Quantum Field Theory

A Selection of Papers in Memoriam Kurt Symanzik

Editors: A. Jaffe, H. Lehmann, and G. Mack

With 36 Figures

Springer-Verlag Berlin Heidelberg New York Tokyo

Professor Dr. Arthur Jaffe
Harvard University, Department of Physics,
Lyman Laboratory of Physics,
Cambridge, MA 02138, USA

Professor Dr. Harry Lehmann
Professor Dr. Gerhard Mack
II. Institut für Theoretische Physik der Universität Hamburg,
D-2000 Hamburg 50, Fed. Rep. of Germany

This book originally appeared as the journal
Communications in Mathematical Physics, Vol. 97, Number 1 and 2

ISSN 0010-3616 Springer Berlin, Heidelberg, New York 1985

ISBN-13:978-3-540-15260-6 e-ISBN-13:978-3-642-70307-2
DOI: 10.1007/978-3-642-70307-2

2153/3130-543210

Publisher's Preface

Kurt Symanzik was certainly one of the most outstanding theoretical physicists of our time. For thirty years, until his untimely death in 1983, he helped to shape the present form of quantum field theory and its application to elementary particle physics. In memoriam of Kurt Symanzik leading scientists present their most recent results, giving, at the same time, an overview of the state of the art.

This collection was originally published in Vol. *97*, 1/2 (1985) of Communications in Mathematical Physics. They range over various interrelated topics of interest to Kurt Symanzik. We hope that making this collection available in an accessible and inexpensive way will benefit the physics community.

The Publisher

Contents

Kurt Symanzik

Kurt Symanzik was born November 23, 1923 in Lyck, East Prussia. He grew up in Königsberg, but because of the war he could only begin to study physics at the age of 23, when he entered the Technical University of Munich. He shortly moved to Göttingen and became a student of Heisenberg. There Symanzik encountered two young colleagues, H. Lehmann and W. Zimmermann, with whom he developed both close friendship and scientific collaboration. This group was later dubbed the "Feldverein" by W. Pauli, when it had become an important influence in theoretical physics.

In 1954, Symanzik completed his doctoral thesis, "On the Schwinger functional in quantum field theory." The deep insights in this work and the technical skill in their implementation set the scene for a series of classic papers in diverse fields of theoretical physics; all these papers share conceptual clarity combined with overwhelming technical ability. The best known work from the period in Göttingen was the famous LSZ "reduction formula" to express scattering cross sections in terms of vacuum expectation values of quantum fields. Today this formula can be found in most books on elementary particles or quantum fields.

From 1955 to 1962, Symanzik worked in many departments in both the United States and in Europe, including the Institute for Advanced Study, the University of Chicago, Göttingen, Hamburg, Stanford, Princeton, UCLA, and CERN. Two themes during this period were a study of dispersion relations and the analysis of how Green's functions reflect the many-particle structure of quantum fields.

In 1962, Symanzik accepted a professorship at the Courant Institute, where he remained for 6 years. While there he developed Euclidean quantum field theory, surely one of his greatest achievements. He recognized that field theory could be reduced to the structure of classical statistical mechanics. He proposed that integral equations, correlation inequalities, Markovian properties, interacting random paths, and other aspects of classical statistical physics had an interpretation in quantum field theory. Originally Symanzik was motivated by his attempt to solve the existence question for scalar quantum fields by this method, culminating in his 1968 Varenna lectures. Later these ideas led to the reconstruction theorem for quantum theory from Euclidean fields, and they became an integral part of constructive field theory. Ultimately this approach made possible the computations based on high temperature series or computer simulation in lattice gauge theories based on the renormalization group. Furthermore this point of view led to the noninteraction theorems for quartic scalar field theories. Euclidean field theory today is an indispensible starting point for the study of many problems in particle physics.

In 1968, Symanzik returned to Germany as a research Professor at DESY. Here his interests turned in a different direction, and the Callan-Symanzik equation was another high point of his career. This renormalization group equation gave impetus to the discovery of asymptotically free quantum field theories. Symanzik found a first model. Soon thereafter it was recognized that nonabelian gauge theories are asymptotically free. This was a precondition for the development of Quantum Chromodynamics, the currently accepted model for hadronic interactions.

In 1981 the German Physical Society presented Kurt Symanzik the Max Planck Medal, its highest honor for scientific achievement.

For many colleagues and young scientists, Symanzik was a physicist whom one visited in order to learn by conversation. His shyness, his penetrating insight, and his dislike for redundancy in communication often made it difficult to establish personal contact with him. But those who did get to know him closely remember not only an extraordinary intellect, but also a loyal and generous friend. He enjoyed contacts with colleagues and young scientists both at DESY and elsewhere. It was usual for Symanzik to perform long calculations and to write long letters to encourage the work of others as well as to explain his own unique and original insights.

He enjoyed with equal gusto unscientific activities including swimming, attending ballet and dancing. Friends and colleagues watched with amusement and affection as he tried to execute dance steps as complicated as the equations in his papers!

Kurt Symanzik's last papers were devoted to lattice gauge theory. They show that he was in full command of his creative force until the end when he died of cancer on October 25, 1983.

A. Jaffe, H. Lehmann, and G. Mack

Publications of Kurt Symanzik

Kaskaden im Atomkern. In: Heisenberg, W.: Kosm. Strahlung, 2. Aufl., S. 164. Berlin: Springer 1953

Praktisch wichtige Formeln aus der Relativitätskinematik. In: Heisenberg, W.: Kosm. Strahlung, S. 558

Zur renormierten einzeitigen Bethe-Salpeter-Gleichung. Nuovo Cimento **11**, 88–91 (1953)

Über das Schwingersche Funktional in der Feldtheorie. Z. Naturforsch. **9a**, 809–824 (1954)

Zur Formulierung quantisierter Feldtheorien. Nuovo Cimento **1**, 205–225 (1955), with H. Lehmann, W. Zimmermann

Zur Vertexfunktion in quantisierten Feldtheorien. Nuovo Cimento **2**, No. 3, 425–432 (1955), with H. Lehmann, W. Zimmermann

Derivation of dispersion relations for forward scattering. Phys. Rev. **105**, 743–749 (1957)

On scattering at very high energies. Nuovo Cimento **5**, 659–665 (1957)

On the formulation of quantized field theories. II. Nuovo Cimento **6**, 319–333 (1957), with H. Lehmann, W. Zimmermann

On the renormalization of the axial vector β-decay coupling. Nuovo Cimento **11**, 269–277 (1959)

Dispersion relations and vertex properties in perturbation theory. Progr. Theor. Phys. **20**, 690–702 (1958)

The asymptotic condition and dispersion relations. In: Lectures on field theory and the many-body problem, pp. 67–96. Caianiello, E.R. (ed.). New York: Academic Press 1961

On the many-particle structure of Green's functions in quantum field theory. J. Math. Phys. **1**, 249–273 (1960)

Green's functions and the quantum theory of fields. In: Lectures in theoretical physics. Vol. III, pp. 490–531. Brittin, W.E., Downs, B.W., Downs, J. (eds.). New York: Interscience 1961

Green's functions method and renormalization of renormalizable field theories. In: Lectures on high energy physics, Zagreb 1961, pp. 485–517 (reprinted, New York: Gordon and Breach 1966)

Grundlagen und gegenwärtiger Stand der feldgleichungsfreien Feldtheorie. In: Werner Heisenberg und die Physik unserer Zeit, pp. 275–298. Braunschweig: Vieweg 1961

Application of functional integrals to Euclidean quantum field theory. In: Analysis in function space, pp. 197–206. Martin, W.T., Segal, I. (eds.). Cambridge, MA: MIT Press 1964

A modified model of Euclidean quantum field theory. Techn. Rep. IMM-NYU 321 (June 1964)

Many particle structure of Green's functions. In: Symposia on theoretical physics, Vol. 3, pp. 121–170. Ramakrishnan, A. (ed.). New York: Plenum Press 1967

Proof and refinements of an inequality of Feynman. J. Math. Phys. **6**, 1155–1156 (1965)

Euclidean quantum field theory. I. Equations for a scalar model. J. Math. Phys. **7**, 510–525 (1966)

A method for Euclidean quantum field theory. In: Mathematical theory of elementary particles, pp. 125–140. Goodman, R., Segal, I. (eds.). Cambridge, MA: MIT Press 1966

Schwinger functions and the classical limit of equilibrium quantum statistical mechanics. Nuovo Cimento **45**, 269–272 (1966)

Euclidean proof of the Goldstone theorem. Commun. Math. Phys. **6**, 228–232 (1967)

Euclidean quantum field theory. In: Local quantum field theory, pp. 152–226. Jost, R. (ed.). New York: Academic Press 1969 (Varenna lectures)

Euclidean quantum field theory. In: Fundamental interactions at high energy, pp. 19–32. Gudehus, T., Kaiser, G., Perlmutter, A. (eds.). New York: Gordon and Breach 1969

Renormalization of models with broken symmetry. In: Fundamental interactions at high energy, pp. 263–278. Perlmutter, A., Iverson, G.J., Williams, R.M. (eds.). New York: Gordon and Breach 1970

Renormalization of certain models with PCAC. Lett. Nuovo Cimento **2**, 10–12 (1969)

Renormalizable models with simple symmetry breaking. I. Symmetry breaking by a source term. Commun. Math. Phys. **16**, 48–80 (1970)

Small-distance behaviour analysis and power counting. Commun. Math. Phys. **18**, 227–246 (1970)

Small-distance behaviour in field theory. Springer Tracts Mod. Phys. **57**, 222–236 (1971)
Lectures in Lagrangian quantum field theory. Interner Bericht DESY T-71/1, Febr. 1971
Renormalization of theories with broken symmetry. In: Cargèse lectures in physics, pp. 179–237. Bessis, J.D. (ed.). New York: Gordon and Breach 1972
Small-distance-behaviour analysis and Wilson expansions. Commun. Math. Phys. **23**, 49–86 (1971)
On computations in conformal invariant field theories. Lett. Nuovo Cimento **3**, 734–738 (1972)
Currents, stress tensor and generalized unitarity in conformal invariant quantum field theory. Commun. Math. Phys. **27**, 247–281 (1972), with G. Mack
A field theory with computable large-momenta behaviour. Lett. Nuovo Cimento **6**, 77–80 (1973)
Infrared singularities in theories with scalar massless particles. Acta Phys. Austriaca, Suppl. **XI**, 199–240 (1973)
On theories with massless particles. In: Renormalization of Yang-Mills fields and applications to particle physics. C.N.R.S. Marseille, 72, p. 470, pp. 221–230
Infrared singularities and small-distance behaviour analysis. Commun. Math. Phys. **34**, 7–36 (1973)
Short review of small-distance-behaviour analysis. In: Renormalization and invariance in quantum field theory, pp. 225–246. Caianiello, R. (ed.). New York: Plenum Press 1974
Massless ϕ^4 theory in $4-\varepsilon$ dimensions. Lett. Nuovo Cimento **8**, 771–774 (1973)
Massless ϕ^4 theory in $4-\varepsilon$ dimensions. Cargèse lectures in physics. Brézin, E. (ed.). New York: Gordon and Breach 1973 (unpublished)
New trends in field theory. J. Phys., Suppl. **10**, T. 34, pp. C1-117–126
Small-distance behaviour in quantum field theory. In: Particles, quantum fields, and statistical mechanics. Alexanian, M., Zepeda, A. (eds.). Berlin, Heidelberg, New York: Springer 1975
Renormalization problem in nonrenormalizable massless ϕ^4 theory. Commun. Math. Phys. **45**, 79–98 (1975)
Renormalization problem in a class of nonrenormalizable theories. Proceedings VI GIFT Seminar on Theoretical Physics, June 1975
Renormalization problem in massless $(\phi^4)_{4+\varepsilon}$ theory. Suppl. Acta Austriaca **XVI**, 177–184 (1976)
Regularized quantum field theory. In: New developments in quantum field theory and statistical mechanics, pp. 265–280. Lévy, M., Mitter, P. (eds.). New York: Plenum Press 1977
$1/N$ expansions in $P(\phi^2)_{4-\varepsilon}$ theory. I. Massless theory, $0<\varepsilon<2$ (DESY 77/05) (unpublished)
Cutoff dependence in lattice ϕ_4^4 theory. In: Recent developments in gauge theories, pp. 313–330. 't Hooft, G., et al. (eds.). New York: Plenum Press 1980
Anomalies of the free loop wave equation in the WKB approximation. Nucl. Phys. B **173**, 365–396 (1980), with M. Lüscher, P. Weisz
Schrödinger representation and Casimir effect in renormalizable quantum field theory. Nucl. Phys. B **190** [FS3], 1–44 (1981)
Some topics in quantum field theory. In: Mathematical problems in theoretical physics. Conference Berlin 1981, pp. 44–58. Schrader, R., Seiler, R., Uhlenbrock, D.A. (eds.). Berlin, Heidelberg, New York: Springer 1982
Improved lattice actions for non-linear sigma model and non-abelian gauge theory. Workshop on non-perturbative field theory and QCD, Trieste, Dec. 1982 (to be published by World Publishing Company, Singapore)
Improved continuum limit in the lattice O(3) non-linear sigma model. Phys. Lett. **126** B, 467 (1983), with B. Berg, I. Montvay, S. Meyer
Concerning the continuum limit in some lattice theories. In: 21st international conference on high energy physics, pp. C3; 254–259. Petiau, P., Porneuf, M. (eds.). Paris: Editiones de Physique 1982
Continuum limit and improved action in lattice theories. I. Principles and ϕ^4-theory. Nucl. Phys. B **226**, 187–204 (1983)
Continuum limit and improved action in lattice theories. II. O(n)-nonlinear sigma model in perturbation theory. Nucl. Phys. B **226**, 205–227 (1983)

Monte Carlo Simulations for Quantum Field Theories Involving Fermions

M. Karowski[1]*, R. Schrader[2]*, and H. J. Thun[3]

1 Institut für Theoretische Physik, Universität Göttingen, Bunsenstrasse 9, D-3400 Göttingen, Federal Republic of Germany
2 State University of New York at Stony Brook, Stony Brook, NY 11794, USA
3 Institut für Theorie der Elementarteilchen, Freie Universität Berlin, Arnimallee 14, D-1000 Berlin 33

Abstract. We present a new variant of a Monte Carlo procedure for euclidean quantum field theories with fermions. On a lattice every term contributing to the expansion of the fermion determinant is interpreted as a configuration of self-avoiding oriented closed loops which represent the fermionic vacuum fluctuations. These loops are related to Symanzik's polymer description of euclidean quantum field theory. The method is extended to the determination of fermionic Green's functions. We test our method on the Scalapino-Sugar model in one, two, three, and four dimensions. Good agreement with exactly known results is found.

1. Introduction

In recent years Monte Carlo simulations for euclidean lattice models have been of considerable help in improving our understanding of those relativistic quantum field theories, which are supposed to describe high energy particle physics. This includes in particular gauge theories. Now any realistic model for particle interactions includes fermionic fields like quark fields. It is, therefore, important to simulate systems with fermionic degrees of freedom. There have been several proposals to deal with this problem, see e.g. [1–16], or [17–19] for a review. However, all these methods require extensive computing time and some of them only work for two-dimensional models or are only approximations from the beginning: quenched approximation, hopping parameter expansion etc. In particular for the interesting case of four-dimensional lattices no way has yet been found to perform Monte Carlo calculations including fermions as efficiently as they can be done when bosonic fields only are present.

It is the aim of this paper to propose a new numerical method which basically treats all fields on the same footing during the upgrading procedure. This new way of treating fermions applies to all lattice models known to the authors and may

---

* On leave of absence from Freie Universität Berlin

easily be combined with the standard methods already used for bosonic fields. We now briefly describe our method, the details of which will be given in the next sections. To deal with the fermionic degrees of freedom we introduce an auxiliary statistical ensemble, the elements of which are labelled by a subset of the set of permutations on the fermionic degrees of freedom. We give a graphical representation of such an element of the statistical ensemble in terms of self-avoiding loops and relate it to Symanzik's polymer description of euclidean quantum field theory [20, 21]. This procedure guarantees that Pauli's exclusion principle for the fermions is automatically taken care of. The antisymmetry in the fermionic degrees of freedom is obtained by equating the Green's functions of the physical model with certain (ratios of) Green's functions of the auxiliary statistical ensemble. This is explained in Sect. 2.

In Sect. 3 we present a local heat bath method for a fermionic action in the form originally introduced by Kogut and Susskind in a Hamiltonian context [22]. As just mentioned, each element of the statistical ensemble, which is a certain permutation, has a graphical representation. The heat bath is local, i.e. most of the fermionic degrees of freedom are frozen, because we apply localized permutations which upgrade a given permutation only. locally. The intricate part of our procedure is to show that the principle of detailed balance is satisfied. This combined with the ergodicity of the upgrading ensures that the Gibbs distribution for the auxiliary statistical ensemble is the unique equilibrium distribution for the upgrading procedure.

In Sect. 4 we consider the standard, simple model introduced by Scalapino and Sugar [3], which has often been used to test methods for dealing with fermions numerically. We compare our Monte Carlo results with exact values obtained by numerical Fourier summation. Finally, Sect. 5 contains some conclusions.

2. Statistical Ensembles for Fermions

In this section we will discuss the auxiliary statistical ensemble, in terms of which the fermionic degrees of freedom may be described so as to give the Green's functions of the physical theory. We begin by recalling the standard formulation of lattice theories involving fermions. We will restrict our attention to (finite) cubical lattices in d dimensions. The question of how to choose the right boundary conditions will not be relevant in this section.

Consider a lattice action of the form

$$S = S(\psi^+, \psi, \phi). \tag{2.1}$$

ψ^+ and ψ denote fermionic fields and are considered to be Grassmann variables. ϕ stands for all other fields, which are supposed to be bosonic. In all applications so far, the fermionic fields are defined on the vertices of the lattice. Thus the lattice site x may be used to label the fermionic degrees of freedom. The fields ϕ may live on vertices (ordinary Bose fields) or on links (lattice gauge fields). Our method will also allow for more complicated situations, where for example ϕ lives on higher dimensional cells.

For the purpose of explaining our method, we first look at the simplified situation, where

(a) the fermionic fields carry no indices (like flavour or colour) other than the vertex index,

(b) the action is quadratic in the fermionic fields.

It is easy to extend our method to situations, where these restrictions are removed and at the end of the section we will briefly indicate how this is done. With these restrictions the action S may be decomposed as

$$S(\psi^+, \psi, \phi) = S_F(\psi^+, \psi, \phi) + S_B(\phi), \tag{2.2}$$

where

$$S_F(\psi^+, \psi, \phi) = \sum_{x, y \in \text{Lattice}} \psi^+(x) A_{xy}(\phi) \psi(y). \tag{2.3}$$

Here the matrix $A = A(\phi) = \{A_{xy}(\phi)\}$ with complex valued entries is indexed by the vertices of the lattice and is a functional of the bosonic fields ϕ, and $S_B(\phi)$ is that part of the action which does not involve fermionic fields. The "partition" function of the theory is therefore

$$Z_{FB} = \int d\phi d\psi^+ d\psi e^{-S}. \tag{2.4}$$

The integration over the fermionic fields is in the sense of Berezin [23], and $d\phi$ describes the integration over the bosonic fields ϕ. By

$$\langle X \rangle = Z_{FB}^{-1} \int d\phi d\psi^+ d\psi X e^{-S} \tag{2.5}$$

we denote the expectation value in this model. If we perform the fermionic integration first, Eq. (2.4) may be written as

$$Z_{FB} = \int d\phi \det A(\phi) e^{-S_B(\phi)}. \tag{2.6}$$

To obtain Green's functions involving fermionic fields, we consider the typical example $\langle \psi(v) \psi^+(u) \rangle$. For given u, v let $A^{(u,v)}$ denote the matrix given by

$$A_{xy}^{(u,v)} = \begin{cases} A_{xy} & \text{if} \quad x \neq u \quad \text{and} \quad y \neq v. \\ 1 & \text{if} \quad x = u \quad \text{and} \quad y = v, \\ 0 & \text{otherwise}. \end{cases} \tag{2.7}$$

Then we have

$$\langle \psi(v) \psi^+(u) \rangle = \frac{\int d\phi \det A^{(u,v)}(\phi) e^{-S_B(\phi)}}{\int d\phi \det A(\phi) e^{-S_B(\phi)}}. \tag{2.8}$$

Of course, $\det A^{(u,v)}$ is (up to a sign) equal to the determinant of the submatrix of A obtained by deleting the uth row and v^{th} column. More generally, any (higher order) Green's function in the fermionic fields may be obtained in this way from determinants of suitable submatrices. Formulas (2.6) and (2.8) are obtained from Berezin's integration theory, by using the standard formula

$$\det A = \sum_{\pi} \text{sgn}(\pi) \prod_{x} A_{x\pi(x)}, \tag{2.9}$$

where π runs through the set of all permutations of lattice points.

Let us now neglect the ϕ dependence for a moment. Each term on the right-hand side of (2.9), which is labelled by π, may be given a graphical presentation as follows. If $\pi(x) \neq x$, we draw an oriented straight line in the lattice from the point x to the point $\pi(x)$. By definition its length is 1. Now every permutation may be written as a product of (nontrivial) cyclic permutations

$$\pi = \pi_1 \dots \pi_s , \tag{2.10}$$

and this description is unique up to the ordering of the cyclic factors π_r, $1 \leqq r \leqq s$. If ℓ_r denotes the order of π_r (such that $\ell_r = 2$ if π_r is a transposition), we have

$$\operatorname{sgn}(\pi) = \prod_{r=1}^{s} (-1)^{\ell_r + 1} . \tag{2.11}$$

Using this graphical presentation each cyclic permutation π_r corresponds to a closed oriented polygonal loop $\mathscr{L}_r$ of length ℓ_r. These loops $\mathscr{L}_1, \dots, \mathscr{L}_s$ are nonintersecting in the sense that each vertex in the lattice is the endpoint of at most one oriented straight line and in that case it is also the starting point of exactly one line.

Conversely to each such family of nonintersecting oriented polygonal loops corresponds a unique π and hence a unique contribution to $\det A$ in the sense of Eq. (2.9). Note that we only need to consider those π which are contained in the set

$$\mathscr{C}(A) = \left\{ \pi ; \prod_x A_{x\pi(x)} \neq 0 \right\} . \tag{2.12}$$

Similarly, the only nonvanishing contributions to $\det A^{(u,v)}$ are among those π for which $\pi(u) = v$. By deleting the particular straight line going from u to v, each such contribution to $\det A^{(u,v)}$ can be graphically described by a set of nonintersecting loops plus an additional "propagator", i.e. an open polygonal line, going from v to u, which is nonselfintersecting and not intersecting the other loops in the sense just described.

It is important to note that this nonintersecting property is a *local property*, i.e. it is only necessary to test all vertices individually to see whether a given set of polygonal loops correspond to a permutation π or not.

In most applications the matrix A will have the following additional property: A is said to be *local*, if $A_{xy} = 0$ unless $\operatorname{dist}(x, y) \leqq 1$. In the context of our graphical presentation this means that all loops are built out of links. Also the nonintersecting property stated above is now the usual nonintersecting property of curves. For local A, we say $\pi \in \mathscr{C}(A)$ is a "dimer" if π is cyclic of order 2. π is said to contain a dimer if at least one of its cyclic factors π_r is of order two. Now for local A, the number $|\mathscr{C}(A)|$ of elements in $\mathscr{C}(A)$ is bounded above by $(2d+1)^{(\text{volume of lattice})}$. On the other hand, since the density of states for the dimer problem is explicitly known for $d = 2$ at zero temperature [24–27], it is easy to obtain the lower bound $\alpha^{\text{volume}/2}$ on $|\mathscr{C}(A)| \left(\alpha = \exp \frac{2G}{\pi} = 1.791\dots, G = \text{Catalan's} = 0.915\dots \right)$, if A is such that $A_{xy} \neq 0$ whenever $\operatorname{dist}(x, y) \leqq 1$. Indeed, this follows by considering the subset of $\mathscr{C}(A)$ consisting of those π with dimers only which point in directions parallel to a given plane. In Sect. 3 we will also give a discussion on how to compute $|\mathscr{C}(A)|$ numerically for the case of such a local A.

The graphical presentation we have given for general (constant) matrix A is related to Symanzik's polymer presentation [20, 21], see also [28, 29], as follows. For simplicity, let A be of the special form

$$A = \mathbb{1} + \Gamma, \tag{2.13}$$

with $\Gamma_{xx} = 0$ and Γ_{xy} small. Then

$$\begin{aligned} \det A &= \exp[\operatorname{tr} \ln A] \\ &= \exp\left[\sum_{n=1}^{\infty} \frac{(-1)^{n-1}}{n} \operatorname{tr} \Gamma^n\right]. \end{aligned} \tag{2.14}$$

Since

$$\operatorname{tr} \Gamma^n = \sum_{x_1, \ldots, x_n} \Gamma_{x_1 x_2} \ldots \Gamma_{x_n x_1} . \tag{2.15}$$

each term of the right-hand side of (2.15) can be viewed graphically as an oriented polygonal loop of length n. In this formulation, however, there is no nonintersecting condition, i.e. each vertex may be the endpoint of more than one oriented straight line. The converse holds also in this case: To each such loop of length n corresponds a unique contribution to $\operatorname{tr} \Gamma^n$ via (2.15). The Taylor expansion of the exponent in Eq. (2.14) then gives a combinatorial description (and proof) of how these terms in (2.14)–(2.15), each of which is presented by one oriented loop, combine to the set of terms in (2.9), each of which is presented by a family of nonintersecting oriented loops. A similar graphically equivalent description may be given for the Green's functions, e.g. $\langle \psi(v) \psi^+(u) \rangle$.

We now turn to a construction of the statistical ensemble. Let the matrix $|A|$ be given by

$$|A|_{xy} = |A_{xy}| , \tag{2.16}$$

and consider the permanent of $|A|$,

$$\operatorname{Per}(|A|) = \sum_{\pi \in \mathscr{C}(A)} \prod_x |A|_{x\pi(x)} . \tag{2.17}$$

Now $\mathscr{C}(A)$ may be viewed as a Gibbs statistical ensemble with energies given by

$$E(\pi, A) = \sum_x E_x(\pi, A) = -\sum_x \ln(|A|_{x\pi(x)}) , \tag{2.18}$$

such that the permanent becomes the corresponding partition function

$$\operatorname{Per}(|A|) = \sum_{\pi \in \mathscr{C}(A)} \exp[-E(\pi, A)] . \tag{2.19}$$

If we denote by

$$\bar{X} = \frac{1}{\operatorname{Per}(|A|)} \cdot \sum_\pi X(\pi) \cdot e^{-E(\pi, A)} \tag{2.20}$$

averages in this statistical ensemble, we have the following relation for the determinant of A

$$\det A = \sum_{\pi} \operatorname{sgn}(\pi, A) e^{-E(\pi, A)} = \operatorname{Per}(|A|) \cdot \overline{\operatorname{sgn}(A)}, \tag{2.21}$$

where $\operatorname{sgn}(\pi, A)$ is defined on $\mathscr{C}(A)$ by

$$\operatorname{sgn}(\pi, A) = \operatorname{sgn}(\pi) \cdot \prod_{x} \frac{A_{x\pi(x)}}{|A_{x\pi(x)}|}. \tag{2.22}$$

To obtain the Green's function $\langle \psi(v) \psi^{+}(u) \rangle$ we have to determine [cf. Eq. (2.8)]

$$\det A^{(u,v)} = \operatorname{Per}(|A^{(u,v)}|) \cdot \overline{\operatorname{sgn}(A^{(u,v)})}, \tag{2.23}$$

where now the configurations of $\mathscr{C}(A^{(u,v)})$ have to be taken into account.

We remark that the following modification of this construction leads to the theory of the noninteracting polymer gas: Take A to be local with $A_{xx} = 1$ and $A_{xy} = \text{const}$ for $\operatorname{dist}(x, y) = 1$. The statistical ensemble is defined to be the subset of $\mathscr{C}(A)$ consisting of all π's without dimers. In fact, the numerical results in [42] were obtained by a corresponding modification of the upgrading procedure, to be explained in Sect. 3.

Let us return to the case, where bosonic fields ϕ are present. Again we define $\mathscr{C}(A)$ by (2.12), where the condition $\prod_{x} A_{x\pi(x)} \neq 0$ is now understood in the sense of functionals. Also in (2.18) we set $E_x(\pi, A(\phi)) = \infty$ for any value of ϕ for which $A_{x\pi(x)}(\phi) = 0$. We are now in a position to describe the auxiliary statistical ensemble and express physical Green's functions as expectation values in this theory. In fact, the thermal average values are now

$$\bar{X} = Z^{-1} \int d\phi \sum_{\pi \in \mathscr{C}(A)} X(\pi, \phi) e^{-S_B(\phi) - E(\pi, A(\phi))}, \tag{2.24}$$

with

$$Z = \int d\phi \sum_{\pi \in \mathscr{C}(A)} e^{-S_B(\phi) - E(\pi, A(\phi))}. \tag{2.25}$$

The set of configurations is now the product $\mathscr{C}(A)$ times the set of the usual bosonic configurations. Consider first an observable F which only depends on ϕ. Then

$$\begin{aligned} \langle F(\phi) \rangle &= \frac{\int d\phi F(\phi) \det A(\phi) e^{-S_B(\phi)}}{\int d\phi \det A(\phi) e^{-S_B(\phi)}} \\ &= \frac{\int d\phi \sum_{\pi \in \mathscr{C}(A)} F(\phi) \operatorname{sgn}(\pi, A(\phi)) e^{-S_B(\phi) - E(\pi, A(\phi))}}{\int d\phi \sum_{\pi \in \mathscr{C}(A)} \operatorname{sgn}(\pi, A(\phi)) e^{-S_B(\phi) - E(\pi, A(\phi))}} \\ &= \overline{F \cdot \operatorname{sgn}(A)} / \overline{\operatorname{sgn}(A)}. \end{aligned} \tag{2.26}$$

We note a crucial property of this relation, which will allow us to perform *local upgrading procedures* both for the fermionic and the bosonic part: $E(\pi, A(\phi))$ is a sum of local terms $E_x(\pi, A(\phi))$ in the sense that for local theories the latter will depend only on the form of π near x and on the field configurations of ϕ which live near x.

Next we turn to fermionic Green's functions. For the two-point function we obtain as a generalization of Eq. (2.23) [see Eq. (2.8)]

$$\langle \psi(v)\psi^+(u)\rangle = \frac{\overline{\mathrm{sgn}(A^{(u,v)})}}{\overline{\mathrm{sgn}(A)}}. \tag{2.27}$$

Analogously higher order Green's functions may be obtained by augmenting the statistical ensemble appropriately.

The remainder of this section is devoted to a brief outline of the modifications necessary to cover the cases that

(a′) the fermionic fields carry indices ϱ which may include the vertex index x and internal degrees of freedom and

(b′) higher order interactions in the fermionic fields are present.

In the case (a′), which is important for treating e.g. non-abelian lattice gauge theories, A is a matrix $A_{\varrho\varkappa}$. And $\mathscr{C}(A)$ is now defined to be the set of all permutations π of the ϱ's for which $\prod_\pi A_{\varrho\pi(\varrho)}(\phi) \neq 0$. Again there is a graphical representation which now is in d' dimensions ($d' > d$) with the extra $(d'-d)$ dimensions being used to describe the additional degrees of freedom.

As for the case (b′) assume the action contains an extra term which is of fourth order in the fermionic fields

$$S'_F(\psi^-, \psi, \phi) = \sum_{\substack{\varrho \neq \varrho' \\ \varkappa \neq \varkappa'}} \psi^+(\varrho)\psi^+(\varrho')B_{\varrho\varrho'\varkappa\varkappa'}\psi(\varkappa)\psi(\varkappa'). \tag{2.28}$$

Let $\mathscr{I} = (I, I_1, \ldots, I_k)$ $(0 \leqq k)$ be a decomposition of the set of all ϱ's into a subset I and ordered sets I_ℓ $(1 \leqq \ell \leqq k)$ containing two elements each. Then

$$\begin{aligned} &\int d\psi^+ d\psi e^{-S_F(\psi^+, \psi, \phi) - S'_F(\psi^+, \psi, \phi)} \\ &\quad = \sum_{\pi, \mathscr{I}} \mathrm{sgn}(\pi) \prod_{\varrho \in I} A_{\varrho\pi(\varrho)}(\phi) \prod_{\substack{\ell=1 \\ (\varrho', \varrho'') = I_\ell}}^{k} B_{\varrho'\varrho''\pi(\varrho')\pi(\varrho'')}(\phi). \end{aligned} \tag{2.29}$$

The auxiliary statistical ensemble is now labelled by $\pi, \mathscr{I} \in \mathscr{C}(A, B)$, and the states representing the bosonic fields ϕ, where

$$\mathscr{C}(A, B) = \left\{ (\pi, \mathscr{I});\ \prod_{\varrho \in I} A_{\varrho\pi(\varrho)} \prod_{\substack{\ell=1 \\ (\varrho, \varrho') = I_\ell}}^{k} B_{\varrho\varrho'\pi(\varrho)\pi(\varrho')} \neq 0 \right\}. \tag{2.30}$$

Also the energies (2.18) are replaced by

$$E(\pi, \mathscr{I}, A(\phi), B(\phi)) = -\sum_{\varrho \in I} \ln|A_{\varrho\pi(\varrho)}(\phi)| - \sum_{\substack{\ell=1 \\ (\varrho, \varrho') = I_\ell}}^{k} \ln|B_{\varrho\varrho'\pi(\varrho)\pi(\varrho')}(\phi)|. \tag{2.31}$$

Again a graphical presentation may be obtained. It is obvious how to extend this procedure to interactions of order higher than four.

Another strategy would work for local and translation invariant B's. One can introduce an intermediate boson $\varphi(\varrho, \varkappa)$ and replace (2.28) by

$$S_F''(\psi^+, \psi, \varphi, \phi) = -\sum[\varphi(\varrho, \varkappa)B_{\varrho\varrho'\varkappa\varkappa'}\varphi(\varrho', \varkappa') + \psi^+(\varrho)\psi(\varkappa)(B_{\varrho\varrho'\varkappa\varkappa'} + B_{\varrho'\varrho\varkappa'\varkappa})\varphi(\varrho', \varkappa')], \quad (2.32)$$

which again turns out to be an action of case (a').

3. The Heat Bath Method

In this section we explain our heat bath method for the determination of fermion determinants for the case of free massive fermions.

There exist different formulations for fermions on a lattice, e.g. the Wilson [30], Kogut-Susskind [31–36], and Dirac-Kähler [37–39] versions. For the purpose of this paper the second version is the most adequate one. Thus we use it to exemplify our heat bath method in spite of its shortcoming, namely numerical results are reliable only for sufficiently heavy fermions. More precisely, the statistical errors obscure the measurements if the hopping parameter $k = 1/(2am)$ (a = lattice spacing, m = fermion mass) exceeds 0.6, 0.25, 0.15 for Susskind fermions in 2, 3, and 4 dimensions, respectively. In a forthcoming publication we shall extend our method to the case of Wilson fermions, where there is no such restriction on the mass.

We consider a hypercubic lattice in d dimensions with lattice points

$$\mathbf{x} = \sum_{\mu=1}^{d} x_\mu \mathbf{a}_\mu \quad (3.1)$$

labelled by integer components, $x_\mu = 0, 1, 2, \ldots, L-1$, and $\{\mathbf{a}_\mu\}$ orthogonal vectors parallel to the lattice axes, $\mathbf{a}_\mu \cdot \mathbf{a}_\nu = a^2\delta_{\mu\nu}$. The naive lattice version of the euclidean free fermion action reads

$$S = \sum_x \psi^+(x)(\gamma_\mu\partial_\mu + m)\psi(x) \quad (3.2)$$

with central differences, i.e.

$$\partial_\mu\psi(x) = \frac{1}{2a}[\psi(x+a_\mu) - \psi(x-a_\mu)].$$

The Susskind formulation is most easily obtained by the transformation [35, 36]

$$\psi(x) \to \prod_{\mu=1}^{d} \gamma_\mu^{x_\mu}\psi(x); \quad (3.3a)$$

the γ-matrices then become proportional to the unit matrix, albeit x-dependent,

$$\gamma_\mu \to \gamma_\mu(x)\mathbb{1} = \prod_{\nu=1}^{\mu-1} (-1)^{x_\nu}\mathbb{1}. \quad (3.3b)$$

It is then sufficient to consider only one component of a full Dirac fermion at each lattice site, and the matrix which appears in Eq. (2.3) takes the form

$$(\gamma_\mu\partial_\mu + m)_{\mathbf{x},\mathbf{y}} = \gamma_\mu(x)\frac{1}{2a}[\delta_{\mathbf{x}+\mathbf{a}_\mu,\mathbf{y}} - \delta_{\mathbf{x}-\mathbf{a}_\mu,\mathbf{y}}] + m\delta_{\mathbf{x},\mathbf{y}}. \quad (3.4)$$

The fermion determinant is therefore given by

$$\det(\gamma\partial+m)=\sum_{\pi\in S_N}\operatorname{sgn}(\pi)\prod_x(\gamma\partial+m)_{x\pi(x)}, \tag{3.5}$$

with $N=L^d=$number of lattice points.

As explained in Sect. 2 there is a one-to-one correspondence between the permutation $\pi\in S_N$ of the lattice points and loop configurations $c\in\mathscr{C}(\gamma\partial+m)$ on the lattice, therefore,

$$Z_F\equiv\det(\gamma\partial+m)=m^N\sum_{c\in\mathscr{C}}\operatorname{sgn}(c,\gamma\partial+m)k^{\ell(c)} \tag{3.6}$$

[with $k=1/(2m)$, and we set $a=|\mathbf{a}_\mu|=1$ from now on]. The sum runs over all configurations c of self-avoiding closed oriented loops on the lattice made up from single links ("monomers") connecting neighbouring sites.

The integer $\ell(c)$ is the total length in lattice units of all loops in the configuration c. The smallest allowed loops are "dimers" with $\ell=2$, righthanded and lefthanded 1×1 square loops with $\ell=4$, etc. There is a direct geometrical interpretation of $\operatorname{sgn}(c,\gamma\partial+m)$, which has been defined in Eq. (2.21). For a single loop $\operatorname{sgn}(c,\gamma\partial+m)=(-1)^{n_i}$, where n_i is the number of internal lattice points in a surface (made up of plaquettes) whose boundary is the loop c. This implies that in $d=1$ dimension there are no negative contributions to Eq. (3.6) at all, whereas in $d=2$ dimensions the simplest ones are due to 2×2 square loops with $n_1=1$, $\ell=8$, and in $d\geqq3$ they arise from certain nonplanar loops with $\ell=6$. Such negative contributions give rise to large statistical errors in the evaluation of expectation values like (2.20) and (2.22) by Monte Carlo methods. This fact is responsible for the empirically determined "critical values" for k mentioned at the beginning of this section.

Next we introduce the partition function of an auxiliary statistical mechanics problem of a gas of self-avoiding closed orientated loops,

$$Z(k)=\sum_{c\in\mathscr{C}}k^{\ell(c)}, \tag{3.7}$$

where $k=e^{-\beta}$ now is the "monomer" fugacity ($\beta=$inverse temperature). In terms of the thermal average of an observable X for this auxiliary problem,

$$\bar{X}=Z^{-1}\sum_{c\in\mathscr{C}}X(c)k^{\ell(c)}, \tag{3.8}$$

we then can rewrite the fermion determinant Eq. (3.6) as

$$Z_F=m^N\cdot Z\cdot\overline{\operatorname{sgn}(\gamma\partial+m)}, \tag{3.9}$$

and the quantum field vacuum expectation value is

$$\langle X\rangle=Z_F^{-1}\int d\psi^+d\psi X\cdot e^{-S}=\overline{X\cdot\operatorname{sgn}}/\overline{\operatorname{sgn}}. \tag{3.10}$$

The partition function Z itself can be obtained from the average energy $\bar{\ell}$. In fact, Eq. (3.7) gives

$$k\frac{\partial}{\partial k}\ln Z(k)=\bar{\ell}(k), \tag{3.11}$$

and by integration

$$Z(k)=\exp\int_0^k \frac{dk'}{k'}\bar{\ell}(k') . \tag{3.12}$$

In particular, we can evaluate Eq. (3.12) for $k=1$ to get the number of all configurations in $\mathscr{C}$ [cf. Eq. (3.7)].

Similarly, the fermion determinant is computable from $\langle \ell \rangle$,

$$Z_F(k)=m^N \exp\int_0^k \frac{dk'}{k'}\langle \ell \rangle(k') , \tag{3.13}$$

which is obtained by differentiating Eq. (3.6) with respect to $k=1/2m$,

$$k\frac{\partial}{\partial k}\ln Z_F(k)=\overline{\ell\cdot \mathrm{sgn}}/\overline{\mathrm{sgn}}-N=\langle \ell \rangle -N . \tag{3.14}$$

It is now clear that it is sufficient to know the average values $\bar{\ell}$, $\overline{\mathrm{sgn}}$, $\overline{\ell\cdot\mathrm{sgn}}$, etc. for the auxiliary problem. They can be determined numerically by means of a Monte Carlo simulation [40, 41]. In the remainder of this section we shall explain in some detail our variant of the heat bath method which generates samples of equilibrium ensembles of configurations in $\mathscr{C}(\gamma\partial+m)$.

Starting from an arbitrary allowed old configuration c_0 of loops with total length ℓ_0 a new configuration c_n with length ℓ_n is proposed. We accept the new one with probability

$$w=\frac{w_n}{w_0+w_n} , \tag{3.15}$$

where

$$w_{0,n}=k^{\ell_{0,n}} \tag{3.16}$$

are the Boltzmann weights of the old and new configurations, respectively. In fact, we actually take the new configuration c_n if a pseudo-random number (equally distributed in the unit interval) is less than w, otherwise, we retain the old configuration c_0. In our updating procedure for the creation of new configurations from old ones we systematically sweep the plaquettes of the entire lattice. At each of the $N=L^d$ lattice points we look in turn at $d(d-1)/2$ plaquettes. In a given plaquette we then perform tentative local changes of the old configuration.

In practice we do not use the components (3.1) to specify a lattice point, but number them lexicographically by the integers

$$x=\sum_{\mu=1}^{d} L^{\mu-1}x_\mu , \tag{3.17}$$

i.e. $x=0, 1, \ldots, N-1$.

As explained above each configuration of loops $c\in\mathscr{C}$ corresponds to a permutation $\pi\in S_N$ and thus gives rise to a term in the expansion of the fermion determinant, Eq. (3.5). It is instructive to think of the permutations in the following way. Imagine e.g. a "dog" sitting at each (lattice) site. In the identical permutation every dog is at home, $\pi(x)=x$. Because of chains of unit length the dogs cannot go beyond their next neighbor's sites. The neighbor then has to look for another place, too. By a permutation π we mean that a dog moves from x to $\pi(x)$, therefore, $\pi^{-1}(x)$ is the site wherefrom the dog, now sitting at x, has come. It is convenient to parametrize the configuration by

$$p(x)=\pi^{-1}(x)-x\,, \tag{3.18}$$

which is the distance (in site numbers) of the home site of a dog from its actual position. Because of the nearest neighbor property of the matrix $\gamma\partial+m$, i.e. the chains, the function $p(x)$ is restricted to $(2d+1)$ values

$$p(x)=0,\ \pm 1,\ \pm L,\ \ldots,\ \pm L^{d-1} \tag{3.19}$$

for allowed configurations. In this picture the links of the self-avoiding closed loops are represented by the chains of those dogs which are not at home.

Let us now explain in more detail how we generate a new configuration from an allowed old one by a rearrangement within a given plaquette. We work with a set of 14 local operations described in terms of permutations and defined on a plaquette, $q_1, q_2, \ldots, q_{14}$. They fall into four classes of cyclic permutations. Elements in one class are obtained from one another by rotations and reflections. Figure 1 gives a representative of each class.

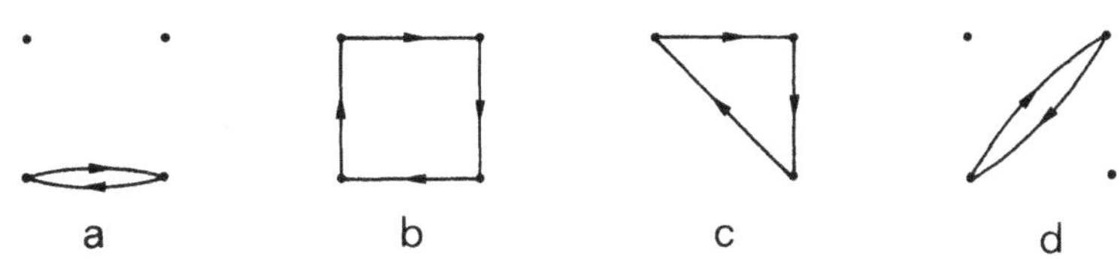

Fig. 1a–d. Representatives of the four classes of local permutations

So these four classes contain 2, 2, 8, and 2 operations, respectively.

A tentative new permutation π_n is obtained from an old one π_0 by

$$\pi_0 \to \pi_n = q_j \circ \pi_0\,; \qquad j=1,2,\ldots,14\,, \tag{3.20}$$

which implies

$$p_n(x)=p_0(q_j^{-1}(x))+q_j^{-1}(x)-x\,. \tag{3.21}$$

The new permutation is allowed, i.e. compatible with the nearest neighbor property, if and only if $p_n(x)$ takes permissible values, cf. (3.19). This has to be checked only for the points in the plaquette since they are the only ones affected by the rearrangement. For $p_n(x)$ unequal to the specific values (3.19) the operation would tear a dog too far from its home site and the corresponding matrix element in that contribution to the right-hand side of Eq. (3.5) would vanish.

We take into consideration the different possibilities $j=1,2,\ldots,14$ by the following strategy. First, we deal with the creation or annihilation of "dimers" corresponding to Fig. 2a

Fig. 2a and b. Dimer creation and annihilation

which result from the application of the two-cycle operation q_1 of Fig. 1a. If the new configuration is allowed we accept it with probability given by Eqs. (3.15) and (3.16). We repeat this procedure with the operation q_2 (a rotated version of q_1).

If after these manipulations the plaquette indeed contains a dimer, we immediately turn to the next plaquette. Otherwise, we try to apply one of the remaining operations $q_3, \ldots, q_{14}$ corresponding to Fig. 1b–d. The resulting four types of transitions are depicted in Fig. 3.

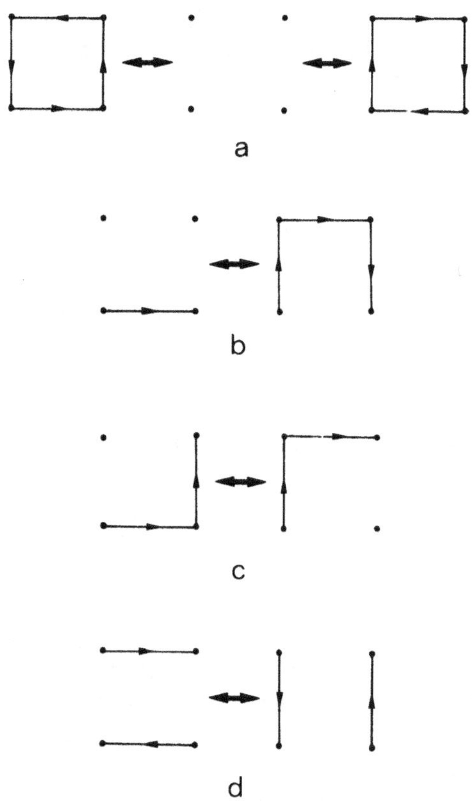

Fig. 3a–d. Local changes of loop configurations not involving dimers

Type (a) is generated by four-cycle operations q_3 and q_4 corresponding to Fig. 1b, types (b) and (c) are due to three-cycles $q_5, \ldots, q_{12}$, cf. Fig. 1c, and finally, type (d) originates from diagonal two-cycles q_{13}, q_{14}, cf. Fig. 1d. If it so happens that more than one of the operations q_j ($j=3, \ldots, 14$) are applicable to the old configuration we generally select the one with lowest index j. The reader may convince himself that we have ordered these operations such that the firstcoming successful one does not produce any dimers. As a rule there is at most one q_j which does not produce dimers. The only exception to the latter statement results from the situation, where the plaquette is empty (all four dogs at home). Then both q_3 and q_4 lead to an allowed new permutation, since they can make a clockwise or counter-clockwise permutation, respectively, cf. Fig. 3a. Now the reason for excluding a further production of dimers by one of the operators $q_3 \ldots q_{14}$ is the following. Whenever possible we prefer a simple alternative (retain c_0 or take c_n) in the generation of configurations (but no more competing new configurations $c'_n, c''_n, \ldots$). Namely, for the case of a simple alternative the probability formula (3.15) implies detailed balance, i.e. for an ensemble of configurations with Gibbs distribution $\propto e^{-\beta\ell}$ the flux from any configuration to another one is equal to the reverse one. In our algorithm there are always only simple alternatives except for the operations q_3 and q_4 as was mentioned above. If they are to be applied to an empty plaquette we first toss a coin between q_3 and q_4, i.e. with weight 1/2 we choose randomly the direction of the cyclic permutation, and only afterwards apply the probability formula (3.15) to decide whether we actually accept the new configuration. For the reverse process of annihilating a 1×1 square loop we have to reduce the probability (3.15) by a factor of 1/2 in order to preserve detailed balance.

The reader may convince himself that "ergodicity" of our upgrading procedure (apparently) is fulfilled: any allowed loop configuration[1] on a d-dimensional hypercubic lattice can in principle be constructed (and again decomposed) after sufficiently many iterations of local changes as depicted in Figs. 2 and 3. Hence we expect that the so obtained Gibbs distribution is the unique equilibrium distribution for the upgrading procedure. Furthermore, starting from an arbitrary initial configuration we expect to reach this equilibrium after an appropriately long "warming up" period. From the sequence of configurations generated by iterations of the procedure just explained we then select a subset of configurations which are separated by so many single steps that their correlations (in the sense that the configurations may still "look very similar") seem negligible. We have performed a systematic investigation of such correlations only for some particular cases. Therefrom we got the impression that after two or three complete sweeps through the entire lattice we may neglect the correlations.

The thermal average $\bar{X}$ of a variable X, cf. Eq. (3.8), is approximated by

$$\bar{X} \approx \frac{1}{n} \sum_{i=1}^{n} X(c_i), \tag{3.22}$$

1 We were unable to find a counterexample. Compare, however, also the remarks after Eq. (3.23)

where the sum now extends over a sufficiently large sample of equilibrium configurations c_i generated according to the above prescription. Practically, our measurements at different values of the parameters m, respectively k, are performed in so-called thermal cycles. Starting with the empty configuration at a low "temperature" (i.e. small hopping parameter k, large mass m) we increment k by a series of small steps up to a maximal value and then similarly return to the minimal value. The idea behind this is that equilibrium configurations do not vary too much for only gradually different temperatures. Of course, an appropriate warming up (or cooling down) period has to be included before an average (3.22) is taken at the new value.

Let us include at this stage some remarks on boundary conditions. In most of our calculations we take an "average" of periodic and antiperiodic boundary conditions, e.g.

$$\det = \tfrac{1}{2}(\det_{\text{periodic}} + \det_{\text{antiperiodic}}), \tag{3.23}$$

the reason being that in this combination the contributions of loops which are topologically nontrivial (e.g. wrapping around the entire lattice) cancel out. Such nontrivial loops cannot be generated locally. We have also done some computations including such loops but in the present applications of our method they seem to be negligible.

At the end of this section we would like to explain how our heat bath method can be adapted for the calculation of fermionic Green's functions $\langle \psi_{x_0} \ldots \psi_{x_n}^+ \rangle$. For simplicity we consider here only the two-point function

$$\langle \psi_v \psi_u^+ \rangle = (\gamma\partial + m)_{vu}^{-1}, \tag{3.24}$$

the generalizations being obvious. We consider a matrix $(\gamma\partial + m)^{(u)}$ which is equal to $\gamma\partial + m$, except that the row u has been replaced by a row of 1's, i.e.

$$(\gamma\partial + m)_{xy} = (1 - \delta_{x,u})\,(\gamma\partial + m)_{xy} + \delta_{x,v}, \tag{3.25}$$

which is a generalization of (2.7). The determinant of the modified matrix is easily calculated by expansion with respect to the row u,

$$\det(\gamma\partial + m)^{(u)} = \det(\gamma\partial + m) \sum_v (\gamma\partial + m)_{vu}^{-1}. \tag{3.26}$$

The auxiliary statistical mechanics system corresponding to $(\gamma\partial + m)^{(u)}$ again is a gas of self-avoiding closed oriented loops. In general, however, the configurations contain a particular closed loop which consists of a special line from the point u to any point v (which need not be a nearest neighbor of u), and then is closed by a path of ordinary links from v to u. Since we have chosen all entries in the modified row to be equal to 1, the energy of the special line is independent of its geometrical length. Therefore, we may disregard it altogether and say the configurations contain in addition to closed loops also an open line starting anywhere in the lattice and ending at the prescribed point u. In terms of our picture this means that the dog with home site u has been replaced by a "cat" which is not confined to its neighborhood by a chain, hence it may displace a dog from any site. The cat, however, can also easily be confined to a certain range I of points. In our computations we considered three different cases of intervals: the whole lattice

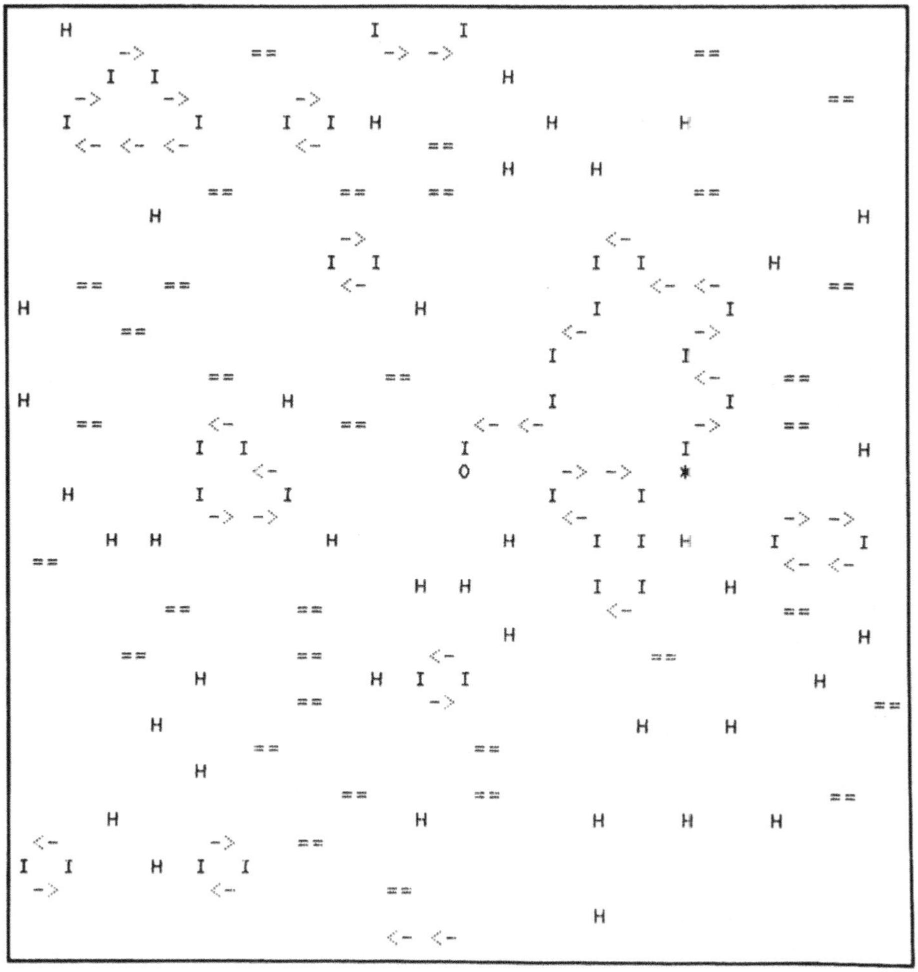

Fig. 4. A typical loop configuration including an open line starting at $*$ and ending at 0. Dimers in the vertical or horizontal direction are represented by H and $==$, respectively. The configuration was obtained in a Monte Carlo run for a free fermion with mass $m=1/(2\kappa)$ for $k=0.6$ on a 20×20 lattice

$I=(0,N-1)$, a ray parallel to an axis, e.g. $I=(0,L-1)$, or a half-ray e.g. $I=(u+1,L-1)$. To this end all entries in row u of $(\gamma\partial+m)^{(u,I)}_{xy}$ which correspond to forbidden sites, $y\notin I$, have to be set equal to zero, and Eqs. (3.25) and (3.26) have to be modified appropriately. In the process of generating new configurations we obviously can distinguish from each other the configurations which contribute to different terms on the right-hand side of Eq. (3.26), cf. Fig. 4. In terms of the characteristic function

$$h(v,u)=\begin{cases}1 & \text{if the open line starts at } v,\\ 0 & \text{otherwise},\end{cases} \tag{3.27}$$

we obtain

$$\langle \psi_v \psi_u^+ \rangle \propto \overline{h(v,u) \cdot \operatorname{sgn}(\gamma\partial + m)^{(u,I)}} \quad \text{for} \quad v \in I . \tag{3.28}$$

The normalization factor may be determined from $\det(\gamma\partial + m)$, cf. Eq. (3.14),

$$\begin{aligned} m \sum_x \langle \psi_x \psi_x^+ \rangle &= m \operatorname{tr}(\gamma\partial + m)^{-1} \\ &= m \frac{\partial}{\partial m} \ln \det(\gamma + m) \\ &= N - <\langle \ell \rangle . \end{aligned} \tag{3.29}$$

The upgrading procedure is modified as follows. Before every new sweep through the lattice we give the starting point of the open line (i.e. the cat) twice the chance to hop one step in any direction within the allowed range. We take the new configuration with probability $w_n/(w_0 + w_n)$. The ratio of the Boltzmann weights is given by $w_n/w_0 = k^{\pm 1}$ if the length of the open line increases or decreases, respectively. Only afterwards we sweep through the entire lattice with our local rearrangement operators but then we keep the starting point of the open line fixed.

4. Application of the Method

In this section we apply our method to the Scalapino-Sugar model [3]. This model is exactly soluble but nontrivial for a Monte Carlo simulation. Since all interesting quantities are exactly known, we thus obtain a good test of the efficiency of our method. The interaction of a (static) scalar real Bose field ϕ and a massive Fermi field ψ is governed by the Lagrangian

$$\mathscr{L}_{F,B} = \psi^+(\gamma\partial + m + g\phi^2)\psi + \phi^2 . \tag{4.1}$$

By integrating out the Bose degrees of freedom one realizes that this interacting field theoretic model is equivalent to a free fermion model with effective mass

$$m_{\text{eff}} = m + g/2 . \tag{4.2}$$

Indeed, one obtains

$$\begin{aligned} Z_{FB}(m,g) &= \int d\psi^+ d\psi d\phi e^{-\int dx \mathscr{L}_{FB}} \\ &= Z_{FB}(m_{\text{eff}}, 0) = \pi^{N/2} Z_F(m_{\text{eff}}) , \end{aligned} \tag{4.3}$$

with Z_F given by Eq. (3.6). The vacuum expectation value of ϕ^2 can be expressed in terms of $\psi\psi^+$:

$$\langle \phi_x^2 \rangle = \tfrac{1}{2}(1 + g\langle \psi_x \psi_x^+ \rangle) . \tag{4.4}$$

The fermion two-point function (on a d-dimensional hypercubic lattice, $L^d = N$) can be evaluated by Fourier sums, e.g.

$$\begin{aligned} \langle \psi_x \psi_y^+ \rangle &= (\gamma\partial + m_{\text{eff}})_{xy}^{-1} \\ &= \frac{1}{N} \sum_{k_\mu = 0}^{L-1} e^{\frac{2\pi i}{L} \mathbf{k} \cdot (\mathbf{x} - \mathbf{y})} \left[m_{\text{eff}} + i\gamma_\mu \sin \frac{2\pi}{L} k_\mu \right]^{-1} \end{aligned} \tag{4.5}$$

for periodic boundary conditions (for the antiperiodic case k_μ in the propagator has to be replaced by $k_\mu+\frac{1}{2}$). In $d=1$ dimension all quantities can be expressed in closed form in terms of Gegenbauer polynomials, e.g. for $L\geqq 4$ and L even one obtains

$$\begin{aligned} \det(\partial+m) &= 2^{1-L}\cosh L\varphi\,, \\ \langle\psi\psi^+\rangle &= \tanh L\varphi/\cosh\varphi\,, \\ \langle\ell\rangle &= L(1-\tanh\varphi\cdot\tanh L\varphi)\,, \end{aligned} \tag{4.6}$$

where $\sinh\varphi=m_{\mathrm{eff}}$.

Let us now address ourselves to the application of the heat bath method to this model. For the interacting case the formula (3.6) has to be replaced by [cf. Eqs. (2.6) and (2.14)]

$$\begin{aligned} Z_{FB} &= \int d\phi \det(\gamma\partial+m+g\phi^2)e^{-\sum_x \phi_x^2} \\ &= 2^{-N}\int d\phi \sum_{c\in\mathscr{C}} \mathrm{sgn}(c,\gamma\partial+m+g\phi^2) \prod_{p(x)=0} (2m+2g\phi_x^2)e^{-\sum_x \phi_x^2}\,, \end{aligned} \tag{4.7}$$

and the partition function Z for the auxiliary problem is obtained by omitting the factor $\mathrm{sgn}(\cdot,\cdot)$ in the last equation. A fermion configuration $c\in\mathscr{C}$ again consists of self-avoiding closed loops made up of simple links (i.e. the chains of the displaced dogs as was explained in Sect. 3). The fermionic Boltzmann weight at a point x is given by

$$w_F(x)=\begin{cases} 2m+2g\phi^2(x) & \text{for} \quad p(x)=0\,, \\ 1 & \text{for} \quad p(x)\neq 0\,, \end{cases} \tag{4.8}$$

i.e. we obtain a factor $(2m+2g\phi^2(x))$ for all those dogs which are at home, $\pi(x)=x$, cf. Eq. (3.18). On the other hand, the local weight to be used in the upgrading of the bosons is

$$w_B(x)=e^{-\phi^2(x)}w_F(x)\,. \tag{4.9}$$

The local upgrading procedure for the fermions is completely analogous to the free case explained in Sect. 3. The probability for accepting a new configuration is still given by Eq. (3.15) with, however, the relative weights w_F^n and w_F^0 given by $\prod_x w_F(x)$, where the product is only over the points which are affected by the tentative local change. After one Monte Carlo step for the fermions all $d(d-1)/2$ plaquettes emanating from the lattice point x we upgrade the Bose field $\phi(x)$ by a modified Metropolis algorithm [40]. We propose a new field,

$$\phi_n(x)=\phi_0(x)+r\,, \tag{4.10}$$

with a random number r uniformly distributed in the interval $(-1,1)$. The new field is accepted with probability

$$w=\max(1,w_B^n(x)/w_B^0(x))\,. \tag{4.11}$$

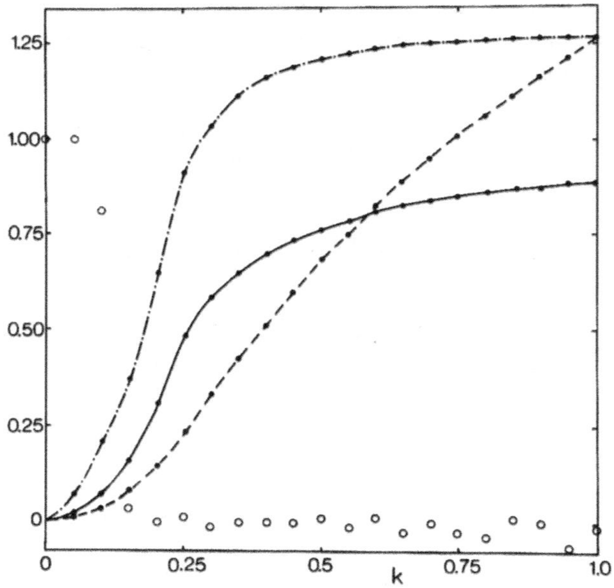

Fig. 5. Monte Carlo results for the thermal average values per lattice site of the length $\bar{\ell}/N$ (——), $(\log Z)/N$ (–·–·), and the entropy S/N (---) versus $k=1/(2m)=e^{-1/T}$ for the auxiliary statistical loop gas system corresponding to a free fermion on a 8^4 lattice. In addition $\overline{\text{sgn}}$ (○) is plotted

We repeat this bose upgrading five times before passing to the next point.

Let us finally report our results. We considered hypercubic lattices in 1, 2, 3, and 4 dimensions with linear sizes $L=100$–1000, 4–40, 4–16, and 4–10, respectively. By means of Monte Carlo simulations in thermal cycles $0 \leqq k \leqq k_{\max}$ for various numbers of intermediate steps and values for $k_{\max}$, we generated samples of equilibrium ensembles of fermion-boson configurations (c, ϕ) in two series of runs:

(a) $$(c, \phi) \in \mathscr{C}(\gamma\partial + m + g\phi^2) \otimes \mathbb{R}^N,$$

i.e. c is a configuration of closed self-avoiding oriented loops.

(b) $$(c, \phi) \in \mathscr{C}(\gamma\partial + m + g\phi^2)^{(u,I)} \otimes \mathbb{R}^N,$$

i.e. c contains in addition to case (a) one self-avoiding open line starting at some $v \in I$ and ending at u, which also avoids the loops.

For the case (a) we "measured" using Eq. (3.22) the auxiliary quantities $\bar{\ell}$, $\overline{\text{sgn}(A)}$, $\overline{\ell\,\text{sgn}(A)}$, and $\overline{\phi^2\,\text{sgn}(A)}$ for the matrix $A=\gamma\partial+m+g\phi^2$. From the average energy $\bar{\ell}$ at temperature $T=-1/\ln k$ (for a sufficiently large number of k-values with $0 \leqq k \leqq k_{\max}$) we can calculate other thermodynamic quantities for our statistical system. For the free fermion case, $g=0$, we then obtain by numerical integration of Eq. (3.12) the partition function Z and therefrom the entropy $S=\ln Z+\bar{\ell}/T$, see Fig. 5. For infinite temperature some results for the entropy per lattice site are listed in Table 1 for various dimensions and lattice sizes. Note that finite size effects seem negligible and a good fit is $S/N \approx (d+1)/4$. Therefore, the

Table 1. Monte Carlo results for the entropy per lattice site S/N at infinite temperature (or mass $=1/2$) for loop gas systems corresponding to free fermion field theories on various lattices L^d

d	1		2				3			4		
L	100	1000	4	10	20	40	4	8	16	4	6	8
S/N	0.479	0.478	0.764	0.763	0.762	0.764	1.02	1.03	1.03	1.25	1.26	1.26

total number of loop configurations for dimensions $d \leqq 4$ is approximately given by $\exp(N \cdot (d+1)/4)$.

The "physical" quantities $\langle \psi\psi^+ \rangle$ and $\langle \phi^2 \rangle$ (see Fig. 6) are obtained by Eqs. (3.10), (3.14), and (3.29), and the fermion determinant $Z_F = \det(\gamma\partial + m)$ for the free case $g=0$ by Eq. (3.13) or by Eqs. (3.7) and (3.9).

We have two types of checks of our Monte Carlo results,

i) consistency checks e.g.

$$\langle \phi^2 \rangle = (1 + g\langle \psi\psi^+ \rangle)/2\,, \quad \text{cf. (4.4)}\,,$$

$$Z_F = m^N Z\,\overline{\mathrm{sgn}}\,, \quad \text{cf. (3.9)}\,,$$

ii) comparison with numbers obtained by numerical Fourier summation, cf. Eq. (4.5).

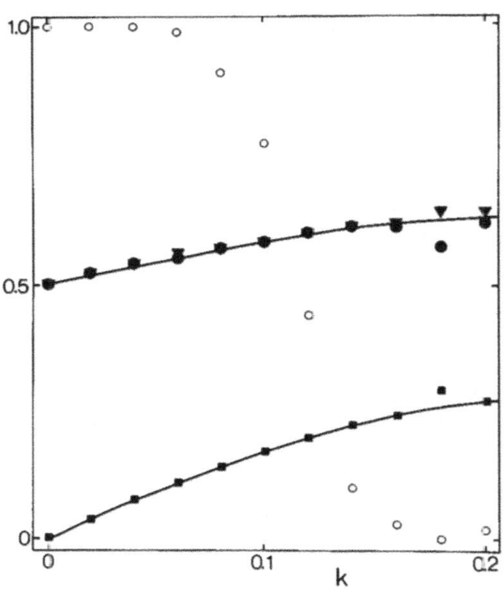

Fig. 6. Monte Carlo results for $\langle \psi\psi^+ \rangle$ (■) and $\langle \phi^2 \rangle$ (●) versus the hopping parameter $k = 1/(2m)$ for the Scalapino-Sugar model with coupling $g=1$ on a 10^4 lattice. The curves are drawn through the values obtained by Fourier summation. For a check of Eq. (4.4) the quantity $(1 + g\langle \psi\psi^+ \rangle)/2$ (▼) is included, too. The auxiliary quantity $\overline{\mathrm{sgn}}$ (○) is also plotted

Table 2. Monte Carlo results for the fermion determinant Z_F in terms of $A=\ln(Z_F^{MC}/m^N)/N$, as a check $B=\ln(Z^{MC}\overline{(\mathrm{sgn})}^{MC})/N$, and the "physical" quantity $\langle\psi\psi^+\rangle^{MC}$ in comparison with Fourier results $\langle\psi\psi^+\rangle^{Fou}$ for free fermions with mass $m=1/(2k)$ on various lattices L^d

d	L	k	A	B	$\langle\psi\psi^+\rangle^{MC}$	$\langle\psi\psi^+\rangle^{Fou}$
1	100	0.5	0.188	0.188	0.704	0.707
1	100	5	1.71	1.71	1.05	1.00
2	20	0.2	0.0727	0.0727	0.346	0.347
2	20	1	0.786	0.753	0.50	0.51
3	16	0.1	0.0288	0.0288	0.191	0.189
3	16	0.25	0.158	0.157	0.359	0.368
4	8	0.1	0.0381	0.0381	0.187	0.185
4	8	0.15	0.0808	0.0818	0.256	0.255

Table 3. Monte Carlo results and checks for the Scalapino-Sugar model with mass $m=1/(2k)$ and coupling $g=1$ on various lattices L^d

d	L	k	$\langle\psi\psi^+\rangle^{MC}$	$\langle\psi\psi^+\rangle^{Fou}$	$\langle\phi^2\rangle^{MC}$	$(1+g\langle\psi\psi^+\rangle)/2$
1	1000	0.5	0.5548	0.5547	0.7769	0.7774
1	1000	5	0.8581	0.8575	0.9284	0.9290
2	20	0.2	0.3010	0.3008	0.6486	0.6505
2	20	1.0	0.5130	0.5136	0.7554	0.7565
3	16	0.1	0.1737	0.1733	0.5868	0.5866
3	16	0.25	0.3227	0.3246	0.6614	0.6623
4	10	0.1	0.1698	0.1706	0.5842	0.5849
4	10	0.2	0.2749	0.2739	0.623	0.637

In $d=1$ dimensions we have also the exact formulae (4.6). In Tables 2 and 3 we list some comparable quantities for various dimensions and lattice sizes for $g=0$ and $g=1$, respectively. The suffixes MC and Fou are employed to distinguish Monte Carlo from Fourier results. As a rule we find agreement within a few percent or better. The errors become large (see Fig. 6) if $\overline{\mathrm{sgn}}$ starts fluctuating around zero. This phenomenon determines the critical values of k which restricts the applicability of our Monte Carlo method to small k, as was already mentioned at the beginning of Sect. 3.

For the case (b) we "measured" the two point function $\langle\psi_v\psi_u^+\rangle$ using Eq. (3.28). The results are in good agreement with the Fourier results, cf. Figs. 7 and 8. The exponential decay can be fitted by the ansatz

$$\langle\psi_v\psi_u^+\rangle \sim e^{-\mu|\mathbf{v}-\mathbf{u}|}. \tag{4.20}$$

In Table 4 we have collected the corresponding mass parameters. Again the agreement is satisfactory. Note that the parameter μ differs from the effective mass of the model if the latter is not small compared to the cut off mass 1/(lattice spacing) which has been set equal to 1.

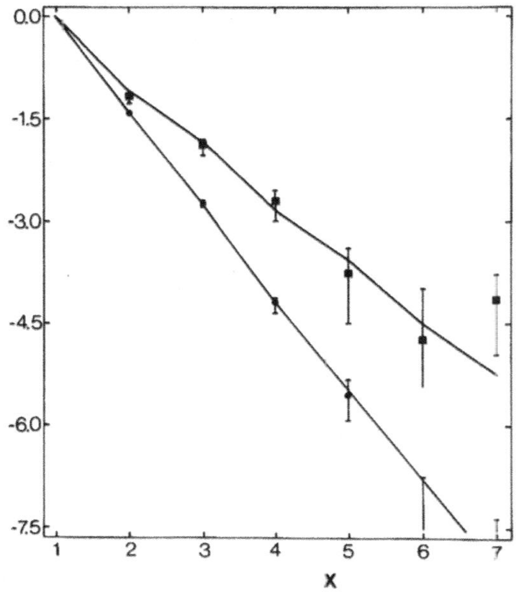

Fig. 7. Monte Carlo results for $\log(\langle\psi(x)\psi^+(0)\rangle/\langle\psi(1)\psi^+(0)\rangle)$ versus the distance x for a free fermion with mass $m=1/(2k)$ and $k=0.3$ (●), and $k=0.6$ (■) on a 20×20 lattice. The solid lines connect the values obtained by Fourier summation. A typical configuration for $k=0.6$ is shown in Fig. 4

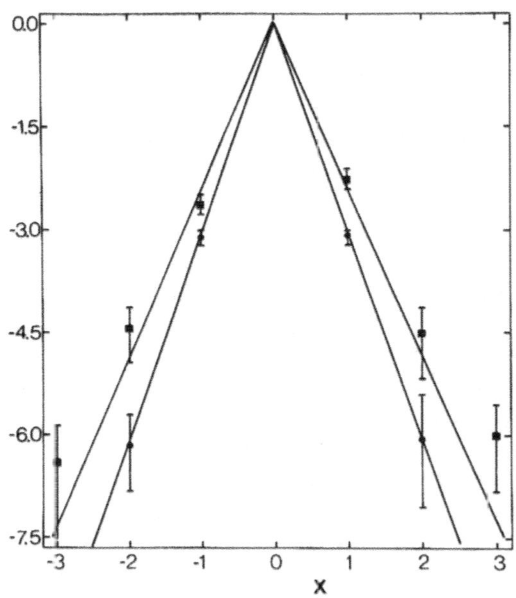

Fig. 8. Monte Carlo results for $\log(\langle\psi(x)\psi^+(0)\rangle/\langle\psi(0)\psi^+(0)\rangle)$ versus the distance x for the Scalapino-Sugar model with $k=1/(2m)=0.05$ (●) and 0.1 (■) and coupling $g=1$ on a 10^4 lattice. The solid lines connect the values obtained by Fourier summation

Table 4. Monte Carlo results for the mass parameter μ obtained from the fit Eq. (4.20) in comparison with Fourier results for the Scalapino-Sugar model with mass $m=1/(2k)$ and coupling g on various lattices L^d

d	L	g	k	m_{eff}	μ^{MC}	μ^{Fou}
1	100	0	1	0.5	0.487	0.481
1	100	0	2	0.25	0.253	0.247
1	100	1	1	1.0	0.878	0.881
1	100	1	2	0.75	0.693	0.693
2	20	0	0.3	1.67	1.36	1.35
2	20	0	0.6	0.83	1.0	0.9
2	20	1	0.3	2.17	1.58	1.56
2	20	1	0.6	1.33	1.18	1.18
3	16	0	0.1	5	2.38	2.33
3	16	0	0.2	2.5	1.26	1.70
3	16	1	0.1	5.5	2.33	2.41
3	16	1	0.2	3	1.75	1.85
4	10	0	0.05	10	2.89	3.00
4	10	0	0.1	5	2.30	2.35
4	10	1	0.05	10.5	3.10	3.05
4	10	1	0.1	5.5	2.42	2.41

5. Conclusions

In this paper we have introduced a new method for numerical investigations of quantum field theories with fermions. The essential point is the construction of a "heat bath for fermion determinants" on a lattice. For simplicity we restricted our discussion to the case of Susskind fermions and hypercubic lattices. The key observation relies on the fact that every term contributing to the expansion of a determinant may be graphically represented by a configuration of self-avoiding oriented loops. A direct application of importance sampling by a Monte Carlo procedure is inhibited by the different signs with which these configurations contribute to the determinant. We therefore introduce an auxiliary thermodynamical system of a gas of self-avoiding oriented loops with Boltzmann weights proportional to the modulus of the above contributions. The sign is then subsumed in the observables. Therefore, the quantum field expectation values turn out to be ratios of thermodynamic average values.

We have devised a local upgrading procedure for generating equilibrium ensembles of loop configurations. The boson fields can be treated as usual in Monte Carlo simulations. This allows us to evaluate bosonic physical quantities in quantum field theories involving fermions. By introducing one open line besides the loops we calculate fermionic Green's functions.

In most previous proposals the fermions are either not considered as dynamical variables at all or they require a nonlocal upgrading. In contrast we include all fermionic vacuum fluctuations, treat fermions and bosons on the same footing and work with a local upgrading procedure. The computer times for the fermionic and the bosonic degrees of freedom are comparable. In the application to the Scalapino-Sugar model we found good agreement with known results.

We plan to extend our calculations to more interesting field theoretic models involving fermions, including gauge theories. Instead of Susskind fermions we then shall use Wilson fermions in order to suppress the extra light fermions which are responsible for large statistical errors in some of our present computations.

In Sect. 2 we also mentioned how by a modification of our techniques, one obtains a description and an upgrading procedure for the statistical ensemble of a polymer gas. In fact, this method was used in [42] to obtain numerical information on its critical behaviour, see also [9].

By extending the ideas developed there, a numerical discussion of the theory of self-avoiding random surfaces is given in [43], which differs from the upgrading procedure used in [44] (see also [45]). We emphasize that although we designed our method with local lattice theories in mind, it could also have other applications like calculating Slater determinants in nuclear and condensed matter physics. Finally, [46, 47] give a discussion of "euclidean" lattice gravity using Regge calculus. There it is pointed out that already this lattice version contains a Faddeev-Popov determinant which necessitates the introduction of fermionic and bosonic ghost fields to restore locality. In principle our method should also be able to treat these fermionic fields numerically.

Acknowledgements. We thank B. Schroer for discussions. One of us (M.K.) thanks C. N. Yang for financial support for a visit to and one of us (R.S.) for kind hospitality at the Institute of Theoretical Physics (SUNY), Stony Brook. This work was supported by the Deutsche Forschungsgemeinschaft under contracts Schr 160/7-1 and Schr 4/9-2 and by NSF Grant Phy 81 09110 A 01. The numerical calculations were performed on the VAX 11/780 computers of Fachbereich Physik der Freien Universität Berlin and Gesellschaft für wissenschaftliche Datenverarbeitung mbH, Göttingen.

References

1. Weingarten, D.H., Petcher, D.N.: Monte Carlo integration for lattice gauge theories with fermions. Phys. Lett. **99** B, 333 (1981)
2. Fucito, F., Marinari, E., Parisi, G., Rebbi, C.: A proposal for Monte Carlo simulations of fermionic systems. Nucl. Phys. B **180**, 369 (1981)
3. Scalapino, D.J., Sugar, R.L.: Method for performing Monte Carlo calculations for systems with fermions. Phys. Rev. Lett. **46**, 519 (1981)
4. Hamber, H.: Monte Carlo simulations of systems with fermions. Phys. Rev. D **24**, 951 (1981)
5. Marinari, E., Parisi, G., Rebbi, C.: Monte Carlo simulation of the massive Schwinger model. Nucl. Phys. B **190**, 734 (1981)
6. Duncan, F., Furman, A.: Monte Carlo calculations with fermions: the Schwinger model. Nucl. Phys. B **190**, 767 (1981)
7. Blankenbecler, R., Scalapino, D.J., Sugar, R.L.: Monte Carlo calculations of coupled boson-fermion systems. I. Phys. Rev. D **24**, 2278 (1981)
8. Hirsch, J., Scalapino, D.J., Sugar, R.L., Blankenbecler, R.: Efficient Monte Carlo procedure for systems with fermions. Phys. Rev. Lett. **47**, 1628 (1981)
9. Berg, B., Foerster, D.: Random paths and random surfaces on a digital computer. Phys. Lett. **106** B, 323 (1981)
10. Hamber, H., Parisi, G.: Numerical estimates of hadronic masses on a pure $SU(3)$ gauge theory. Phys. Rev. Lett. **47**, 1792 (1981)
Hamber, H., Marinari, E., Parisi, G., Rebbi, C.: Spectroscopy in a latter gauge theory. Phys. Lett. **108** B, 314 (1982)

11. Weingarten, D.H.: Monte Carlo evaluation of hadron masses in lattice gauge theories with fermions. Phys. Lett. **109** B, 57 (1982)
12. Hasenfratz, A., Hasenfratz, P., Kunszt, Z., Lang, C.B.: Hopping parameter expansion for the meson spectrum in $SU(3)$ lattice QCD. Phys. Lett. **110** B, 289 (1982)
13. Kuti, J.: Stochastic method for the numerical study of lattice fermions. Phys. Rev. Lett. **49**, 183 (1982)
14. Anthony, S.J., Llewellyn Smith, C.H., Wheater, J.F.: A new proposal for Monte Carlo simulation of fermions on a lattice. Phys. Lett. **116** B, 287 (1982)
15. Duffy, W., Guralnik, G., Weingarten, D.: Error estimate for the valence approximation to lattice QCD. Bloomington preprint (1983)
16. Joos, H., Montvay, I.: The screening of colour charge in numerical hopping parameter expansion. Nucl. Phys. B **225** [FS9], 565 (1983)
17. Bhanot, J.: Lattice gauge theory: the Monte Carlo approach. CERN preprint TH. 3507 (1983)
18. Creutz, M., Jacobs, L., Rebbi, C.: Monte Carlo computations in lattice gauge theories. Brookhaven preprint BNL 32438 (1983)
19. Rebbi, C.: Lattice gauge theories and Monte Carlo simulations. Singapore: World Scientific 1983
20. Symanzik, K.: Euclidean quantum field theory. I. Equations for a scalar model. J. Math. Phys. **7**, 510 (1966)
21. Symanzik, K.: In, Proc. Intern. School of Physics "Enrico Fermi", Varenna Course XLV, Jost, R. (ed.). New York: Academic Press 1969
22. Kogut, J., Susskind, L.: Hamiltonian formulation of Wilson's lattice gauge theories. Phys. Rev. D **11**, 395 (1975)
23. Berezin, F.A.: The method of second quantization. New York, San Francisco, London: Academic Press 1966
24. Fisher, M.E.: Statistical mechanics of dimers on a plane lattice. Phys. Rev. **124**, 1664 (1961)
25. Temperley, H.N.V., Fisher, M.E.: Dimer problem in statistical mechanics – an exact result. Philos. Mag. **6**, 1061 (1961)
26. Kasteleyn, P.W.: The statistics of dimers on a lattice. I. The number of dimer arrangements on a quadratic lattice. Physica **27**, 1209 (1961)
27. Fisher, M.E., Stephenson, J.: Statistical mechanics of dimers on a plane lattice. II. Dimer correlations and monomers. Phys. Rev. **132**, 1411 (1963)
28. Brydges, D., Fröhlich, J., Spencer, T.: The random walk representation of classical spin systems and correlation inequalities. Commun. Math. Phys. **83**, 123 (1982)
29. Aizenman, M.: Geometric analysis of ϕ^4 fields and Ising models. Parts I and II. Commun. Math. Phys. **86**, 1 (1982)
30. Wilson, K.: Confinement of quarks. Phys. Rev. D **10**, 2445 (1974)
Wilson, K.: Quarks on a lattice, or, the colored string model. Phys. Reports **23**, 331 (1975)
Wilson, K.: In: New phenomena in subnuclear physics, Zichichi, A. (ed.). New York: Plenum Press 1977
31. Kogut, J., Susskind, L.: Hamiltonian formulation of Wilson's lattice gauge theories. Phys. Rev. D **11**, 395 (1975)
32. Banks, T., Raby, S., Susskind, L., Kogut, J., Jones, D.R.T., Scharbach, P., Sinclair, D.: Strong-coupling calculations of the hadron spectrum of quantum chromodynamics. Phys. Rev. D **15**, 1111 (1977)
33. Susskind, L.: Lattice fermions. Phys. Rev. D **16**, 3031 (1977)
34. Sharatchandra, H.S., Thun, H.J., Weisz, P.: Susskind fermions on a Euclidean lattice. Nucl. Phys. B **192**, 205 (1981)
35. Kawamoto, N., Smit, J.: Effective lagrangian and dynamical symmetry breaking in strongly coupled lattice QCD. Nucl. Phys. B **192**, 100 (1981)
36. Kluberg-Stern, H., Morel, A., Napoly, O., Petersson, B.: Flavours of Lagrangean Susskind fermions. Nucl. Phys. B **220** [FS8], 447 (1983)
37. Becher, P.: Dirac fermions on the lattice – a local approach without spectrum degeneracy. Phys. Lett. **104** B, 221 (1981)

38. Becher, P., Joos, H.: The Dirac-Kähler equation and fermions on a lattice. Z. Phys. C **15**, 343 (1982)
39. Mitra, P.: Geometry of non-degenerate Susskind fermions. Nucl. Phys. B **227**, 349 (1983)
40. Metropolis, N., Rosenbluth, A.W., Rosenbluth, M.N., Teller, A.H., Teller, E.: Theory of the effects of exchange on the nuclear fine structure in the paramagnetic resonance spectra of liquids. J. Chem. Phys. **27**, 1087 (1953)
41. Friedberg, R., Cameron, J.: Test of the Monte Carlo method: Fast simulation of a small Ising lattice. J. Chem. Phys. **52**, 6049 (1970)
42. Karowski, M., Thun, H.J., Helfrich, W., Rys, F.: Numerical study of self-avoiding loops on *d*-dimensional hypercubic lattices. J. Phys. A **16**, 4073 (1983)
43. Karowski, M., Thun, H.J.: Monte Carlo simulations for self-avoiding surfaces (in preparation)
44. Sterling, T., Greensite, J.: Entropy of self-avoiding surfaces on the lattice. Phys. Lett. **121** B, 345 (1983)
45. Schrader, R.: On the Euler characteristics of random surfaces. Stony Brook preprint ITP-SB-84-44 (1984)
46. Cheeger, J., Müller, W., Schrader, R.: In: Unified theories of elementary particles (Heisenberg Symposium 1981), Breitenlohner, P., Dürr, H.P. (eds.). Lecture Notes in Physics, Vol. 160. Berlin, Heidelberg, New York: Springer 1982
47. Cheeger, J., Müller, W., Schrader, R.: On the curvature of piecewise flat spaces. Commun. Math. Phys. **92**, 405 (1984)

Communicated by G. Mack

Received February 7, 1984

SU(2) Lattice Gauge Theory: Standard Action Versus Symanzik's Tree-Improved Action

B. Berg[1], A. Billoire[2], S. Meyer[3]*, and C. Panagiotakopoulos[4]

1 II. Institut für Theoretische Physik der Universität Hamburg, Luruper Chaussee 149, D-2000 Hamburg 50, Federal Republic of Germany
2 CEN Saclay, Service de Physique Théorique, Orme des Merisiers, F-91191 Gif-sur-Yvette Cedex, France
3 Fachbereich Physik der Universität Kaiserslautern, Erwin-Schrödinger-Strasse, D-6750 Kaiserslautern, Federal Republic of Germany
4 Physics Department (Theory), The Rockefeller University, 1230 York Avenue, New York, NY 10021-6399, USA

Dedicated to the memory of Kurt Symanzik

Abstract. We carry out Monte Carlo simulations of the $4d$ SU(2) lattice gauge theory. The standard action and the Symanzik tree-improved action are used. Results for the string tension, glueball masses, and energy-momentum dispersion are reported. In case of the standard action our results are a finite size study extending previous investigations.

I. Introduction

Symanzik [1, 2] has pointed out that corrections to continuum theory stemming from finite lattice spacing can be systematically diminished by use of a judiciously chosen lattice action. A number of Monte Carlo (MC) studies [3] have been carried out using Symanzik improved action. In this paper we report high statistics results for $4d$ SU(2) lattice gauge theory.

In the continuum limit each physical quantity is proportional to an appropriate power of the correlation length (inverse mass gap) ξ with an universal coefficient. For non-zero lattice spacing $a \neq 0$ this "scaling" is violated by non-universal terms of order $(a^2/\xi^2)\ln(a/\xi)$. Symanzik improved actions allow us to reduce these violations to order $(a^2/\xi^2)^2 \ln(a/\xi)$, to all orders of perturbation theory, by including in the lattice action suitable chosen irrelevant terms.

In principle the coefficients of these irrelevant terms can be calculated up to any given order of perturbation theory. For $4d$ SU(n) lattice gauge theories the tree-level improved action (TIA) has been determined [4–6]. A motivated ansatz [5] for the improved action includes Wilson loops up to length 6 and the result

* Supported by Deutsche Forschungsgemeinschaft

[4, 6] for the TIA becomes

$$S^{\mathrm{TI}} = -\frac{4}{g^2}\left[\frac{5}{3}\sum_{\square}\mathrm{Re}(\mathrm{Tr}\,\square) - \frac{1}{12}\sum_{\square\!\square}\mathrm{Re}(\mathrm{Tr}\,\square\!\square)\right]. \tag{I.1}$$

Here $\square\!\square$ represents the planar rectangular double plaquettes of size 1×2.

We carry out a high statistics MC simulation using

a) the standard action (SA) on an 8^4 lattice,

b) the TIA on $5^3\cdot 8$ and 8^4 lattices.

We are mainly interested in calculating the glueball mass spectrum[1] for both actions. We use a variant of the MC variational (MCV) method, which was pioneered in [8–10]. The Wilson loops involved in our calculations are numerated in Fig. I.1. Following the classification of [11] we construct states in various irreducible representations of the cubic group. For more details about the glueball calculations see Sect. II.2.

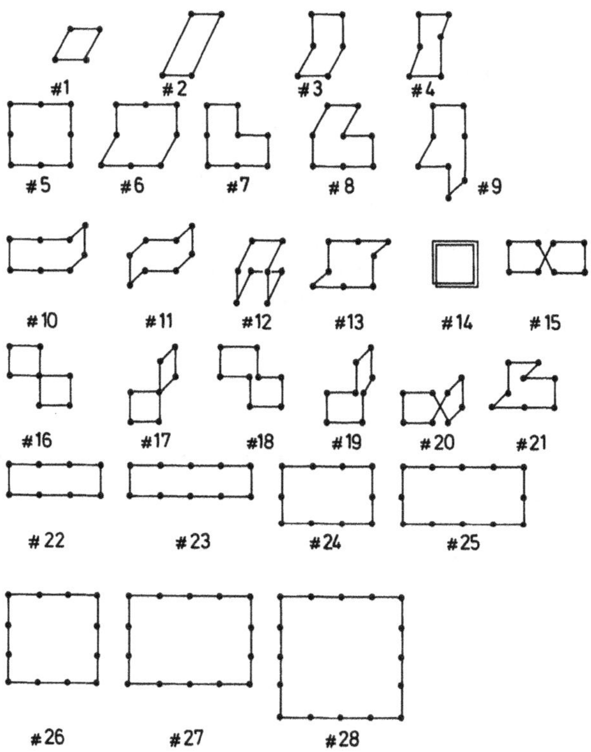

Fig. I.1. Wilson loops as used in this paper

For some operators we also calculate the energies of momentum eigenstates and consider the relativistic energy-momentum dispersion. In the $2d$ O(3) non-linear σ-model such an analysis was previously carried out in [12]. Improvement was found for the 1-loop improved action, but not for the TIA. In the present case the TIA behaves even worse than the SA.

1 For a recent review of glueball calculations in lattice gauge theories, see [7]

In case of the SA a scaling window for the 0^+ glueball (mass gap) was first obtained in [8]. The MCV method was used on a $4^3 \cdot 16$ lattice, leading to the estimate

$$m(0^+)=(170 \pm 30)\Lambda_L . \tag{I.2}$$

In this paper we present a finite size study of this and other results [10] obtained on small lattices. Previous [13, 14] investigations on an 8^4 lattice are improved and corrected. Our *final* $m(0^+)$ estimate is well consistent with the estimate (I.2).

As a byproduct we obtain high statistics data for Creutz [15] ratios up to $\chi(4,4)$. This allows an improvement of previous string tension estimates (for instance [15, 16]) which were mainly based on $\chi(3,3)$. Creutz ratios $\chi(I,J)$ are not stable under increasing (I,J) from $(3,3)$ to $(4,4)$, and the string tension has a decreasing tendency. The potential analysis of [17] on a 15^4 lattice and measurements of Polyakov loop correlation functions [18, 19] indicate that this process continues also beyond the estimate one obtains from $\chi(4,4)$.

In the case of the TIA we have assembled very large statistics on a $5^3 \cdot 8$ lattice, on which our glueball estimates are based. This improves previously reported results [20]. It has to be emphasized [21] that the TIA has severe problems with the transfer matrix. The transfer matrix is not positive definite, not even symmetric. It connects three time-layers. Therefore, the MCV method has problems, and the smallest distance, where we can expect to get meaningful results is $t=2$.

As in [12] we distinguish in this paper scaling in the general sense and asymptotic scaling. Scaling in general means scaling with respect to the (unknown) lattice β-function, and asymptotic scaling means scaling with respect to the *universal* part of the β-function. Asymptotic scaling implies proportionality of a physical quantity with respect to a standard two-loop Λ_L scale. The improvement program [1, 2, 4–6] is made for general scaling. Interesting quantities are mass ratios for which we have

$$\frac{m_1}{m_2} \simeq \left(\frac{a^2}{\xi_r^2}\right) \ln\left(\frac{a}{\xi_r}\right) \to \left(\frac{a^2}{\xi_r^2}\right)^2 \ln\left(\frac{a}{\xi_r}\right) \tag{I.3}$$

up to the considered order of the coupling g^2. In Eq. (I.3) we have introduced the "relevant range of interaction" [7] $\xi_r = \alpha\xi$. The parameter α should be adjusted, such that the proportionality constant in (I.3) is numerically close to 1. Typically $\alpha \approx 4$ for $4d$ lattice gauge theories. This emphasizes the meaning of the correlation length ξ as a counting parameter for (a^2/ξ^2) corrections to scaling and indicates that improvement may already work at a correlation length $\xi < 1$.

The string tension for the SU(2) TIA was first studied in [22, 23]. The MC statistics of these investigations, particularly of [22], is rather poor. Therefore, we did a calculation with a more reasonable MC statistics on an 8^4 lattice. We obtain results for Creutz ratios up to $\chi(4,4)$. Also improved Creutz ratios [6] are studied and found to converge better with respect to the loop size. This indicates nicely that improvement is sensible in the considered $\beta = 4/g^2$-region. However, the potential analysis [17] indicates again a further lowering of the string tension beyond the result as obtained from $\chi(4,4)$.

Our paper is organized as follows: Section II contains our SA results, Sect. III the TIA results, and finally, summary and conclusions are presented in Sect. IV.

Table II.1. Rectangular Wilson loops

β	2.20	2.25	2.30	2.35	2.40	2.50
Sweeps (Eq)	40,000 (1,200)	60,000 (1,200)	60,000 (1,200)	60,000 (1,200)	60,000 (1,200)	60,000 (1,200)
Measurements (Bins)	20,000 (20)	24,000 (20)	24,000 (20)	20,000 (16)	20,000 (20)	20,000 (20)
W_{11}	0.56928 ±0.00008	0.58619 ±0.00008	0.60215 ±0.00008	0.61690 ±0.00011	0.63025 ±0.00008	0.65241 ±0.00005
W_{12}	0.34126 ±0.00012	0.36385 ±0.00012	0.38575 ±0.00013	0.40646 ±0.00017	0.42551 ±0.00014	0.45733 ±0.00008
W_{22}	0.13644 ±0.00013	0.15767 ±0.00014	0.17976 ±0.00017	0.20186 ±0.00022	0.22315 ±0.00021	0.25971 ±0.00014
W_{13}	0.20682 ±0.00013	0.22862 ±0.00012	0.25049 ±0.00014	0.27173 ±0.00019	0.29178 ±0.00016	0.32594 ±0.00010
W_{14}	0.12562 ±0.00012	0.14403 ±0.00011	0.23071 ±0.00013	0.18228 ±0.00018	0.20082 ±0.00016	0.23325 ±0.00011
W_{23}	0.05716 ±0.00011	0.07192 ±0.00010	0.08848 ±0.00015	0.10626 ±0.00020	0.12436 ±0.00020	0.15706 ±0.00016
W_{24}	0.02426 ±0.00007	0.03330 ±0.00007	0.04424 ±0.00011	0.05691 ±0.00015	0.07057 ±0.00017	0.09683 ±0.00015
W_{33}	0.01747 ±0.00008	0.02518 ±0.00007	0.03505 ±0.00011	0.04683 ±0.00010	0.05992 ±0.00018	0.08566 ±0.00018
W_{34}	0.00554 ±0.00005	0.00915 ±0.00005	0.01437 ±0.00008	0.02140 ±0.00010	0.03003 ±0.00013	0.04867 ±0.00015
W_{44}	0.00137 ±0.00005	0.00271 ±0.00003	0.00501 ±0.00005	0.00855 ±0.00008	0.01363 ±0.00011	0.02613 ±0.00015

II. The Standard Action

In this section we report our MC results as obtained with the SA. We use the icosaeder SU(2) subgroup [24], our simulation is carried out on an 8^4 lattice. Our main interest is concerned with the glueball spectrum. Therefore, we concentrate on a few important β-values and collect at each β-value very high statistics, as is obvious from Table II.1. As in [11] we use random upgrading and a sweep is defined by upgrading each link once in the mean. After two sweeps ($\beta \leqq 2.30$), three sweeps ($\beta \geqq 2.35$) measurements are performed. As analyzed in [11] this is expected to save computer time as compared with calculating up to a similar accuracy by doing measurements after each sweep. We have used the Metropolis method with 5 trials per upgrading. A typical acceptance rate for 5 trials is 70%.

For each measurement we select a time direction (which is cyclically rotated for the next measurement etc.) and calculate in each t=const plane the vacuum expectation values for all Wilson loops up to length 8 and additionally for the rectangular loops up to size $4 \cdot 4$. The Wilson loops are numerated in Fig. I.1. Our

Table II.2. Creutz ratios

β	2.20	2.25	2.30	2.35	2.40	2.50
$\chi(1,1)$	0.5634 ± 0.0002	0.5341 ± 0.0002	0.5072 ± 0.0002	0.4830 ± 0.0002	0.4616 ± 0.0002	0.42708 ± 0.00007
$\chi(1,2)$	0.5117 ± 0.0003	0.4769 ± 0.0002	0.4453 ± 0.4453	0.4172 ± 0.0003	0.3928 ± 0.0002	0.35526 ± 0.00011
$\chi(2,2)$	0.4051 ± 0.0005	0.3594 ± 0.0004	0.3181 ± 0.0004	0.2827 ± 0.0005	0.2526 ± 0.0005	0.21058 ± 0.00027
$\chi(1,3)$	0.5008 ± 0.0003	0.4646 ± 0.0002	0.4317 ± 0.0003	0.4027 ± 0.0003	0.3773 ± 0.0003	0.33871 ± 0.00014
$\chi(1,4)$	0.4986 ± 0.0004	0.4620 ± 0.0002	0.4287 ± 0.0003	0.3993 ± 0.0003	0.3736 ± 0.0003	0.33460 ± 0.00016
$\chi(2,3)$	0.3693 ± 0.0007	0.3205 ± 0.0004	0.2769 ± 0.0005	0.2390 ± 0.0006	0.2074 ± 0.0005	0.1643 ± 0.0004
$\chi(2,4)$	0.3583 ± 0.0011	0.3084 ± 0.0006	0.2640 ± 0.0006	0.2251 ± 0.0007	0.1929 ± 0.0007	0.1491 ± 0.0005
$\chi(3,3)$	0.3154 ± 0.0023	0.2649 ± 0.0011	0.2169 ± 0.0008	0.1778 ± 0.0007	0.1456 ± 0.0009	0.1032 ± 0.0007
$\chi(3,4)$	0.2910 ± 0.0044	0.2421 ± 0.0018	0.1979 ± 0.0015	0.1586 ± 0.0011	0.1242 ± 0.0011	0.0818 ± 0.0007
$\chi(4,4)$	0.251 ± 0.029	0.2115 ± 0.0081	0.1657 ± 0.0046	0.1350 ± 0.0042	0.0992 ± 0.0028	0.0563 ± 0.0015

notation for expectation values of rectangular loops of size $I \cdot J$ is W_{IJ} and our normalization is $\lim_{\beta \to \infty} W_{IJ}(\beta) = 1$. We obtain high precision expectation values for these loops. The results for rectangular loops are collected in Table II.2. In most of the cases the statistical analysis is done with respect to 20 bins considered as typical "independent events". This means that a typical "independent event" is an average over 1000 measurements done over 2000–3000 sweeps.

At $\beta = 2.2$, 2.3, 2.4, and 2.5 some values can be compared with those of [16], where the full SU(2) group and the heat bath method was used. Within statistical errors (Table 5 of [16], where all Wilson loops (including timelike directions) were measured after each sweep (systematic upgrading), has to be used) we find agreement up to $\beta = 2.4$. At $\beta = 2.5$ a small systematic lowering is observed. Using the measured Wilson loops we obtain high statistics results for the string tension relying in Creutz ratios. These results are presented in Sect. II.1.

For each measured Wilson loop (in each $t = \text{const}$ plane) we construct various irreducible representations of the cubic group (see [11] for details) and keep on-diagonal correlations up to distance $t = 2$ for excited states and up to distance $t = 3$ for the 0^+ state. For some operators we also calculate momentum $\mathbf{p} \neq 0$ eigenstates (see Sect. II.5). Our spectrum results are presented in Sects. II.2–II.4: The $m(0^+)$ mass gap in Sect. II.2, excited glueball states in Sect. II.3 and finally the energy-momentum dispersion is studied in Sect. II.4.

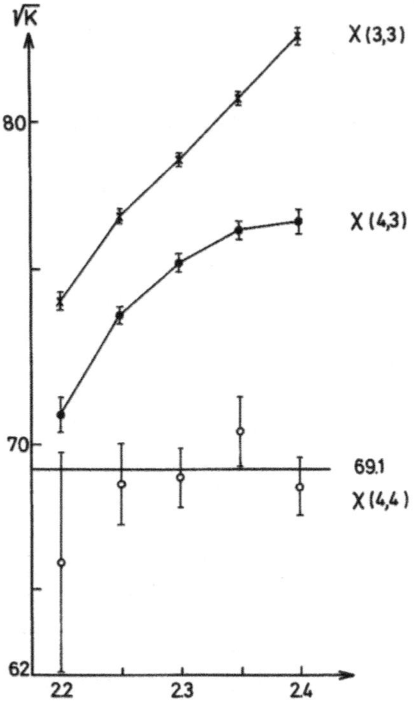

Fig. II.1. String tension estimate (in units of Λ_L^{SA}) from Creutz ratios

II.1. The String Tension

From our measured rectangular Wilson loops (Table II.1) we obtain high precision results for the Creutz [15] ratios

$$\chi(I,J) = -\ln\left(\frac{W_{IJ}\cdot W_{I-1,J-1}}{W_{I-1,J}\cdot W_{I,J-1}}\right) \tag{II.1}$$

up to $I=J=4$. These results are listed in Table II.2. Previous MC investigations (for instance [15, 16]) had only a sufficient statistics for obtaining reliable results up to $\chi(3,3)$ and led to continuum estimates of the string tension of order

$$\sqrt{K} \approx 76\,\Lambda_L^{SA}. \tag{II.2}$$

We have now a nearly 100 times larger statistics than in [16]. The new very precise data for $\chi(3,3)$ are within statistical errors consistent with the early data. In Fig. II.1 we plot our present results for $\chi(3,3)$, $\chi(4,3)$, and $\chi(4,4)$ in physical units of Λ_L. On the detailed scale of this figure we realize clear deviations from asymptotic scaling for $\chi(3,3)$. The situation is improving for $\chi(4,3)$ and $\chi(4,4)$. Within statistical errors the $\chi(4,4)$ data are consistent with asymptotic scaling and lead to the continuum estimate

$$\sqrt{K} \approx 69\,\Lambda_L^{SA}. \tag{II.3}$$

There are, however, evidences that the string tension is in fact lower. The first one comes from a measurement of Wilson loops on a 15^4 lattice [17]. Fitting the potential to a Coulomb plus linear form, results in a considerable lowering of the string tension estimate. The final $\sqrt{K}$ estimate is

$$\sqrt{K} \approx 56\,\Lambda_L^{SA}. \tag{II.4}$$

As asymptotic scaling deviations are also reported, the meaning of this number for the continuum limit is, however, obscure. Further, clarification is desirable.

The second evidence that the string tension is in fact lower than II.3 comes from investigations of Polyakov loop correlations [18]. As a Polyakov loop wraps around the whole lattice, this method gives a better estimate of the quark anti-quark potential. After projection on zero momentum, the estimate of the string tension becomes very stable with respect to the distance between the two loops. Results are consistent with asymptotic scaling leading to the estimate

$$\sqrt{K} \approx 52\,\Lambda_L^{SA}.$$

Comparing our $\chi(4,4)$ results with those of [17] (on a 15^4 lattice) we find agreement except for $\beta=2.4$, where finite lattice effects may well be important for a 4×4 loop (our lattice is 8^4). For the other β-values finite size effects on $\chi(4,4)$ are small. Therefore, our scaling for $\chi(4,4)$ might be a finite size effect.

II.2. The Mass Gap $m(0^+)$

The aim of this section is to investigate the previous MCV mass gap estimate [8][2] relying on calculations on a $4^3\cdot 16$ lattice, with respect to finite lattice size effects. Further, the use of a larger lattice will allow us to extend the scaling window beyond $\beta=2.25$ up to $\beta=2.4$. We project our 28 Wilson loops into the A_1^+ representation and measure on-diagonal correlations up to distance $t=3$. For each of these operators we define distance t glueball masses by means of

$$m_i(t_2,t_1)=\frac{-1}{t_2-t_1}\ln\left(\frac{\langle 0_i(0)0_i(t_2)\rangle}{\langle 0_i(0)0_i(t_1)\rangle}\right) \qquad (t_2>t_1). \tag{II.5}$$

In the infinite statistics limit all these numbers become upper bounds for the real mass gap $m(0^+)$. Due to technical reasons (disk space and tape handling) we did not measure off-diagonal correlations and therefore minimization in the sense of [8–10] is not possible. In the spirit of minimization we take our final estimates from the single operators, which give the lowest results for $m_i(1,0)$ ($i=1,\ldots,28$). By definition these operators are called the "*best*" operators. In view of our experience [8, 11] that at distance $t=2$, for good operators, statistical noise is already overwhelming the effect of minimization, we expect rather reliable results. To take directly the smallest value $m(2,1)$ ($i=1,\ldots,28$) is not correct, because we would obtain a too small result due to statistical fluctuations. In fact, the pattern of operators which give low results at distance $t=2$ is rather unstable with respect to β, whereas it is stable at distance $t=1$.

2 Similar estimates were obtained by other authors. For a review see [7]

Table II.3. $m(0^+)$ results from the best three operators

β	OP	$m(1,0)$	$m(2,0)$	$m(2,1)$
2.20	6	1.79 ± 0.03	1.75 ± 0.07	1.72 ± 0.14
2.20	13	1.78 ± 0.03	1.73 ± 0.07	1.68 ± 0.14
2.20	21	1.79 ± 0.03	1.73 ± 0.06	1.67 ± 0.14
2.25	6	1.69 ± 0.02	1.53 ± 0.03	1.37 ± 0.07
2.25	13	1.68 ± 0.02	1.54 ± 0.03	1.40 ± 0.08
2.25	21	1.71 ± 0.02	1.56 ± 0.03	1.47 ± 0.08
2.30	6	1.64 ± 0.03	1.42 ± 0.03	1.20 ± 0.08
2.30	13	1.63 ± 0.03	1.41 ± 0.03	1.20 ± 0.09
2.30	21	1.65 ± 0.03	1.42 ± 0.03	1.19 ± 0.09
2.35	6	1.73 ± 0.03	1.55 ± 0.04	1.36 ± 0.09
2.35	13	1.72 ± 0.03	1.54 ± 0.04	1.36 ± 0.09
2.35	21	1.78 ± 0.03	1.60 ± 0.04	1.42 ± 0.10
2.40	6	1.80 ± 0.03	1.52 ± 0.04	1.23 ± 0.09
2.40	13	1.78 ± 0.03	1.50 ± 0.04	1.21 ± 0.09
2.40	24	1.78 ± 0.03	1.44 ± 0.04	1.11 ± 0.09
2.50	24	2.03 ± 0.03	1.64 ± 0.05	1.25 ± 0.11
2.50	25	2.06 ± 0.04	1.62 ± 0.04	1.18 ± 0.10
2.50	27	2.11 ± 0.04	1.59 ± 0.04	1.08 ± 0.10

In Table II.3 we have collected for the three best operators the results up to distance $t=2$. For all three operators the $m(2,1)$ results are always identical within statistical errors. Only for $\beta \geqq 2.4$ operators with length >8 are among the best operators.

Previously a finite size study, also on an 8^4 lattice, was attempted by Ishikawa et al. [13, 14]. Unfortunately, these authors used "momentum smeared" wave functions to improve the MC statistics, then the unknown $\mathbf{p}^2$-contribution was subtracted by introducing one new parameter. In Fig. II.2 we compare their results

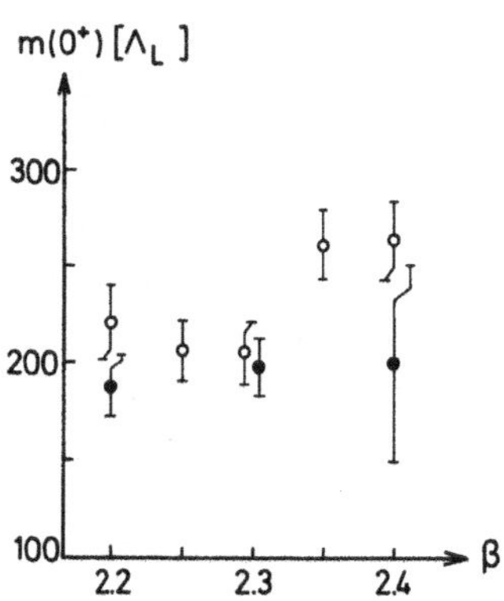

Fig. II.2. Our momentum $\mathbf{p}=0$ results (⌽) in comparison with results (⏀) from [13, 14] as obtained from "momentum smeared" wave functions

Table II.4. $m(0^+)$ distance $t=3$ results from the best three operators

β	OP	$m(3,0)$	$m(3,1)$	$m(3,2)$
2.20	6	$1.84\pm^{1.03}_{0.20}$	$1.87\pm^{1.55}_{0.35}$	$2.01\pm^{3.20}_{0.79}$
2.20	13	$1.85\pm^{0.77}_{0.22}$	$1.88\pm^{1.17}_{0.33}$	$2.08\pm^{2.43}_{0.78}$
2.20	21	$1.94\pm^{\infty}_{0.30}$	$2.01\pm^{\infty}_{0.39}$	$2.35\pm^{\infty}_{0.88}$
2.25	6	1.41 ± 0.07	$1.27\pm^{0.11}_{0.09}$	$1.17\pm^{0.24}_{0.21}$
2.25	13	1.41 ± 0.07	1.27 ± 0.10	$1.14\pm^{0.24}_{0.22}$
2.25	21	1.42 ± 0.07	1.28 ± 0.10	$1.15\pm^{0.25}_{0.23}$
2.35	6	1.37 ± 0.08	1.18 ± 0.12	$0.96\pm^{0.29}_{0.25}$
2.35	13	$1.39\pm^{0.10}_{0.07}$	$1.22\pm^{0.16}_{0.12}$	$1.04\pm^{0.34}_{0.29}$
2.35	21	$1.45\pm^{0.10}_{0.07}$	$1.28\pm^{0.17}_{0.13}$	$1.10\pm^{0.37}_{0.31}$
2.40	6	1.27 ± 0.05	1.00 ± 0.09	0.78 ± 0.22
2.40	13	1.25 ± 0.05	0.98 ± 0.08	0.76 ± 0.20
2.40	24	1.24 ± 0.05	0.97 ± 0.09	0.83 ± 0.21
2.50	24	1.36 ± 0.06	1.03 ± 0.09	0.80 ± 0.23
2.50	25	1.36 ± 0.06	1.01 ± 0.11	0.84 ± 0.23
2.50	26	1.35 ± 0.06	0.98 ± 0.10	0.88 ± 0.23

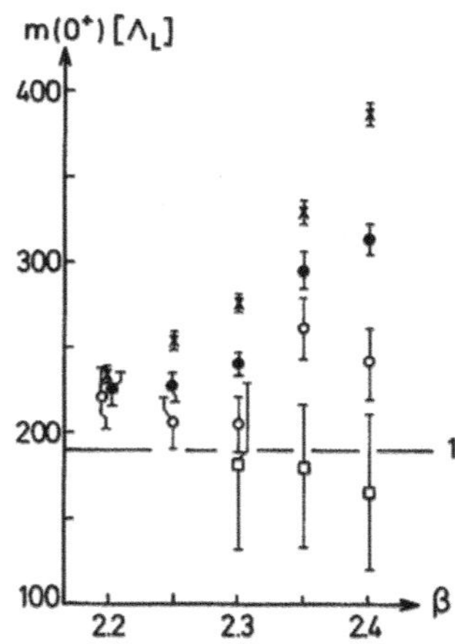

Fig. II.3. Scaling for the 0^+ mass gap, obtained from $m(1)$ ( ), $m(2)$ ( ), $\hat{m}(2)$ ( ), and $m(3,2)$ ( )

(after $\langle \mathbf{p}^2 \rangle$ subtraction) with our $\mathbf{p}=0$ zero results from the best operator of length $\leqq 8$. We do not take into account our length 10 operators, or $t=3$ results, in order to work in a similar approximation as the authors of [13, 14]. In this approximation we see no scaling beyond $\beta=2.3$. We, therefore, attribute their result of [13, 14] to be due to the subtraction of the unknown $\langle \mathbf{p}^2 \rangle$.

With our high statistics we obtain also signals at distance $t=3$. For the three best operators these results are collected in Table II.4. In Fig. II.3 we exhibit at each β-value distance $t=1,2,3$ results for the best operator. Our results at distance $t=3$ extend the scaling window up to $\beta=2.4$. A conventional continuum limit estimate is

$$m(0^+)=(190\pm30)\,\Lambda_L^{SA}\,. \tag{II.6}$$

Table II.5. Critical length for a deconfinement temperature $T_c = 43\,\Lambda_L$

β	T_c	$L_{\text{crit}} = 1/T_c$
2.1	0.42	2.4
2.2	0.33	3.1
2.3	0.25	3.9
2.4	0.20	5.1
2.5	0.15	6.5

Table II.6. $m(2^+)$ MC results from the best operator and $m(2^+)$ SC results [26, 27] (last two orders)

β	OP	$m(1)$	$m(2)$	$\hat{m}(2)$	$SC, O(u^6)$	$SC, O(u^8)$
2.20	7	3.18 ± 0.04	$3.29 \pm^{\infty}_{0.45}$	$3.40 \pm^{\infty}_{0.93}$	3.42	3.17
2.25	7	3.04 ± 0.04	$2.72 \pm^{0.23}_{0.16}$	$2.40 \pm^{0.51}_{0.35}$	3.37	3.09
2.30	7	3.03 ± 0.04	$3.09 \pm^{0.58}_{0.26}$	$3.15 \pm^{1.19}_{0.55}$	3.33	3.01
2.35	24	2.89 ± 0.04	$2.54 \pm^{0.32}_{0.20}$	$2.18 \pm^{0.69}_{0.43}$	3.30	2.93
2.40	24	2.78 ± 0.03	2.38 ± 0.10	$1.97 \pm^{0.25}_{0.20}$	3.26	2.86
2.50	24	2.73 ± 0.04	$2.74 \pm^{0.30}_{0.19}$	$2.75 \pm^{0.62}_{0.40}$	3.20	2.70

This is slightly higher than the previous result $m(0^+) = (170 \pm 30)\,\Lambda_L^{\text{SA}}$ [8], obtained on an $4^3 \cdot 16$ lattice. Even more notable finite size effects are found by comparing correlations for identical operators on the $4^3 \cdot 16$ and the 8^4 lattice, even at distance $t=1$ clear discrepancies exist. This is argued to be due to the finite temperature-like phase transition for the small spacelike part of the $4^3 \cdot 16$ lattice. Indeed the estimate $T_c = 43\,\Lambda_L^{\text{SA}}$ [26] would give critical length scales as given in Table II.5.

The experience from string tension measurements shows, however, that more accurate data at distance $t \geqq 3$ would be very desirable. There is some hope to achieve such results by using source methods (linear response together with the Langevin equation [27], study of boundary effects [28] or a "cold wall" method [29]).

II.3. Excited States

We have considered the excited spin states 0^-, 2^+, and 1^+. The irreducible representations of the cubic group on the Wilson loops up to length 8 are classified in [11]. According to this classification we have constructed the $A_1^-(0^-)$, $E^+(2^+)$ and $T_1^+(1^+)$ representations, whenever possible for a given Wilson loop.

In contrary to [14] our lowest excited state is the 2^+ tensor. For this state our results from the best operators are collected in Table II.6. As before in Sect. II.1 the best operator is defined to be the one which gives lowest mass values at distance $t=1$. The $m(2,0)$ and $m(2,1)$ values are, of course, better upper bounds than the $m(1,0)$ values. Within the large statistical noise it is, however, not possible to give a

Table II.7. Mass ratios $m(2^+)/m(0^+)$

β	$m(1)$	$m(2)$	$\hat{m}(2)$
2.20	1.79	1.90	2.02
2.25	1.81	1.77	1.71
2.30	1.86	2.19	2.63
2.35	1.68	1.65	1.60
2.40	1.56	1.65	1.77

Table II.8. Best results for 0^- and mass ratios $0^-/0^+$

β	OP	$m(1)$	$0^-/0^+$ $(t=1)$
2.20	21	$4.42\pm^{0.21}_{0.17}$	2.48
2.25	8	$4.44\pm^{0.14}_{0.13}$	2.64
2.30	8	$4.16\pm^{0.19}_{0.16}$	2.55
2.35	8	$4.08\pm^{0.39}_{0.28}$	2.37
2.40	8	$3.88\pm^{0.11}_{0.10}$	2.18

Table II.9. Best results for 1^+ and mass ratios $1^+/0^+$

β	OP	$m(1)$	$1^+/0^+$ $(t=1)$
2.20	13	$5.04\pm^{0.22}_{0.18}$	2.83
2.25	10	$5.03\pm^{0.19}_{0.15}$	3.05
2.30	8	5.00 ± 0.13	3.07
2.35	21	$4.82\pm^{0.21}_{0.17}$	2.80
2.40	21	$4.83\pm^{0.22}_{0.19}$	2.71

reliable estimate of the expected lowering. Comparing the distance $t=2$ results for $\beta=2.25$, 2.30 and $\beta=2.35$ we find that even with our high statistics we do not obtain reliable error bars for these results. See Sect. III.3 for a related more detailed discussion.

In Table II.6 we have also given the two last orders from the strong coupling (SC) expansion [30, 31] in the character variable u. There is amazingly good agreement between the SC u^8 and the MC $m(1)$ results, indicating that the E^+ SC series may – in contrast to the A_1^+ SC series – still be convergent in the considered β-region or at least in part of it. As the cubic symmetry is an exact symmetry of the lattice such a possibility exists. For the convenience of the reader we collect in Table II.7 mass ratios $m(2^+)/m(0^+)$ as obtained by dividing the best 2^+ and 0^+ results. If we optimistically *assume* that the 2^+ state likes to scale for $\beta \geqq 2.2$ we would obtain the order of magnitude

$$m(2^+) \approx 1.8\, m(0^+) \tag{II.7}$$

Table II.10. Considered momentum eigenstates

n_i^1	n_i^2	n_i^3	i
0	0	0	1
1	0	0	2
0	1	0	3
0	0	1	4
1	1	0	5
1	0	1	6
0	1	1	7
1	−1	0	8
1	0	−1	9
0	1	−1	10
1	1	1	11
2	0	0	12
0	2	0	13
0	0	2	14

from analyzing Table II.7. We have to emphasize that the low $m(1)$-ratios for $\beta \geqq 2.35$ are meaningless, because at these values the mass estimates for 0^+ at distance $t=1$ are already in the spin wave region [8]. The order of magnitude (II.7) is consistent with MC data of previous investigations [14] and also with results [32] as obtained using Manton's action.

Finally, Table II.8 contains our best 0^- and the Table II.9 our best 1^+ results. Only $m(1)$ values are given, because results at distance $t=2$ are statistical noise. We also give the corresponding $0^-/0^+$ and $1^+/0^+$ mass ratios. Taking into account β-values up to $\beta=2.30$ we find the orders of magnitude

$$m(0^-) \approx 2.5\, m(0^+)\,, \tag{II.8}$$

$$m(1^+) \approx 3\, m(0^+)\,. \tag{II.9}$$

Even more than for the 2^+ state, it is obscure whether these numbers have any meaning for the continuum limit $a(\beta) \to 0$.

II.4. Energy-Momentum Dispersion

In this section we consider momentum states for the $A_1^+(0^+)$ and the $E^+(2^+)$ states. In case of the A_1^+ representation we construct momentum eigenstates for all Wilson loops up to length 6, whereas for the E^+ representation we only take the 1-plaquette operator. In each case our momenta are $\mathbf{K}_i = \frac{\mathbf{n}_i}{2\pi L}$ $(L=8)$, with $\mathbf{n}_i = (n_i^1, n_i^2, n_i^3)$, and $i=1, \ldots, 14$ as given in Table II.10. As in [33] we construct the considered irreducible representations of the cubic group on appropriate spacelike cubes $C_j(\mathbf{x}, t)$, and perform the Fourier transform for these cube operators.

$$\tilde{C}_j(\mathbf{K}, t) = \sum_{\mathbf{x}} e^{i\mathbf{K}\mathbf{x}} C_j(\mathbf{x}, t)\,.$$

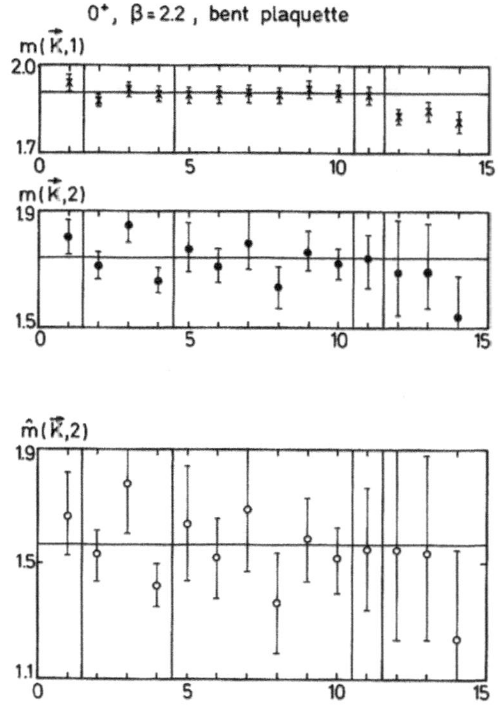

Fig. II.4. Results for 0^+ momentum states at $\beta = 2.2$

Different irreducible representations of the cubic group do not mix up to order $\mathbf{K}^2$. This means we obtain good projections on J^P eigenstates for small momenta and short distances. Otherwise, 0^+ will finally win.

Let us define

$$m(\mathbf{K}, t) = \sqrt{E(\mathbf{K}, t, 0)^2 - \mathbf{K}^2} \tag{II.10a}$$

and

$$\hat{m}(\mathbf{K}, t) = \sqrt{E(\mathbf{K}, t, 1)^2 - \mathbf{K}^2}\,, \tag{II.10b}$$

where $E(\mathbf{K}, t_2, t_1)$ is in analogy to Eq. (II.5) a finite distance definition of the energy of the momentum $\mathbf{K}$ state:

$$E(\mathbf{K}, t_2, t_1) = \frac{-1}{t_2 - t_1} \ln \left(\frac{\mathrm{Re}\langle \overline{\tilde{C}_j(\mathbf{K}, t_2) \tilde{C}_j(\mathbf{K}, 0)} \rangle}{\mathrm{Re}\langle \overline{\tilde{C}_j(\mathbf{K}, t_1) \tilde{C}_j(\mathbf{K}, 0)} \rangle} \right) \quad (t_2 > t_1)\,. \tag{II.11}$$

Using these definitions we find for the 0^+ state restoration of Lorentz invariance in the considered β-region ($\beta \geqq 2.2$). This is illustrated in Fig. II.4. From the now considered operators the bent plaquette (operator #3 of Fig. I.1) is the best operator at $\beta = 2.2$. At distance $t = 2$ and already $t = 1$ we find restoration of Lorentz invariance for the 11 lowest momenta ($i = 1, \ldots, 11$ as defined in Table

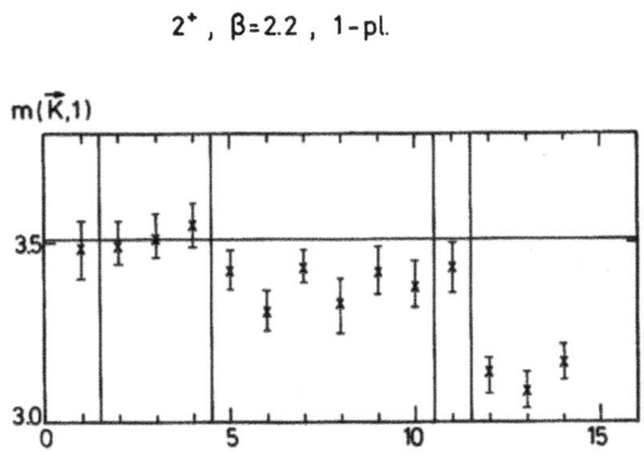

Fig. II.5. Results for 2^+ momentum state at $\beta = 2.2$

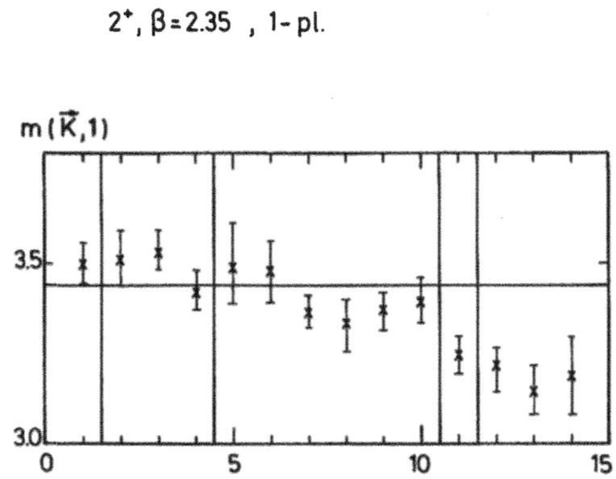

Fig. II.6. Results for 2^+ momentum states at $\beta = 2.35$

II.10). The horizontal line is the mean value calculated from these 11 momenta. The last three momenta deviate due to the finite lattice spacing. Indeed replacing

$$\mathbf{K}^2 \to \sum_{i=1}^{3} (2 - 2\cos K_i) \tag{II.12}$$

gives also consistency with Lorentz invariance for these momenta. [For the first 10 momenta the change under the replacement (II.12) is numerically very small.]

In Fig. II.4 momenta within vertical lines have identical $\mathbf{K}^2$ and hence identical $E(\mathbf{K}, t_2, t_1)$.

For all β we find their energies to be statistically rather independent. Therefore, we may considerably increase the MC statistics by measuring momentum eigenstates for the low lying momenta and by using Lorentz invariance. In the present case we would win a factor $\lesssim 11$. Independently a similar observation has been made in [34].

For the 2^+ state the situation is depicted in Figs. II.5 and II.6. Only distance $t = 1$ correlations are sufficiently out of the statistical noise to allow clear conclusions. At $\beta = 2.2$ we find consistency with Lorentz invariance only for the first 4 momenta, whereas at $\beta = 2.35$ it is found for the first 10 momenta. The

Table III.1. Rectangular Wilson loops on the 8^4 lattice

β	1.5	1.6	1.7	1.8	1.9	2.0
W_{11}	0.5617 ±0.0002	0.6003 ±0.0004	0.6347 ±0.0003	0.6629 ±0.0002	0.6856 ±0.0002	0.7043 ±0.0001
W_{12}	0.3118 ±0.0003	0.3629 ±0.0007	0.4114 ±0.0006	0.4529 ±0.0003	0.4866 ±0.0004	0.5146 ±0.0002
W_{22}	0.1002 ±0.0003	0.1437 ±0.0008	0.1920 ±0.0009	0.2381 ±0.0003	0.2771 ±0.0007	0.3101 ±0.0003
W_{13}	0.1744 ±0.0003	0.2220 ±0.0007	0.2707 ±0.0007	0.3145 ±0.0003	0.3512 ±0.0005	0.3827 ±0.0002
W_{14}	0.0977 ±0.0002	0.1362 ±0.0006	0.1787 ±0.0007	0.2193 ±0.0003	0.2546 ±0.0006	0.2853 ±0.0002
W_{23}	0.0337 ±0.0002	0.0604 ±0.0005	0.0959 ±0.0009	0.1341 ±0.0003	0.1688 ±0.0008	0.1997 ±0.0003
W_{24}	0.0114 ±0.0001	0.0258 ±0.0003	0.0487 ±0.0008	0.0769 ±0.0004	0.1049 ±0.0008	0.1310 ±0.0003
W_{33}	0.0070 ±0.0001	0.0187 ±0.0003	0.0390 ±0.0007	0.0655 ±0.0004	0.0929 ±0.0009	0.1187 ±0.0003
W_{43}	0.0017 ±0.0001	0.0060 ±0.0001	0.0164 ±0.0005	0.0333 ±0.0004	0.0532 ±0.0008	0.0732 ±0.0003
W_{44}	0.00012 ±0.00015	0.0015 ±0.0001	0.0059 ±0.0002	0.0154 ±0.0005	0.0289 ±0.0007	0.0432 ±0.0007
β	1.45	1.55	1.65	1.75	1.85	1.95
$\beta^2 \partial_\beta W_{11}$	0.856 ±0.006	0.927 ±0.013	0.937 ±0.014	0.864 ±0.013	0.777 ±0.011	0.711 ±0.012

horizontal lines are the mean values as calculated from the corresponding momentum eigenstates. As the momentum $\mathbf{K}=0$ 2^+ mass is rather high, we find only small changes under the replacement (II.12).

III. Symanzik Tree-Improved Action

We now report our MC results as obtained with the TIA. As before we use the icosaeder subgroup [24]. Our high statistics glueball calculations are done on a $5^3 \cdot 8$ lattice. Additionally, we calculate with a rather mode rate statistics the string tension from Creutz ratios on a 8^4 lattice.

III.1. The String Tension

At each β-value we have done first 200 sweeps without measurements for reaching equilibrium and then 2000 sweeps with measurements. For the relevant values $1.5 \leqq \beta \leqq 2.0$, our Wilson loop results are collected in Table III.1. Error bars are calculated with respect to 10 bins. Because of correlations between successive bins

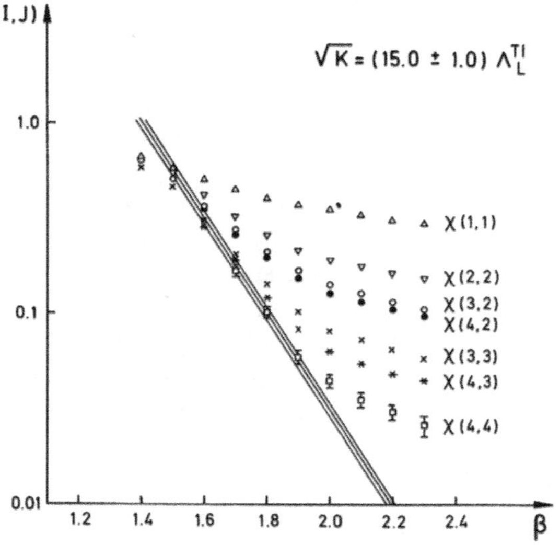

Fig. III.1. Conventional estimate of the string tension from Creutz ratios

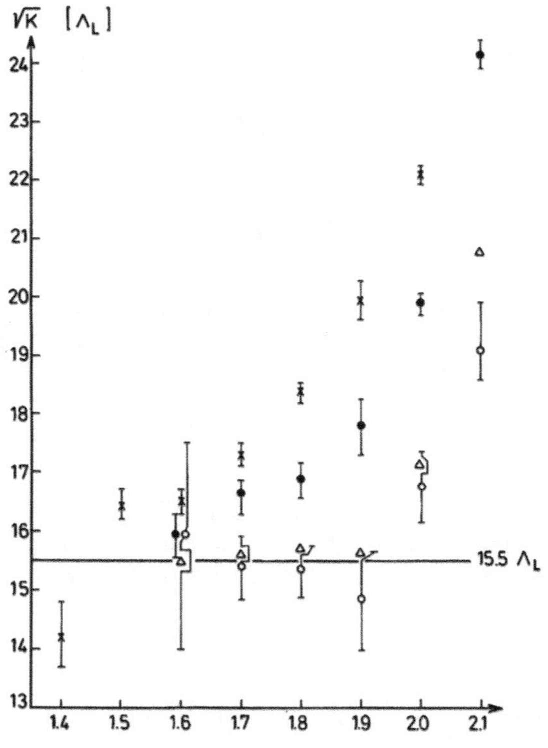

Fig. III.2. Relevant Creutz ratios on a precise physical scale: $\chi(3,3)$ ⌶, $\chi(4,3)$ ●, $\chi(4,4)$ ○, and improved $\hat{\chi}(3,3)$ △

true error bars may be larger (see [11, 16] for a more detailed analysis). From W_{11} we can easily calculate the specific heat by numerical differentiation. For the convenience of the reader, we collect in Table III.1 also the specific heat as obtained from neighbour β-values. The peak is reached at $\beta \approx 1.65$.

Using Creutz ratios we give in Fig. III.1 a conventional plot for the scaling behaviour of the string tension. A scaling window is found and the continuum estimate is obtained from the envelope. The accuracy becomes much more obvious from Fig. III.2. These data in physical units of Λ_L^{TI} are plotted on a very precise scale. By a similar interpretation as in Sect. II.1 the $\chi(4,4)$ results indicate the upper bound

$$\sqrt{K} \approx 15.5\, \Lambda_L^{\mathrm{TI}}. \tag{III.1}$$

This is lower than the previous results $\sqrt{K} \approx 17.9\Lambda_L^{\mathrm{TI}}$ [22] and $\sqrt{K} \approx 16.4\Lambda_L^{\mathrm{TI}}$ [23]. On the other hand, the potential analysis of [17] finds again (asymptotic) scaling deviations and a considerably lower value

$$\sqrt{K} \approx 12.5\, \Lambda_L^{\mathrm{TI}}. \tag{III.2}$$

Perturbative calculations [6, 35] of the Λ-ratio give

$$r = \Lambda_L^{\mathrm{TI}}/\Lambda_L^{\mathrm{SA}} = 4.13. \tag{III.3}$$

It is amazing to note that our results for $\sqrt{K^{\mathrm{TI}}}/\sqrt{K^{\mathrm{SA}}}$ and those of [17] are both consistent with the asymptotic estimate of Eq. (III.3).

A consistent string tension measurement using the improved action should involve improved loop operators. To the tree level order, improved Creutz ratios are given by [6]

$$\hat{\chi}(I,J) = \sum_{m,n} C_{mn} \ln[W(I+m, J+n)]. \tag{III.4}$$

Here $C_{mn} = C_{nm}$ and the non-zero C_{mn} are listed in the following:

$$c_{11} = y + \tfrac{1}{3}, \quad c_{00} = 3y + \tfrac{5}{4}, \quad c_{-1-1} = 1 - 3y, \tag{III.5a}$$

$$c_{-2-2} = -y - \tfrac{1}{12}, \quad c_{10} = -2y - \tfrac{1}{3}, \quad c_{-1-2} = 2y, \tag{III.5b}$$

$$c_{1-1} = y, \quad c_{0-2} = -y + \tfrac{1}{12}, \quad c_{0-1} = -1. \tag{III.5c}$$

y is a free parameter. (Note that the standard Creutz ratios are obtained with $c_{00} = 1$, $c_{0-1} = -1$, $c_{-1-1} = -1$, and all other $c_{mn} = 0$.)

We tried three choices:

$$\text{(a) } y = -\tfrac{1}{2}, \quad \text{(b) } y = 0, \quad \text{(c) } y = -\tfrac{1}{3}. \tag{III.6}$$

(a) and (b) are close to the standard Creutz ratios and (c) avoids W_{44}, which is the noisiest Wilson loop. Together with $\chi(3,3)$ and $\chi(4,4)$ we collect the results for $\hat{\chi}(3,3)$ (a, b, c) in Table III.2.

There is y-independence up to an excellent precision and the $\hat{\chi}(3,3)$ values are well compatible with $\chi(4,4)$. This shows that the *improvement works* in a consistent way. In Fig. III.2 the triangles are mean values from our three considered $\hat{\chi}(3,3)$ definitions, and there is good consistency with asymptotic scaling.

Table III.2. $\chi(3,3)$, $\chi(4,4)$, and improved string tensions $\hat{\chi}(3,3)$

β	1.6	1.7	1.8	1.9	2.0	2.1
$\chi(3,3)$	0.3052	0.2058	0.1416	0.1020	0.0799	0.0726
$\chi(4,4)$	0.2837	0.1625	0.0987	0.0566	0.0440	0.0346
$\chi(3,3)$ (a)	0.2669	0.1670	0.1043	0.0634	0.0468	0.0414
$\chi(3,3)$ (b)	0.2689	0.1671	0.1054	0.0640	0.0480	0.0420
$\chi(3,3)$ (c)	0.2607	0.1666	0.1010	0.0613	0.0429	0.0398

Table III.3. Rectangular spacelike Wilson loops on a $5^3 \cdot 8$ lattice

β	W_{11}	W_{12}	W_{22}	Sweeps
1.45	0.5412 ± 0.0001	0.2864 ± 0.0002	0.0818 ± 0.0002	40,000
1.50	0.5613 ± 0.0001	0.3112 ± 0.0002	0.0997 ± 0.0902	40,000
1.55	0.58125 ± 0.00008	0.3371 ± 0.0002	0.1206 ± 0.0002	122,000
1.60	0.6003 ± 0.0002	0.3628 ± 0.0002	0.1437 ± 0.0003	80,000
1.65	0.61888 ± 0.00008	0.3891 ± 0.0002	0.1698 ± 0.0002	164,000
1.70	0.63599 ± 0.00008	0.4141 ± 0.0002	0.1968 ± 0.0002	160,000
1.75	0.6513 ± 0.0002	0.4369 ± 0.0003	0.2227 ± 0.0004	42,000
1.80	0.6643 ± 0.0002	0.4560 ± 0.0003	0.2443 ± 0.0004	40,000
1.90	0.68617 ± 0.00004	0.48824 ± 0.00006	0.2811 ± 0.0001	160,000
2.00	0.70446 ± 0.00007	0.5152 ± 0.0002	0.3121 ± 0.0003	40,000

III.2. The Mass Gap $m(0^+)$

Our glueball calculations are carried out on a $5^3 \cdot 8$ lattice. In view of the double plaquette involved in the Symanzik TIA a $5^3 \cdot 8$ lattice seems to be the smallest feasible lattice. As in [11, 20] our MC calculation is based on 21 Wilson loops W_i $(i = 1, \dots, 21)$ of length $\leqq 8$. We only measure spacelike Wilson loops. In Table III.3 we give mean values for the considered rectangular loops. As is obvious from this table we have collected a very high statistics of up to 160,000 sweeps at some β-values. This was mainly done for hunting the 2^+ state (see next section). We have done first 1200 sweeps for thermalization and carried out measurements after every second sweep.

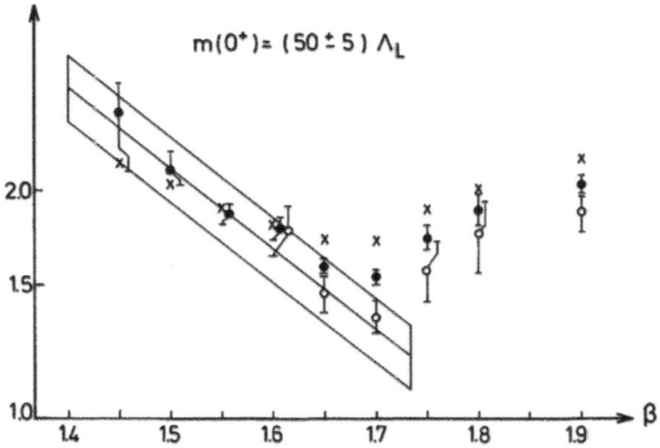

Fig. III.3. Scaling for the 0^+ mass gap

Finite distance glueball masses and best operators are defined as in Sect. II.2. We only consider correlations up to distance $t=2$. In Table III.4 we have for the three best operators collected the mass results. These three operators give always nearly identical numerical values. As depicted in Fig. III.3 we find a window $1.45 \leqq \beta \leqq 1.70$, where our distance $t=2$ data are in agreement with asymptotic scaling. The data of Fig. III.3 are always taken from the best operator of Table III.4.

As discussed in the introduction, the MCV method is in trouble[3] in the case of the improved action. This may be illustrated by the lowest order SC expansion for connected plaquette-plaquette correlations in the A_1^+ representation of the cubic group. Up to a positive proportionality constant one obtains[4] the following leading behaviour:

$$\langle \square(0)\,\square(t)\rangle_c = \begin{cases} (\frac{5}{3}\beta/4)^4 & (t=1) \\ (-\frac{1}{12}\beta/4)^4 & (t=2) \\ (\frac{5}{3}\beta/4)^4(-\frac{1}{12}\beta/4)^4 & (t=3) \\ (-\frac{1}{12}\beta/4)^8 & (t=4) \end{cases} \tag{III.7}$$

and $\langle \square(0)\,\square(t)\rangle_c = \text{const}$,

$$\begin{cases} (c_1\beta)^2(c_2\beta)^2(c_2\beta)^{2K}(c_1\beta)^2\,, & t=2K+1\,, \quad K \geqq 2\,, \\ (c_1\beta)^2(c_2\beta)^2(c_2\beta)^{2K-1}(c_1\beta)^2\,, & t=2K\,, \quad K \geqq 3\,, \end{cases}$$

with $c_1 = \frac{5}{3}$, $c_2 = -\frac{1}{12}$ (const > 0).

For $t \to \infty$ we obtain from the *absolute value* of the correlation the leading order glueball mass

$$m(0^+) = -\ln\beta\,. \tag{III.8}$$

3 In the case of Manton's action the negative eigenvalues of the transfer matrix are not a severe problem for a MCV calculation [32]

4 We would like to thank G. Münster for clarifying discussions on this subject

Table III.4. $m(0^+)$ results from the best three operators

β	OP	$m(1)$	$m(2)$	$\hat{m}(2)$
1.45	8	2.17 ± 0.03	$2.29\pm^{0.25}_{0.17}$	$2.40\pm^{0.53}_{0.36}$
1.45	10	2.19 ± 0.03	$2.26\pm^{0.24}_{0.17}$	$2.33\pm^{0.51}_{0.35}$
1.45	21	2.19 ± 0.03	$2.24\pm^{0.20}_{0.14}$	$2.30\pm^{0.41}_{0.31}$
1.50	6	2.07 ± 0.03	$2.15\pm^{0.14}_{0.11}$	$2.23\pm^{0.50}_{0.24}$
1.50	8	2.05 ± 0.03	$2.16\pm^{0.13}_{0.09}$	$2.28\pm^{0.26}_{0.22}$
1.50	21	2.03 ± 0.03	$2.12\pm^{0.13}_{0.09}$	$2.22\pm^{0.26}_{0.21}$
1.55	6	1.89 ± 0.02	1.86 ± 0.05	1.83 ± 0.11
1.55	13	1.89 ± 0.02	1.87 ± 0.05	1.85 ± 0.12
1.55	21	1.89 ± 0.02	1.87 ± 0.05	1.84 ± 0.12
1.60	6	1.80 ± 0.03	1.78 ± 0.06	1.77 ± 0.13
1.60	13	1.80 ± 0.03	1.78 ± 0.05	1.77 ± 0.13
1.60	21	1.82 ± 0.03	1.81 ± 0.06	1.81 ± 0.13
1.65	6	1.73 ± 0.02	1.59 ± 0.04	1.46 ± 0.08
1.65	13	1.72 ± 0.02	1.51 ± 0.03	1.46 ± 0.08
1.65	21	1.77 ± 0.02	1.64 ± 0.04	1.50 ± 0.09
1.70	6	1.73 ± 0.02	1.55 ± 0.03	1.38 ± 0.07
1.70	13	1.72 ± 0.01	1.54 ± 0.03	1.36 ± 0.06
1.70	21	1.79 ± 0.01	1.60 ± 0.03	1.41 ± 0.06
1.75	6	1.89 ± 0.04	1.74 ± 0.06	1.60 ± 0.14
1.75	7	1.94 ± 0.04	1.74 ± 0.06	1.54 ± 0.14
1.75	13	1.89 ± 0.04	1.73 ± 0.06	1.57 ± 0.14
1.80	6	2.03 ± 0.04	1.91 ± 0.09	$1.78\pm^{0.21}_{0.18}$
1.80	7	2.08 ± 0.04	1.88 ± 0.07	$1.68\pm^{0.19}_{0.16}$
1.80	13	2.01 ± 0.03	1.89 ± 0.09	$1.76\pm^{0.21}_{0.18}$
1.90	6	2.22 ± 0.02	2.06 ± 0.05	1.91 ± 0.11
1.90	13	2.20 ± 0.02	2.04 ± 0.05	1.87 ± 0.10
1.90	21	2.29 ± 0.02	2.14 ± 0.06	1.99 ± 0.14
2.00	5	2.31 ± 0.04	$2.15\pm^{0.14}_{0.11}$	$2.00\pm^{0.31}_{0.25}$
2.00	6	2.34 ± 0.04	$2.11\pm^{0.12}_{0.09}$	$1.89\pm^{0.16}_{0.12}$
2.00	13	2.32 ± 0.04	2.12 ± 0.12	$1.91\pm^{0.28}_{0.23}$

At small distances t the fall-off of the correlation function is, however, rather different.

As TIA and SA are in the same universality class, one expects the problem to be less important in the continuum limit $\beta\to\infty$. In our β-region we find for all considered operators positive correlations at distance $t=1,2$. At our low β-values ($\beta=1.45, 1.55$) $m(2,0)$ results are systematically higher than $m(1,0)$ results. Because of the interaction range going over two lattice spacings, one should discard the $m(1,0)$ results for the purpose of estimating the mass gap. Our selection of best operators is based on distance $t=1$ correlations, and one may now doubt this procedure. But our high MC statistics allow us also to select directly the lowest $m(2,0)$ result. With only one exception ($\beta=1.75$) we obtain always one of our three best operators (Table III.4), and at $\beta=1.75$ the result we find remains within

Table III.5. $m(2^+)$ results for the best two operators

β	OP	$m(1)$	$m(2)$	$\hat{m}(2)$
1.45	2	3.35 ± 0.06	$4.55 \pm^{\infty}_{1.44}$	$5.75 \pm^{\infty}_{2.34}$
1.45	7	3.34 ± 0.04	∞	∞
1.50	7	3.18 ± 0.06	$3.19 \pm^{\infty}_{0.42}$	$3.20 \pm^{\infty}_{0.88}$
1.50	8	3.16 ± 0.04	∞	∞
1.55	7	3.12 ± 0.03	$3.14 \pm^{0.42}_{0.23}$	$3.15 \pm^{0.87}_{0.47}$
1.55	8	3.15 ± 0.03	$3.49 \pm^{1.26}_{0.33}$	$3.83 \pm^{2.54}_{1.68}$
1.60	7	2.94 ± 0.03	$2.84 \pm^{0.20}_{0.14}$	$2.73 \pm^{0.42}_{0.30}$
1.60	8	3.00 ± 0.03	$2.95 \pm^{0.35}_{0.21}$	$2.90 \pm^{0.72}_{0.44}$
1.65	7	2.93 ± 0.03	$2.93 \pm^{0.24}_{0.16}$	$2.93 \pm^{0.50}_{0.34}$
1.65	8	2.98 ± 0.01	$3.13 \pm^{0.38}_{0.21}$	$3.28 \pm^{0.78}_{0.44}$
1.70	5	2.93 ± 0.03	$2.61 \pm^{0.13}_{0.10}$	$2.29 \pm^{0.27}_{0.23}$
1.70	7	2.89 ± 0.03	$2.72 \pm^{0.16}_{0.12}$	$2.55 \pm^{0.34}_{0.27}$
1.75	5	2.90 ± 0.04	$2.68 \pm^{0.25}_{0.17}$	$2.46 \pm^{0.54}_{0.37}$
1.75	7	2.86 ± 0.04	$2.92 \pm^{0.47}_{0.24}$	$2.99 \pm^{0.97}_{0.52}$
1.80	5	2.94 ± 0.04	$2.75 \pm^{0.31}_{0.19}$	$2.57 \pm^{0.65}_{0.42}$
1.80	7	2.91 ± 0.04	$2.78 \pm^{0.37}_{0.21}$	$2.65 \pm^{0.77}_{0.45}$
1.90	5	2.81 ± 0.02	2.60 ± 0.09	2.39 ± 0.18
1.90	7	2.82 ± 0.02	2.72 ± 0.10	$2.61 \pm^{0.24}_{0.19}$
2.00	5	2.76 ± 0.03	$2.50 \pm^{0.13}_{0.10}$	$2.25 \pm^{0.28}_{0.23}$
2.00	7	2.81 ± 0.03	$2.57 \pm^{0.18}_{0.14}$	$2.32 \pm^{0.40}_{0.30}$

statistical errors unchanged. In summary, our procedure is consistent up to distance $t=2$. From Fig. III.3 our final estimate is

$$m(0^+) = (50 \pm 5)\,\Lambda_L^{\mathrm{TI}}. \tag{III.9}$$

Comparing this with Eq. (II.6) gives us the Λ-ratio, $r_{mg} \approx 3.8$, in good agreement with universality.

III.3. Excited States

As for the SA we consider 0^-, 2^+, and 1^+ states. The warnings from Sect. III.2 with respect to the MCV method apply again. Within our practical restrictions we obtain results as presented in this section. Our lowest state is again the 2^+ tensor. For this state mass values from the best two operators at each β are collected in Table III.5, and results obtained from the best operators are also graphically depicted in Fig. III.4. There is no signal for asymptotic scaling, hence also no signal for scaling of the mass ratio $m(2^+)/m(0^+)$. Up to $\beta=1.8$ distance $t=2$ results are (within their errors) even well compatible with the $t=1$ results. After drastically increasing the MC statistics our previous [20] indication of asymptotic scaling in this β-region (from $t=2$ correlations) disappeared.

We illustrate this by drawing "continuum estimate" lines into Fig. III.4, which are very different from those of [20]. This demonstrates the arbitrariness, but does

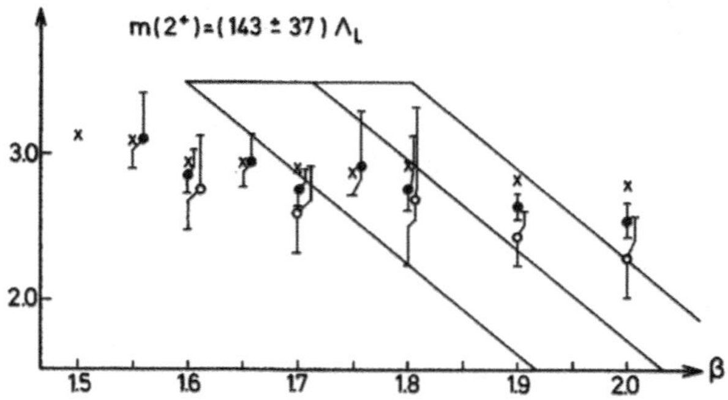

Fig. III.4. 2^+ state results

not rule out the previous estimate. In particular, it should be mentioned in this connection, that the finite temperature phase transition is around $\beta \approx 1.6$–1.7 (see also [17]).

Results of [20] are based on 40,000 sweeps at each β-value. We added new β-values and increased the statistics up to 160,000 sweeps at some selected β-values. Independently of the total number of sweeps we calculate at each β error bars by dividing our complete set of data into 20 bins. Even in the range 40,000–160,000 sweeps the $m(2)$ and $\hat{m}(2)$ error bars do not obey the asymptotic $1/\sqrt{N_{\text{sweeps}}}$ behaviour. This indicates the existence of correlations in computer time over an enormous number of sweeps and/or very rare but important events.

For the convenience of the reader we collect in Table III.6 mass ratios $m(2^+)/m(0^+)$ as obtained by dividing the best 2^+ and 0^+ results. $m(1)$ ratios are of particular interest because of their statistical reliability. The reached maximum is lower than in the case of the SA. (See the corresponding Table II.7.) Also the maximum is reached at a much smaller correlation length, because the $m(0^+)$ scaling window is shifted to a smaller correlation length as compared with the SA. The correlation length $1/m(0^+)$ as obtained from our mass gap estimates (II.6) and (II.9) is given in Table III.7 for the relevant β-values of both actions. Assuming that

Table III.6. Mass ratios $m(2^+)/m(0^+)$

β	$m(1)$	$m(2)$	$\hat{m}(2)$
1.45	1.54	∞	∞
1.50	1.56	∞	∞
1.55	1.65	1.69	1.72
1.60	1.63	1.60	1.54
1.65	1.70	1.84	2.01
1.70	1.68	1.77	1.88
1.75	1.51	1.69	1.90
1.80	1.45	1.47	1.51
1.90	1.28	1.27	1.28
2.00	1.19	1.16	1.13

Table III.7. Correlation for SA and TIA

SA		TIA	
β	ξ	β	ξ
2.20	0.69	1.55	0.53
2.25	0.79	1.60	0.60
2.30	0.89	1.65	0.68
2.35	1.01	1.70	0.76
2.40	1.14	1.75	0.86

Table III.8. $m(0^-)$ results from the best operator

β	OP	$m(1)$	$0^-/0^+$ $(t=1)$
1.45	21	$4.78\pm^{0.23}_{0.19}$	2.20
1.50	21	$4.32\pm^{0.25}_{0.20}$	2.13
1.55	6	4.21 ± 0.13	2.23
1.60	21	4.12 ± 0.11	2.29
1.65	6	3.91 ± 0.06	2.27
1.70	21	3.93 ± 0.06	2.28
1.75	6	3.78 ± 0.13	2.00
1.80	6	3.45 ± 0.12	2.34
1.90	21	3.60 ± 0.05	2.06
2.00	21	3.59 ± 0.08	1.88

Table III.9. $m(1^+)$ results from the best operator

β	OP	$m(1^+)$	$1^+/0^+$ $(t=1)$
1.45	11	$5.53\pm^{0.43}_{0.30}$	2.54
1.50	19	$5.17\pm^{0.30}_{0.23}$	2.55
1.55	13	5.07 ± 0.14	2.68
1.60	21	4.91 ± 0.12	2.73
1.65	13	4.92 ± 0.11	2.86
1.70	13	4.76 ± 0.08	2.77
1.75	19	4.46 ± 0.15	2.36
1.80	21	4.70 ± 0.15	2.34
1.90	8	4.54 ± 0.09	2.06
2.00	9	4.36 ± 0.12	1.89

the rather smooth behaviour of $m(1)$ ratios up to the maximum at $\xi\approx0.68$ gives an indication about the continuum limit behaviour, we would obtain for the TIA

$$m(2^+)\approx1.6\,m(0^+) \qquad \text{(III.10)}$$

as compared with $m(2^+)\approx1.8\,m(0^+)$ in case of the SA.

In Tables III.8 and III.9 we give our $m(1)$ results from the best operators for 0^- and 1^+ states. For these states distance $t=2$ results are noise. We also give the mass ratios $0^-/0^+$ and $1^+/0^+$. By taking into account ratios up to the maximum at $\beta=1.65$, we find the order of magnitude

$$m(0^-)\approx2.2\,m(0^+), \qquad \text{(III.11)}$$

$$m(1^+)\approx2.8\,m(0^+). \qquad \text{(III.12)}$$

As for the 2^+ state these ratios are somewhat lower than the corresponding ratios (II.8) and (II.9) in case of the SA.

III.4. Energy-Momentum Dispersion

In this section we consider momentum eigenstates for the $A_1^+(0^+)$ and the $E^+(2^+)$ states. Details, in particular the considered operators, are as described for the SA in

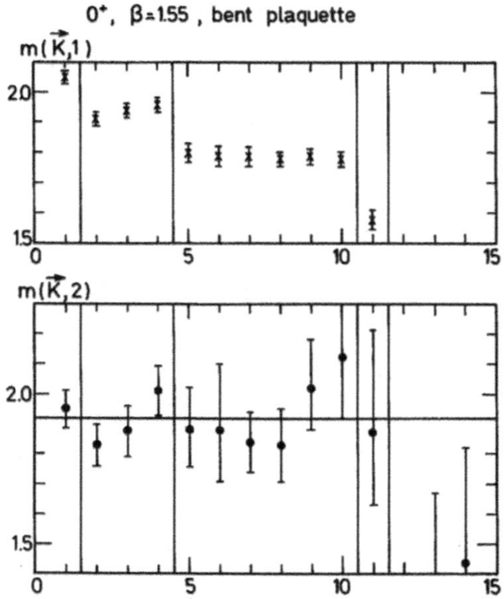

Fig. III.5. Results for 0^+ momentum eigenstates. (Equation II.12 is used for $\mathbf{K}^2$)

Sect. II.4. For $i=1, \ldots, 14$ the momenta are defined in Table II.10. Of course, we have now $L=5$ and the MC statistics of Table III.3.

As compared with the SA the energy-momentum dispersion is *not* improved. Typical results for the 0^+ state are depicted in Fig. III.5. As $L=5$ is rather small, we always replace $\mathbf{K}^2$ by means of Eq. (II.12) for the figures of this section. For all considered higher momenta the relativistic energy-momentum dispersion is clearly violated at distance $t=1$. This has to be seen in contrast to the corresponding SA results of Fig. II.4. We attribute the bad behaviour of the TIA to the already discussed problems with the transfer matrix, implying strong reservations against all distance $t=1$ results. At distance $t=2$ we find restoration of Lorentz invariance within the statistical errors.

The 2^+ state results, as for a typical example depicted in Fig. III.6, look similar. At distance $t=2, 2^+$ data are, however, so noisy that no clear conclusions are possible. The lack of Lorentz invariance at distance $t=1$ is particularly ugly, if one thinks about using higher momentum eigenstates for improving the 2^+ state MC statistics at distance $t=2$. In contrast to the SA no self-consistent procedure would be possible. For both actions calculations of the 2^+ state using momentum states may suffer from the 0^+ mixing.

IV. Summary and Conclusions

Within the limits of our approximations we obtain the following results: $\sqrt{K} \approx 69\, \Lambda_L^{\mathrm{SA}}$ (II.3), $m(0^+) \approx 190\, \Lambda_L^{\mathrm{SA}}$ (II.6) for the SA and $\sqrt{K} \approx 15.5\, \Lambda_L^{\mathrm{TI}}$ (III.1),

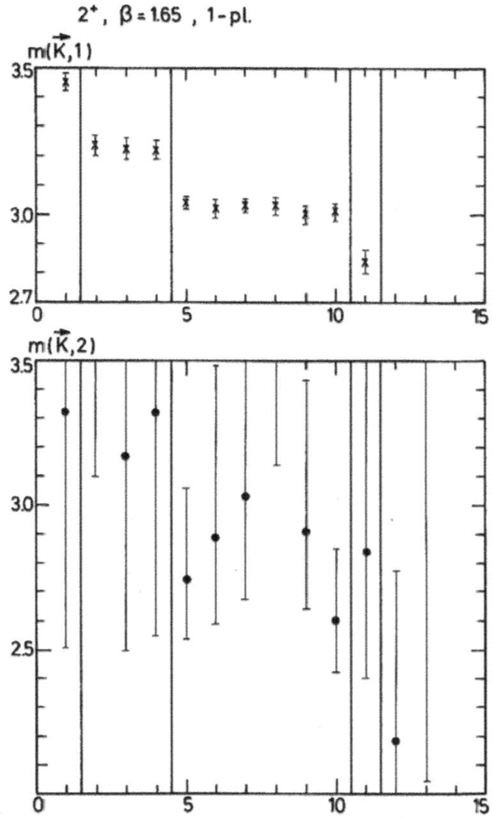

Fig. III.6. Results for 2^+ momentum eigenstates. (Equation II.12 is used for $\mathbf{K}^2$)

$m(0^+) \approx 50\,\Lambda_L^{\rm TI}$ (III.9) for the TIA. This gives for the Λ-ratio $r = \Lambda_L^{\rm TI}/\Lambda_L^{\rm SA} = 4.13$ [6, 35] the MC estimates $r_{\sqrt{K}} \approx 4.5$ and $r_{m(0^+)} \approx 3.8$ in consistency with universality.

Results for the excited states $2^+, 0^-, 1^+$, due to lack of clear scaling signals, are rather inconclusive. Very tentative orders of magnitude are: $m(2^+) \approx 1.6$–$1.8\,m(0^+)$, $m(0^-) \approx 2.2$–$2.5\,m(0^+)$, and $m(1^+) \approx 2.8$–$3.0\,m(0^+)$. The lower values corresponds always to the TIA and the higher value to the SA.

Finally, we have calculated energies for momentum eigenstates. In case of the 0^+ state we find the relativistic energy-momentum dispersion restored for both actions. The SA behaves nicer than the TIA, as the restoration of Lorentz invariance already happens at distance $t = 1$. This allows us to establish such a result also for the 2^+ state.

A future MCV calculation of the glueball spectrum should be based on important operators only, and use – at least for the lowest state – restoration of Lorentz invariance for improving the MC statistics. For the different actions and states the important ("best" at some values of β in the notation of Sects. II and III),

Table IV.1. Important operators

SA, 0^+:	6,		13,		21, 24.
SA, 2^+:	7,				21, 24.
SA, 0^-:	8,				21.
SA, 1^+:	8,	10,	13,		21.
TIA, 0^+:	5, 6, 7, 8,	10,	13,		21.
TIA, 2^+:	2, 5, 7, 8,				
TIA, 0^-:	6,				21.
TIA, 1^+:	8, 9,	11,	13,	19,	21.

Table IV.2. Size of scaling windows

	β_1	β_2	ξ_1	ξ_2	ξ_2/ξ_1
$\sqrt{K}$, SA	2.20[a]	2.50[a]	0.69	1.47	2.1
$m(0^+)$, SA	2.00[b]	2.4 [2.25[b]]	0.42	1.14 [0.78]	2.7 [1.9]
$\sqrt{K}$, TIA	1.5	1.9	0.47	1.24	2.6
$m(0^+)$, TIA	1.45	1.7	0.42	0.76	1.8

[a] Incorporates also results of [16]

[b] Relies on [8], where a $4^3 \cdot 16$ lattice was used

operators are collected in Table IV.1. (The numbers are defined by means of Fig. I.1.) Doing a more detailed analysis, particularly emphasizing the 0^+ and 2^+ states, we find in conclusion everything well covered by measuring only correlations of the operators 6, 7, 13, 21, and 24.

The crucial question is, of course: does the improvement work? In Table IV.2 we collect approximate estimates for the beginning and end of the scaling windows obtained so far. Scaling (with reservations coming from [17]) of the TIA string tension sets on at a smaller correlation length than for the SA. Otherwise, scaling windows, if compared on similar lattices, look similar. Therefore, from a practical point of view, the improvement is modest. An important consistency check is, of course, the observation that the improved definitions [6] of Creutz ratios work well (see Sect. III.1).

For spectrum calculations the SA has the decisive advantage of a positive definite transfer matrix. Together with the greater computational simplicity, this is an important feature in favour of the SA.

Acknowledgements. We would like to thank M. Lüscher, E. Marinari, I. Montvay, G. Münster, and P. Weisz for useful discussions. We benefitted particularly from the advice of K. Symanzik during major parts of this work. It is a pleasure to thank the staff of the University of Karlsruhe Computer Center for providing us with the necessary computer time on the CYBER 205.

Note added. After finishing the manuscript we became aware of an extension [36] of the work of [22].

References

1. Symanzik, K.: In: Mathematical problems in theoretical physics. Schrader et al. (eds.). Lecture Notes in Physics, Vol. 153. Berlin, Heidelberg, New York: Springer 1982
2. Symanzik, K.: Continuum limit and improved action in lattice theories. I. Principles and ϕ^4 theory. Nucl. Phys. B **226**, 187 (1983)
3. Berg, B., Meyer, S., Montvay, I., Symanzik, K.: Improved continuum limit in the lattice $O(3)$ non-linear sigma model. Phys. Lett. **126** B, 467 (1983)
   Falcioni, M., Martinelli, G., Paciello, M.L., Taglienti, B., Parisi, G.: A Monte Carlo simulation with an "improved" action for the $O(3)$ non-linear sigma model. Nucl. Phys. B **225**, (FS9), 313 (1983)
   For recent reviews see: Berg, B.: Preprint, DESY 83-120, published in the proceedings of the "International Symposium on the Theory of Elementary Particles," Ahrenshoop (DDR), 1983
   Meyer, S.: Talk presented at the Santa Barbara Workshop, 20.6.–1.7.83
4. Curci, G., Menotti, P., Paffuti, G.P.: Symanzik's improved lagrangian for lattice gauge theory. Phys. Lett. **130** B, 205 (1983)
5. Weisz, W.: Continuum limit improved lattice action for pure Yang-Mills theory. I. Nucl. Phys. B **212**, 1 (1983)
6. Weisz, P., Wohlert, R.: Nucl. Phys. B **236**, 137 (1984)
7. Berg, B.: Cargèse lectures 1983
8. Berg, B., Billoire, A., Rebbi, C.: Monte Carlo estimates of the $SU(2)$ mass gap. Ann. Phys. (NY) **142**, 185 (1982) and Addendum **146**, 470 (1983)
9. Falcioni, M., Marinari, E., Paciello, M.L., Parisi, G., Rapuano, F., Taglienti, B., Zhang, Yi-Chang: On the masses of the glueballs in pure $SU(2)$ lattice gauge theory. Phys. Lett. **110** B, 295 (1982)
10. Ishikawa, K., Schierholz, G., Teper, M.: The glueball mass spectrum in QCD: First results of a lattice Monte Carlo calculation. Phys. Lett. **110** B, 399 (1982)
11. Berg, B., Billoire, A.: Glueball spectroscopy in $4d$ $SU(3)$ lattice gauge theory. I. Nucl. Phys. B **221**, 109 (1983) and: Glueball spectroscopy in $4d$ $SU(3)$ lattice gauge theory. II. Nucl. Phys. B **226**, 405 (1983)
12. Berg, B., Meyer, S., Montvay, I.: Nucl. Phys. B **235** [FS11], 149 (1984)
13. Ishikawa, K., Schierholz, G., Teper, M.: Renormalization group behaviour of 0^+ and 2^+ glueball masses in $SU(2)$. Z. Phys. C **16**, 69 (1982)
14. Ishikawa, K., Schierholz, G., Teper, M.: Calculation of the glueball mass spectrum of $SU(2)$ and $SU(3)$ non-Abelian lattice gauge theories. I. Introduction and $SU(2)$. Z. Phys. C **19**, 327 (1983)
15. Creutz, M.: Asymptotic-freedom scales. Phys. Rev. Lett. **45**, 313 (1980)
16. Berg, B., Stehr, J.: $SU(2)$ lattice gauge theory and Monte Carlo calculations. Z. Phys. C **9**, 333 (1981)
17. Gutbrod, F., Montvay, I.: Phys. Lett. **136** B, 411 (1984)
18. Billoire, A., Marinari, E.: Phys. Lett. **139** B, 399 (1984)
19. Karsch, F., Lang, C.B.: Phys. Lett. **138** B, 176 (1984)
20. Berg, B., Billoire, A., Meyer, S., Panagiotakopoulos, C.: Monte Carlo calculation of $SU(2)$ glueball states with Symanzik's tree-level improved action. Phys. Lett. **133** B, 359 (1983)
21. Lüscher, M., Weisz, P.: On-shell improved lattice gauge theories. Commun. Math. Phys. **97**, 59–77 (1984)
22. Belforte, S., Curci, G., Menotti, P., Paffuti, G.P.: Monte Carlo calculations with Symanzik's improved action for QCD. Phys. Lett. **131** B, 423 (1983)
23. Fukugita, M., Kaneko, T., Niuya, T., Ukawa, A.: Measurement of the string tension in $SU(2)$ lattice gauge theory with action including six-link loops. Phys. Lett. **130** B, 73 (1983) and Addendum
24. Rebbi, C.: Phase structure of non-Abelian lattice gauge theories. Phys. Rev. D **21**, 3350 (1980)
    Petcher, D., Weingarten, D.: Monte Carlo calculations and a model of the phase structure for gauge theories on discrete subgroups of $SU(2)$. Phys. Rev. D **22**, 2465 (1980)
    Bhanot, G., Lang, C., Rebbi, C.: Comp. Phys. Commun. **25**, 275 (1982)

25. Stack, J.D.: Heavy quark potential in $SU(2)$ lattice gauge theory. Phys. Rev. D **27**, 412 (1983)
26. Engels, J., Karsch, F., Montvay, I., Satzs, H.: Finite size effects in euclidean lattice thermodynamics for non-interacting Bose and Fermi systems. Nucl. Phys. B **205** [FS5], 545 (1982)
27. Falcioni, M., Marinari, E., Paciello, M.L., Parisi, G., Taglienti, B., Zhang, Yi-Chang: Large-distance correlation functions for an $SU(2)$ lattice gauge theory. Nucl. Phys. B **215** [FS5], 265 (1983)
28. Mütter, K.H., Schilling, K.: Glueball mass from variant actions in lattice Monte Carlo simulations. Phys. Lett. **121** B, 267 (1983)
Parisi, G.: Private communication
29. de Forcrand, Ph.: Work in progress
30. Münster, G.: Strong coupling expansions for the mass gap in lattice gauge theories. Nucl. Phys. B **190**, 439 (1981); Erratum B **205** [FS5], 545 (1982)
31. Seo, K.: Glueball mass estimate by strong coupling expansion in lattice gauge theories. Nucl. Phys. B **209**, 200 (1982)
32. Berg, B., Billoire, A., Koller, K.: Monte Carlo calculation of $SU(2)$ glueball states with Manton's action. Nucl. Phys. B **233**, 50 (1984)
33. Kimura, N., Ukawa, A.: Energy-momentum dispersion of glueballs and the restoration of Lorentz invariance in lattice gauge theories. Nucl. Phys. B **205** [FS5], 637 (1982). See also: Berg, B., Panagiotakopolous, C.: Photon in $U(1)$ lattice gauge theory. Phys. Rev. Lett. **52**, 94 (1984)
34. Schierholz, G., Teper, M.: Phys. Lett. **136** B, 64, 69 (1984)
35. Bernreuther, W., Wetzel, W.: The relation between the Λ parameters of the standard and of the continuum limit improved lattice action for pure Yang-Mills theory. Phys. Lett. **132** B, 382 (1983)
Iwasaki, Y., Sakai, S.: Phys. Lett. **136** B, 73 (1984)
36. Belforte, S., Curci, G., Menotti, P., Paffuti, G.: Phys. Lett. **137** B, 207 (1984)

Communicated by G. Mack

Received February 7, 1984

On-shell Improved Lattice Gauge Theories

M. Lüscher[1] and P. Weisz[2]*

1 Deutsches Elektronen-Synchrotron DESY, D-2000 Hamburg,
Federal Republic of Germany
2 II. Institut für Theoretische Physik der Universität Hamburg, D-2000 Hamburg 50,
Federal Republic of Germany

Abstract. By means of a spectrum conserving transformation, we show that one of the 3 coefficients in Symanzik's improved action can be chosen freely, if only spectral quantities (masses of stable particles, heavy quark potential etc.) are to be improved. In perturbation theory, the other 2 coefficients are however completely determined and their values are obtained to lowest order.

1. Introduction

Symanzik's improvement programme [1–6] for lattice gauge theories (and other lattice theories) is designed to systematically reduce the cutoff dependence of on- and off-shell amplitudes near the continuum limit. Mainly, this is achieved by choosing an improved lattice action, which, at the first level of improvement, is equal to the standard one-plaquette action plus a linear combination of 3 operators[1] of dimension 6 with perturbatively calculable coefficients $c_i(g_0^2)$ (g_0: bare coupling constant, $i=1, 2, 3$). In addition, gauge invariant lattice operators are in general also *intrinsically* cutoff dependent and, in order to obtain improved correlation functions, must therefore be corrected by subtracting a combination of higher dimensional operators. The necessity of such subtractions has been explicitly demonstrated by Symanzik in the case of the non-linear σ-model [3. Sect. 4].

For the computation of the coefficients $c_i(g_0^2)$, the intrinsic cutoff dependence of operators is a potential source of difficulty, because it must be carefully disentangled from the "dynamical" cutoff effects, which are to be cancelled by improving the action. This problem can however be entirely avoided, if only the improvement of spectral quantities (e.g. the static quark-antiquark potential at physical distances) is required. Such quantities are independent of the choice of "interpolating" operator field and it therefore makes no difference whether the operators one uses have or have not been corrected.

In this paper, we show that the requirement of on-shell improvement places only two constraints on the coefficients $c_i(g_0^2)$ so that without loss one may choose $c_3(g_0^2)=0$, for example. $c_1(g_0^2)$ and $c_2(g_0^2)$ are then completely fixed and can be

---

* Heisenberg foundation fellow

1 By abuse of notation, we use the word "operator" for any Euclidean ($\mathbb{C}$-number) field, which can be composed from the fundamental gauge field

computed straightforwardly by evaluating two distinct spectral quantities in perturbation theory. We are presently performing such a calculation to one-loop order and shall here only quote the tree level result (Sect. 4).

In view of the rather non-trivial conceptual structure of the improvement programme, we include two preparatory sections, which are slightly more extensive than absolutely necessary. First, we explain what is meant by the (classical) dimension of lattice operators and present an efficient method to compute it (Sect. 2). Improved lattice gauge theories are introduced subsequently, emphasizing the necessity to formulate an improvement condition. In Sect. 4, we establish the main result announced above by constructing a transformation, which shifts the parameters in the improved action, but leaves the spectrum unchanged up to $O(a^4)$ (a: lattice spacing). Apparent discrepancies with previous calculations of the coefficients $c_i(g_0^2)$ are resolved in Sect. 5 and conclusions are drawn in Sect. 6.

2. Classical Dimension of Local Lattice Operators

Consider SU(N) lattice gauge fields $U(n,\mu)$ living on a four-dimensional hypercubic lattice with sites $n\in\mathbb{Z}^4$. A simple example for the kind of operators to be studied below is the plaquette operator

$$\mathcal{O}(n)=\sum_{\mu,\nu}\mathrm{Tr}\{\mathbb{1}-U(n,\mu)U(n+\hat{\mu},\nu)U(n+\hat{\nu},\mu)^{-1}U(n,\nu)^{-1}\} \tag{1}$$

($\hat{\mu}$: unit vector in the positive μ-direction). Some characteristic features of $\mathcal{O}(n)$ are

a) $\mathcal{O}(n)$ is a gauge invariant polynomial of link variables $U(m,\nu)$ located near n.

b) $\mathcal{O}(n)$ moves covariantly under space-time translations of the lattice gauge field.

All operators that we shall consider later have these general properties, and no further properties are required to unambiguously define the classical dimension of an operator.

Any given smooth continuum gauge field $A_\mu(x)$, $x\in\mathbb{R}^4$, can be arbitrarily well approximated by lattice gauge fields $U(n,\mu)$ in the following way. Let "a" denote a small distance and superimpose a lattice on $\mathbb{R}^4$ with sites $x=an$, $n\in\mathbb{Z}^4$. The (continuum) parallel transporter from $an+a\hat{\mu}$ to an in the field $A_\mu(x)$ is then given by the familiar path ordered exponential[2]

$$U(n,\mu)=T\exp a\int_0^1 dt\,A_\mu(an+a\hat{\mu}-ta\hat{\mu}). \tag{2}$$

For sufficiently small spacing a, $U(n,\mu)$ will be a slowly varying lattice gauge field "triangulating" A_μ, and one expects that lattice operators $\mathcal{O}(n)$ evaluated in this field approximate local (continuum) operators formed from A_μ. Indeed, inspecting Eq. (2), one easily establishes the existence of an asymptotic expansion

$$\mathcal{O}(0)\underset{a\to 0}{\simeq}\sum_{k=0}^{\infty}a^k O_k(0), \tag{3}$$

2 We use a notation, where A_μ is an anti-hermitian, traceless $N\times N$-matrix

where $O_k(x)$ is a polynomial of $A_\mu(x)$ and its derivatives. Due to the gauge covariance of Eq. (2) and the gauge invariance of $\mathcal{O}(0)$, the O_k's are actually gauge invariant, too.

We now define the dimension of a lattice operator $\mathcal{O}(n)$ to be the smallest k for which $O_k \neq 0$. For example, in case of the plaquette operator we have

$$\mathcal{O}(0) = -\frac{a^4}{2} \sum_{\mu,\nu} \mathrm{Tr}\{F_{\mu\nu}(0) F_{\mu\nu}(0)\} + O(a^5), \tag{4}$$

$$F_{\mu\nu} = \partial_\mu A_\nu - \partial_\nu A_\mu + [A_\mu, A_\nu], \tag{5}$$

so that its dimension is equal to 4. We emphasize that the dimension of lattice operators does not depend on how precisely one approximates continuum gauge fields by lattice fields. In particular, Eq. (2) could be replaced by

$$U(n,\mu) = \exp a A_\mu(an)$$

without affecting the leading term in the expansion (3). For the actual computation of the dimension of complicated lattice operators, Eq. (2) is preferable, however, because it allows us to take full advantage of gauge invariance.

When $\mathcal{O}(n)$ transforms as a scalar field under rotations and reflections of the lattice, the operators O_k that can occur in the small "a" expansion (3) are severely restricted. In particular, $O_k = 0$ for odd k, and the lowest operators assume the general form

$$O_4 = r_0 \sum_{\mu,\nu} \mathrm{Tr}(F_{\mu\nu} F_{\mu\nu}), \tag{6}$$

$$\begin{aligned} O_6 = {} & r_1 \sum_{\mu,\nu} \mathrm{Tr}(D_\mu F_{\mu\nu} D_\mu F_{\mu\nu}) + r_2 \sum_{\mu,\nu,\varrho} \mathrm{Tr}(D_\mu F_{\nu\varrho} D_\mu F_{\nu\varrho}) \\ & + r_3 \sum_{\mu,\nu,\varrho} \mathrm{Tr}(D_\mu F_{\mu\varrho} D_\nu F_{\nu\varrho}) + \sum_\mu \partial_\mu G_\mu . \end{aligned} \tag{7}$$

Here, the numbers r_i depend on the operator $\mathcal{O}(n)$ considered and

$$D_\mu F_{\nu\varrho} = \partial_\mu F_{\nu\varrho} + [A_\mu, F_{\nu\varrho}] . \tag{8}$$

G_μ is a linear combination of the fields

$$\sum_{\nu,\varrho} \mathrm{Tr}(F_{\nu\varrho} D_\mu F_{\nu\varrho}), \ \sum_{\nu,\varrho} \mathrm{Tr}(F_{\mu\nu} D_\varrho F_{\varrho\nu}), \ \sum_{\nu} \mathrm{Tr}(F_{\mu\nu} D_\mu F_{\mu\nu}),$$

which we shall never need to work out explicitly. The coefficients r_i, on the other hand, will be required frequently, and the remainder of this section is therefore devoted to the question of how to compute them efficiently.

Let us begin with the small 'a' expansion of the simple operator

$$\mathcal{O}_{\mu\nu}(n) = \mathrm{Tr}\{U(n,\mu) U(n+\hat{\mu},\nu) U(n+\hat{\nu},\mu)^{-1} U(n,\nu)^{-1}\} . \tag{9}$$

For any given fixed pair of indices $\mu \neq \nu$, choose the gauge

$$\begin{aligned} & A_\mu(x) = 0 \quad \text{for all } x, \\ & A_\nu(x) = 0 \quad \text{for all } x \text{ with } x_\mu = 0 . \end{aligned} \tag{10}$$

$\mathcal{O}_{\mu\nu}(0)$ then reduces to

$$\mathcal{O}_{\mu\nu}(0) = \mathrm{Tr}\left\{ T \exp a \int_0^1 dt\, A_\nu(a\hat{\mu} + (1-t) a\hat{\nu}) \right\} . \tag{11}$$

Furthermore, at $x=0$, we have

$$\partial_\nu^p \partial_\mu^{q+1} A_\nu = D_\nu^p D_\mu^q F_{\mu\nu}, \tag{12}$$

so that

$$A_\nu(a\hat{\mu}+(1-t)a\hat{\nu}) = aF_{\mu\nu} + \frac{1}{2}a^2[D_\mu + 2(1-t)D_\nu]F_{\mu\nu}$$
$$+\frac{1}{6}a^3[D_\mu^2 + 3(1-t)D_\nu D_\mu + 3(1-t)^2 D_\nu^2]F_{\mu\nu} + O(a^4). \tag{13}$$

Expanding the exponential in Eq. (11), noting $\mathrm{Tr}A_\nu = 0$ and using (13), quickly yields the result

$$\mathcal{O}_{\mu\nu}(0) = N + \frac{1}{2}a^4\mathrm{Tr}(F_{\mu\nu}F_{\mu\nu}) + \frac{1}{2}a^5\mathrm{Tr}[F_{\mu\nu}(D_\mu + D_\nu)F_{\mu\nu}]$$
$$+\frac{1}{6}a^6\mathrm{Tr}\Big\{F_{\mu\nu}F_{\mu\nu}F_{\mu\nu} + F_{\mu\nu}(D_\mu^2 + \frac{3}{2}D_\mu D_\nu + D_\nu^2)F_{\mu\nu}$$
$$+\frac{3}{4}(D_\mu + D_\nu)F_{\mu\nu}(D_\mu + D_\nu)F_{\mu\nu}\Big\} + O(a^7). \tag{14}$$

This is a relation between gauge invariant quantities and it therefore holds for all gauge fields A_μ, including those which do not satisfy the gauge condition (10).

More complicated operators than $\mathcal{O}_{\mu\nu}$ can be treated similarly by writing them as a sum of $\mathcal{O}_{\mu\nu}$'s plus traces of products of gauge covariant plaquette matrices such as

$$P_{\mu\nu}(n) = U(n,\mu)\,U(n+\hat{\mu},\nu)\,U(n+\hat{\nu},\mu)^{-1}\,U(n,\nu)^{-1} - \mathbb{1}. \tag{15}$$

Proceeding as above, these can be expanded and one obtains, for example,

$$P_{\mu\nu}(0) = a^2F_{\mu\nu} + \frac{1}{2}a^3(D_\mu + D_\nu)F_{\mu\nu}$$
$$+\frac{1}{6}a^4\Big\{3F_{\mu\nu}F_{\mu\nu} + (D_\mu^2 + \frac{3}{2}D_\mu D_\nu + D_\nu^2)F_{\mu\nu}\Big\} + O(a^5). \tag{16}$$

Because $P_{\mu\nu}$'s are always multiplied by other $P_{\varrho\sigma}$'s, it is not necessary to go beyond $O(a^4)$ in Eq. (16). In this way, all the operators encountered later can be expanded up to order a^6 with little effort (see Sect. 3.3 for explicit results).

3. Improved Lattice Gauge Theories

3.1. $O(a^2)$ Scaling Violations

Suppose we are given an $\mathrm{SU}(N)$ lattice gauge theory with action $S[U]$ and bare coupling constant g_0. Euclidean expectation values of gauge invariant combinations Q of the field $U(n,\mu)$ are defined by

$$\langle Q\rangle = \frac{1}{Z}\int \mathcal{D}[U]\,Q[U]\,e^{-S[U]}, \tag{17}$$

where Z is a normalizing factor such that $\langle 1\rangle = 1$ and

$$\mathscr{D}[U] = \prod_{n,\mu} dU(n,\mu) \tag{18}$$

[dU denotes the normalized Haar measure on SU(N)].

Physically interesting quantities M such as the masses of stable particles and their scattering amplitudes[3] can (in principle) be extracted from the Euclidean correlation functions of suitable local lattice operators. Initially, when expressed in lattice units, these quantities only depend on g_0 and a set of external space-like lattice momenta collectively denoted by $k(-\pi \leqq k_i \leqq \pi$ for each component $k_i)$[4]. In order to study the continuum limit, it is convenient to introduce a lattice spacing "a" by defining scaled quantities m through

$$m(p, g_0, a) = a^{-\delta} M(ap, g_0), \tag{19}$$

where δ is the engineering dimension of M and $p = k/a$ a "physical" momentum.

In a lattice gauge theory with perfect scaling it would be possible to find a function $\bar{g}_0(a)$ such that

$$a\frac{d}{da} m(p, \bar{g}_0(a), a) = 0 \tag{20}$$

for *all* quantities m. For a generic action this will however not be possible and, according to perturbation theory, the best one can achieve is approximate scaling near the continuum limit, i.e. for $a \to 0$ we then have[5]

$$a\frac{d}{da} m(p, \bar{g}_0(a), a) = O(a^2). \tag{21}$$

Note that there are many functuions $\bar{g}_0(a)$, which lead to approximate scaling. For what follows, it is helpful to remove this ambiguity by choosing once and for all a physical condition, which fixes $\bar{g}_0(a)$ uniquely. For example, one may require that the mass gap is equal to 1 GeV for all a. Or, more suitable for perturbation theory, that the force between heavy quarks is equal to $Cg^2/4\pi\,\mathrm{GeV}^2$ at a distance of $1\,\mathrm{GeV}^{-1}$ [g a fixed small number, the "renormalized" coupling, and $C = (N^2-1)/2N$]. Ultimately, it does not matter what condition is chosen, but we shall assume that it is the same for all theories considered and that it has the general form

$$m_0(\bar{g}_0(a), a) = \hat{m}_0, \tag{22}$$

where m_0 is some spectral physical quantity and $\hat{m}_0$ a fixed value.

It is possible that different lattice actions S and S' yield the same physical amplitudes within an error of order a^{2k}. In more precise terms, this means the following. Any given physical quantity can be computed in the two theories so that one has two sets of functions $m(p, g_0, a)$ and $m'(p, g_0, a)$ (the bare coupling constant g_0 is just a parameter in the action and we are free to denote it by the same symbol

3 I.e. full propagator amputated, connected n-point functions evaluated on the (lattice) mass shell

4 Other external length scales may occur but are omitted here for simplicity

5 We use a sloppy notation, where $O(a^2)$ may also stand for a term of order $a^2(\ln a)^r$

in all theories). In particular, $\bar{g}_0(a)$ and $\bar{g}'_0(a)$ may be determined and the approximate physical equivalence of S and S' is then expressed through

$$m(p,\bar{g}_0(a),a)=m'(p,\bar{g}'_0(a),a)+O(a^{2k}). \qquad (23)$$

Due to universality, a large class of actions are expected to have identical continuum limits and are hence equivalent up to $O(a^2)$. Examples of actions S' that are physically indistinguishable from a given action S [i.e. no error term in Eq. (23)] are also easily constructed by local, gauge covariant substitutions of the field variables $U(n,\mu)$. Such transformations affect correlation functions, but do not change spectral quantities. This fact will be exploited in Sect. 4.

3.2. Improvement Conditions

The aim of Symanzik's improvement programme is to reduce the $O(a^2)$ scaling violations in Eq. (21) down to $O(a^4)$ by choosing an improved lattice action. Actually, in his study of the ϕ^4-theory [2], Symanzik was able to show that not only spectral quantities can be improved, but also the (properly normalized) n-point functions of the fundamental field in momentum space. In the 2-dimensional non-linear σ-model, the situation turned out to be not so simple, in particular, the n-point functions of the lattice spin field $\boldsymbol{\phi}$ could not be improved by merely choosing an improved action [3]. However, the n-point functions of a corrected field operator

$$\boldsymbol{\phi}'=\boldsymbol{\phi}+\bar{c}_1(g_0)(\boldsymbol{\phi}\cdot\Box\boldsymbol{\phi})\boldsymbol{\phi}+\bar{c}_2(g_0)\Box\boldsymbol{\phi}, \qquad (24)$$

can be improved, provided the coefficients $\bar{c}_i$ are chosen appropriately and provided one subtracts certain contact terms from the correlation functions. For gauge theories, no in depth analysis of $O(a^2)$ scaling violations has yet been made and it is not known whether e.g. the improvement of uncorrected (but normalized) Wilson loops is possible to all orders of perturbation theory.

An improvement condition is a statement specifying which quantities are to be improved. In general, different improvement conditions require different improved actions. Moreover, some do not determine the improved action uniquely and still others may be inconsistent as discussed above. Here we shall adopt the "minimal" improvement condition, which requires that the error term in Eq. (21) be reduced to $O(a^4)$ for all low lying energy values m[6]. This includes, in particular, the force between heavy quarks at physical distances and the spectrum of the field excitations in a periodic (space-) box of physical size [8]. Because of the relation between finite size effects on the mass spectrum and scattering amplitudes [9], we expect that the improvement of the energy spectrum implies the improvement of the latter, too. In any case, lattice theories satisfying our condition will be called "on-shell improved."

The scalar field theories for which Symanzik has worked out the improved action are also on-shell improved. His improvement conditions can however be

6 Energy values can be read off from the exponential decay of correlation functions at large times. Equivalently, they can be identified with the eigenvalues λ of the transfer matrix $\mathbb{T}$ of [7] through $m=-\frac{1}{a}\ln(\lambda/\Lambda)$ (Λ: ground state eigenvalue of $\mathbb{T}$)

Table 1. List of the sets $\mathscr{S}_i$ of elementary loops $\mathscr{C}$ on the lattice. Only the loops of Fig. 1 occur in the improved action. The other loops are generated by the field transformation discussed in Sect. 4

Set	Description of elements	No. of elements/site
$\mathscr{S}_0$	Plaquettes (Fig. 1a)	6
$\mathscr{S}_1$	Rectangles (Fig. 1b)	12
$\mathscr{S}_2$	Parallelograms (Fig. 1c)	16
$\mathscr{S}_3$	Bent rectangles (Fig. 1d)	48
$\mathscr{S}_4$	Double plaquettes (Fig. 2a)	6
$\mathscr{S}_5$	Twisted rectangles (Fig. 2b)	12
$\mathscr{S}_6$	Bent twisted rectangles (Fig. 2c)	48

more restrictive. For example, in the ϕ^4-theory, the requirement that all n-point functions of the field be improved completely determines the coefficients on the improved action, while some of them remain free, if only on-shell improvement is required (cf. Sect. 4).

3.3. Ansatz for the Improved Action

Following [4] the improved action is written as[7]

$$S[U]=\frac{2}{g_0^2}\sum_{i=0}^{3} c_i(g_0^2)\sum_{\mathscr{C}\in\mathscr{S}_i}\mathscr{L}(\mathscr{C}), \tag{25}$$

where the $\mathscr{S}_i$'s denote sets of elementary loops $\mathscr{C}$ on the lattice as described by Table 1. A given set $\mathscr{S}_i$ contains all loops of a definite shape that can be drawn on the lattice. Loops $\mathscr{C}$ that differ by orientation only are considered equal and the weight $\mathscr{L}(\mathscr{C})$ is defined by

$$\mathscr{L}(\mathscr{C})=\operatorname{Re}\operatorname{Tr}[\mathbb{1}-U(\mathscr{C})], \tag{26}$$

$U(\mathscr{C})$ being the ordered product of the link variables $U(n,\mu)$ along $\mathscr{C}$. Finally, the coefficients $c_i(g_0^2)$ are regular at $g_0^2=0$ and are normalized such that

$$c_0(g_0^2)+8c_1(g_0^2)+8c_2(g_0^2)+16c_3(g_0^2)=1. \tag{27}$$

This is merely a convention. It could always be enforced by renormalizing g_0, if it should not hold initially.

The structure of the improved action becomes more transparent if we rewrite it in terms of local lattice operators. Thus, pick a set of operators $\mathscr{O}_i(n)$, scalar under rotations and reflections, such that

$$\sum_{\mathscr{C}\in\mathscr{S}_i}\mathscr{L}(\mathscr{C})=\sum_n \mathscr{O}_i(n). \tag{28}$$

7 Our notation differs from [4–6] by an interchange of c_2 with c_3

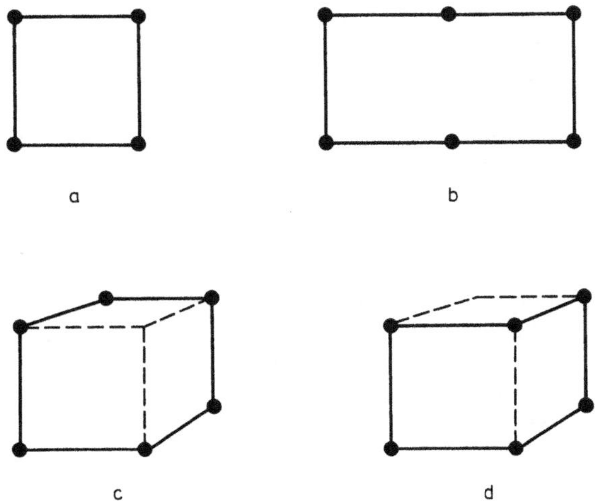

Fig. 1a–d. Elementary loops $\mathscr{C}$ on the lattice. **a** and **b** are planar loops, while **c** and **d** extend in 3 dimensions. Dashed lines are drawn to guide the eye

Table 2. List of the coefficients r_i [Eqs. (6) and (7)] that occur in the small "a" expansion (3) of the operators $\mathcal{O}_i(n)$ defined in Sect. 3.3

	r_0	r_1	r_2	r_3
$\mathcal{O}_0$	$-\frac{1}{4}$	$\frac{1}{24}$	0	0
$\mathcal{O}_1$	-2	$\frac{5}{6}$	0	0
$\mathcal{O}_2$	-2	$\frac{1}{3}$	$\frac{1}{6}$	$\frac{1}{6}$
$\mathcal{O}_3$	-4	$\frac{1}{6}$	0	$\frac{1}{2}$
$\mathcal{O}_4$	-1	$\frac{1}{6}$	0	0
$\mathcal{O}_5$	0	$-\frac{1}{2}$	0	0
$\mathcal{O}_6$	-4	$\frac{7}{6}$	0	$-\frac{1}{2}$

Following the method explained in Sect. 2, the classical small "a" expansion (3) of the $\mathcal{O}_i$'s can be worked out. The results are listed in Table 2.[8] It follows from these that $\mathcal{O}_0$ is an operator of dimension 4 and that the operators

$$\mathcal{O}'_1 = \mathcal{O}_1 - 8\mathcal{O}_0, \tag{28a}$$

$$\mathcal{O}'_2 = \mathcal{O}_2 - 8\mathcal{O}_0, \tag{28b}$$

$$\mathcal{O}'_3 = \mathcal{O}_3 - 16\mathcal{O}_0, \tag{28c}$$

8 Note that the coefficients r_i are independent of the choice of the operators $\mathcal{O}_i$ as long as (27) holds. The divergence terms $\sum_\mu \partial_\mu G_\mu$, on the other hand, are dependent and we shall (and may) assume that $\sum_\mu \partial_\mu G_\mu = 0$ for all operators $\mathcal{O}_i$

are all of dimension 6. In terms of these operators the improved action reads

$$S=\frac{2}{g_0^2}\sum_n\left\{\mathcal{O}_0(n)+\sum_{i=1}^{3}c_i(g_0^2)\mathcal{O}_i'(n)\right\}. \tag{29}$$

In other words, this is the Wilson action plus a linear combination of 3 operators of dimension 6.

Actually, the operators $\mathcal{O}_i'$, $i=1,2,3$, from a basis in the space of all operators of dimension 6, which transform as a scalar field under rotations and reflections. Namely, any such operator $\mathcal{O}$ can be represented by

$$\mathcal{O}=\sum_{i=1}^{3}\lambda_i\mathcal{O}_i'+\sum_\mu\partial_\mu\mathcal{G}_\mu+\mathcal{R}, \tag{30}$$

where $\mathcal{G}_\mu$ and $\mathcal{R}$ are lattice operators of dimension 5 and 8 respectively and ∂_μ denotes a lattice derivative ($\partial_\mu f(n)=f(n+\hat{\mu})-f(n)$). Thus, up to operators of dimension 8, the Ansatz (29) is the most general one can write down.

As already pointed out, the improvement programme has not yet been shown to work in the case of gauge theories. However, in view of the experience with scalar theories, we are confident (and shall here assume) that there is at least one choice of the coefficients $c_i(g_0^2)$ such that the theory described by the action (25) is on-shell improved. There is also a further structural property of improvement, which we expect to hold and which we shall make use of in Sect. 4. Namely, suppose S is an on-shell improved action and suppose we add a term

$$\Delta S=\varepsilon\frac{2}{g_0^2}\sum_n\mathcal{O}(n), \tag{31}$$

where ε is infinitesimal and $\mathcal{O}(n)$ a dimension 6 lattice operator scalar under rotations and reflections. In general, this leads to $O(a^2)$ scaling violations in the energy spectrum, and we expect to be able to cancel these effects by adding another term of the form

$$\tilde{\Delta}S=-\varepsilon\frac{2}{g_0^2}\sum_n\sum_{i=1}^{3}\Delta_i^{\mathcal{O}}(g_0^2;S)\mathcal{O}_i'(n), \tag{32}$$

where the coefficients $\Delta_i^{\mathcal{O}}(g_0^2;S)$ are computable in perturbation theory (as we shall see in Sect. 4, they are however not uniquely determined). Decomposing the operator $\mathcal{O}(n)$ as in Eq. (30), we may take

$$\Delta_i^{\mathcal{O}}(g_0^2;S)=\lambda_i+\Delta_i^{\mathcal{R}}(g_0^2;S).$$

Now, $\mathcal{R}$ is an operator of dimension 8 and it therefore does not induce $O(a^2)$ effects at the tree level of perturbation theory. Such effects may however show up at one-loop order, but to cancel these, it is sufficient to choose $\Delta_i^{\mathcal{R}}(g_0^2;S)=O(g_0^2)$, so that altogether we may assume

$$\Delta_i^{\mathcal{O}}(g_0^2;S)=\lambda_i+O(g_0^2). \tag{33}$$

We emphasize again that although the properties just described are highly plausible they really have the status of an unproven hypothesis at present.

3.4. Applicability of Ordinary Perturbation Theory

The only practical way known today to calculate the coefficients $c_i(g_0^2)$ is by weak coupling expansion about the (locally) pure gauge configurations. Also, all structural statements [Eq. (33) for example] ultimately rely on perturbation theory. Now it can happen that for an action S of the form (25) the classical vacuum $U(n,\mu)=1$ is a false vacuum, i.e. that there are other field configurations with $S<0$. In such a case the perturbation expansion about the pure gauge configurations is not relevant to the continuum limit and any conclusions based on it may be completely misleading. For example, we have found an action, which is on-shell improved at tree level of ordinary (false) perturbation theory, but turns out to be unimproved when expanded about the absolute minima of the action.

Perturbation theory enters our analysis at many places and the conclusions reached therefore apply only in cases where the improved action S is positive for $g_0 \to 0$ in the sense that

$$S[U]>0, \tag{34}$$

for all lattice gauge fields $U(n,\mu)$, except for (locally) pure gauge configurations. This condition restricts the admissable tree level coefficients $c_i(0)$ to some convex domain P in the plane defined by Eq. (27). P could be determined numerically, but it is easier to establish some inequalities, which in many cases of interest are sufficient to decide whether the improved action is positive or not. For example, if

$$c_0+\frac{8}{3}c_2<0 \quad \text{or} \quad c_0+8c_3<0, \tag{35}$$

one can show quickly that the action is not positive by exhibiting a field configuration with $S<0$ (details are given in Appendix A). On the other hand, if

$$c_0+8\hat{c}_1+\frac{80}{3}\hat{c}_2+32\hat{c}_3>0, \tag{36}$$

where for all i, $\hat{c}_i=\frac{1}{2}(c_i-|c_i|)$, one can prove that the action is positive and that ordinary perturbation theory is hence trustworthy. Note, however, that the domain P is certainly larger than the region characterized by the inequality (36).

4. Isospectral Transformation of Improved Actions

In order to determine the coefficients c_i in the improved action (25), we have computed the energies of a number of excited states of the lattice gauge field enclosed in an $L\times L\times L$ space box with twisted periodic boundary conditions (the twist is introduced for purely technical reasons to simplify the perturbative calculations). We found that in all cases considered the $O(a^2)$ scaling violations could be removed at tree level provided,

$$c_1(0)-c_2(0)-c_3(0)=-\frac{1}{12}, \tag{37}$$

$$c_2(0)=0. \tag{38}$$

To this order, Eq. (37) also insures the improvement of the heavy quark potential at physical distances [4]. Equation (38) is however a new result (cf. Sect. 5).

While our computations prove that the relations (37) and (38) are necessary for on-shell improvement, it is not immediately clear that they are also sufficient, because we have only considered a subset of spectral values. This question is resolved by our main result, which is summarized by the following

Theorem. *Suppose S is an on-shell improved action of the form* (25). *Then there exist coefficients $\Delta_i(g_0^2; S)$ such that*

$$S+\varepsilon\frac{2}{g_0^2}\sum_n\sum_{i=1}^{3}\Delta_i(g_0^2; S)\mathcal{O}_i'(n) \tag{39}$$

is an on-shell improved action to first order in ε, and such that

$$\Delta_1(0; S)=1\,, \tag{40a}$$

$$\Delta_2(0; S)=0\,, \tag{40b}$$

$$\Delta_3(0; S)=1\,, \tag{40c}$$

for all actions S.

Postponing the proof of the theorem, we note that at one-loop order (and beyond) the coefficients $\Delta_i(g_0^2; S)$ must be expected to depend on S so that a non-linear differential equation must be solved to integrate up the infinitesimal transformation (39). According to Eqs. (40), this differential equation however degenerates to a linear one at $g_0^2=0$ and it immediately follows that *all* actions S with coefficients $c_i(g_0^2)$ satisfying (37) and (38) are tree-level improved. Thus, to lowest order g_0^2, the most general on-shell improved action is given by

$$c_0(0)=\frac{5}{3}-24x\,, \tag{41a}$$

$$c_1(0)=-\frac{1}{12}+x\,, \tag{41b}$$

$$c_2(0)=0\,, \tag{41c}$$

$$c_3(0)=x\,, \tag{41d}$$

where x is a free parameter restricted only by the requirement that the action be positive (cf. Sect. 3.4). In particular, it is necessary that

$$x<\frac{5}{72}. \tag{42}$$

On the other hand, all values x with

$$|x|<\frac{1}{16}, \tag{43}$$

yield a positive action and are hence allowed. For example, $x=0$ is possible, but $x=\frac{1}{12}$ [which would imply $c_1(0)=0$] violates (42) and is therefore excluded.

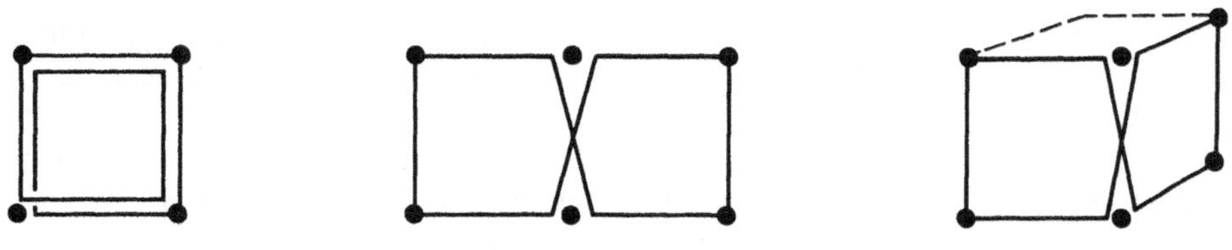

Fig. 2a–c. Elementary loops $\mathscr{C}$ generated from the one plaquette loop by the field transformation (46). **a** winds twice around a single plaquette

Our theorem is also useful at higher orders of g_0^2. For suppose we have already calculated the c_i's up to the $(\ell-1)^{\text{th}}$ loop order and suppose we wanted to compute the next order coefficients $c_i^{(\ell)}$. Then, by choosing the parameter ε in Eq. (39) proportional to $g_0^{2\ell}$, one deduces that these are only determined up to a shift

$$\begin{aligned} c_1^{(\ell)} &\to c_1^{(\ell)} + y\,, \\ c_2^{(\ell)} &\to c_2^{(\ell)} \\ c_3^{(\ell)} &\to c_3^{(\ell)} + y\,, \end{aligned} \tag{44}$$

where y is arbitrary. In particular, we may choose

$$c_3(g_0^2) = 0\,, \tag{45}$$

to all orders in perturbation theory. This choice is the most convenient one for weak coupling calculations, but we emphasize that for Monte Carlo simulations it may be advantageous to take $c_3 \neq 0$ to keep away from phase transitions at intermediate values of g_0^2. Incidentally, we note that Wilson [10] also chose $c_3 = 0$ for his improved action for a reason not disclosed to the reader.

We now proceed to prove the theorem. Essentially, the idea is to invent a transformation Φ of lattice gauge fields $U(n, \mu)$, which does not affect the energy spectrum but which changes the improved action by an additive term of dimension 6. Allowing for an error of order a^4, the transformed improved action may then be replaced by the standard improved action (25) with new coefficients c_i so that altogether we have constructed a transformation, which shifts the c_i's, but which does not generate $O(a^2)$ scaling violations.

Φ maps any given lattice gauge field $U(n, \mu)$ onto a new field $\tilde{U}(n, \mu)$. In order to preserve the basic structure of the action, this mapping should be local and invariant under rotations and reflections of the lattice. We are thus led to make the Ansatz

$$\tilde{U}(n, \mu) = e^{\varepsilon X_\mu(n)} U(n, \mu)\,, \tag{46}$$

where ε is infinitesimal and $X_\mu(n)$ the anti-hermitian traceless part of another field $Y_\mu(n)$:

$$X_\mu(n) = Y_\mu(n) - Y_\mu(n)^+ - 1/N \,\mathrm{Tr}[Y_\mu(n) - Y_\mu(n)^+]\,, \tag{47a}$$

$$\begin{aligned} Y_\mu(n) = \frac{1}{4} \sum_\nu \{ & U(n, \nu) U(n+\hat{\nu}, \mu) U(n+\hat{\mu}, \nu)^{-1} U(n, \mu)^{-1} \\ & - U(n, \mu) U(n-\hat{\nu}+\hat{\mu}, \nu)^{-1} U(n-\hat{\nu}, \mu)^{-1} U(n-\hat{\nu}, \nu) \}\,. \end{aligned} \tag{47b}$$

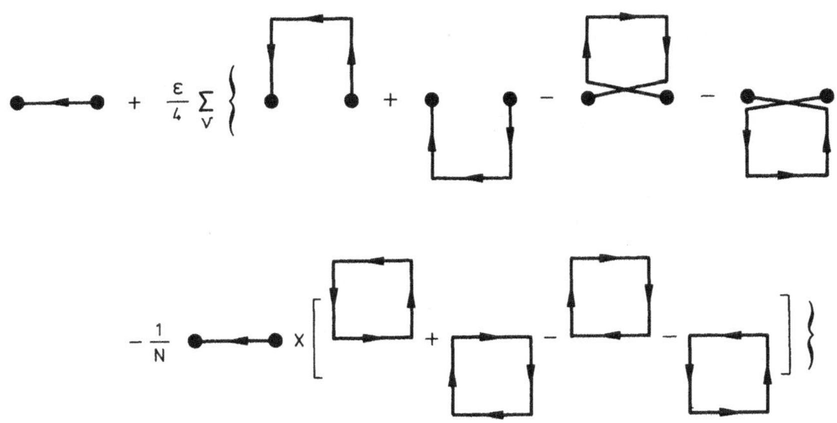

Fig. 3. Graphical representation of $\tilde{U}(n, \mu)$ [cf. Eq. (46)]. The vertices n and $n+\hat{\mu}$ are indicated by dots, link variables by arrows and the whole drawing is in the (μ, ν)-plane

The transformed field $\tilde{U}(n, \mu)$ is shown graphically in Fig. 3. Note that it depends on the old field $U(n, \mu)$ in a gauge covariant fashion.

In the classical continuum limit one finds (cf. Sect. 2)

$$X_\mu = \frac{1}{2} a^3 \sum_\nu D_\nu F_{\nu\mu} + O(a^4) \tag{48}$$

and Φ reduces to

$$A_\mu \to A_\mu + \frac{\varepsilon}{2} a^2 \sum_\nu D_\nu F_{\nu\mu} + O(a^3).$$

It follows that for any lattice operator $\mathcal{O}$ with dimension d we have $\mathcal{O}[\tilde{U}] = \mathcal{O}[U] + \varepsilon \mathcal{P}[U]$, where $\mathcal{P}$ is an operator of dimension $d+2$. In particular, the improved action (29) transforms as

$$S[\tilde{U}] = S[U] + \varepsilon \frac{2}{g_0^2} \sum_n \left\{ \mathcal{P}_0(n) + \sum_{i=1}^{3} c_i(g_0^2) \mathcal{P}_i(n) \right\}, \tag{49}$$

$\mathcal{P}_0$ and the $\mathcal{P}_i$'s being operators of dimension 6 and 8, respectively.

We are now ready to complete the proof of the theorem. Thus, suppose S is an improved action of the form (25). As discussed at the end of Sect. 3.3, we may replace S by

$$S - \varepsilon \frac{2}{g_0^2} \sum_n \left\{ \mathcal{P}_0(n) + \sum_{i=1}^{3} c_i(g_0^2) \mathcal{P}_i(n) - \sum_{i=1}^{3} \tilde{\Delta}_i(g_0^2; S) \mathcal{O}_i'(n) \right\} \tag{50}$$

without affecting the energy spectrum except perhaps at order a^4. In Eq. (50), the coefficients $\tilde{\Delta}_i$ are given by

$$\tilde{\Delta}_i(g_0^2; S) = \Delta_i^{\mathcal{P}_0}(g_0^2; S) + \sum_{k=1}^{3} c_k(g_0^2) \Delta_i^{\mathcal{P}_k}(g_0^2; S). \tag{51}$$

Secondly, we perform a substitution of integration variables in the functional integral (17) replacing $U(n, \mu)$ by $\tilde{U}(n, \mu)$. Because this change of variables is local, the exponential decay of correlation functions of local operators is not affected and the energy spectrum is exactly the same as before. The action, however, changes according to Eq. (49), and in addition there is a Jacobian, which is calculated in Appendix B. Collecting all terms, the new action reads

$$S+\varepsilon\frac{2}{g_0^2}\sum_n\left\{-2Cg_0^2\mathcal{O}_0(n)+\sum_{i=1}^{3}\tilde{\Delta}_i(g_0^2;S)\mathcal{O}_i'(n)\right\}, \tag{52}$$

where $C=(N^2-1)/2N$. Finally, we replace g_0^2 by $g_0^2-\varepsilon 2Cg_0^4$, after which the action (52) assumes the form (39) with

$$\Delta_i(g_0^2;S)=\tilde{\Delta}_i(g_0^2;S)+2Cg_0^2\left(1-g_0^2\frac{\partial}{\partial g_0^2}\right)c_i(g_0^2). \tag{53}$$

Since all transformations applied conserve the energy spectrum up to at most an error of order a^4, we have thus generated a new on-shell improved action. It remains to show that the coefficients $\Delta_i(g_0^2;S)$ satisfy Eqs. (40). Using Eq. (33), we have

$$\Delta_i(0;S)=\Delta_i^{\mathcal{P}_0}(0;S)=\lambda_i, \tag{54}$$

where the λ_i's are to be determined from the classical expansion (30) of the operator $\mathcal{P}_0$. To this end, we first recall that $\mathcal{P}_0$ is obtained from the single plaquette operator $\mathcal{O}_0$ by replacing $U(n,\mu)$ by $\tilde{U}(n,\mu)$ and subtracting the old $\mathcal{O}_0$. Thus, with the help of Fig. 3, one finds

$$\mathcal{P}_0=\frac{1}{2}\{\mathcal{O}_1+\mathcal{O}_3-2\mathcal{O}_4-\mathcal{O}_5-\mathcal{O}_6\}, \tag{55}$$

where here and below divergence terms and operators of dimension 8 are neglected. Next, from Table 2 it follows that

$$\mathcal{O}_4=4\mathcal{O}_0, \tag{56a}$$

$$\mathcal{O}_5=8\mathcal{O}_0-\mathcal{O}_1, \tag{56b}$$

$$\mathcal{O}_6=32\mathcal{O}_0-\mathcal{O}_3, \tag{56c}$$

and hence $\mathcal{P}_0=-24\mathcal{O}_0+\mathcal{O}_1+\mathcal{O}_3$. In other words, $\lambda_1=\lambda_3=1$ and $\lambda_2=0$ as required.

5. Comparison with Earlier Calculations of Improved Actions

A computation of the coefficients $c_i(g_0^2)$ to lowest order was first attempted in [4] and the following two relations were obtained:

$$c_1(0)=-\frac{1}{12}, \tag{57}$$

$$c_2(0)+c_3(0)=0. \tag{58}$$

Later, Curci et al. [6] provided an argument for

$$c_2(0)=c_3(0)=0 \tag{59}$$

by studying the heavy quark potential at one-loop order. Equation (59) was also derived from a purely classical consideration in [5]. Taken together, the conclusion was that the choice

$$c_0(0)=\frac{5}{3}, \quad c_1(0)=-\frac{1}{12}, \quad c_2(0)=c_3(0)=0 \tag{60}$$

is necessary for improvement.

At first sight, there seems to be a discrepancy with our result, which asserts that there is a whole one-parameter family of tree level improved actions of which (60) is but one member [cf. Eqs. (41)]. The origin of this mismatch is that the authors of [4–6] use improvement conditions, which differ from the one adopted here (Sect. 3.2). It is instructive to trace in each case how the improvement condition influences the outcome of the calculations. In [4], for example, the requirement was to improve all physical size Wilson loops. This includes in particular the heavy quark potential at physical distances L, which to lowest order of perturbation theory is given by

$$V(L)=-C\frac{g_0^2}{4\pi L}\left\{1+3\frac{a^2}{L^2}\left[c_1(0)-c_2(0)-c_3(0)+\frac{1}{12}\right]+O\left(\frac{a^4}{L^4}\right)\right\} \tag{61}$$

($C=(N^2-1)/2N$ as before). Now we can see that while the improvement of all Wilson loops leads to Eqs. (57) and (58), the single relation (37) already insures the improvement of the quark potential and this would in fact be the only result that can be extracted from the calculations of [4] once one adopts our on-shell improvement condition.

In their work, Curci et al. consider the Fourier transform $\tilde{V}(\mathbf{p})$ of the heavy quark potential at one-loop order. Besides other $O(a^2)$ contributions, there are two terms proportional to

$$a^2 \ln(a^2\mathbf{p}^2) \quad \text{and} \quad a^2(\mathbf{p}^2)^{-2} \sum_{i=1}^{3} p_i^4 \ln(a^2\mathbf{p}^2).$$

If one requires $\tilde{V}(\mathbf{p})$ to be improved, the coefficients of both terms must vanish and, in conjunction with Eqs. (57) and (58), this yields $c_2(0)=c_3(0)=0$. However, from the point of view of our improvement condition, we do not expect $\tilde{V}(\mathbf{p})$ to be improved, because it is a sum of energy values $V(L)$, which also includes energies of the order of the cutoff ($V(L)$ for $L=a$, for example). On the other hand, for physical L the $O(a^2)$ contributions to the quark potential at one-loop order are proportional to $a^2L^{-3}\ln L/a$ and a^2L^{-3}. Now it turns out that the first term disappears as soon as (37) holds, and no further relation for the tree-level coefficients $c_i(0)$ is obtained from the second term, because it can be made to vanish by adjusting the one-loop coefficients. Summing up, we have found that also at one-loop order Eq. (37) is in fact the only constraint on the coefficients $c_i(0)$, which is implied by the improvement of the quark potential at physical distances. In the framework of our improvement condition, Eq. (38) is therefore a new result.

In [5] a purely classical argument in favour of Eqs. (60) was presented. The starting point was the small "a" expansion of the improved action S, which reads

$$S=-\frac{1}{2g_0^2}\int d^4x\left\{\sum_{\mu,\nu}\mathrm{Tr}(F_{\mu\nu}F_{\mu\nu})-\frac{2}{3}c_2a^2\sum_{\mu,\nu,\varrho}\mathrm{Tr}(D_\mu F_{\nu\varrho}D_\mu F_{\nu\varrho})\right.$$
$$-\left(2c_1-2c_3+\frac{1}{6}\right)a^2\sum_{\mu,\nu}\mathrm{Tr}(D_\mu F_{\mu\nu}D_\mu F_{\mu\nu})$$
$$\left.-\left(\frac{2}{3}c_2+2c_3\right)a^2\sum_{\mu,\nu,\varrho}\mathrm{Tr}(D_\mu F_{\mu\varrho}D_\nu F_{\nu\varrho})+O(a^4)\right\}. \tag{62}$$

The classical action may be considered a generating functional for the vertex functions at tree level and one may therefore expect them to be improved, if the $O(a^2)$ terms in Eq. (62) are absent. This is the case if and only if the coefficients $c_i(0)$ assume the values (60). On the other hand, for the on-shell improved actions (41), Eq. (62) reduces to

$$S=-\frac{1}{2g_0^2}\int d^4x\left\{\sum_{\mu,\nu}\mathrm{Tr}(F_{\mu\nu}F_{\mu\nu})-2c_3(0)a^2\sum_{\mu,\nu,\varrho}\mathrm{Tr}(D_\mu F_{\mu\varrho}D_\nu F_{\nu\varrho})+O(a^4)\right\},$$

so that at the classical level on-shell improvement apparently means the improvement of the action for solutions of the equations of motions only.[9]

We fin ally remark that Symanzik was quite aware of the possibility to shift parameters in the improved action by making use of the Schwinger-Dyson equations. He has, however, never considered doing so at the tree level, probably because he primarily wanted to improve the correlation functions of the fundamental lattice field without explicit subtractions of the field operator.

6. Concluding Remarks

From a practical point of view, our main results are

a) that one coefficient in the improved action remains free if only spectral quantities are to be improved, and

b) that the most general tree level on-shell improved action is characterized by

$$c_1(0)-c_3(0)=-\frac{1}{12}\quad\text{and}\quad c_2(0)=0\,.$$

(We remind the reader that the improved action must also be positive. See Sect. 3.4.)

One particularly natural choice of the undetermined parameter in the action is

$$c_3(g_0^2)=0$$

not only because perturbative calculations are greatly simplified, but also because in addition to the spectral quantities many other quantities get improved as well, at least to lowest order of perturbation theory.

One may also try to fix the free coefficient in the action by adding a further improvement condition, for example that all physical size (and properly norma-

9 We do not know of an independent argument to this effect and we are therefore unable to prove Eqs. (37) and (38) by inverting the above reasoning

lized) Wilson loops should be improved. However, one must be very careful not to run into inconsistencies, especially so beyond the tree level. The danger is that at some stage one gets more independent equations than there are free parameters in the action. For this reason we prefer to stick to spectral quantities, which are anyhow the only objects of physical significance.

Appendix A: Positivity of Improved Actions

We first show that the improved action S cannot be positive if $c_0+\frac{8}{3}c_2<0$. To this end, choose

$$U(n,\mu)=\begin{bmatrix} (-1)^{s_\mu} & & & 0 \\ & (-1)^{s_\mu} & & \\ & & 1 & \\ 0 & & & \ddots 1 \end{bmatrix}, \quad s_\mu=\sum_{\nu=0}^{\mu-1} n_\nu . \tag{A1}$$

The sums

$$\sum_i = \sum_{\mathscr{C}\in\mathscr{S}_i} \mathscr{L}(\mathscr{C}) \tag{A2}$$

are then easily evaluated and one finds

$$\sum_{\mathrm{C}}=24N_s, \quad \sum_1=\sum_3=0, \quad \sum_2=64N_s,$$

where N_s is the number of sites in the lattice (we assume periodic boundary conditions for convenience). It follows that

$$S=\frac{2}{g_0^2}24N_s\left(c_0+\frac{8}{3}c_2\right)<0,$$

i.e. the action is not positive.

Next, suppose $c_0+8c_3<0$. Then, replacing s_μ in (A1) by

$$s_\mu=\delta_{\mu 3}\sum_{\mu=0}^{2} n_\nu, \tag{A3}$$

we have for this configuration

$$S=\frac{2}{g_0^2}12N_s(c_0+8c_3)<0,$$

and it follows again that the action is not positive.

Finally, we prove that the inequality (36) implies the positivity of the improved action. The basic ingredient in the proof is the following corollary of the Cauchy-Schwarz inequality:

Lemma. *Suppose U and V are two unitary matrices. Then*

$$\mathrm{Re}\,\mathrm{Tr}(1-UV)\leqq 2\mathrm{Re}\,\mathrm{Tr}(1-U)+2\,\mathrm{Re}\,\mathrm{Tr}(1-V), \tag{A4}$$

where the equality holds if and only if $U=V=1$.

Now let $\mathscr{C}$ be any rectangular loop as shown in Fig. 1b. Writing it as product of two plaquette loops $\mathscr{C}_1$ and $\mathscr{C}_2$, the lemma yields $\mathscr{L}(\mathscr{C}) \leqq 2\mathscr{L}(\mathscr{C}_1)+2\mathscr{L}(\mathscr{C}_2)$. Summing over all rectangles, we have

$$\sum_1 \leqq 8 \sum_0 . \tag{A5}$$

Similarly, one derives

$$\sum_2 \leqq \frac{80}{3} \sum_0 , \quad \sum_3 \leqq 32 \sum_0 . \tag{A6}$$

These inequalities together with $\sum_i \geqq 0$ imply

$$S \geqq \frac{2}{g_0^2}\left(c_0 + 8\hat{c}_1 + \frac{80}{3}\hat{c}_2 + 32\hat{c}_3\right)\sum_0 . \tag{A7}$$

Thus, if (36) holds, the improved action is bounded from below by a positive multiple of the Wilson action and is hence positive.

Appendix B: Computation of the Jacobian of the Transformation Φ

The calculation of the Jacobian of Φ can be greatly simplified with the help of a lemma from the general theory of Lie group (e.g. [11, Chap. X]). Thus, let G be a Lie group with Lie algebra $\mathfrak{G}$, $d\mu(g)$ a (right-) invariant measure on G, and $\Phi: G \to G$ a differentiable invertible mapping. For every $g \in G$ a linear mapping $\varphi_g: \mathfrak{G} \to \mathfrak{G}$ can be defined through

$$\Phi(e^{tZ}g) \cdot \Phi(g)^{-1} = \exp\{t\varphi_g(Z) + O(t^2)\} \tag{B1}$$

for all $Z \in \mathfrak{G}$. The lemma alluded to above then reads

Lemma. *For every integrable function $f: G \to \mathbb{C}$ the following substitution rule holds:*

$$\int_G d\mu(g) f(g) = \int_G d\mu(g) |\det \varphi_g| f(\Phi(g)) . \tag{B2}$$

In our case, the group G is equal to the set of all lattice gauge fields, i.e. it is isomorphic to $[\mathrm{SU}(N)]^4 N_s$, where N_s is the number of sites in the lattice. Correspondingly, the Lie algebra $\mathfrak{G}$ is identified with the set of all fields $Z_\mu(n)$ of traceless anti-hermitian $N \times N$-matrices. For example, if the group element g stands for the field $U(n,\mu)$, we have the correspondence

$$e^{tZ} \cdot g \leftrightarrow e^{tZ_\mu(n)} \cdot U(n,\mu) . \tag{B3}$$

Inserting this into the definition (46) of the transformation Φ, it is trivial to compute φ_g, and one finds to first order in ε,

$$\det \varphi_g = 1 + \varepsilon \frac{2}{N}(N^2-1) \sum_{\mathscr{C} \in \mathscr{S}_0} [\mathscr{L}(\mathscr{C}) - N] . \tag{B4}$$

This is in fact the Jacobian we are looking for, because the integral (B2) has exactly the form of the functional integral (17), if we identify $d\mu(g)$ with the measure $\mathscr{D}[U]$ [Eq. (18)].

References

1. Symanzik, K.: Some topics in quantum field theory. In: Mathematical problems in theoretical physics. Schrader, R. et al. (eds.). Lecture Notes in Physics, Vol. 153, Berlin, Heidelberg, New York: Springer 1982
2. Symanzik, K.: Continuum limit and improved action in lattice theories. I. Principles and ϕ^4 theory. Nucl. Phys. **B226**, 187 (1983)
3. Symanzik, K.: Continuum limit and improved action in lattice theories. II. $O(N)$ nonlinear sigma model in perturbation theory. Nucl. Phys. **B226**, 205 (1983)
4. Weisz, P.: Continuum limit improved lattice action for pure Yang-Mills theory. I. Nucl. Phys. **B212**, 1 (1983)
5. Weisz, P., Wohlert, R.: Continuum limit improved lattice action for pure Yang-Mills theory. II. Nucl. Phys. **B236**, 397 (1984)
6. Curci, G., Menotti, P., Paffuti, G.: Symanzik's improved lagrangian for lattice gauge theory. Phys. Lett. **130** B, 205 (1983)
7. Lüscher, M., Weisz, P.: Definition and general properties of the transfer matrix in continuum limit improved lattice gauge theories. DESY-preprint 84-018 (1984). Nucl. Phys. B (to be published)
8. Lüscher, M.: Some analytic results concerning the mass spectrum of Yang-Mills gauge theories on a torus. Nucl. Phys. B **219**, 233 (1983)
9. Lüscher, M.: On a relation between finite size effects and elastic scattering processes. Cargèse lecture notes, DESY-preprint 83-116 (1983). New York: Plenum Press (to be published)
10. Wilson, K.G.: Monte-Carlo calculations for lattice gauge theory. In: Recent developments in gauge theories, G. 't Hooft et al. (eds.). (Cargèse 1979), New York: Plenum Press 1980
11. Helgason, S.: Differential geometry and symmetric spaces. New York: Academic Press 1962

Communicated by G. Mack

Received April 5, 1984

On the Modular Structure of Local Algebras of Observables

Klaus Fredenhagen*

Centre de Physique Théorique, CNRS, Luminy, Case 907, F-13288 Marseille Cedex 9, France

Abstract. An asymptotic expression for the modular operators of local algebras of observables in quantum field theory is given. In an asymptotically scale invariant theory this leads to an identification of the spectra of all modular operators with $\mathbb{R}_+$. So in this case the local algebras are algebras of type III, and only factors of type III_1 can occur in a central decomposition.

1. Introduction

A characteristic feature of relativistic quantum field theory is the existence of vacuum fluctuations for all local observables. Actually, the vacuum state, restricted to a bounded region $\mathcal{O}$, has many features of an equilibrium state at nonzero temperature. Mathematically, this analogy has been made precise by the Tomita-Takesaki theory [1]. The vacuum vector Ω is cyclic and separating for each local algebra $\mathfrak{A}(\mathcal{O})$ according to the Reeh-Schlieder theorem [2]. Therefore, the theory of Tomita and Takesaki implies the existence of a positive invertible operator $\Delta_{\mathcal{O}}$ such that $\Delta_{\mathcal{O}}^{it}\mathfrak{A}(\mathcal{O})\Delta_{\mathcal{O}}^{-it}\subset\mathfrak{A}(\mathcal{O})$ for all $t\in\mathbb{R}$, and the vacuum is an equilibrium state with temperature 1 if $\Delta_{\mathcal{O}}^{-it}$ is considered as the time translation operator.

It is not clear from the beginning whether the "dynamics" which generates these time translations has a physical interpretation. There is an important special case, where such an interpretation is possible. Bisognano and Wichmann [3] have shown that for the wedge $W=\{x\in\mathbb{R}, |x^0|<x^1\}$ the operators Δ_w^{-it} coincide with the unitary operators $U(\Lambda_{2\pi t})$ which implement the Lorentz transformation

$$\Lambda_{2\pi t}=\begin{pmatrix}\cosh 2\pi t & \sinh 2\pi t & & \\ \sinh 2\pi t & \cosh 2\pi t & & 0 \\ & & 1 & \\ & 0 & & 1\end{pmatrix}. \tag{1.1}$$

* Heisenberg fellow

For an uniformly accelerated observer on the world line $\{\Lambda_{2\pi t}(0,\varrho,0,0)_{t\in\mathbb{R}},\ \varrho>0\}$, $\Lambda_{2\pi\varrho\tau}$ is a translation by proper time τ, hence he finds that the vacuum has temperature $(2\pi\varrho)^{-1}$, a phenomenon, similar to the Hawking radiation of black holes [4], which has first been discussed by Unruh [5]; the connection with the Bisognano-Wichmann result has been observed by Sewell [6]. Recently, Haag, Narnhofer, and Stein [7] have shown that the Hawking temperature can be explained using only local properties of the theory. They consider the scaling limit of the theory and require that this limit is, in every admissible state and at every point of space time, a Wightman field theory, fulfilling the spectrum condition. Using this property of the scaling limit for points in the intersection of the future and the past horizon (in the case of the accelerated observer considered before the 2-plane $x^0=x^1=0$) they find that the value of the Hawking temperature is uniquely fixed.

These results strongly suggest that the modular operators of finitely extended regions $\mathcal{O}$ might have a physical meaning, too. A purely geometric interpretation of $\Delta_{\mathcal{O}}^{-it}$, however, seems to be possible only under special conditions [8]. Since $\Delta_{\mathcal{O}}^{-it}$ induces automorphisms of the algebra of observables and therefore preserves locality, the associated transformation of space time should preserve the causal structure. So only conformal transformations can occur. In fact, Hislop and Longo [9] showed that in the case of the free massless scalar field the modular operator of a double cone induces conformal transformations

$$\Delta_{\mathcal{O}}^{-it}\varphi(x)\Delta_{\mathcal{O}}^{it}=\varphi(K_{2\pi t}), \tag{1.2}$$

where (K_t) is a 1-parameter subgroup of the conformal group which has a timelike generator and leaves the double cone $\mathcal{O}$ invariant.

To compare this result with that of Bisognano and Wichmann one may look at a double cone $\mathcal{O}\subset W$ which contains the origin in its closure. Then for $x\in\mathcal{O}$, x sufficiently near to the origin, the actions of K_t and Λ_t approximately coincide,

$$|x|^{-1}(K_t-\Lambda_t)x\to 0\,, \tag{1.3}$$

where $|x|^2=(x^0)^2+|\underline{x}|^2$. In the short distance limit, a general theory is expected to exhibit conformal invariance; therefore, one may conjecture that the approximate coincidence of $\Delta_{\mathcal{O}}^{-it}$ and $U(\Lambda_{2\pi t})$ in the neighbourhood of the origin is a general phenomenon.

Actually, using only the fact that $\Delta_w^{-it}\mathfrak{A}(\lambda\mathcal{O})\Delta_w^{it}\subset\mathfrak{A}(\mathcal{O})$ for $|t|\leqq(2\pi)^{-1}|\ln\lambda|$, $\lambda<1$, which is a consequence of the identification of Δ_w^{-it} with $U(\Lambda_{2\pi t})$ and the Lorentz covariance of the net of local algebras, one arrives at the following result.

Theorem 1.1. *For each $f\in\mathscr{L}_1(\mathbb{R})$ and $0<\lambda<1$ there exists a constant $c_f(\lambda)>0$ such that:*

(i) $$c_f(\lambda)\to 0 \quad as \quad \lambda\to 0\,, \tag{1.4}$$

(ii) $$\left\|\int dt f(t)\{\Delta_{\mathcal{O}}^{-it}-\Delta_w^{-it}\}A\Omega\right\|^2\leqq c_f(\lambda)\{\|A\Omega\|^2+\|A^*\Omega\|^2\} \tag{1.5}$$

for all $A\in\mathfrak{A}(\lambda\mathcal{O})$. $c_f(\lambda)$ neither depends on the details of the theory nor on the size of $\mathcal{O}$.

The proof of this theorem will be given in Sect. 2.

We would like to proceed from there to show that the spectra of $\Delta_{\mathcal{O}}$ and Δ_w coincide. If the theory is dilation invariant this follows simply because in that case we have a group of unitary operators $D(\lambda)$, representing the dilations, with the properties:

$$D(\lambda)\Omega=\Omega\,,\qquad D(\lambda)\mathfrak{A}(\mathcal{O})D(\lambda)^{-1}=\mathfrak{A}(\lambda\mathcal{O})\,,\qquad D(\lambda)\Delta_w=\Delta_w D(\lambda)\,. \tag{1.6}$$

The last relation follows from the fact that the dilations commute with the Lorentz boosts (it would hold even without the identification of the modular transformations with Lorentz boosts since the wedge is dilation invariant). Then

$$b_f(\lambda)=\sup_{A\in\mathfrak{A}(\lambda\mathcal{O})}\|\int dt f(t)\Delta_w^{-it}A\Omega\|\{\|A\Omega\|^2+\|A^*\Omega\|^2\}^{-1/2}\equiv b_f \tag{1.7}$$

is independent of λ. If we choose f such that the support of its Fourier transform $\tilde{f}$ lies in the complement of the spectrum of $(-\ln\Delta_{\mathcal{O}})$, then the first term of the left-hand side of inequality (1.5) vanishes, (1.5) reduces to

$$b_f(\lambda)^2\leqq c_f(\lambda) \tag{1.8}$$

for any λ, and thus by (1.4) $b_f=0$ which means that the support of $\tilde{f}$ lies also in the complement of the spectrum of $(-\ln\Delta_w)$. Since the spectrum of $(-\ln\Delta_w)$ is the whole real axis, its complement is empty and thus the spectrum of $(-\ln\Delta_{\mathcal{O}})$ must also be the whole real line.

We shall show in Sect. 3 that an analogous argument with the same conclusion can be given if the theory is asymptotically dilation invariant in the following sense: there is a scalar Wightman field φ associated to the net of local algebras of observables such that the Wightman functions have a scaling limit.

Can one determine also the spectra of modular operators associated to other normal states of $\mathfrak{A}(\mathcal{O})$? This question is related to the type of $\mathfrak{A}(\mathcal{O})$ in the classification of Connes [10]. In a dilation invariant theory it follows immediately from the general results on asymptotically abelian systems [11, 12] that the spectra of all modular operators coincide with $\mathrm{sp}\,\Delta_{\mathcal{O}}=\mathbb{R}_+$. In Sect. 4 we shall show that this remains true under the assumption of asymptotic scale invariance of the Wightman field φ. In particular, this implies that the local algebras are von Neumann algebras of type III; moreover, in a central decomposition only factors of type III_1 can occur. In the case of dilation invariant theories this improves a result of Driessler [13].

2. Comparison of Modular Operators

Let $\mathcal{N}_1\subset\mathcal{N}_2$ be von Neumann algebras in a Hilbert space with a common cyclic and separating vector Ω. For $j=1,2$, S_j denotes the closure of the antilinear operator $S_j^{(0)}$

$$S_j^{(0)}A\Omega=A^*\Omega\,,\qquad A\in\mathcal{N}_j\,, \tag{2.1}$$

and the modular operator Δ_j is the square of the positive operator occurring in the polar decomposition of S_j,

$$S_j=J_j\Delta_j^{1/2}\,. \tag{2.2}$$

$\mathcal{N}_1 \subset \mathcal{N}_2$ implies $\mathcal{D}(S_1) \subset \mathcal{D}(S_2)$, and from (2.1)

$$S_2 \upharpoonright \mathcal{D}(S_1) = S_1 . \tag{2.3}$$

$\mathcal{D}(S_2)$, equipped with the scalar product

$$(\Phi, \Psi)_2 = (\Phi, \Psi) + (S_2\Psi, S_2\Phi), \qquad \Phi, \Psi \in \mathcal{D}(S_2) \tag{2.4}$$

is a Hilbert space which will be denoted by $\mathcal{H}_2$, with the identification map $I: \mathcal{H}_2 \to \mathcal{D}(S_2) \subset \mathcal{H}$. For $\Phi, \Psi \in \mathcal{D}(S_1)$, from (2.3)

$$(\Phi, \Psi)_2 = (\Phi, \Psi) + (S_1\Psi, S_1\Phi) . \tag{2.5}$$

Hence as S_1 is closed, $\mathcal{H}_1 = I^{-1}\mathcal{D}(S_1)$ is a closed subspace of $\mathcal{H}_2$. Let P_1 denote the orthogonal projection from $\mathcal{H}_2$ onto $\mathcal{H}_1$. The operators $V_j = I^{-1}(1+\Delta_j)^{-1/2}$ are isometries from $\mathcal{H}$ onto $\mathcal{H}_j$, with adjoints $V_j^* = (1+\Delta_j)^{1/2} I P_j, j = 1, 2\ (P_2 = 1)$. Thus $IV_j = (1+\Delta_j)^{-1/2}$ and

$$IP_jI^* = IV_jV_j^*I^* = (1+\Delta_j)^{-1}, \quad j = 1, 2 . \tag{2.6}$$

Equation (2.6) implies in particular that for an increasing sequence of von Neumann algebras $\mathcal{N}^{(n)}$, $\mathcal{N}_1 \subset \mathcal{N}^{(n)} \subset \mathcal{N}_2$, $\{\cup \mathcal{N}^{(n)}\}'' = \mathcal{N}_2$, the modular operators converge in the sense of strong resolvent convergence [14]. We shall use Eq. (2.6) to estimate the difference

$$[(1+\Delta_2)^{-1} - (1+\Delta_1)^{-1}]A\Omega, \qquad A \in \mathcal{N}_1 , \tag{2.7}$$

by the time spent by A in $\mathcal{N}_1$ under the time evolution corresponding to Δ_2.

Proposition 2.1. *Let $A \in \mathcal{N}_1$ such that $\Delta_2^{it} A \Delta_2^{-it} \in \mathcal{N}_1$ for $|t| \leqq \tau$. Then*

$$\|[(1+\Delta_2)^{-1} - (1+\Delta_1)^{-1}]A\Omega\| \leqq \frac{2}{\pi} \tanh^{-1} e^{-\pi\tau} \{\|A\Omega\|^2 + \|A^*\Omega\|^2\}^{1/2} .$$

Proof. From (2.6) $I^* = I^{-1}(1+\Delta_2)^{-1}$. We have

$$I^{-1}(1+\Delta_2)^{-1} = \underset{\varepsilon \searrow 0}{s\text{-lim}}\, I^{-1}\Delta_2^{\varepsilon}(1+\Delta_2)^{-1} = \underset{\varepsilon \searrow 0}{s\text{-lim}}\, I^{-1} \int dt f_\varepsilon(t) \Delta_2^{it}$$

with $f_\varepsilon(t) = \frac{i}{2}[\sinh(t+i\varepsilon)\pi]^{-1}$, and therefore,

$$\begin{aligned}[(1+\Delta_2)^{-1} - (1+\Delta_1)^{-1}]A\Omega &= I(1-P_1)I^{-1}(1+\Delta_2)^{-1}A\Omega \\ &= s\text{-lim} \int dt f_\varepsilon(t) I(1-P_1)I^{-1}\Delta_2^{it}A\Omega . \end{aligned} \tag{$*$}$$

For $|t| \leqq \tau$ one has $\Delta_2^{it}A\Delta_2^{-it} \in \mathcal{N}_1$, hence $\Delta_2^{it}A\Omega \in \mathcal{D}(S_1)$, $I^{-1}\Delta_2^{it}A\Omega \in \mathcal{H}_1$ and $(1-P_1)I^{-1}\Delta_2^{it}A\Omega = 0$. Thus one can restrict the t-integration in ($*$) to values $|t| > \tau$. The limit $\varepsilon \searrow 0$ can then be performed under the integral sign, and the proposition follows with $\|I^{-1}\Delta_2^{it}A\Omega\| = \{\|A\Omega\|^2 + \|A^*\Omega\|^2\}^{1/2}$ for all t and $\int_{|t| \geqq \tau} dt \frac{1}{2} |\sinh \pi t|^{-1} = \frac{2}{\pi} \tanh^{-1} e^{-\pi\tau}$. q.e.d.

If one replaces Δ_j by $e^u\Delta_j$ in Proposition 2.1, $j=1,2$, $u\in\mathbb{R}$, one obtains, by the same method, the estimate

$$\|[1+e^u\Delta_2)^{-1}-(1+e^u\Delta_1)^{-1}]A\Omega\| \leqq \frac{2}{\pi}\tanh^{-1}e^{-\pi\tau}\{\|A\Omega\|^2+e^u\|A^*\Omega\|^2\}^{1/2}. \tag{2.8}$$

The next lemma gives a connection between the resolvants $(1+e^u\Delta)^{-1}$ and the unitaries Δ^{it} for a positive invertible operator Δ.

Lemma 2.2. *Let* $\varphi\in\mathcal{S}(\mathbb{R})$ *with* $\int\varphi(u)du=0$. *Then*

$$\int du\varphi(u)(1+e^u\Delta)^{-1}=\int dt\frac{i}{2}[\sinh\pi t]^{-1}\tilde{\varphi}(t)\Delta^{it}$$

with $\tilde{\varphi}(t)=\int du\varphi(u)e^{itu}$.

Proof. Using the spectral representation of $\ln\Delta$, the lemma follows from the string of identities

$$\begin{aligned}
&\int dt\frac{i}{2}[\sinh\pi t]^{-1}\tilde{\varphi}(t)e^{ixt}\\
&=\int dt\int du\frac{i}{2}[\sinh\pi t]^{-1}\varphi(u)e^{it(x+u)}\\
&=\lim_{\varepsilon\searrow 0}\int du\varphi(u)\int dt\frac{i}{2}[\sinh\pi(t+i\varepsilon)]^{-1}e^{it(x+u)}\\
&=\lim_{\varepsilon\searrow 0}\int du\varphi(u)e^{\varepsilon(x+u)}(1+e^{x+u})^{-1}\\
&=\int du\varphi(u)(1+e^{x+u})^{-1}. \quad \text{q.e.d.}
\end{aligned}$$

Now we have the means to prove Theorem 1.1.

Proof of Theorem 1.1. Let $\mathfrak{A}(\mathcal{O})=\mathcal{N}_1$, $\mathfrak{A}(W)=\mathcal{N}_2$, $\Delta_{\mathcal{O}}=\Delta_1$, and $\Delta_w=\Delta_2$. For $g\in\mathcal{D}(\mathbb{R})$, we define

$$\varphi_g(u)=(2\pi)^{-1}\int dtg(t)(-2i)\sinh(\pi t)e^{-itu}.$$

We have the estimates

$$\begin{aligned}
&\|\int dtf(t)(\Delta_2^{it}-\Delta_1^{it})A\Omega\|\\
&\leqq\int dt|f(t)-g(t)|\cdot 2\|A\Omega\|\\
&\quad+\int du|\varphi_g(u)|\,\|[(1+e^u\Delta_2)^{-1}-(1+e^u\Delta_1)^{-1}]A\Omega\|\\
&\leqq\left\{\int dt|f(t)-g(t)|+\int_{|u|\geqq c}du|\varphi_g(u)|\right\}\cdot 2\|A\Omega\|\\
&\quad+\int du|\varphi_g(u)|e^{c/2}\frac{2}{\pi}\tanh^{-1}\lambda^{1/2}\{\|A\Omega\|^2+\|A^*\Omega\|^2\}^{1/2},
\end{aligned}$$

which give the theorem with

$$c_f(\lambda)=\inf_{\substack{g\in\mathcal{D}(\mathbb{R})\\ c>0}}\left\{2\int dt|f-g|+2\int_{|u|\geqq c}du|\varphi_g|+\frac{2}{\pi}\tanh^{-1}\lambda^{1/2}e^{c/2}\int du|\varphi_g|\right\}. \quad \text{q.e.d.}$$

3. The Spectrum of $\Delta_{\mathcal{O}}$

For a determination of sp$\Delta_{\mathcal{O}}$ by Theorem 1.1 in a dilationally invariant theory one needs some information on the quantity $b_f(\lambda)$ defined in Eq. (1.7). In a certain sense the local algebras $\mathfrak{A}(\lambda\mathcal{O})$ must be large enough for arbitrarily small λ. We, therefore, restrict ourselves to asymptotically dilation invariant theories. Since there is no satisfying intrinsic definition of asymptotic dilation invariance available, we use the following characterization: there exists a scalar hermitian Wightman field φ which is associated to the local algebras of observables such that the closure $\varphi(f)^-$ of $\varphi f)$ is affiliated to $\mathfrak{A}(\mathcal{O})$ if supp $f \subset \mathcal{O}$. Furthermore, there exists a monotone function $N: \mathbb{R}_+^* \to \mathbb{R}_+^*$ such that the scaled Wightman functions W_n^λ,

$$W_n^\lambda(x_1, \ldots, x_n) = N(\lambda)^n(\Omega, \varphi(\lambda x_1) \ldots \varphi(\lambda x_n)\Omega), \tag{3.1}$$

converge for $\lambda \to 0$ in the sense of tempered distributions with $\lim W_2^\lambda \equiv W_2^0$ different from zero. Such a behaviour is expected for renormalizable theories with an ultraviolet fixed point.

It is convenient to smear the field φ only over trajectories $\{\Lambda_t x\}$ with $x \in W$. According to a result of Borchers [16] this is sufficient to get well defined operators. We introduce the operators

$$\varphi_\varrho(g) = N(\varrho) \int dt g(t) \varphi(\varrho \sinh 2\pi t, \varrho \cosh 2\pi t, 0, 0), \tag{3.2}$$

with $\varrho > 0$ and $g \in \mathcal{D}(\mathbb{R})$. If supp$g \subset (-L, L)$ then $\varphi_\varrho(g)^-$ is affiliated to $\mathfrak{A}(\mathcal{O}_R)$, $R = \varrho e^L$, where $\mathcal{O}_R$, $R \in \mathbb{R}_+$, denotes the double cone with center $(0, R, 0, 0)$ and radius R. φ_ϱ may be considered as a one-dimensional field which transforms covariantly under the Lorentz transformations Λ_t,

$$U(\Lambda_{2\pi t}) \varphi_\varrho(\tau) U(\Lambda_{2\pi t})^{-1} = \varphi_\varrho(\tau + t). \tag{3.3}$$

The Fourier transforms of the two-point functions of these fields

$$\{\int dt e^{-ik\tau}(\Omega, \varphi_\varrho(0)\varphi_\varrho(\tau)\Omega)\} dk = d\nu_\varrho(k) \tag{3.4}$$

are positive measures. They are tempered, as a consequence of the temperedness of the field φ; their support is the real line; this follows for instance by a representation as an integral over mass values with the Källen-Lehmann weight μ of the two-point function of the field φ,

$$d\nu_\varrho(k) = \{\int d\mu(m^2) B_{\varrho m}(k)\} dk (N(\varrho)/\varrho)^2, \tag{3.5}$$

where $B_{\varrho m}(k)$ is an analytic function with nonnegative values for $m \geqq 0$ [1]; furthermore, $d\nu_\varrho(-k) = e^{-k} d\nu_\varrho(k)$; this is the expression of the fact that Ω induces a KMS-state on the field φ_ϱ with temperature 1 and implies in particular that multiplication by e^{-k} does not destroy the temperedness of ν_ϱ; and finally, from the scaling assumption on φ, the measures converge for $\varrho \to 0$ in the sense of tempered distributions to a tempered measure ν_0 which has the same properties.

The fields φ_ϱ can be used for an investigation of the function $b_f(\lambda)$. Let $f \in \mathcal{L}_1(\mathbb{R})$ and $g \in \mathcal{D}(\mathbb{R})$. Then for $\lambda > 0$ and ϱ sufficiently small $\varphi_\varrho(g)^-$ is affiliated

---

1 Explicitly, $B_m(k) = 2m^2\{K_{1+ik/2\pi}(m) K_{1-ik/2\pi}(m) - K_{ik/2\pi}(m) K_{-ik/2\pi}(m)\}$ for $m \neq 0$, where K_n denotes the modified Bessel function of the third kind of order n [17]

to $\mathfrak{A}(\lambda\mathcal{O})$, hence

$$b_f(\lambda) \geqq \|\int dt f(t) \Delta_w^{-it} \varphi_\varrho(g)\Omega\|^2 \{\|\varphi_\varrho(g)\Omega\|^2 + \|\varphi_\varrho(g)^*\Omega\|^2\}^{-1}. \tag{3.6}$$

Because of the Lorentz covariance of the field φ_ϱ (3.3) and the KMS-property of the measure v_ϱ, the right-hand side of inequality (3.6) is equal to

$$\frac{\int \{dv_\varrho(k) + dv_\varrho(-k)\} |\tilde{g}(k)|^2 |\tilde{f}(k)|^2 (1+e^{-k})^{-1}}{\int \{dv_\varrho(k) + dv_\varrho(-k)\} |\tilde{g}(k)|^2}. \tag{3.7}$$

In the limit $\varrho \to 0$ the support restriction for g disappears. So varying g freely in $\mathcal{D}(\mathbb{R})$, one finds that the lowest upper bound of the term in (3.7) is $\sup\{|\tilde{f}(k)|^2(1+e^{-k})^{-1},\ k \in \mathbb{R}\}$. This number, however, is equal to $\|\int dt f(t) \Delta_w^{-it}(1+\Delta_w)^{-1/2}\|^2$, which is an upper bound for the right-hand side of inequality (3.6), since from the definition of the modular operator

$$\|\int dt f(t) \Delta_w^{-it}(1+\Delta_w)^{-1/2}\|^2 = \sup_{A \in \mathfrak{A}(W)} \frac{\|\int dt f(t) \Delta_w^{-it} A\Omega\|^2}{\|A\Omega\|^2 + \|A^*\Omega\|^2}. \tag{3.8}$$

Hence $b_f(\lambda)$ is independent of λ also in the case of asymptotic dilation invariance, and one concludes that $\mathrm{sp}\,\Delta_{\mathcal{O}} = \mathbb{R}_+$, in the same way as in the dilation invariant case.

4. Spectra of all Modular Operators

There are general theorems asserting that the spectra of all modular operators coincide for asymptotically abelian systems [11, 12]. In a Wightman field theory the scale transformations of the fields act in a similar way as the automorphisms in an asymptotically abelian system; in fact, in a dilation invariant theory the analogy is exact. The intuitive reason for the independence of the spectrum of the modular operator from the choice of the normal faithful state is that the unitary cocycles relating different modular automorphism groups approximately commute with operators $A \in \mathfrak{A}(\lambda\mathcal{O})$ for λ sufficiently small; only these operators have been used to identify $\mathrm{sp}\,\Delta_{\mathcal{O}}$.

It is not known whether the general structure of quantum field theory implies that the center of $\mathfrak{A}(\mathcal{O})$ is always contained in the center of the whole algebra of observables [this is known to be the case e.g. for the algebra $\mathfrak{A}(W)$]. Actually, Buchholz [18] has given simple examples of local nets, where the local algebras have nontrivial centers, whereas the center of the global algebra is trivial; these examples have all properties usually required with the exception of duality and the time slice axiom.

Therefore, it is reasonable to consider also the modular operators associated to normal states which are not necessarily faithful. If ω is a normal state of a von Neumann algebra $\mathcal{N}$ there is a minimal projection $E \in \mathcal{N}$ with $\omega(E) = 1$. The restriction ω_E of ω to the reduced algebra $\mathcal{N}_E = E\mathcal{N}E$ is faithful, so one can introduce the modular operator Δ_ω as the modular operator associated to the pair $(\omega_E, \mathcal{N}_E)$. The intersection

$$S'(\mathcal{N}) = \cap\, \mathrm{sp}\,\Delta_\omega, \tag{4.1}$$

where ω ranges over the set of normal states of $\mathcal{N}$ is an algebraic invariant of $\mathcal{N}$ [19]. It coincides with the invariant S of Connes [10] if $\mathcal{N}$ is a factor.

For the determination of $\mathrm{sp}\,\Delta_\omega$ we use the following criterion which is a slight variation of a criterion of Stormer [12].

Proposition 4.1. *Let ω denote a normal state of a von Neumann algebra $\mathcal{N}$, and let $\lambda > 0$. Then $\lambda \in \mathrm{sp}\,\Delta_\omega$ iff for each $\varepsilon > 0$ there is some $A \in \mathcal{N}$ with $\omega(A^*A) \geqq 1$ such that for all $B \in \mathcal{N}$,*

$$|\omega(AB) - \lambda\omega(BA)| \leqq \varepsilon\{\omega(B^*B) + \lambda\omega(BB^*)\}^{1/2}.$$

Proof. Let $\mathcal{N}$ act on a Hilbert space $\mathcal{H}$ such that ω is induced by the cyclic vector Ω. Let E denote the minimal projection in $\mathcal{N}$ with $\omega(E) = 1$. If $\lambda \in \mathrm{sp}\,\Delta_\omega$ there exists, for each $\varepsilon > 0$, some $A \in \mathcal{N}_E$ with $\|A\Omega\| = 1$ and $\|(\Delta_\omega^{1/2} - \lambda^{1/2})A\Omega\| < \varepsilon/\sqrt{2}$. Then for all $B \in \mathcal{N}$,

$$\begin{aligned}
|\omega(AB) - \lambda\omega(BA)| &= |(\Delta_\omega^{1/2} - \lambda^{1/2})A\Omega, (\Delta_\omega^{1/2} + \lambda^{1/2})EB^*\Omega)| \\
&\leqq \frac{\varepsilon}{\sqrt{2}} \|(\Delta_\omega^{1/2} + \lambda^{1/2})EB^*\Omega\| \\
&\leqq \varepsilon\{\|\Delta_\omega^{1/2}EB^*\Omega\|^2 + \lambda\|EB^*\Omega\|^2\}^{1/2} \\
&= \varepsilon\{\|EB\Omega\|^2 + \lambda\|EB^*\Omega\|^2\}^{1/2} \\
&\leqq \varepsilon\{\omega(B^*B) + \lambda\omega(BB^*)\}^{1/2}.
\end{aligned}$$

If, on the other hand, for some $\varepsilon > 0$ there exists an $A \in \mathcal{N}$ with $\omega(A^*A) \geqq 1$ and $|\omega(AB) - \lambda\omega(BA)| \leqq \varepsilon\{\omega(B^*B) + \lambda\omega(BB^*)\}^{1/2}$ for all $B \in \mathcal{N}$, then for $B \in \mathcal{N}_E$

$$\begin{aligned}
|((\Delta_\omega^{1/2} - \lambda^{1/2})EA\Omega, (\Delta_\omega^{1/2} + \lambda^{1/2})B^*\Omega)| &\leqq \varepsilon\{\|\Delta_\omega^{1/2}B^*\Omega\|^2 + \lambda\|B^*\Omega\|^2\}^{1/2} \\
&\leqq \varepsilon\|(\Delta_\omega^{1/2} + \lambda^{1/2})B^*\Omega\|,
\end{aligned}$$

and therefore, $\|(\Delta_\omega^{1/2} - \lambda^{1/2})EA\Omega\| \leqq \varepsilon$. The norm of $EA\Omega$ can be estimated by inserting $B = A^*(1-E)$ into the criterion. Because of $(1-E)\Omega = 0$, this gives

$$\lambda\|(1-E)A\Omega\|^2 \leqq \varepsilon\lambda^{1/2}\|(1-E)A\Omega\|$$

and thus $\|(1-E)A\Omega\| \leqq \varepsilon\lambda^{-1/2}$ and $\|EA\Omega\|^2 \geqq 1 - \varepsilon^2\lambda^{-1}$. For $\varepsilon < \lambda^{1/2}$, one gets

$$\mathrm{dist}(\mathrm{sp}\,\Delta_\omega^{1/2}, \lambda^{1/2}) \leqq \frac{\varepsilon}{(1 - \varepsilon^2\lambda^{-1})^{1/2}};$$

so if the criterion is fulfilled for all $\varepsilon > 0$ one finds $\lambda \in \mathrm{sp}\,\Delta_\omega$. q.e.d.

Theorem 4.2. *Let $\mathcal{N}$ be a von Neumann algebra with $\mathfrak{A}(\mathcal{O}) \subset \mathcal{N} \subset \mathfrak{A}(W)$. If a Wightman field φ with a scale invariant limit is associated to the local net then $S'(\mathcal{N}) = \mathbb{R}_+$.*

Proof. We shall use the Wightman field φ to construct, for each $\lambda > 0$, $\varepsilon > 0$, a uniformly bounded sequence $A_n \in \mathfrak{A}\left(\frac{1}{n}\mathcal{O}\right)$ such that

$$\|A_n\Omega\| \geqq 1 \quad \text{and} \quad \|(j(A_n^*) - \lambda^{1/2}A_n)\Omega\| \leqq \varepsilon$$

for all $n \in \mathbb{N}$, where j denotes the modular involution of $(\mathfrak{A}(W), \Omega)$, i.e. j is the PCT-transformation composed with a rotation by π in the x_2-x_3-plane.

Provided these sequences exist, the argument goes as follows. Let ω be a normal state of $\mathcal{N}$, and let $\Phi \in \mathcal{H}$ be a vector inducing ω. (Φ exists since $\mathcal{N}$ has a separating vector; the argument below, however, works equally well with density matrices.) Since the weak limit points of the sequences $(|A_n|^2)$ and $(|j(A_n^* - \lambda^{1/2}A_n|^2$ are localized in the origin they are multiples of the identity; hence for some n, $\|A_n\Phi\| \geqq 1$ and $\|(j(A_n^*) - \lambda^{1/2}A_n)\Phi\|$, $\|(A_n^* - \lambda^{1/2}j(A_n))\Phi\| \leqq \varepsilon$. But then the criterion of Proposition 4.1 is satisfied:

$$\begin{aligned}|\omega(A_nB) - \lambda\omega(BA_n)| &= |([A_n^* - \lambda^{1/2}j(A_n)]\Phi, B\Phi) + \lambda^{1/2}(j(A_n)\Phi, B\Phi)\\ &\quad - \lambda^{1/2}(B^*\Phi, [\lambda^{1/2}A_n - j(A_n^*)]\Phi) - \lambda^{1/2}(B^*\Phi, j(A_n^*)\Phi)|\\ &\leqq \varepsilon\{\|B\Phi\| + \lambda^{1/2}\|B^*\Phi\|\}\\ &\leqq \varepsilon\sqrt{2}\{\omega(B^*B) + \lambda\omega(BB^*)\}^{1/2}.\end{aligned}$$

It remains to establish the existence of the sequences (A_n). We choose

$$A_n = [1 + c\varphi_{1/n+k}(f)^- \varphi_{1/n+k}(f)^*]^{-1}\varphi_{1/n+k}(f)^-$$

with $f \in \mathcal{D}(\mathbb{R})$, $c > 0$ and $k \in \mathbb{N}$ such that the Fourier transform of f is concentrated around $\ln\lambda$, $A_n \in \mathfrak{A}\left(\frac{1}{n}\mathcal{O}\right)$ and c is sufficiently small. More precisely, we choose f such that $\|\varphi_{1/n}(f)\Omega\| \geqq 1 + \varepsilon/3$ and $\|(\Delta_W^{1/2} - \lambda^{1/2})\varphi_{1/n}(f)\Omega\| \leqq \varepsilon/3$ for all sufficiently large n. This is possible since the measures $(1 + e^{-k})\nu_\varrho$ are tempered and converge for $\varrho \to 0$. The constant c will be chosen such that $(1 + \lambda^{1/2})\|[A_n - \varphi_{1/n}(f)]\Omega\|$ and $\|\Delta_w^{1/2}(A_n - \varphi_{1/n}(f))\Omega\|$ are bounded by $\varepsilon/3$. This is possible uniformly in n because of the estimates

$$\|(A_n - \varphi_{1/n}(f))\Omega\| \leqq (\sqrt{c}/2)\|\varphi_{1/n}(f)^*\varphi_{1/n}(f)\Omega\|,$$

$$\|\Delta_w^{1/2}(A_n - \varphi_{1/n}(f))\Omega\| \leqq (\sqrt{c}/2)\|\varphi_{1/n}(f)\varphi_{1/n}(f)^*\Omega\|,$$

and the assumed scaling properties of the 4-point function. Since $\Delta_w^{1/2}A_n\Omega = j(A_n^*)\Omega$, the sequence (A_n) has the properties required above. q.e.d.

If a von Neumann algebra $\mathcal{N}$ with $\mathfrak{A}(\mathcal{O}) \subset \mathcal{N} \subset \mathfrak{A}(W)$ is a factor, then the Connes invariant $S(\mathcal{N})$ [10] coincides with $S'(\mathcal{N})$, which is equal to $\mathbb{R}_+$ according to Theorem 4.2. Hence $\mathcal{N}$ is a factor of type III_1. Moreover, if E is a projection in the center of $\mathcal{N}$ such that $\mathcal{N}_E = E\mathcal{N}E$ is a factor, one has $S(\mathcal{N}_E) = S'(\mathcal{N}_E) \supset S'(\mathcal{N}) = \mathbb{R}_+$ and finds that $\mathcal{N}_E$ is a factor of type III_1. An analogous statement holds for factors occurring in a direct integral central decomposition of $\mathcal{N}$.

Proposition 4.3. *Let $\mathcal{N}$ be a von Neumann algebra in a separable Hilbert space. If the invariant $S'(\mathcal{N})$ defined in Eq. (4.1) is equal to $\mathbb{R}_+$, only type III_1-factors can occur in the central decomposition of $\mathcal{N}$.*

Proof. Let (σ_t) denote the modular automorphism group associated to a faithful normal state of $\mathcal{N}$, $\left(\mathcal{N} \underset{\sigma}{\times} R\right)$ the crossed product of $\mathcal{N}$ by σ, and Θ the dual action of $\mathbb{R}$ on the crossed product. Then by a theorem of Connes and Takesaki [20]

$$e^s \in S'(\mathcal{N}) \Leftrightarrow \Theta_s \text{ acts trivially on the center of } \left(\mathcal{N} \underset{\sigma}{\times} \mathbb{R}\right).$$

Now let $\mathcal{N} = \int^{\oplus} \mathcal{N}(r) d\mu(r)$ denote the central decomposition of $\mathcal{N}$. According to Sutherland [21] there are corresponding direct decompositions of the modular automorphisms, $\sigma_t = \int^{\oplus} \sigma_t^{(r)} d\mu(r)$, of the crossed product, of the center of the crossed product and of the dual action, $\Theta_s = \int^{\oplus} \Theta_s^{(r)} d\mu(r)$. $S'(\mathcal{N}) = \mathbb{R}_+$ implies then that, for all $s \in \mathbb{R}$, $\Theta_s^{(r)}$ is trivial on the center of $\left(\mathcal{N}(r) \underset{\sigma^{(r)}}{\times} \mathbb{R}\right)$ for μ-almost every r. Hence, for μ-almost every r, $\Theta_s^{(r)}$ is trivial on the center of $\left(\mathcal{N}(r) \underset{\sigma^{(r)}}{\times} \mathbb{R}\right)$ for all $s \in \mathbb{Q}$ and, because of continuity, for all $s \in \mathbb{R}$. Thus $S'(\mathcal{N}(r)) = \mathbb{R}_+$, i.e. $\mathcal{N}(r)$ is a type III_1-factor for μ-almost every r. q.e.d.

The results in this section show in particular that there exist no intermediate type I factors between $\mathfrak{A}(\mathcal{O})$ and $\mathfrak{A}(W)$ in contrast to the situations considered recently by Doplicher and Longo [22]. The physical reason is that the commutant $\mathfrak{A}(W)'$ of $\mathfrak{A}(W)$ is the algebra associated to the left wedge W' which has a common boundary point with $\mathcal{O}$. In such a situation product states, i.e. states without correlations between measurements in W' and $\mathcal{O}$, cannot exist, a fact which was known from the work of Roberts in the dilation invariant case [23].

Acknowledgements. The ideas for this work arose from discussions with Rudolf Haag and Daniel Kastler on Ref. [7]. These discussions are gratefully acknowledged. I also would like to thank Detlev Buchholz, Joachim Cuntz, Daniel Kastler, and Roberto Longo for valuable suggestions on the meaning of the invariant S' which finally led to the proof of Proposition 4.3, Rudolf Haag for useful hints concerning the presentation of the results and Daniel Kastler for the hospitality extended to me at the Centre de Physique Théorique of the C.N.R.S., Luminy.

References

1. Takesaki, M.: Tomita's theory of modular Hilbert algebras. Berlin, Heidelberg, New York: Springer 1970
2. Reeh, H., Schlieder, S.: Bemerkungen zur Unitäräquivalenz von Lorentzinvarianten Feldern. Nuovo Cimento **22**, 1051 (1961)
   See also: Streater, R.F., Wightman, A.S.: PCT, spin and statistics and all that. New York: Benjamin 1964
3. Bisognano, J.J., Wichmann, E.H.: On the duality for a hermitian scalar field. J. Math. Phys. **16**, 985 (1975)
4. Hawking, S.W.: Particle creation by black holes. Commun. Math. Phys. **43**, 199 (1975)
5. Unruh, W.G.: Notes on black-hole evaporation. Phys. Rev. D **14**, 870 (1976)
6. Sewell, G.L.: Relativity of temperature and the Hawking effect. Phys. Lett. **79** A, 23 (1980)
7. Haag, R., Narnhofer, H., Stein, U.: On quantum field theory in gravitational background. Commun. Math. Phys. **94**, 219–238 (1984)
8. Haag, R.: Private communication
9. Hislop, P.D., Longo, R.: Modular structure of the local algebras associated with the free massless scalar field theory. Commun. Math. Phys. **84**, 71 (1982)
10. Connes, A.: Une classification des facteurs de type III. Ann. Ec. Norm. Sup. **6**, 133 (1973)
11. Araki, H.: Remarks on spectra of modular operators of von Neumann algebras. Commun. Math. Phys. **28**, 267 (1972)
12. Størmer, E.: Spectra of states and asymptotically abelian C^*-algebras. Commun. Math. Phys. **28**, 279 (1972) [corrections Commun. Math. Phys. **38**, 341 (1974)]

13. Driessler, W.: On the type of local algebras in quantum field theory. Commun. Math. Phys. **53**, 295 (1977)
14. Longo, R.: A simple proof of the existence of modular automorphisms in approximately finite dimensional von Neumann algebras. Pac. J. Math. **75**, 199 (1978) and private communication
15. Longo, R.: Algebraic and modular structure of von Neumann algebras of physics. Proc. Symp. Pure Math. **38**, Part 2, 551 (1982)
16. Borchers, H.-J.: Field operators as C^∞-functions in spacelike directions. Nuovo Cimento **38**, 1600 (1964)
17. Erdelyi, A., Magnus, W., Oberhettinger, F.: Higher transcendental functions. New York, Toronto, London: McGraw-Hill 1953
18. Buchholz, D.: Private communication
19. Pedersen, G.K.: C^*-algebras and their automorphism groups, LMS Monographs 14. London, New York: Academic Press 1979
20. Connes, A., Takesaki, M.A.: The flow of weights on factors of type III. Tôhoku Math. J. **29**, 473 (1977)
21. Sutherland, C.: Crossed products, direct integrals, and Connes' classification of type III factors. Math. Scand. **40**, 209 (1977)
22. Doplicher, S.: Local aspects of superselection rules. I. Commun. Math. Phys. **85**, 73 (1982)
Doplicher, S., Longo, R.: Local aspects of superselection rules. II. Commun. Math. Phys. **88**, 399 (1983)
23. Roberts, J.E.: Some applications of dilatation invariance to structural questions in the theory of local observables. Commun. Math. Phys. **37**, 273 (1974)

Communicated by G. Mack

Received April 26, 1984

The Intersection of Brownian Paths as a Case Study of a Renormalization Group Method for Quantum Field Theory

Michael Aizenman*

Departments of Mathematics and Physics, Rutgers University, New Brunswick, NJ 08903, USA

Abstract. A new approach is presented for the study of the probability that the random paths generated by two independent Brownian motions in $\mathbb{R}^d$ intersect or, more generally, are within a short distance a of each other. The well known behavior of that function of a-above, below, and at the critical dimension $d=4$, as well as further corrections, are derived here by means of a single renormalization group equation. The equation's derivation is expected to shed some light on the β-function of the $\lambda\phi_d^4$ quantum field theory.

1. Introduction

The Brownian motion in addition to its own attraction has been a source of insight in various branches of mathematical analysis and physics. In particular, its properties have been quoted in the analysis of various "random walk" and "random surface", expansions used in statistical mechanics and quantum field theory. Since it is often the case that the insight is based on an analogy – rather than a complete reduction, the interest is not confined to the results, but extends also to the methods by which these can be proven. Thus we expect that something may still be gained from new derivations of old results (some of which are improved here).

With these comments in mind, we present here a new method for the study of the probabilities of intersection, and almost-intersection, for the *paths* of two independent Brownian motions: above, below and *at* the critical dimension $d=4$. In contrast with the previously available proofs, the analysis is based on a single equation – whose derivation requires only the more basic properties of the Bronian motion. In particular, a stronger use is made of the scale covariance – by means of which a "β-function" is derived for this problem.

---

* Sloan Foundation Research Fellow. Research supported in part by NSF grant PHY-8301493

The above terminology is used in recognition of the similarity of the main equation with the one which appears in the Callan-Symanzik equations for the $\lambda\phi_d^4$ quantum field theory, where it is widely believed to describe Wilson's "renormalization group flow" of the coupling constant. While the equations are similar, their derivations are quite different. In fact, part of the motivation for this study is the desire to develop some insight which might be useful for a non-perturbative analysis of quantum-field theoretical and statistic-mechanical issues (see Sect. 8).

A relation between the Brownian motion and the more complex scalar quantum field theory, for which $d=4$ is the critical dimension, first appeared, more than a decade and a half ago, in the work of Professor K. Symanzik. He also discovered the renormalization group equation whose analog is considered here. Recent works on the subject, including this one, derive from, and were inspired by his deep insights – which we are only beginning to understand. I should like to present this work as a small tribute to Professor Symanzik's scientific contributions.

The main problem studied here, and the organization of the paper, are described in the next section.

2. Brownian Motion

With the term Brownian motion we refer to the stochastic process with *continuous paths* $b(t):[0,\infty)\to\mathbb{R}^d$, which is described by the Wiener measure. $b(t)$ has independent increments, and its transition probabilities are:

$$\int \mu_x(db)\delta(b(t)-y)=e^{t\Delta}(x,y)=(2\pi t)^{-d/2}e^{-(x-y)^2/(2t)}, \tag{2.1}$$

where $\mu_x(db)$ is probability measure for paths which start at $x\in\mathbb{R}^d$ (with $\int \mu_x(db)=1$), and Δ is the Laplace operator. The measure μ defined by the above properties can be formally described by the following expression

$$\mu_x(db)=\delta(b(0)-x)\prod_{t\in[0,\infty)} db(t)e^{-\frac{1}{2}\int\left|\frac{db(s)}{ds}\right|^2 ds}/\text{Normalization}. \tag{2.2}$$

The realizability of the measure μ on the space of continuous functions is, of course, not a matter of definition – but a well known theorem. The Brownian motion has an alternative description as the continuum limit of a simple random walk (with a proper scaling).

For each $b(t)$ we denote by ω its "path" in $\mathbb{R}^d$. Specifically:

$$\omega_{[t_1,t_2]}=b([t_1,t_2])=\{y\in\mathbb{R}^d | y=b(s) \text{ for some } s\in[t_1,t_2]\}, \tag{2.3}$$

and $\omega=\omega_{[0,\infty)}$. From this point on, by the term *path* we refer only to the set ω (not $b(\cdot)$).

We shall discuss the behavior, in the limit $a\searrow 0$, of the probability that the paths of two independent Brownian motions are within the distance a of each other, i.e.

$$\begin{aligned}&\text{Prob}(\text{dist}(\omega^1_{[0,t_1]},\omega^2_{[0,t_2]})\leqq a|0,x)\\ &\overset{\text{def}}{=}\int \mu_0(db^1)\mu_x(db^2)\chi[\text{dist}(\omega^1_{[0,t_1]},\omega^2_{[0,t_2]})\leqq a].\end{aligned} \tag{2.4}$$

The symbol $\chi[\mathscr{C}]$ will always denote a characteristic function which is 1 if the condition $\mathscr{C}$ is satisfied, and 0 otherwise.[1]

Since the motion is continuous, the probability that the paths actually intersect is also well defined, and for *finite* t_i it equals the limit of (2.4) as $a \searrow 0$.

The main intersection properties (but not all, see Sect. 8) of Brownian paths are known. The pioneering works of Kakutani [1] and Dvoretzky, Erdös, and Kakutani [2] showed the following.

Proposition 2.1 [1, 2]. *For independent pairs of d-dimensional Brownian motions:*

$$\mathrm{Prob}(\omega^1_{[0,t_1]} \cap \omega^2_{[0,t_2]} \neq \emptyset | 0, x) = \begin{cases} 0 & d \geqq 4 \\ f(x, t_1, t_2) \neq 0 & d < 4 \end{cases} \tag{2.5}$$

for each $x \neq 0$ *and* $t_1, t_2 > 0$.

One may be puzzled by the fact that the critical dimension is $d = 4$, since ω^i are the images, under the continuous maps b^i, of the *one dimensional* $\mathbb{R}^+$. Nevertheless, the Hausdorff dimensions of ω^i are ("almost surely") 2.

Each of the previously available proofs of (2.5) deals separately with the two cases distinguished there, and (except for $d > 4$) employs some property of the Brownian motion which is a not quite trivial consequence of its definition. Our main purpose is to present a simple, and unified proof. As a bonus, we hope to shed some light on a renormalization-group equation which has appeared (by quite a different reasoning) in quantum field theory.

For simplicity we shall focus our attention on the single function,

$$g(a) = \int_{\mathbb{R}^d} dx \int_0^\infty dt_1 e^{-t_1} \int_0^\infty dt_2 e^{-t_2} \,\mathrm{Prob}(\ \mathrm{dist}(\omega^1_{[0,t_1]}, \omega^2_{[0,t_2]}) \leqq a | 0, x\ ), \tag{2.6}$$

which can be shown to describe also the limiting behavior of the probability (2.4), as $a \to 0$, at each fixed $x \neq 0$ and $t_1, t_2 > 0$. [It is easy to see that $g(a) < \infty$; e.g. by the argument used in (6.13).]

The arguments presented here offer a simple proof that, as $a \to 0$, $g(a)$ exhibits the following behavior (in a strong sense – see Proposition 5.2):

$$g(a) \cong \begin{cases} c_d \geqq C(4-d) & d < 4, \\ C(d-4)a^{d-4}(1 + 0(a^{d-4+\min\{0,6-d\}})) & d > 4 \\ & \neq 6, \\ C(d-4)a^{d-4}(1 + 0(|\ln a| a^{d-4})) & d = 6 \end{cases} \tag{2.7}$$

and, in the critical dimension $d = 4$;

$$\frac{c}{1 + |\ln a|} \geqq g(a) \geqq \frac{c'}{1 + |\ln a|} \tag{2.8}$$

Here, *and in the rest of the paper C* stands for various constants, which we can bound from above and below. Furthermore, these bounds have natural extensions to non-integral values of d – which are regular (i.e. neither vanish nor diverge for finite d).

---

1 In this work the superscipts on b^i and ω^i should always be read as indices, not powers

The *leading* behavior in (2.7) and (2.8) has in effect been known. Previous results, and methods, are briefly described in the next section. The emphasis in this paper is on the *β-function formalism*, which is introduced in Sect. 4. Our bounds on the β-function, $W(g)$, are summarized in Proposition 4.1. Their derivation is postponed to Sects. 6 and 7. In Sect. 5 the results on $W(g)$ are used to present a simple picture of the d-dependence of $g(0)$, and for the derivation of the above asymptotics for $g(a)$. Some open problems which call for an improvement and extensions of this approach are mentioned in Sect. 8.

Section 7 contains also some useful *direct bounds* on $g(a)$. Yet another point of view is described in the appendix.

3. Previous Arguments

The earliest analysis of the intersection properties of Brownian paths in $d=4$ and $d<4$ dimensions [2] was based on the fact that the probability of a d-dimensional Brownian motion to hit a set $A\subset\mathbb{R}^d$ does not vanish if and only if the sets' Newtonian capacity is non-zero. Furthermore, an additional basic – yet not entirely trivial – result of potential theory has been invoked for the case $d=3$; and for $d=4$ use was made of the somewhat sophisticated (yet by now standard) uniform estimate of Levy[2]

$$|b(s)-b(t)|\leqq M(b,T)\sqrt{|t-s|\ln|t-s|^{-1}}, \qquad s,t\in[0,T]. \tag{3.1}$$

Closely related questions were considered in [3] with regard to intersection properties of discrete random walks (see Sect. 8). The stopping-time argument used there, led to an analog of the $|\ln a|^{-1}$ law for $d=4$ dimensions [see (2.8)]. An error of method (which led to incorrect results in other cases) was pointed out by Lawler [4], who proved this law by rather detailed considerations, combining stopping time arguments with another property which will not be described here.

Our quest for simpler methods is motivated by the desire to eventually be able to analyze the interaction in physical models which admit stochastic geometrical (interacting-) random walk or random surface representations, or expansions. Needless to say, for such models one does not yet have an analog of the potential theory, nor detailed information like (3.1).

The one case for which a simple argument has been known is the proof of the non-intersection of Brownian paths in $d>4$ dimensions (first derived in [1]). In Sect. 7 we recall a specially simple upper bound of Spencer [5] and derive, by a related method, a rather good lower bound for all d. An analog of Spencer's method was used for some of the results mentioned in 8iii.

---

2 In fact, Levy's estimate can be used more effectively by considering the stopping time value of the function $f(b(t_1))$, where

$$f(z)=\int_0^{t_1} dt|z-b_1(t)|^{2-d}, \qquad d=4$$

4. A Renormalization Group Equation

a) A Heuristic Argument

A key idea in our approach to the study of the function $g(a)$ is to consider its changes under scaling. In particular, we shall make use of the Brownian motion's *self similarity:*

$$b(t) \overset{\mathscr{D}}{=} \alpha b(\alpha^{-2}t) \text{ with respect to } \mu_0(db), \tag{4.1}$$

$\mathscr{D}$ meaning that the two functions have the same distribution. The well known scaling relation (4.1) is clearly seen in the formal expression (2.2), as well as in (2.1).

The scale covariance is, of course, broken by the constraint that $\text{dist}(\omega^1_{[0,t_2]}, \omega^2_{[0,t_2]}) \leqq a$. Let us first offer a *heuristic analysis* of the resulting behavior of $g(a)$.

For an integer n, $g\left(\frac{a}{n}\right)$ may be related to $g(a)$ by the change of scale: $x = \hat{x}/n$, $t = \hat{t}/n^2$. When presented in terms of $\hat{x}$ and $\hat{t}$, in (4.1), $g\left(\frac{a}{n}\right)$ expresses the integral of the probability of pairs of *longer* paths to come within the distance a. The volume integral is $dx = n^{-d}d\hat{x}$, and the change in the time scale can be accounted for by replacing each path by a chain of n^2 linked paths – whose time duration is again of the order 0(1). The required contact would occur if in any of the $n^2 \times n^2$ pairs the paths are within the distance a. Therefore, by the inclusion-exclusion principle:

$$g\left(\frac{a}{n}\right) = n^{2\cdot 2}n^{-d}g(a) - R(a,n), \tag{4.2}$$

where R is a (positive) multiple-intersection correction, which for a fixed n we may expect it to behave as $O(g(a)^2)$.

Under iteration (4.2), with some more specific information on $R(a,n)$, leads to Theorem 1. We shall now present a more careful, and complete, argument in which (4.2) is replaced by a differential equation.

b) The Main Result

It is evident from (2.6) that $g(a)$ has the following properties

i) $g(a)$ is a strictly increasing function of a,

ii) $\lim\limits_{a\to\infty} g(a) = \infty$,

iii) $\lim\limits_{a\to 0} g(a) = g(0)$ (by the monotone convergence theorem).

The differentiability of the function $g(a)$ is certainly not manifest in (2.6). The situation may however be remedied by means of the scaling relation (4.1), with $\alpha = a$. One gets:

$$\begin{aligned} g(a) &= \int_{\mathbb{R}^d} dx \iint_0^\infty dt_1 dt_2 e^{-(t_1+t_2)} \int \mu_0(db^1)\mu_{x/a}(db^2)\chi[\text{dist}(\omega^1_{[0,t_1/a^2]}, \omega^2_{[0,t_2/a^2]}) \leqq 1] \\ &= a^{d+4} \int_{\mathbb{R}^d} dx \iint_0^\infty dt_1 dt_2 e^{-a^2(t_1+t_2)} \int \mu_0(db^1)\mu_x(db^2)\chi[\text{dist}(\omega^1_{(0,t_1]}, \omega^2_{[0,t_2]}) \leqq 1]\,. \end{aligned} \tag{4.3}$$

Equation (4.3) shows that $g(a)$ is infinitely differentiable in $a \in (0,\infty)$, and allows one to obtain an expression for its derivative, which is given in Lemma 6.1, below. The above listed properties of $g(a)$ allow one to define implicitly the function $W(g)$, on $(g(0,\infty) \subset \mathbb{R}$, to which we refer in the next proposition.

Proposition 4.1. *The function $g(a)$ obeys the following differential equation*

$$\frac{dg}{d\ln a^{-1}} = -W(g(a)), \tag{4.4}$$

where W (the "β-function") is a continuous function, defined uniquely in the regime $[g(0), \infty)$, having the form

$$W(g) = (d-4)g + 4\hat{F}(g), \tag{4.5}$$

with a non-linear term which, for $g \leqq g(1)$ in the lower bound, satisfies

$$c_+ g^2 \leqq \hat{F}(g) \leqq c_\delta g^{1+\frac{2}{d+2+\delta}} \quad (\text{for any } \delta > 0). \tag{4.6}$$

Furthermore, $F(a) = \hat{F}(g(a))$ (given explicitly by (6.2)) obeys the following bounds (for $a \leqq 1$)

$$\begin{cases} C|d-4|^2 & d \neq 4 \\ c(1+|\ln a|)^2 & d=4 \end{cases} \leqq F(a)/[a^{2(d-4)+\min(0,6-d)}] \leqq \begin{cases} C|6-d|^{-1} & d \neq 6 \\ c(1+|\ln a|) & d=6. \end{cases} \tag{4.7}$$

With the restriction of the lower bound to $g \approx g(0)$, (4.6) holds with:

$$c_+(0) = \sqrt{\pi} \Big/ \left[(2\sqrt{2\pi})^d \Gamma\left(\frac{d+1}{2}\right) \right], \quad c_\delta = \left(\frac{(d+\delta)(d+2+\delta)}{2\sqrt{2\pi}^d} \Gamma\left(\frac{\delta}{2}\right) \right)^{\frac{2}{d+2+\delta}}. \tag{4.8}$$

Summary of the Proof. The main results here are the bounds on the β-function. The exact expression for $W(g(a))$ is derived in Lemma 6.1, where $F(a)$ is defined explicitly. The continuity of $W(g)$ *at* $g(0)$ follows from the obvious monotonicity of $F(a)$ [see (6.2)]. For the bounds (4.6) and (4.7) we use two different sets of arguments. (4.6) is proven by Lemmas 6.2 and 6.3; and (4.7) is derived, by direct estimates, in Sect. 7c. □

Of the two inequalities in (4.6) the more faithful one is the lower bound. The other may in fact be improved using (4.7). However, (4.6) represents our a priori estimates which allow one to establish the main elements of the picture described next.

Let us precede the derivation of the results described above by a presentation of their implications.

5. Analysis of the Renormalization-Group Inequalities

The function $W(g)$ and a single value, say $g(1)$, contain all the information about $g(a)$ – the question of the vanishing of $g(0)$ being determined by $W(g)$ for $g \ll 1$. Notice that in (4.5) we give W's leading linear term exactly, and determine that the non-linear part is positive. In the role it plays, and its dependence on g and the dimension d (Fig. 1), W is quite similar to the β-function of the ϕ_d^4 euclidean-field [6], which was mentioned in the introduction. Its analysis belongs to the standard repertoire of physics literature (see, e.g. [7]).

For $g > g(0)$, $W(g)$ is positive and, by (6.2), it converges to a limit as $g \searrow g(0)$. (4.4) shows that this limit has to be zero. For convenience, we shall refer to $W(\cdot)$ as if it is defined for all $g \in [0, \infty)$. Clearly, $W(g)$ has such extensions which have all the properties listed in Proposition 4.1.

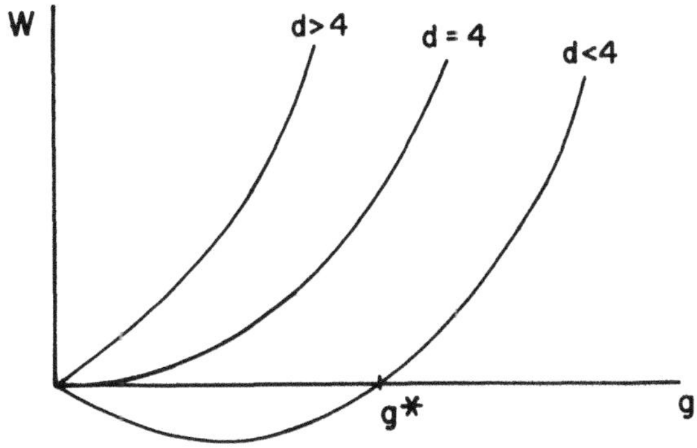

Fig. 1. The shape of the function $W(g)$

Proposition 5.1. *In any dimension d, g(a) takes values only in the region* $[g^*, \infty)$, *where*

$$g^* = \sup\{g | W(g) = 0\} \tag{5.1}$$

Furthermore, $g(0) = g^*$ *and it satisfies:*

$$\begin{cases} g(0) = 0 & \text{for} \quad d = 4 \quad (5.2) \\ g(0) > \dfrac{2\sqrt{2\pi}\, d(4-d)^{(d+4)/2}}{(d+2)(d+4)} & \text{for} \quad d < 4 . \quad (5.3) \end{cases}$$

Remark. While (5.4) shows the main result for $d<4$, its d dependence in the "$d \to 4$ limit" is not optimal, see the direct bound (7.12).

Proof. g^* is well defined since W is a continuous function, with

$$W(0) = 0 \quad \text{and} \quad \lim_{g \to \infty} W(g) = \infty . \tag{5.4}$$

The continuity of $g(a)$, and its properties i) and ii) (Sect. 4b) imply that $g(a) \in (g^*, \infty)$ for $a > 0$, where $W(g) > 0$.

As $a \to 0$ $g(a)$ which obeys Eq. (4.4), is driven all the way to g^*, since on the logarithmic scale of $\ln a^{-1}$, the interval $a \in (0, 1]$ is mapped onto $[0, \infty)$.

The properties (5.2) and (5.3) of g^* are simple consequences of (4.5) for which we choose $\delta = 2$. □

For dimensions $d \geqq 4$, where $g(0)$ vanishes, one is also interested in the actual behavior of $g(a)$ for small a.

Proposition 5.2. i) *For dimensions* $d > 4$:

$$g(a) = c_d a^{d-4}(1 + O(a^{\min(d-4,2)}\{1 + |\ln a| \chi[d=6]\})) \tag{5.5}$$

in the strong sense that $0 < c_d \leqq g(1)$ *and the lower order term is also of not less than the indicated order, up to a logarithmic correction for* $d = 6$.

ii) *In* $d = 4$ *dimensions*

$$g(a) \leqq \frac{1}{g(1) + 4c_+ \ln a^{-1}} \tag{5.6}$$

for any $a \leqq 1$. *(An opposite bound of the same order is provided by (7.12).)*

Proof. i) $d \geqq 4$. Notice that the simple bound

$$\frac{dg}{d\ln a^{-1}} \leqq -(d-4)g \tag{5.7}$$

[which is a consequence of the positivity of $\hat{F}(g)$, in (4.5)] implies that

$$g(a) \leqq g(1)a^{d-4}, \quad \text{for} \quad a \leqq 1. \tag{5.8}$$

For more complete information, let us write down the equation for $\hat{g}(a) = g(a)/a^{d-4}$:

$$\frac{d\hat{g}(a)}{d\ln a^{-1}} = -F(a)a^{4-d} = -O(a^{\min(2,d-4)}\{1+|\ln a|\chi[d=6]\}) \tag{5.9}$$

[by (4.5) and (4.7)]. Hence

$$\hat{g}(a) = g(1) - \int_0^{\ln a^{-1}} dz O(e^{-\min(2,d-4)z}\{1+z\chi[d=6]\}). \tag{5.10}$$

Splitting

$$\int_0^{\ln a^{-1}} dz \ldots = \int_0^{\infty} dz \ldots - \int_{\ln a^{-1}}^{\infty} dz \ldots,$$

we get

$$g(a)/a^{d-4} = \text{const} + O(a^{\min(2,d-4)}\{1+|\ln a|\chi[d=6]\}), \tag{5.11}$$

which proves (5.5). Notice that the *lower* bound in (4.7) shows that the second term in the right-hand side is in fact of the indicated order of magnitude, up to logarithmic corrections for $d=6$ (and $d=4$).

ii) $d=4$. In this case $W(g)$ has no linear term, see (4.5). To integrate (4.6) we cast it in the following form:

$$\frac{dg^{-1}}{d\ln a^{-1}} \geqq 4c_+ \quad \text{and} \quad \frac{dg^{-\frac{1}{3+\delta}}}{d\ln a^{-1}} \leqq \frac{4}{3+\delta}c_{2\delta}, \tag{5.12}$$

which directly leads to (5.6), and to a lower bound which looks like the right-hand side of (5.6) raised to the power $3+\delta$. [A better lower bound is given in (7.12).] □

Remark. An alternative way to integrate bounds on $W(g)$ [but not those on $W(g(a))$] is to invert and integrate (4.4), by which:

$$\ln a^{-1} = \int_{g(a)}^{g(1)} dg W(g)^{-1}. \tag{5.13}$$

This is the expression used in physics literature when one has a perturbative expansion for $W(g)$.

We have seen here how the problem introduced in Sect. 2 can be solved by the "renormalization group" equations (4.4) and (4.5). It still remains to prove the inequalities which lead to the bounds (4.6) and (4.7). This is done in the next two sections.

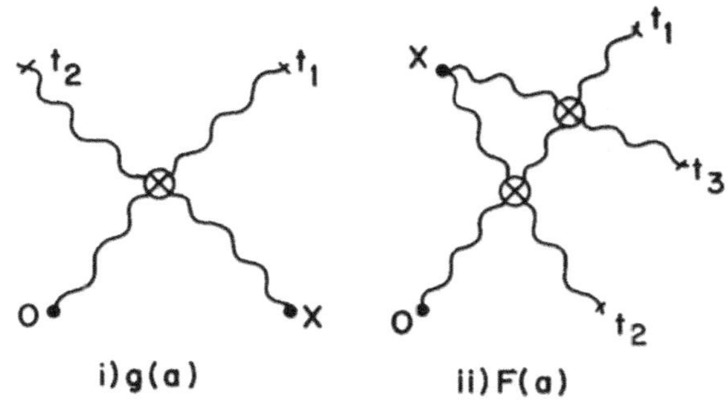

Fig. 2. A diagramatic representation of the functions $g(a)$ and $F(a)$. Each wiggled line represents a Brownian path, and a circle indicates that the paths come to within the distance a. g and F are obtained by integrating over x and over the corresponding Brownian motions with an exponential killing time

6. Proof of the Bounds on the β-Function

a) An Exact Expression for $W(g(a))$

We start by differentiating the expression (4.3) for $g(a)$, producing in effect an exact version of the heuristic argument which led to (4.2).

Lemma 6.1. *The function $g(a)$ is infinitely differentiable in $(0, \infty)$, with the derivative*

$$a\frac{dg(a)}{da} = (d-4)g(a) + 4F(a), \tag{6.1}$$

where

$$F(a) = \int_{\mathbb{R}^d} dx \iiint_0^\infty \prod_{k=1}^{3} (dt_k e^{-t}) \int \mu_0(db^1)\mu_x(db^2)\mu_x(db^3) \cdot \chi[\mathrm{dist}(\omega^1_{[0,t_1]}, \omega^i_{[0,t_i]}) \leqq a;\ i=2 \text{ and } 3] \tag{6.2}$$

(see Fig. 2).

Proof. The C^∞ property follows directly from the expression (4.3) for $g(a)$. Differentiating (4.3) [which is symmetric in (t_1, t_2)], and then changing back to the scale of (2.6), we obtain:

$$\frac{dg}{da} = \frac{d+4}{a} g - 2 \cdot \frac{2}{a} I, \tag{6.3}$$

with

$$I = \int_{\mathbb{R}^d} dx \iint_0^\infty dt_1 dt_2 e^{-(t_1+t_2)} t_2 \int \mu_0(db^1)\mu_x(db^2)\chi[\mathrm{dist}(\omega^1_{[0,t_1]}, \omega^2_{[0,t_2]}) \leqq a]. \tag{6.4}$$

The integration $\int_0^\infty dt_2 e^{-t_2} t_2 \ldots$ in (6.4) may be replaced by

$\iint_0^\infty ds_1 ds_2 e^{-(s_1+s_2)} \ldots$ with $t_2 = s_1 + s_2$. For each (s_1, s_2), we split the path

$$\omega^2_{[0,t_2]} = \omega^2_{[0,s_1]} \cup \omega^2_{[s_1,s_1+s_2]}. \tag{6.5}$$

For a specified $b(s_1) = y$ the two paths in the above splitting are independent.

Furthermore, translation invariance permits us to replace $\int dx$ by $\int dy$. Thus:

$$I = \int_{\mathbb{R}^d} dy \int_0^\infty dt_1 e^{-t_1} \iint_0^\infty ds_1 ds_2 e^{-(s_1+s_2)} \int \mu_0(db^1)\mu_y(db^2)\mu_y(db^3) \cdot \chi[\operatorname{dist}(\omega^1_{[0,t_1]}, \omega^2_{[0,s_1]} \cup \omega^3_{[0,s_2]}) \leqq a] \tag{6.6}$$

(where we also used a reflection in time, for ω^2).

By the simplest case of the inclusion-exclusion principle,

$$\chi[\operatorname{dist}(\omega^1, \omega^2 \cup \omega^3) \leqq a] = \chi[\operatorname{dist}(\omega^1, \omega^2) \leqq a] + \chi[\operatorname{dist}(\omega^1, \omega^3) \leqq a] - \chi[\operatorname{dist}(\omega^1, \omega^i) \leqq a;\ i=2 \text{ and } 3]. \tag{6.7}$$

Using it in (6.6), we get

$$I = 2g - F(a), \tag{6.8}$$

which, when combined with (6.3), yields the claimed (6.2) □

The right-hand side of (6.1) provides us with an exact expression for $W(g(a))$. In order to derive the bounds (4.6) one has now to produce bounds on $F(a)$, in terms of $g(a)$. This is done next.

b) The Lower Bound on $W(g)$

The lower bound is quite natural. Since $F(a)$ represents events with two intersections (see Fig. 2) one may expect it to be related to the square of $g(a)$. The key tool for the following result is just the Schwarz inequality.

Lemma 6.2. *In any dimension, for $a \leqq \alpha$:*

$$F(a) \geqq c_+(\alpha) g(a)^2 \tag{6.9}$$

with a dimension dependent constant $c_+(\alpha) > 0$ for which $c_+(0)$ is given in (4.8).

Proof. Let us denote by $\mathscr{C}_i$ the condition

$$\sup_{0 \leqq s \leqq t_i} \pm \frac{x \cdot (b^i(s) - b^i(0))}{|x|} \geqq \frac{|x| - \alpha}{2}, \text{ with } + \text{ for } i=1, \text{ and } - \text{ for } i=2.$$

Clearly

$$F(a) \geqq \int_{\mathbb{R}^d} dx \int_0^\infty dt_1 e^{-t_1} \int \mu_0(db^1) \chi[\mathscr{C}_1] \cdot \left\{ \int_0^\infty dt_2 e^{-t_2} \int \mu_x(db^2) \chi[\operatorname{dist}(\omega^1_{[0,t_1]}, \omega^2_{[0,t_2]}) \leqq a] \right\}^2, \tag{6.10}$$

since without the characteristic function $\chi[\mathscr{C}_1]$ we have an equality. Applying the Schwarz inequality we obtain.

$$F(a) \geqq \left(\int_{\mathbb{R}^d} dx \int_0^\infty dt_1 e^{-t_1} \int \mu_0(db^1) \chi[C_1] \right)^{-1} \cdot \left\{ \int_{\mathbb{R}^d} dx \int_0^\infty dt_1 e^{-t_1} \int d\mu_0(b^1) \int_0^\infty dt_2 e^{-t_2} \int \mu_x(db^2) \chi[\mathscr{C}_1] \cdot \chi[\operatorname{dist}(\omega^1_{[0,t_1]}, \omega^2_{[0,t_2]}) \leqq a] \right\}^2. \tag{6.11}$$

By the symmetry of the integral in the brackets $\{-\}$, one may replace $\chi[\mathscr{C}_1]$ there with

$$\tfrac{1}{2}(\chi[\mathscr{C}_1]+\chi[\mathscr{C}_2]) \geqq \tfrac{1}{2}, \tag{6.12}$$

the inequality being subject to the second constraint which is present in (6.11).

Without $\chi[\mathscr{C}_1]$, the bracketed expression is just g. Therefore we have derived (6.9), with the following value for $c_+(\alpha)$:

$$c_+(\alpha)^{-1} = 4 \int_{\mathbb{R}^d} dx \int_0^\infty dt e^{-t} \int \mu_0(dx) \chi\left[\sup_{0 \leqq s \leqq t} \frac{x \cdot b(s)}{|x|} \geqq \frac{|x|}{2} \right].$$

To evaluate the integral, we use the reflection principle for the one dimensional projection of $b(s)$. One gets

$$\begin{aligned} c_+(\alpha)^{-1} &= 2 \int_{\mathbb{R}^d} dx \int_0^\infty dt e^{-t} \int \mu_0(dx) \chi\left[\frac{x \cdot b(t)}{|x|} \geqq \frac{|x|-\alpha}{2} \right] \\ &= 2 \int_{\mathbb{R}^d} dx \int_0^\infty dt e^{-t} \int_{|x-\alpha|_+/2} dz \frac{1}{\sqrt{2\pi t}} e^{-z^2/(2t)} \end{aligned} \tag{6.13}$$

[using (2.1)]. The value of $c_+(0)$ is given in (4.7). □

c) The Upper Bound

Lemma 6.3. *For any* $\delta > 0$

$$F(a) \leqq c_\delta g^{1+\frac{2}{d+2+\delta}} \tag{6.14}$$

with the constant given in (4.8).

Although the above inequality completes the derivation of the qualitatively appealing picture described in Sects. 4 and 5, it is less satisfactory than its counterpart (6.9). Note that any *upper* bound on $F(a)$ leads to a *lower* bound on $g(a)$. It turns out, however, that $g(a)$ may be bounded below by means of a more direct argument – with a better result for the critical dimension $d=4$ (see Sect. 7). For these reasons, the proof of (6.14) would be described here in a somewhat abbreviated form.

The main difficulty in bounding F from *above* by const g^2 is that the two "intersections" required in $F(a)$ (see Fig. 2) can be highly correlated – if both occur in the vicinity of x. In fact, the contribution from the simplest of such events has the following consequence [see (7.22)]

$$F(a) \geqq \frac{\text{const}}{a^{d-2}} \geqq \text{const}\, a^{(6-d)} g(a)^2 . \tag{6.15}$$

Hence $F(a)/g(a)^2 \to \infty$, as $a \to 0$, in $d > 6$ dimensions.

The above problem does not exist in estimating the quantity

$$T(a) - \int_{\mathbb{R}^d} dx \int_{\mathbb{R}^d} dy T_{x,y}(a),$$

where

$$T_{x,y}(a)=\iiint_0 \prod_{n=1}^{3}(dt_n e^{-t_n})\int \mu_0(db^1)\mu_x(db^2)\mu_y(db^3)\chi[\mathscr{D}_2]\chi[\mathscr{D}_3] \tag{6.16}$$

with $\mathscr{D}_i$ denoting the events, $\mathrm{dist}(\omega^1_{[0,t_1]}, \omega^i_{[0,t_i]})\leqq a$.

The event $\mathscr{D}_2\cap\mathscr{D}_3$ is the non-exclusive union of two sequences of events. In one, the trajectory of $b^1(t)$ comes to within distance a from $\omega^3_{[0,t_3]}$ *after* reaching such a neighborhood of $\omega^2_{[0,t_2]}$. In the other sequence, the roles of ω^2 and ω^3 are reversed. By a simple stopping time argument, the contribution to $T(a)$ of each of these scenarios is exactly g^2. Thus one gets:

$$(g(a)^2\leqq)T(a)\leqq 2g(a)^2\,. \tag{6.17}$$

Proof of Lemma 6.3. Returning to $F(a)$, one may combine the integrations over the two paths b^2 and b^3, in (6.2) into a single-path integral [reversing the step (6.5)], and a single time integral. This leads to the following representation

$$F(a)=\int_{\mathbb{R}^d} dx\int_0^\infty dt_1 e^{-t_1}\int_0^\infty dt e^{-t}\int \mu_0(db^1)\mu_x(db^2)$$
$$\cdot\,\chi[\mathrm{dist}(\omega^1_{[0,t_1]},\omega_{[0,t]})\leqq a]\times(t_{\mathrm{L}}-t_{\mathrm{F}})\,, \tag{6.18}$$

where t_{F} and t_{L} are the values of the *first* and the *last* times $s\in[0,t]$ at which $\mathrm{dist}(\omega^1_{[0,t_1]}, b(s))\leqq a$.

Without the term $(t_{\mathrm{L}}-t_{\mathrm{F}})$, the right-hand-side of (6.18) would be just $g(a)$. By the Hölder inequality, we get for each $p>1$

$$F(a)\leqq g(a)^{1-\frac{1}{p}}B(a)^{\frac{1}{p}}\,, \tag{6.19}$$

where $B(a)$ is the quantity obtained by replacing $(t_{\mathrm{L}}-t_{\mathrm{F}})$ in (6.25) with the factor:

$$(t_{\mathrm{L}}-t_{\mathrm{F}})^p=\mathrm{p}(\mathrm{p}-1)\int_0^\infty du u^{p-2}\int_0^\infty ds\chi[s\geqq t_F\,; s+u\leqq t_{\mathrm{L}}]\,. \tag{6.20}$$

Reversing the steps which led from (6.2) to (6.18), and using (in addition to the strong Markov property) Eq. (2.1), one gets

$$B(a)=p(p-1)\int_0^\infty du u^{p-2}e^{-u}\int_{\mathbb{R}^d} dx\int_{\mathbb{R}^d} dy e^{u\Delta}(x,y)T_{x,y}(a)\,. \tag{6.21}$$

By (6.17), the uniform bound $e^{u\Delta}(x,y)\leqq(2\pi u)^{-d/2}$ [see (2.1)], and the choice of $p=1+(d+\delta)/2$, we have

$$B(a)\leqq\left[\frac{(d+\delta)(d+\delta+1)}{2(2\pi)^{d/2}}\Gamma\left(\frac{\delta}{2}\right)\right]g(a)^2\,. \tag{6.22}$$

The substitution of (6.22) in (6.19) (with the above choice of p) results in the claimed bound (6.14) on $F(a)$. □

7. Direct Estimates of $F(a)$ and $g(a)$

In this section we derive the bounds (4.7) on $F(a)$ (part 7c). It is also shown (in 7a and 7b) how $g(a)$ may be estimated by a more direct approach, which yields the *leading* behavior of $g(a)$ (in a qualitative sense) in all the cases *except* for the important logarithmic *upper* bound in $d=4$ dimensions (derived in Proposition 5.2). Note that the logarithmic *lower* bound for $d=4$, presented in 7b, is better than what we got via the a priori estimate of the β-function of Lemma 6.3 [in a remark after (5.12)].

The upper bounds of 7a and 7b are derived by an adaptation of a specially simple (and quite versatile) argument of Spencer [5] for the proof of non-intersection in $d>4$ dimensions. The crux of it was an upper bound on the *probability* of intersection of the paths of lattice random walks, provided by the expected value of the *size* of their intersection. For the opposite bound, of 7b, we evaluate also the second moment, and use the Schwarz inequality.

The direct lower bound on $g(a)$ has also an alternative derivation, due to Sokal, which is given in [14]. I am grateful to Felder and Fröhlich for reminding me of the relevance of this estimate for the problem discussed here.

a) An Upper Bound for $d>4$ (Following Spencer [5])

The basic ingredient needed for the direct estimates is the exact expression for the *ball-hitting probability* (which has an easy derivation)

$$\int_0^\infty dt \int \mu_x(db)\chi[\text{dist}(\omega_{[0,t]}, y) \leqq a] = \frac{G(x,y)}{G(0,a)} \tag{7.1}$$

for $|x-y| \geqq a$, where $G(x,y)$ (the "free propagator") is the kernel, $G(x,y) = (-\Delta+1)^{-1}(x,y)$, (and $G(0,a)$ stands for $G(0,z)$ with $|z|=a$).

The *Fourier-transform* of $G(0,x)$ is $(1+p^2)^{-1}$. We shall also use its following properties

$$\text{i) } G(0,x)/[d|B(1)|\, x|^{2-d}] \leqq 1, \quad \text{and} \quad \to 1 \quad \text{as} \quad |x| \to 0, \tag{7.2}$$

$|B(r)|$ being the volume of the ball of radius r, and

$$\text{ii) } \int dx G(0,x) = 1\,. \tag{7.3}$$

Let us denote

$$\int \nu(db^1 db^2)\ldots = \int_{\mathbb{R}^d} dx \int_0^\infty dt_1 e^{-t_1} \int_0^\infty dt_2 e^{-t_2} \int \mu_0(db^1)\mu_x(db^2)\ldots \tag{7.4}$$

and

$$A_i = \omega^i_{[0,t_i]}\,. \tag{7.5}$$

Thus ν is the measure over *pairs of paths*, with which

$$g(a) = \int \nu(db^1 db^2)\chi[\text{dist}(A_1, A_2] \leqq a]\,. \tag{7.6}$$

Trivial geometric considerations show that

$$\chi[|V_{a/2}(A_1, A_2)| \neq 0] \leqq \chi[\text{dist}(A_1, A_2) \leqq a] \leqq \frac{|V_{2a}(A_1, A_2)|}{|B(a)|}, \tag{7.7}$$

where $V_r(A_1, A_2) = \{z \in \mathbb{R}^d | \text{dist}(A_i, z) \leqq r\}$, and $|V|$ is the d-dimensional volume of the set V.

We shall now apply (7.7) – the main point being that while it is difficult to evaluate the integral (7.6), it is easy to evaluate the corresponding integral of $|V_a|$ (as well as to estimate its higher moments).

Proposition 7.1. *In any dimension*

$$g(a) \leqq \hat{c}a^{d-4}(1+o(1))(\text{as } a \to 0) \tag{7.8}$$

with $\hat{c} = [4^{2-d}d^2|B(1)|^3]^{-1}$.

Proof. Substituting (7.7) in (7.6) one obtains

$$\begin{aligned} g(a) &\leqq \int_{\mathbb{R}^d} dz \int \nu(db^1 db^2)\chi[z \in V_{2a}(A_1, A_2)]/|B(a)| \\ &= \int dz \int \nu(db^1 db^2) \prod_{i=1}^{2} \chi[\operatorname{dist}(\omega^i_{[0,t_i]}, z) \leqq 2a]/|B(a)| \\ &= |B(a)|^{-1} G(0,2a)^{-2} \int_{\mathbb{R}^d} dx \int_{\mathbb{R}^d} dz G(0,z) G(0,x) \\ &= |B(a)|^{-1} G(0,2a)^{-2}, \end{aligned} \tag{7.9}$$

where use was made of (7.1) and (7.3). By (7.2), the last expression reduces to (7.8). □

Thus the simple estimate (7.7) leads to a rather good bound for $d>4$, although the above result does not, by itself, convey any information about $d<4$, and not much about the critical dimension $d=4$.

b) Direct Lower Bounds for $g(a)$

Using the left-hand side of (7.7), and then the Schwartz inequality we obtain the following lower bound for $g(a)$.

$$g(a) \geqq \int \nu(db^1 db^2)\chi[|V_{a/2}(A_1, A_2)| \neq 0] \geqq \frac{(\int \nu(db^1 db^2)\,|V_{a/2}(A_1, A_2)|)^2}{\int \nu(db^1 db^2)\,|V_{a/2}(A_1, A_2)|^2}. \tag{7.10}$$

Evaluating the moments of $|V_{a/2}|$ one gets:

Proposition 7.2. *In any dimension*

$$g(a) \geqq \left(4 \int_{|u| \geqq a/2} du G(0,u)^2 + 4^{2(d-2)} \hat{c} a^{4-d}\right)^{-1} \tag{7.11}$$

(with $\hat{c}$ *of* (7.8)*) and hence, for* $a \leqq 1$,

$$g(a) \geqq \begin{cases} C(d-4)a^{d-4}, & d>4, \\ c[1+|\ln a|]^{-1}, & d=4, \\ \left(\dfrac{4}{(2\pi)^d} \displaystyle\int_{\mathbb{R}^d} dp(1+p^2)^{-2}\right)^{-1} = C(4-d), & d<4. \end{cases} \tag{7.12}$$

Proof. To estimate the denominator in (7.10) we write:

$$\int \nu(db^1 db^2)\,|V_{a/2}(A_1, A_2)|^2 = \int_{\mathbb{R}^d} du \int_{\mathbb{R}^d} dv \int \nu(db^1 db^2)\chi[u, v \in V_{a/2}(A_1, A_2)]. \tag{7.13}$$

There are two possibilities for the order in which the neighborhoods of u and v can be reached by each of the paths $b^i(t)$. Splitting the paths in a natural way, i.e.

conditioning on the site of the first hit, one easily obtains:

$$\int \nu(db^1 db^2)\,|V_{a/2}(A_1, A_2)|^2 = \iint du\,dv \int \nu(db^1 db^2)\chi[u, v \in V_{a/2}(A_1, A_2)]$$
$$\leqq 2 \iiint_{|y-z| \geqq a/2} dx\,dy\,dz\,G(0,y)G(y,z)\,[G(x,y)G(y,z)+G(x,z)G(z,y)]/G(0,a/2)^2$$
$$+ \iiint_{|u-v| \leqq a} dx\,du\,dv\,G(0,u)G(x,u)\;/G(0,a/2)^2$$
$$= 4 \int_{|z|>a/2} dz\,G(0,z)^2\;/G(0,a/2)^4 + |B(1)|a^d/G(0,a)^2\,. \tag{7.14}$$

The *numerator* in (7.10) equals $[G(0,a/2)^{-2}]^2$, by the argument of (7.9). The substitution of (7.14) (with (7.2)) in (7.10) results therefore in the claimed bound (7.11). The transition to (7.12) can be made using the Fourier transform of $G(0,x)$, which is $(1+p^2)^{-1}$. □

c) Direct Estimates of F(a)

The method used in 7a yields rather directly the following upper bound for the quantity $F(a)$, defined by (6.2):

$$F(a) \leqq 2 \int_{\mathbb{R}^d} dx \iint_{|u| \geqq a} du\,dv\,G(0,u)G(0,v)G(x,u)G(u,v)\,[|B(a)|\,G(0,2a)^2]^{-2}$$
$$+ \int_{\mathbb{R}^d} dx\,G(0,x)G(0,2a)^{-1}$$
$$\leqq [\hat{c}\tilde{V}(a)a^{2(d-4)} + \hat{c}|B(1)|a^{d-2}]\,(1+o(1))\,, \tag{7.19}$$

where

$$\tilde{V}(a) = \iint_{|u|>a} du\,dv\,G(0,u)G(u,v)G(v,0) \cong \begin{cases} C\dfrac{1}{6-d}a^{6-d}, & d>6, \\ \text{const}\,|\ln a|, & d=6, \\ \int_{\mathbb{R}^d} dp(1+p^2)^{-3}, & d<6, \end{cases} \tag{7.20}$$

as $a \to 0$.

Therefore

$$F(a) \leqq Ca^{2(d-4)+\min(0,6-d)}\{|6-d|^{-1}\chi[d \neq 6] + c|\ln a|\chi[d=6]\}\,. \tag{7.21}$$

For an opposite bound, let us first make the trivial observation that by (7.1) (with $y=0$)

$$F(a) \geqq \int dx\,G(0,x)/G(0,a) \geqq [d|B(1)|]^{-1}a^{d-2}\,. \tag{7.22}$$

Combining it with (6.9) we have

$$F(a) \geqq [\tfrac{1}{2}d|B(1)|]^{-1}a^{d-2} + \tfrac{1}{2}c_+ g(a)^2\,, \tag{7.23}$$

and thus, using (7.12),

$$F(a) \geqq Ca^{2(d-4)+\min(0,6-d)}\{|d-4|^2\chi[d \neq 4] + c(1+|\ln a|)^{-1}\chi[d=4]\}\,. \tag{7.24}$$

These results are summarized in (4.7) of Proposition 4.1 and used in Sect. 5.

8. Summary and Open Problems

We have seen in Sect. 5 that the "renormalization group" equation described by Proposition 4.1 is an effective tool for the study of the intersection properties of Brownian motion. It would be interesting to see refinements of these techniques, e.g. to resolve the issue mentioned next. An even more important development would be an extension of these methods to some non-Markovian cases. Following are some specific problems.

i) Find the exact behavior, in $d<4$ dimensions, of the quantity

$$\kappa(a)=\int_0^\infty dt_1 e^{-t_1}\int_0^\infty dt_2 e^{-t_2}\,\mathrm{Prob}(\mathrm{dist}(\omega^1_{[0,t_1]},\omega^2_{[0,t_2]})\geqq a|0,x(a)), \tag{8.1}$$

in which $x(a)$ scales, as $a\to 0$, with $|x(a)|=2a$.

We join [4], where such a quantity is reconsidered for the random walks (and an error of [3] is corrected), with the conjecture that for $d=4$

$$\kappa(a)\cong\frac{c}{\sqrt{\ln a^{-1}}},\quad \text{as}\quad a\to 0. \tag{8.2}$$

Such a quantity plays an important role in the physical models which are mentioned in iii).

ii) The probability that Brownian paths come to within distance a is of course analogous, and related to, the probability of an actual intersection of random walks on a lattice of spacing a. Considering the lattice random walks, it is natural to look at the ensemble, Ω_x, of all the paths ω on $\mathbb{Z}^d$, with a given starting point x and the weight $e^{\beta|\omega|}$, where $|\omega|$ is the number of steps. The normalizing factor and the mean square length are

$$S=\sum_{\omega\in\Omega_0} e^{\beta|\omega|},\qquad \xi^2=\sum_{\omega\in\Omega_0}|x(\omega)|^2 e^{\beta|\omega|}/S, \tag{8.3}$$

with $x(\omega)=$ the last site visited by ω.

For simple random walks S and ξ diverge at the critical point $\beta_c=\ln(2d)^{-1}$. The following quantity

$$\hat{g}(\beta)=\sum_{\substack{x\in\mathbb{Z}^d\,\omega^1\in\Omega_0\\ \omega^2\in\Omega_0}} e^{\beta|\omega_1|}e^{\beta|\omega_2|}\chi[\omega^1\cap\omega^2\neq\emptyset]/[S^2\xi^d] \tag{8.4}$$

is analogous to $g(a)$, with $a=\sqrt{\beta_c-\beta}\cong 1/\xi$.

Our methods are easily adaptable to the study of the limiting behavior of $\hat{g}(\beta)$, as $\beta\nearrow\beta_c$, with results which are given by (2.7), with the above correspondence. The problem is to study the analog of $\hat{g}(\beta)$ for *self avoiding* walks. E.g., prove there that

$$\hat{g}\to 0\quad \text{as}\quad \xi\to\infty\quad (\text{i.e. when } \beta\nearrow\beta_c),\text{ in } d\geqq 4 \text{ dimensions}. \tag{8.5}$$

iii) Derive non-perturbative bounds on the β-function of the $\lambda\phi_d^4$ field theory and ferromagnetic spin systems. The renormalized coupling constant in these theories admits a random walk representation which is similar to (8.4), albeit with weights which are more complicated than just $e^{\beta|\omega|}$. Nevertheless, the analysis of the critical behavior in these models has already been greatly aided by the intuition gathered from the study of the Brownian motion [8–10, 15]. (The physical significance of the quantity $\kappa(a)$ for such models is described in [9–12].)

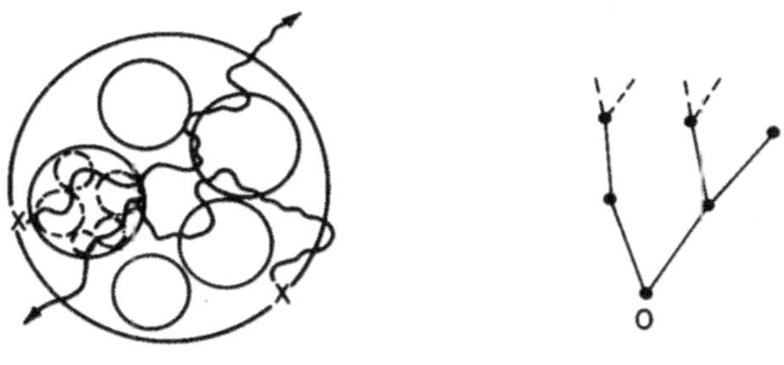

Fig. 3. A random-tree formulation of the intersection problem. (After [9])

Appendix: A Percolation Formulation of the Intersection of Paths

Another approach to the question of intersection of paths was presented in [9]. While it has not yet been useful for any complete argument, it offers yet another point of view. We shall provide here the proof of the main assertion which was made in the appendix of that work.

Since the probability of intersection of the paths of two independent Brownian motions depends only mildly on their starting points (being either zero or nonzero for *all* $x_1 \neq x_2$), it may be studied by means of the following quantity.

$$P = \int_S \sigma(dx) \int_S \sigma(dy)\, \mathrm{Prob}(\omega^1 \cap \omega^2 \cap B \neq \emptyset | 0, x) \Big/ \Big(\int_S \sigma(dx) \Big)^2 , \qquad (A.1)$$

where B is the unit ball in $\mathbb{R}^d$, and $\sigma(dx)$ is the surface measure on $S^d = \partial B$. P is therefore the probability that two paths which start independently on the boundary of a unit ball, intersect inside.

Let $Q_1 = \{q_{1,j}\}_{j \in \mathbb{Z}_-}$ be a partition of the unit ball B into a family of balls, $q_{1,j} \subset B$, whose interiors are disjoint, and which exhaust the volume of B, i.e. for which

$$\sum_{j=1}^{\infty} |q_{1,j}| = |B| . \qquad (A.2)$$

Subpartitioning each of the balls of Q_1, and iteratively repeating this construction, one obtains a hierarchy of partitions which has the structure of a tree, whose "ascending" paths correspond to points of B. The correspondence is $1-1$, up to a set of measure zero (the "cheese") which may be ignored. For each pair of paths ω^1, ω^2 we now mark those points on the tree which correspond to balls (of various generations) which are hit by *both* paths. The marked points, and the edges which connect them form a *random tree.* The problem of paths intersection is thus reformulated as one of "*percolation*" (see Fig. 3, or the more detailed description in [9]).

A classical result states that for trees whose branching numbers are *independent* and identically distributed, the answer to the main question is determined by the single quantity $E(N_1)$, which is the expected value of the number of ascending links from the first vertex (and hence from each vertex which is reached). Specifically, the probability of "percolation" does not vanish if and only if $E(N_1) \geqq 1$ (unless $N_1 \equiv 1$). In the case obtained here there are subtle correlations between the branching numbers. Consequently, we are not able to solve the problem in this form.

Nevertheless, we would like to point out the following striking property of the Brownian motion.

Proposition A.1. *Let $N_1(\omega^1, \omega^2)$ be the number of balls of Q_1 which are hit by both paths, and let $E(-)$ represent the average with respect to the probability measure*

$$\sigma(dx)\sigma(dy)\mu_x(db^2)\mu_y(db^2)\Big/\Big(\int_S \sigma(dz)\Big)^2 .$$

Then

$$E(N_1(\omega^1, \omega^2)) \quad is \quad \begin{cases} <1 & if \quad d>4 \\ =1 & \quad\;\; d=4 \\ >1 & \quad\;\; d<4, \end{cases} \tag{A.3}$$

for every (nontrivial) partition Q_1 of B into balls $\{q_{1,j}\}_{j\in\mathbb{Z}_+}$. Furthermore, the conditional distribution of the number of the next generation balls in $q_{k,j}$ which are intersected by both ω^1 and ω^2 – conditioned on $q_{k,j}$ being intersected, has the same distribution as $N_1(\omega_1, \omega_2)$.

Thus, were we to ignore the existing *interactions* between the numbers of lines which are drawn at the various stages of our construction, we would obtain a solvable model. This "approximation," and Proposition A.1. offer another simple insight (but not yet a proof) into why

$$P \neq 0 \quad \text{if and only if} \quad d<4 . \tag{A.4}$$

Proof. Consider a path which starts at z, with $|z|>1$. The probability of it intersecting the unit ball B is eactly $|z|^{2-d}$. If the intersection does occur then the site of the "first hit" of B has a slightly a-symmetric distribution, which tends to $\sigma(dx)\Big/\int_{S^d} \sigma(du)$ when $|z|\to\infty$. Thus, by the renewal property of the Brownian motion:

$$P = \lim_{|z|\to\infty} \int \mu_z(db^1)\mu_z(db^2)\chi[\omega^1\cap\omega^2\cap B \neq \emptyset]/\int \mu_z(db^1) \cdot \mu_z(db^2)\chi[\omega^i\cap B \neq \emptyset; i=1,2] \tag{A.5}$$

and, similarly,

$$E(N_1(\omega^1, \omega^2)) = \lim_{z\to\infty} \sum_{j=1}^{\infty} \text{Prob}(\omega^i\cap q_{1,j} \neq \emptyset;\, i=1,2\ |z,z)/ \cdot \text{Prob}(\omega^i\cap B \neq \emptyset;\, i=1,2\ |z,z) . \tag{A.6}$$

Let (x_j, r_j) be the centers, and radia, of the balls $q_{1,j}$. Then, by (A.6) and the formula mentioned above

$$E(N_1) = \lim_{z\to\infty} \sum_{j=1}^{\infty} \left(\frac{r_i}{|r_i - z|}\right)^{2(d-2)} \Big/ \left(\frac{1}{|z|}\right)^{2(d-2)} = \frac{1}{|B(1)|} \sum_{j=1}^{\infty} |q_{1,j}| r_j^{d-4} . \tag{A.7}$$

Equation (A.3) follows now from the fact that $\sum_{j=1}^{\infty} |q_{1,j}| = |B(1)|$, and r_j are less than 1. The above argument implies also the other statement made in the proposition. □

Remark. Instead of using balls in Proposition A.1 one could also take any $A \subset \mathbb{R}^d$ which is the closure of an open bounded subset of $\mathbb{R}^d$. For such a set, $\sigma(dx)$ should be replaced with the unique measure which minimizes the capacity integral:

$$C(A)^{-1} = \inf\left\{ \int_A \varrho(dx) \int_A \varrho(dy) \frac{1}{|x-y|^{d-2}} \,\middle|\, \int_A \varrho(dx) = 1 \right\}. \tag{A.8}$$

The general result is that if one partitions A_1 into sets which are similar to A_2, then

$$\left(\frac{\hat{C}(A_1)}{\hat{C}(A_2)}\right)^2 E(N_1) = \sum_{j=1}^{\infty} \left(\frac{|q_{1,j}|}{|A_1|}\right)^{1+\left(1-\frac{4}{d}\right)} \quad \text{which is} \quad \begin{cases} <1 & d>4 \\ =1 & d=4 \\ >1 & d<4, \end{cases} \tag{A.9}$$

where $\hat{C}(A_i) = C(A_i)/|A_i|$. In particular, (A.3) holds in the general self similar case.

The derivation of (A.9) rests on the above argument and the facts that

i) the hitting probability, for large z, behaves as $\frac{C(A)}{z^{d-2}}$, and

ii) for similar sets q and A: $C(q) = C(A)\left(\frac{|q|}{|A|}\right)^{(d-2)/d}$.

Note. After the submission of this article I learned that a β-function approach to the problems discussed here has been advocated by J. Fröhlich [13]. A closely related work of Felder and Fröhlich [14] appears in this issue. I am grateful to T. Hara, T. Hattori, and H. Tasaki for pointing out that Lemma 6.2 requires a restriction on a.

References

1. Kakutani, S.: On Brownian motions in n-space. Proc. Acad. Tokyo **20**, 648 (1944)
2. Dvoretzky, A., Erdös, P., Kakutani, S.: Double points of Brownian paths in n-space. Acta Sci. Math. Szeged **12**, 75 (1950)
3. Erdös, P., Taylor, S.J.: Some intersection properties of random paths. Acta Math. Acad. Sci. Hung. **11**, 231 (1960)
4. Lawler, G.F.: The probability of intersection of Random walks in four dimensions. Commun. Math. Phys. **86**, 539 (1982)
5. Spencer, T.: Private communication
6. Symanzik, K.: Small distance behavior in field theory and power counting. Commun. Math. Phys. **18**, 227 (1970)
   Callan, C.G. Jr.: Broken scale invariance in scalar field theory. Phys. Rev. D **2**, 1541 (1970)
   Wilson, K.G.: Anomalous dimension and the breakdown of scale invariance in perturbation theory. Phys. Rev. D **2**, 1478 (1970)
7. Brézin, E., Le Guillou, J.C., Zinn-Justin, J.: In: Phase transitions and critical phenomena. Domb, C., Green, M.S., (eds.). London, New York, San Francisco: Academic Press 1976
   Itzykson, C., Zuber, J.B.: Quantum field theory. New York: McGraw-Hill 1980
8. Symanzik, K.: Euclidean quantum field theory. I. Equations for a scalar model. J. Math. Phys. **7**, 510 (1966)
9. Aizenman, M.: Geometric analysis of ϕ^4 fields and Ising models. Commun. Math. Phys. **86**, 1 (1982)
10. Fröhlich, J.: On the triviality of $\lambda\phi_d^4$ theories and the approach to the critical point in $d>4$ dimensions. Nucl. Phys. B **200**, [FS4] 281 (1982)
11. Aragão de Carvalho, C., Caraciollo, S., Fröhlich, J.: Polymers and $g|\phi|^4$ theory in four dimensions. Nucl. Phys. B **215** [FS7], 209 (1983)

12. Aizenman, M., Graham, R.: On the renormalized coupling constant and the susceptibility in ϕ_4^4 field theory and the Ising model in four dimensions. Nucl. Phys. B **225** [FS9], 261 (1983)
13. Fröhlich, J.: Quantum field theory in terms of Random walks and Random surfaces. Cargèse (1983) lecture notes
14. Felder, G., Fröhlich, J.: Intersection properties of simple Random walks: a renormalization group approach. Commun. Math. Phys. **97**, 111–124 (1984)
15. Aizenman, M.: Rigorous studies of critical behavior, to appear in the proceedings of the VIII sitges conference. L. Garrido (ed.), Berlin, Heidelberg, New York: Springer 1984

Communicated by A. Jaffe

Received January 9, 1984; in revised form April 23, 1984

Intersection Properties of Simple Random Walks: A Renormalization Group Approach

G. Felder and J. Fröhlich

Theoretical Physics, ETH-Hönggerberg, CH-8093 Zürich, Switzerland

Abstract. We study estimates for the intersection probability, $g(m)$, of two simple random walks on lattices of dimension $d=4$, $4-\varepsilon$ as a problem in Euclidean field theory. We rigorously establish a renormalization group flow equation for $g(m)$ and bounds on the β-function which show that, in $d=4$, $g(m)$ tends to zero logarithmically as the killing rate (mass) m tends to zero, and that the fixed point, g^*, in $d=4-\varepsilon$ is bounded by $\text{const}'\varepsilon \leqq g^* \leqq \text{const}\,\varepsilon$. Our methods also yield estimates on the intersection probability of three random walks in $d=3$, $3-\varepsilon$. For $\varepsilon=0$, these results were first obtained by Lawler [1].

1. Introduction

Two Brownian paths in $\mathbb{R}^d$ starting at different points intersect with positive probability in less than four dimensions, but do never intersect in four or more dimensions [2, 3].

The continuum limit of $g_0|\boldsymbol{\phi}|_d^4$ theory, $\boldsymbol{\phi}=\phi$, or $\boldsymbol{\phi}=(\phi^1,\phi^2)$, $g_0>0$, is an interacting theory in less than four dimensions, in the superrenormalizable regime [4], but is a (generalized) free field in more than four dimensions [5]. Results in four dimensions remain incomplete, but there are strong reasons to expect that the continuum limit is trivial in that case, too.

Symanzik recognized the connection between these two facts in his work [6] on a representation of $g_0|\boldsymbol{\phi}|^4$-theory as a gas of Brownian paths with local, repulsive interaction. Further work on that connection led to a novel, rather intuitive approach to scalar quantum field theory to which several people contributed valuable results, in the past few years. (See e.g. [7] and references therein for reviews of recent results.)

On a more abstract, more heuristic level, much insight into the theory of critical points in lattice field theories and the related problem of constructing continuum limits in dimensions close to four has been accomplished by using renormalization group methods; see e.g. [8] and references given there. In particular, for $g_0|\boldsymbol{\phi}|^4$-theories, perturbative renormalization group equations predict that, in four dimensions, the renormalized coupling constant $g=g(m)$ tends to zero like

$|\ln m|^{-1}$, as the mass m (or, equivalently, the lattice spacing) tends to 0, while in $d=4-\varepsilon$ dimensions $g(m)$ approaches a fixed point $g^*=0(\varepsilon)$, as $m \searrow 0$. Unfortunately, it has not been possible, so far, to find rigorous renormalization group flow equations which would yield mathematical proofs of these facts. It may be a little interesting, therefore, that, in estimating intersection probabilities of simple random walks, one can successfully use rigorous renormalization group flow equations. This is demonstrated in the following. Hence, while it has been shown already that random-walk methods are powerful tools to study field theory, it is now known that field-theoretic methods are useful in the study of random walk problems.

It is worthwhile and amusing to first pose the problem of estimating intersection probabilities of random walks explicitly as a problem in Euclidean field theory. This will show how and why heuristic renormalization group methods can be applied to random-walk problems.

Consider a lattice field theory of two interacting N-component fields, $\boldsymbol{\phi} = (\phi^1, \ldots, \phi^N)$ and $\psi = (\psi^1, \ldots, \psi^N)$, with action

$$\mathscr{A}(\boldsymbol{\phi}, \boldsymbol{\psi}) = \sum_{x \in \mathbb{Z}^d} \left\{ \frac{1}{2} |(\nabla \boldsymbol{\phi})_x|^2 + \frac{1}{2} |(\nabla \boldsymbol{\psi})_x|^2 + \frac{m^2}{2} (|\boldsymbol{\phi}_x|^2 + |\boldsymbol{\psi}_x|^2) \right.$$
$$\left. + \frac{g_0}{4} |\boldsymbol{\phi}_x|^2 |\boldsymbol{\psi}_x|^2 + \frac{g_0'}{4} (|\boldsymbol{\phi}_x|^4 + |\boldsymbol{\psi}_x|^4) \right\}. \tag{1.1}$$

Such models were studied from the point of view of ε-expansions by Brézin et al. [9]. Let $\langle(\cdot)\rangle$ denote the Euclidean vacuum expectation for the theory with action $\mathscr{A}(\boldsymbol{\phi}, \boldsymbol{\psi})$ given by (1.1). We define a susceptibility χ as

$$\chi = \sum_{x \in \mathbb{Z}^d} \langle \phi_0^1 \phi_x^1 \rangle = \sum_{x \in \mathbb{Z}^d} \langle \psi_0^1 \psi_x^1 \rangle, \tag{1.2}$$

the physical mass, m_{phys}, as the exponential decay rate of $\langle \phi_0^1 \phi_x^1 \rangle$ in $|x|$, and physical coupling constants g and g' by

$$g = -\bar{u}_4 \chi^{-2} m_{\text{phys}}^d, \quad g' = -\bar{u}_4' \chi^{-2} m_{\text{phys}}^d, \tag{1.3}$$

where

$$\bar{u}_4 = \sum_{x, y, z} \langle \phi_0^1 \phi_x^1 ; \psi_y^1 \psi_z^1 \rangle,$$

$$\bar{u}_4' = \sum_{x, y, z} \langle \phi_0^1 ; \phi_x^1 ; \phi_y^1 ; \phi_z^1 \rangle,$$

and $\langle A_1 ; A_2 ; \ldots \rangle$ denotes a connected expectation. In [9] renormalization group fixed points for g and g' have been computed as functions of N, to first order in $\varepsilon = 4-d$. The result is:

$$g^* = \frac{(4-N)\varepsilon}{2N(N+8)-16(N-1)}, \quad g'^* = \frac{N\varepsilon}{2N(N+8)-16(N-1)}. \tag{1.4}$$

Thus g'^* and hence g' tend to 0, as $N \to 0$ [in the domain of attraction of the fixed point (4)]. In the limit $N=0$, the flow equation for g as a function of $m_{\text{phys}} = m$ is given by

$$-\frac{dg}{d|\ln m|}=\beta(g), \quad \text{where} \quad \beta(g)=-\varepsilon g+4g^2+0(g^3). \tag{1.5}$$

(Note that the roots of β are $g=0$, $g=g^*$, to first order in ε.)

Surprisingly, the $N\to 0$ limit of the theories considered here is a theory of Brownian paths with a local, repulsive interaction between any two paths. For the lattice theory this is easy to understand: Following [10] one may use a random-walk representation of that theory to show that

$$\lim_{N,g_0'\to 0}\langle \phi_{x_1}^{\alpha_1}\ldots\phi_{x_{2n}}^{\alpha_{2n}}\psi_{y_1}^{\beta_1}\ldots\psi_{y_{2k}}^{\beta_{2k}}\rangle$$

$$=\sum_{p,\;q}\sum_{\substack{\omega_1,\ldots,\omega_n\\ \omega_1',\ldots,\omega_k'}}\int\ldots\int\prod_{i=1}^{n}dv_{\omega_i}^{m}(t^i)\prod_{j=1}^{k}dv_{\omega_j'}^{m}(s^j)\exp\left(-g_0\sum_x\sum_{i,j}t_x^i\cdot s_x^j\right), \tag{1.6}$$

where the (nearest-neighbor) walks ω_i, $i=1,\ldots,n$ have endpoints $x_{p(2i-1)}$, $x_{p(2i)}$ with $\alpha_{p(2i-1)}=\alpha_{p(2i)}$, and, similarly, the walks ω_j' connect $y_{q(2j-1)}$ to $y_{q(2j)}$, where $\beta_{q(2-1)}=\beta_{q(2j)}$; p and q are paitings of $\{1,\ldots,2n\}$, $\{1,\ldots,2k\}$, respectively. The lattice fields $t^i=\{t_x^i\}_{x\in\mathbb{Z}^d}$ and $s^j=\{s_x^j\}_{x\in\mathbb{Z}^d}$ are waiting (or local) times associated with $\boldsymbol{\phi}$ and $\boldsymbol{\psi}$, respectively. Their *a priori*-distributions are given by

$$dv_\omega^m(t)=\prod_{x\in\mathbb{Z}^d}dv_{n_x(\omega)}(t_x)\exp(-(2d+m^2)t_x),$$

$$dv_n(t)=\begin{cases}\delta(t)\,dt, & n=0\\ \dfrac{t^{n-1}}{(n-1)!}\theta(t)\,dt, & n\geqq 1,\end{cases} \tag{1.7}$$

and $n_x(\omega)$ is the total number of visits of ω at the site x.

It should be noted that for the lattice $g_0|\boldsymbol{\phi}|^2|\boldsymbol{\psi}|^2$-theory in finite volume an analytic interpolation in N follows immediately from Symanzik's representation [6, 10], and a rigorous proof of (1.6) is a straight-forward consequence of the formulas in [10]. This has first been noticed in [11]. Important special cases of (1.6) are:

(i)

$$G_m(x,y)\equiv\lim_{N,g_0\to 0}\langle\phi_x^1\phi_y^1\rangle=\lim_{N,g_0\to 0}\langle\psi_x^1\psi_y^1\rangle$$

$$=\sum_{\omega:\,x\to y}\int dv_\omega^m(t)=\sum_{\omega:\,x\to y}\xi^{|\omega|+1}, \tag{1.8}$$

where $\xi\equiv(2d+m^2)^{-1}$, $|\omega|=\#$ jumps of ω. This follows directly from (1.6) and (1.7). The right-hand side of (1.8) is well-known to be the Green's function of $-\Delta+m^2$, where Δ is the finite difference Laplacian. From (1.8) and the simple identity

$$[e^{T(\Delta-m^2)}]_{xy}=\sum_{\omega:\,x\to y}\int dv_\omega^m(t)\,\delta\left(T-\sum_x t_x\right),$$

we infer that $d\nu_\omega^m(t)$ is the waiting-time representation of the discrete analogue of Wiener measure with killing rate m.

(ii)
$$G_{m,g_0}^c(x_1,x_2;y_1,y_2) \equiv \lim_{N,g_0'\to 0} \langle \phi_{x_1}^1, \phi_{x_2}^1; \psi_{y_1}^1, \psi_{y_2}^1 \rangle$$
$$= \sum_{\substack{\omega: x_1\to x_2 \\ \omega': y_1\to y_2}} \iint d\nu_\omega^m(t)\, d\nu_{\omega'}^m(s)\, I_{g_0}(t,s), \tag{1.9}$$

where
$$I_{g_0}(t,s) = \exp\left(-g_0 \sum_x t_x s_x\right) - 1\,. \tag{1.10}$$

Note that
$$I_{g_0}(t,s) = 0\,, \tag{1.11}$$

unless ω and ω' intersect each other somewhere, in which case $-1 < I_{g_0}(t,s) < 0$, i.e. the functions $\exp\left(-g_0 \sum_x t_x s_x\right)$ describe local, repulsive pair-interactions between random walks. By (1.9)–(1.11)

$$g = g(m) = -\sum_x m^d \sum_{y,z} \chi^{-2} G_{m,g_0}^c(0,y;\,x,z), \tag{1.12}$$

with $\chi = \sum_x G_m(0,x) = m^{-2}$, is proportional to the intersection probability between simple random walks with killing rate m. The actual intersection probability is obtained from (1.12) by letting g_0 tend to $+\infty$:

$$\lim_{g_0\to+\infty} -\sum_{y,z} \chi^{-2} G_{m,g_0}^c(0,y;\,x,z) \equiv P(0,x),$$

where, by (1.7), (1.9), and (1.10),

$$P(0,x) = \chi^{-2} \sum_{\substack{\omega: 0\to\cdot \\ \omega': x\to\cdot}} \xi^{|\omega|+|\omega'|+2} \chi(\omega\cap\omega' \neq \phi). \tag{1.13}$$

This is the intersection probability for two simple random walks, ω and ω', starting at $0, x$, respectively. Since $P(0,x)$ is a pure number, and m has the dimension of $[\text{length}]^{-1}$, the average intersection probability is given by

$$P = \sum_x m^d P(0,x) = g(m)|_{g_0\to\infty}\,. \tag{1.14}$$

If (1.5) were rigorously valid we would conclude that

$$g(m) \sim |\log m|^{-1}, \text{ as } m \searrow 0, \quad \text{for } d=4,$$
and
$$g(m) \xrightarrow[m\to 0]{} g^* = \text{const}\,\varepsilon, \quad \text{for } d=4-\varepsilon. \tag{1.15}$$

We note that, in the sense of dimensional analysis [1]), d is a *continuous* parameter, and it is quite clear what is meant by $\mathbb{Z}^{4-\varepsilon}$. More rigorously, one may fix d, e.g. $d=3$, and study intersection probabilities for walks with jumping probabilities

$$P_{xy} = -[(-\Delta + m^2)^{\gamma}]_{xy}, \quad x \neq y, \tag{1.16}$$

with $0<\gamma<1$. We note that $P_{xy}>0$, and $[(-\Delta+m^2)^{-\gamma}]_{xy} = \sum_{\omega: x \to y} P_\omega \xi^{|\omega|+1}$, where $\xi = [(-\Delta + m^2)^{\gamma}]_{xx}^{-1}$, $P_\omega = \prod_{(u,v)\in\omega} P_{uv}$. Then $d=4$, $\gamma=1$ is easily seen to be equivalent to $d=d_0$, $\gamma=d_0/4$, e.g. $d=3$, $\gamma=3/4$, in the sense of dimensional analysis. Moreover $d=4-\varepsilon$, $\gamma=1$ corresponds to $d=3$, $\gamma=(3+\varepsilon)/4$.

We now summarize how much of (1.15) can be proven rigorously, using a rigorous version of the flow equation (1.5):

A) *In four dimensions,*

$$c_1 |\ln m|^{-1} \leqq g(m) \leqq c_2 |\ln m|^{-1} \tag{1.17}$$

for small m.

B) *In* $d=4-\varepsilon$ *dimensions, there exists a non-trivial fixed point* $g^* = \lim_{m \searrow 0} g(m)$, *with*

$$c_3 \varepsilon \leqq g^* \leqq c_4 \varepsilon \tag{1.18}$$

for small ε.

Here $c_1, \ldots, c_4$ are finite, positive constants.

Similar results can be proven for the probability of intersection in one point of three random walks, with $d=4$ replaced by $d=3$; see also [1].

The lower bound in A) is due to Erdös and Taylor [3]. The upper bound in A) was first proven by Lawler [1]. In this paper we present a simple proof of it, motivated by the field-theoretic arguments leading to (1.5).

Our paper is organized as follows: In Sect. 2, we recall the proof of the lower bound on $g(m)$. In Sect. 3, we derive estimates on the β-function which imply the upper bounds on $g(m)$. Our estimates are based on the results of Sect. 2. In Sect. 4, we sketch some extensions of our results and draw conclusions.

2. The Lower Bound

The lower bound (1.17) on the average intersection probability was proven in four dimensions by Erdös and Taylor [3]. We present here a variant of their proof of this result, due to Sokal [12]. [For another proof see also Sect. 4, (a).]

Theorem 2.1. *In four dimensions*

$$g \geqq c_1 \left(\ln \frac{1}{m}\right)^{-1}. \tag{2.1}$$

1 I.e. naive power counting

In $d=4-\varepsilon$ dimensions

$$g \geqq c_3 \varepsilon . \tag{2.2}$$

The proof of this theorem rests on the following ("skeleton"-)inequalities.

Lemma 2.2. *There exist constants d_1 and d_2 such that, for small m,*

$$P(0,x) \leqq d_1 \sum_z G_m(0,z)\, G_m(z,x), \tag{2.3}$$

$$P(0,x) \geqq \frac{d_2}{B(m)} \left(2\sum_z G_m(0,z)\, G_m(z,x) - \frac{1}{B(m)} \sum_{z_1, z_2} G_m(0,z_1)\, G_m(z_1,z_2)^2\, G_m(z_2,x) \right), \tag{2.4}$$

where $B(m)=\sum_z G_m(0,z)^2$.

Proof of Lemma 2.2. In order to prove the upper bound, we note that

$$\chi(\omega \cap \omega' \neq \phi) \leqq \sum_z \chi(\omega \cap \omega' \ni z). \tag{2.5}$$

Splitting the walks ω and ω' at the site z and re-summing over all resulting walks yields

$$P(0,x) \leqq \sum_z \xi^{-2} m^4 \sum_{\substack{\omega_1:\, 0 \to z \\ \omega_2:\, z \to x \\ \omega_3:\, z \to \cdot \\ \omega_4:\, z \to \cdot}} \prod_{i=1}^{4} \xi^{|\omega_i|+1} \leqq \xi^{-2} \sum_z G_m(0,z)\, G_m(z,x), \tag{2.6}$$

and we have used that

$$\chi = m^{-2} = \sum_z G_m(0,z) = \sum_{\omega:\, 0 \to \cdot} \xi^{|\omega|+1}. \tag{2.7}$$

In order to prove the lower bound, we choose some sublattice, L, of $\mathbb{Z}^d$. Then, by the inclusion-exclusion principle,

$$\begin{aligned} P(0,x) &\geqq \sum_{\substack{\omega:\, 0 \to \cdot \\ \omega':\, x \to \cdot}} m^4 \xi^{|\omega|+|\omega'|+2} \chi(\omega \cap \omega' \cap L \neq \phi) \\ &\geqq \sum_{z \in L} \sum_{\substack{\omega:\, 0 \to \cdot \\ \omega':\, x \to \cdot}} m^4 \xi^{|\omega|+|\omega'|+2} \chi(\omega \cap \omega' \cap L \ni z) \\ &\quad - \frac{1}{2!} \sum_{\substack{z_1, z_2 \\ z_1 \neq z_2 \in L}} \sum_{\substack{\omega:\, 0 \to \cdot \\ \omega':\, x \to \cdot}} m^4 \xi^{|\omega|+|\omega'|+2} \chi(\omega \cap \omega' \cap \mathrm{L} \ni z_1, z_2). \end{aligned} \tag{2.8}$$

The probability that a walk ω starting at x hits a site y is given by

$$g_m(x,y) \equiv \sum_{\omega:\, x \to \cdot} m^2 \xi^{|\omega|+1} \chi(\omega \ni y) = G_m(x,y)/G_m(y,y).$$

Thus

$$P(0,x) \geqq \sum_{z \in L} g_m(0,z) g_m(z,x) - \sum_{\substack{z_1, z_2 \in L \\ z_1 \neq z_2}} g_m(0,z_1) g_m(z_1,z_2)^2 (g_m(z_1,x) + g_m(z_2,x)). \tag{2.9}$$

It is convenient to choose L to be a percolation lattice, i.e. we assign a probability p that a site x is in L and a probability $(1-p)$ that $x \notin L$. Taking the expectation value on both sides of (2.9) we obtain

$$P(0,x) \geqq (G_m(0,0)^{-2} p - G_m(0,0)^{-4} p^2 B(m)) \sum_z G_m(0,z) G_m(z,x)$$
$$- G_m(0,0)^{-4} p^2 \sum_{z_1, z_2} G_m(0,z_1) G_m(z_1,z_2)^2 G_m(z_2,x), \tag{2.10}$$

which is equivalent to (2.4) if we choose

$$p = \frac{G_m(0,0)^2}{3B(m)} < 1, \text{ for } m \text{ small enough.} \quad \square \tag{2.11}$$

Proof of Theorem 2.1. Inserting (2.10) in the definition (1.14) of $g(m)$ and using (2.7) we get

$$g(m) > 2 \frac{d^2}{B(m)} m^{d-4}. \tag{2.12}$$

But since

$$B(m) \sim \begin{cases} \ln\left(\frac{1}{m}\right), & d=4 \\ \frac{1}{\varepsilon} m^{-\varepsilon}, & d=4-\varepsilon, \end{cases} \tag{2.13}$$

the theorem is proven. [See also Sect. 4, (a).] $\square$

3. The Upper Bound

We now turn to the main part of this paper: The proof of the upper bound on the intersection probability. Our proof is based on the following lemma, which relates the β-function to the average probability that *two* independent, simple random walks starting at the same point *both* intersect a third walk. We set $g = g(m)$.

Lemma 3.1. *In $d = 4 - \varepsilon$ dimensions, $\varepsilon \geqq 0$,*

$$m \frac{dg}{dm} = -\varepsilon g + 4m^d \sum_x Q(0,x), \tag{3.1}$$

where

$$Q(0,x) = m^6 \sum_{\substack{\omega_1 : 0 \to \cdot \\ \omega_2 : 0 \to \cdot \\ \omega_3 : x \to \cdot}} \prod_{i=1}^{3} \xi^{|\omega_i|+1} \chi(\omega_1 \cap \omega_3 \neq \phi) \chi(\omega_2 \cap \omega_3 \neq \phi). \tag{3.2}$$

Proof. Starting from the definition of g,

$$g = m^{d+4} \sum_{\substack{\omega: 0 \to \cdot \\ \omega': \cdot \to \cdot}} \xi^{|\omega|+|\omega'|+2} \chi(\omega \cap \omega' \neq \phi), \qquad \xi = \frac{1}{2d+m^2}, \tag{3.3}$$

we calculate the derivative, using $\frac{d\xi}{dm} = -2\xi^2 m$. Thus

$$\begin{aligned} m\frac{dg}{dm} &= (d+4)g - 2m^{d+6} \sum_{\substack{\omega: 0 \to \cdot \\ \omega': \cdot \to \cdot}} (|\omega|+|\omega'|+2)\,\xi^{|\omega|+|\omega'|+3} \chi(\omega \cap \omega' \neq \phi) \\ &= (d+4)g - 4m^{d+6} \sum_{\substack{\omega: 0 \to \cdot \\ \omega': \cdot \to \cdot}} (|\omega|+1)\,\xi^{|\omega|+|\omega'|+3} \chi(\omega \cap \omega' \neq \phi), \end{aligned} \tag{3.4}$$

since we can interchange ω and ω'. Note that $|\omega|+1$ is the number of points that ω hits (counted with multiplicity). In each of these points we can split the walk ω into two independent walks ω_1 and ω_2. We thus have (with $\omega' = \omega_3$):

$$m\frac{dg}{dm} = (d+4)g - 4m^{d+6} \sum_z \sum_{\substack{\omega_1: 0 \to z \\ \omega_2: z \to \cdot \\ \omega_3: \cdot \to \cdot}} \prod_{i=1}^{3} \xi^{|\omega_i|+1} \chi(\omega_1 \cap \omega_3 \neq \phi \text{ or } \omega_2 \cap \omega_3 \neq \phi). \tag{3.5}$$

A moment of reflection shows that

$$\begin{aligned} \chi(\omega_1 \cap \omega_3 \neq \phi \text{ or } \omega_2 \cap \omega_3 \neq \phi) &= \chi(\omega_1 \cap \omega_3 \neq \phi) + \chi(\omega_2 \cap \omega_3 \neq \phi) \\ &\quad - \chi(\omega_1 \cap \omega_3 \neq \phi \text{ and } \omega_1 \cap \omega_3 \neq \phi). \end{aligned} \tag{3.6}$$

Inserting this identity on the right-hand side of (3.5) we have, after translation of z to the origin,

$$m\frac{dg}{dm} = (d+4)g - 8g + 4m^{d+6} \sum_{\substack{\omega_1: 0 \to \cdot \\ \omega_2: 0 \to \cdot \\ \omega_3: \cdot \to \cdot}} \prod_{i=1}^{3} \xi^{|\omega_i|+1} \chi(\omega_1 \cap \omega_3 \neq \phi)\ \chi(\omega_2 \cap \omega_3 \neq \phi), \tag{3.7}$$

where the term $-8g$ on the right-hand side of (3.7) results from the first and second term on the right-hand side of (3.6), by re-summing over ω_2, ω_1, respectively, and adding the two contributions.

We now proceed to proving our main estimates.

Theorem 3.2 *In four dimensions,*

$$g(m) \leqq c_2 |\ln m|^{-1}. \tag{3.8}$$

In $d = 4 - \varepsilon$ dimensions,

$$g^* \leqq c_4 \varepsilon, \tag{3.9}$$

where $g^* = \lim_{m \searrow 0} g(m)$.

Proof. Defining

$$\beta(g,m) \equiv m\frac{dg}{dm} = -\frac{dg}{d|\ln m|}, \tag{3.10}$$

we need a lower bound on $\beta(g,m)$, in order to obtain upper bounds on $g(m)$ and g^*, by integrating (3.10). By the Schwarz inequality,

$$\begin{aligned} Q(0,x) &= m^2 \sum_{\omega_3:x\to\cdot} \xi^{|\omega_3|+1}\left(m^2 \sum_{\omega:0\to\cdot} \xi^{|\omega|+1}\chi(\omega\cap\omega_3 \neq \phi)\right)^2 \\ &\geqq \left(m^4 \sum_{\substack{\omega_3:x\to\cdot \\ \omega:0\to\cdot}} \xi^{|\omega_3|+|\omega|+2}\chi(\omega\cap\omega_3 \neq \phi)\right)^2 = P(0,x)^2, \end{aligned} \tag{3.11}$$

and we have used that

$$m^2 = \chi^{-1} = \left(\sum_{\omega:0\to\cdot} \xi^{|\omega|+1}\right)^{-1}.$$

We define $B_R = \{x\in\mathbb{Z}^d \,|\, |x| \leqq R\}$, and use the Schwarz inequality once more. This yields

$$\beta(g,m) \geqq -\varepsilon g + 4m^d \sum_{x\in B_R} P(0,x)^2 \geqq -\varepsilon g + \frac{4m^{-d}}{|B_R|}\left(m^d \sum_{x\in B_R} P(0,x)\right)^2, \tag{3.12}$$

where $|B_R|$ denotes the number of sites in B_R. In order to compare $m^d \sum_{x\in B_R} P(0,x)$ to $g = m^d \sum_x P(0,x)$, we need an *upper bound* on $P(0,x)$, for $|x| > R$: For $x = (x_1, \ldots, x_d)\in\mathbb{Z}^d$, let x_α be the coordinate of x with the largest absolute value. Thus $|x_\alpha| \geqq d^{-1/2}|x|$. Let π_x be the lattice plane perpendicular to the α-axis and intersecting that axis in the point $[x_\alpha/2]e_\alpha$, where $[a]$ is the largest integer $\leqq a$, and e_α is the unit lattice vector in the α-direction. We now recall that $P(0,x)$ is the probability that a random walk ω starting at 0 and a random walk ω' starting at x intersect in a common point of $\mathbb{Z}^d$. This probability is clearly bounded above by the probability, $R(0,x)$, that either ω hits the plane π_x, or ω' hits π_x. (For if neither ω, nor ω' hit π_x, then ω and ω' will always be on different sides of π_x and hence never intersect.) Thus

$$P(0,x) \leqq R(0,x) \leqq c\,T(0,x), \tag{3.13}$$

where

$$T(0,x) = \text{Prob}\{\omega\,|\,\omega:0\to\cdot,\ \omega \text{ hits } \pi_x\}$$

(and $c = 2$ if x_α is even). Now, $T(0,x)$ can easily be calculated explicitly:

$$T(0,x) = \sum_{z\in\pi_x} [G_m(0,z-(\text{sign}\,x_\alpha)e_\alpha) - G_m(0,z+(\text{sign}\,x_\alpha)e_\alpha)]. \tag{3.14}$$

By using the Fourier representation of G_m and the calculus of residues to carry out the momentum space integration we find

$$0 < T(0,x) \leqq c' e^{-c'' m|x_\alpha|} \leqq c' \exp(-c'' d^{-1/2} m|x|). \tag{3.15}$$

Thus $P(0,x) \leqq \tilde{c} \exp(-\tilde{c}\, m|x|)$, from which we get

$$m^d \sum_{|x|>R} P(0,x) \leqq C e^{-\tilde{c} m R}, \tag{3.16}$$

for some finite constants C and $\tilde{c}$.

We now discuss separately the cases $d=4$ and $d=4-\varepsilon$.

(i) For $d=4$, we choose

$$R = (\tilde{c}\, m)^{-1} \ln\left(\frac{2C}{c_1} |\ln m|\right), \tag{3.17}$$

where c_1 is the constant appearing in Theorem 2.1, and $\tilde{c}$, C are the constants of estimate (3.16). Then, by (2.1) and (3.16),

$$m^4 \sum_{|x|>R} P(0,x) \leqq c_1 (2|\ln m|)^{-1} \leqq \frac{1}{2} g, \tag{3.18}$$

and thus

$$m^4 \sum_{x \in B_R} P(0,x) \geqq \frac{1}{2} g. \tag{3.19}$$

With (3.11) and (3.12) this yields

$$m \frac{dg}{dm} = \beta(g,m) \geqq \text{const}\, g^2 (\ln|\ln m|)^{-d}, \tag{3.20}$$

and after integration,

$$g(m)^{-1} - g(m_0)^{-1} \geqq \text{const} \int_{|\ln m_0|}^{|\ln m|} d\tau\, (\ln \tau)^{-d}. \tag{3.21}$$

Since the right-hand side of (3.21) diverges, as $m \searrow 0$, we conclude that

$$0 \leqq g(m) \leqq c_2 |\ln m|^{-1} (\ln|\ln m|)^4, \tag{3.22}$$

in particular $g(m) \to 0$, as $m \to 0$.

With more effort, one can improve (3.22) and show that $g(m) \leqq c_2 |\ln m|^{-1}$.

(ii) For $d = 4 - \varepsilon$, we choose in (3.16)

$$R = (\tilde{c}\, m)^{-1} \ln\left(\frac{2C}{c_3 \varepsilon}\right). \tag{3.23}$$

Then

$$m^d \sum_{|x|>R} P(0,x) \leqq \frac{c_3}{2} \varepsilon \leqq \frac{1}{2} g, \tag{3.24}$$

by Theorem 2.1, and hence

$$m \frac{dg}{dm} \geqq \text{const}\, g^2 |\ln \varepsilon|^{-d} - \varepsilon g, \tag{3.25}$$

or

$$\frac{dg}{d|\ln m|} \leqq \varepsilon g - \text{const}\, g^2 |\ln \varepsilon|^{-d}. \tag{3.26}$$

Hence

$$\lim_{m \searrow 0} g(m) = g^* \leqq c_4\, \varepsilon |\ln \varepsilon|^d. \tag{3.27}$$

Note added in typescript. We now briefly outline how to remove the factors $(\ln|\ln m|)^4$, $|\ln \varepsilon|^d$, respectively, by sharpening the bounds (3.12) and (3.16). Our argument is somewhat inspired by one used by M. Aizenman in his independent work on closely related problems. (We received his preprint after completion of the first draft of this paper.)

The simple estimate

$$Q(0,x) \geqq m^2 \sum_{\omega_1 : 0 \to \cdot} \xi^{|\omega_1|-1} \left(m^2 \sum_{\omega_2 : x \to \cdot} \xi^{|\omega_2|+1} \chi(\omega_1 \cap \omega_2 \neq \phi) \right)^2 \chi(\omega_1 \cap \pi_x \neq \phi), \tag{3.28}$$

where π_x is the plane defined after (3.12), implies, by the Schwarz inequality,

$$\begin{aligned} \beta(g,m) \geqq {} & -\varepsilon g + \left(m^d \sum_x m^2 \sum_{\omega_1 : 0 \to \cdot} \xi^{|\omega_1|+1} \chi(\omega_1 \cap \pi_x \neq \phi) \right)^{-1} \\ & \cdot \left\{ m^d \sum_x m^4 \sum_{\substack{\omega_1 : 0 \to \cdot \\ \omega_2 : x \to \cdot}} \xi^{|\omega_1|+|\omega_2|+2} \chi(\omega_1 \cap \omega_2 \neq \phi) \frac{1}{2} [\chi(\omega_1 \cap \pi_x \neq \phi) + \chi(\omega_2 \cap \pi_x \neq \phi)] \right\}^2, \end{aligned} \tag{3.29}$$

where use was made of the symmetry under interchange of ω_1 and ω_2. The denominator is just $m^d \sum_x T(0,x) \leqq \text{const}$, by (3.15), and since

$$\chi(\omega_1 \cap \omega_2 \neq \phi) \frac{1}{2} [\chi(\omega_1 \cap \pi_x \neq \phi) + \chi(\omega_2 \cap \pi_x \neq \phi)] \geqq \frac{1}{2} \chi(\omega_1 \cap \omega_2 \neq \phi),$$

we get the lower bound

$$\beta(g,m) \geqq -\varepsilon g + C_d g^2 \tag{3.30}$$

(C_d is a positive constant), which, upon integration, yields the claim of the theorem. □

4. Extensions and Conclusions

The following results can be established by using the techniques developed in our paper:

(a) In (1.9) we have defined a connected four-point function for the $g_0 |\phi|^2 |\psi|^2$-theory, for finite values of g_0, in the limit $N \to 0$. We have noted that, for *finite* g_0,

$$P_{g_0}(0,x) \equiv -m^4 \sum_{y,z} G^c_{m,g_0}(0,y;x,z) \tag{4.1}$$

is proportional, but strictly smaller, than the intersection probability, $P(0,x)$. Since the function $-I_{g_0}(t,s)=1-\exp\left(-g_0\sum_x t_x s_x\right)$ is monotone increasing in g_0, we conclude, using (1.9)–(1.11), that

$$P_{g_0}(0,x) \text{ is monotone increasing in } g_0, \text{ with} \qquad \lim_{g_0\searrow 0} P_{g_0}(0,x)=0, \quad \lim_{g_0\nearrow\infty} P_{g_0}(0,x)=P(0,x). \tag{4.2}$$

Clearly

$$-I_{g_0}(t,s)\geqq -I_{g_0'}(t,s)\geqq g_0'\sum_z t_z s_z - \frac{g_0'^2}{2}\sum_{z_1,z_2} t_{z_1}t_{z_2}s_{z_1}s_{z_2}, \tag{4.3}$$

For arbitrary $g_0'\leqq g_0$. By the splitting lemma of [13], this yields

$$\begin{aligned} P_{g_0}(0,x) \geqq {} & g_0'\sum_z G_m(0,z)\,G_m(x,z) \\ & -\frac{g_0'^2}{2}\sum_{z_1,z_2} G_m(0,z_1)\,G_m(z_1,z_2)^2\,G_m(z_1,x) \\ & -\frac{g_0'^2}{2}\sum_{z_1,z_2} G_m(0,z_1)\,G_m(z_1,z_2)^2\,G_m(z_2,x). \end{aligned} \tag{4.4}$$

By choosing $g_0'=B(m)^{-1}g_0$, we obtain

$$\begin{aligned} P_{g_0}(0,x) \geqq {} & \frac{g_0(1-g_0/2)}{B(m)}\sum_z G_m(0,z)\,G_m(x,z) \\ & -\frac{cg_0}{B(m)^2}\sum_{z_1,z_2} G_m(0,z_1)\;G(z_1,z_2)^2\,G_m(z_2,x), \end{aligned} \tag{4.5}$$

with $B(m)$ as in Lemma 2.2.

For g_0 small enough, this lower bound has the desired feature. But, since $P_{g_0}(0,x)$ is monotone increasing in g_0 we have an adequate lower bound for all values of g_0. Thus, Theorem 2.1 holds in this more general situation, too. It may be checked without difficulty that the results in Sect. 3 extend to the full range of values of g_0, as well.

Let $P_\phi(m)$ be the probability that two simple random walks starting at the origin never intersect after the first step. Lawler [1] has studied this quantity which provides a useful notion for the study of triviality of $g\phi_4^4$ theory. He has proven that

$$P_\phi(m)\geqq\begin{cases} c_\varepsilon m^\varepsilon, & d=4-\varepsilon \\ c|\ln \mathrm{m}|^{-1}, & d=4 \end{cases}$$

and

$$P_\phi(m)\leqq\begin{cases} c_\varepsilon' m^{\varepsilon/2}, & d=4-\varepsilon \\ c'|\ln m|^{-1/2}, & d=4. \end{cases}$$

These results can be recovered, in our formalism, from the following straightforward estimates.

$$P_\phi(m)=1-\chi^{-2}P(0,0)\geqq 1-B(m)\,P_\phi(m),$$

providing the lower bound, and

$$P_\phi(m)^2 \leqq \text{const}\, m^{4-d} g(m),$$

which, with the estimates on $g(m)$, yields the upper bound.

In the context of the Edwards model closely related bounds were proven in [14]. It would be interesting to calculate the exact behaviour of $P_\phi(m)$, as $m \searrow 0$. A heuristic argument (based on an approximate flow equation for $P_\phi(m)$) suggests that, in $d = 4 - \varepsilon$, $P_\phi(m) \sim m^\alpha$, with $\frac{\varepsilon}{2} < \alpha < \varepsilon$. We thank A. Sokal and T. Spencer for very instructive discussions on the results reported in this paragraph.

(b) The results of Sects. 1–3 can be generalized to include estimates of the probability that n simple random walks $\omega_1, \ldots, \omega_n$ intersect in a common point of $\mathbb{Z}^d$, $n = 3, 4, \ldots$. This problem, too, may be posed as a problem in Euclidean field theory by considering a lattice theory with action

$$\begin{aligned}\mathscr{A}(\boldsymbol{\phi}_1, \ldots, \boldsymbol{\phi}_n) \\ = \sum_{x \in \mathbb{Z}^d} \left\{ \sum_j \frac{1}{2} (|(\nabla \boldsymbol{\phi}_j)_x|^2 + m^2 |\boldsymbol{\phi}_{j,x}|^2) + 2^{-n} g_0 |\boldsymbol{\phi}_{1,x}|^2 \, |\boldsymbol{\phi}_{2,x}|^2 \ldots |\boldsymbol{\phi}_{n,x}|^2 \right\}.\end{aligned} \tag{4.6}$$

By the methods of [6, 10], the limits

$$G^c_{m,g_0}(x_1, x_1'; \ldots; x_n, x_n') = \lim_{g_0 \to \infty} \lim_{N \to 0} \langle \phi^1_{1,x_1}\, \phi^1_{1,x_1'}; \ldots; \phi^1_{n,x_n}\, \phi^1_{n,x_n'} \rangle \tag{4.7}$$

can be shown to exist, and the average intersection probability $g_n(m)$ for n walks is given by

$$g_n(m) = - \sum_{x_2, \ldots, x_n} (m^d)^{n-1} \sum_{x_1', \ldots, x_n'} \chi^{-n} G^c_{m,g_0}(x_1, x_1'; \ldots; x_n, x_n'). \tag{4.8}$$

From these formulas it follows that the upper critical dimension is $d_u = 2n/(n-1)$, in particular, for $n = 3$, $d_u = 3$. For $d = d_u$, we have

$$g_n(m) \sim |\log m|^{-1}, \quad \text{as} \quad m \searrow 0, \tag{4.9}$$

as was first proven by Lawler [1]. Moreover, for a fixed value of n, one can estimate the behaviour in dimension $d_u - \varepsilon$, with ε small. Our results suggest that there may be no *non-trivial* $g_0 \phi^6$-theories in three dimensions, and that the violations of mean-field theory in tricritical point theory in three dimensions are logarithmic.

(c) Techniques related to the ones we have developed in this paper do also extend to theories like the $-g_0 \operatorname{tr}(\phi^3)$ model, where ϕ is a hermitian $N \times N$ matrix-field, in the limit $N \to \infty$. For this model, one can establish, for example, asymptotic freedom ($\beta < 0$), for $d \leqq 6$. However, the consequences of our results are not terribly spectacular. A more interesting application concerns intersection probabilities of branched polymers without self-interactions: We can prove that two branched polymers without self-interactions do not meet each other, in the limit where the correlation length diverges, with probability 1 in $d \geqq 8$, and we have fairly sharp estimates for the intersection probability in dimension $d = 8 - \varepsilon$. Precise statements and proofs will appear elsewhere.

Conclusions

We believe that the results and techniques reported in this paper have, at the very least, some pedagogical value: They show that renormalization group flow equations can, sometimes, make rigorous mathematical sense and yield precise results. However, the idea to use rigorous renormalization flow inequalities to analyze the critical behaviour of lattice models may turn out to actually have a much wider range of applications than the ones considered in this paper. In fact, this is suggested by results (c), above.

We have been informed that Michael Aizenman has also obtained results on intersection probabilities of random walks, using a β-function approach apparently related to ours.

Acknowledgements. We are indebted to D. Brydges, K. Gawedzki, T. Spencer, and K. Symanzik for very helpful discussions about different topics related to this paper. We are especially indebted to A. Sokal for sharing many insights with us and for permission to describe an unpublished proof of his.

This paper is dedicated to the memory of K. Symanzik, on whose deep ideas it is based.

References

1. Lawler, G.F.: The probability of intersection of independent random walks in four dimensions. Commun. Math. Phys. **86,** 539 (1982); and Preprint, Duke University 1983
2. Dvoretzky, A., Erdös, P., Kakutani, S.: Acta Sci. Math. Szeged. **12,** 75 (1950)
3. Erdös, P., Taylor, S.J.: Acta Math. Acad. Sci. Hung. **11,** 137 and 231 (1960)
4. Glimm, J., Jaffe, A.: Quantum physics. Berlin, Heidelberg, New York: Springer 1981
5. Aizenman, M.: Geometric analysis of ϕ^4 fields and Ising models. Parts I and II. Commun. Math. Phys. **86,** 1 (1982)
   Fröhlich, J.: On the triviality of $\lambda\varphi_d^4$ theories and the approach to the critical point in $d \geqq 4$ dimensions. Nucl. Phys. B**200** [FS4], 281 (1982)
6. Symanzik, K.: Euclidean quantum field theory. In: Local quantum theory. Jost, R. (ed.). New York, London: Academic Press 1969
7. Fröhlich, J.: Quantum field theory in terms of random walks and random surfaces, Cargèse lecture notes 1983 (to be published). Aizenman, M.: Rigorous results on the critical behaviour in statistical mechanics. In: Scaling and Self-Similarity in Physics. Fröhlich, J. (ed.). Boston, Basel, Stuttgart: Birkhäuser 1983
   Brydges, D.: Field theories and Symanzik's polymer representation. In: Gauge theories: fundamental interactions and rigorous results. Dita, P. et al. (eds.). Boston, Basel, Stuttgart: Birkhäuser 1983
8. Amit, D.: Field theory, renormalization group and critical phenomena. New York: McGraw-Hill 1978
9. Brézin, E., Le Guillou, J., Zinn-Justin. J.: Discussion of critical phenomena for general n-vector models. Phys. Rev. B**10,** 892 (1974)
10. Brydges, D., Fröhlich, J., Spencer, T.: The random walk representation of classical spin systems and correlation inequalities. Commun. Math. Phys. **83,** 123 (1982)
11. Brydges, D., Fröhlich, J., Sokal, A.: Unpublished
12. Sokal. A.: Private communication
13. Brydges, D., Fröhlich, J., Sokal, A.: The random walk representation of classical spin systems and correlation inequalities. II. The skeleton inequalities. Commun. Math. Phys. **91,** 117 (1983)
14. Bovier, A., Felder, G., Fröhlich, J.: On the critical properties of the Edwards and the self-avoiding walk model of polymer chains. Nucl. Phys. B**230** [FS 10], 119 (1984)

Communicated by A. Jaffe

Received March 19, 1984; in revised form April 30, 1984

Self-Avoiding Walk in 5 or More Dimensions*

David Brydges[1] and Thomas Spencer[2]

1 Department of Mathematics, University of Virginia, Charlottesville, VA 22903, USA
2 Courant Institute of Mathematical Sciences, 251 Mercer Street, New York, NY 10012, USA

Abstract. Using an expansion based on the renormalization group philosophy we prove that for a T step weakly self-avoiding random walk in five or more dimensions the variance of the endpoint is of order T and the scaling limit is gaussian, as $T \to \infty$.

1. Introduction and Results

We consider walks $\omega(s)$ in $\mathbb{Z}^d$ which start at the origin and consist of $|\omega| = T$ nearest neighbor steps. If each such walk ω is assigned a weight proportional to

$$P_T(\omega) \equiv \prod_{0<s<t\leqq T} (1 - \lambda\delta(\omega(s) - \omega(t))), \qquad 0 < \lambda \leqq 1,$$

we say that the walk is weakly self-avoiding or self-repelling. Here s, t denote non-negative integers and $\delta(j) = 1$ if $j = 0$ and $\delta(j) = 0$ otherwise. When $\lambda = 1$ only walks which strictly self-avoid are counted. Now let us define an expectation of a functional F on paths ω, $|\omega| = T$ by

$$\langle F(\cdot)\rangle_T(\lambda) \equiv \frac{\sum_{|\omega|=T} F(\omega)\, P_T(\omega)}{\sum_{|\omega|=T} P_T(\omega)}.$$

A natural quantity to study is the mean square displacement of $\omega(T)$ defined by

$$R^2(T) \equiv \langle \omega^2(T)\rangle_T(\lambda).$$

In the physics literature $R^2(T)$ is expressed in terms of a critical exponent ν via the relation $R^2(T) \cong C(\lambda)\, T^{2\nu}$ for large T. On the basis of renormalization group

---

Dedicated to the memory of Kurt Symanzik whose profound contributions have guided and inspired us

* Work partially supported by N.S.F. Grant DMR 81-00417

** A. P. Sloan Foundation Fellow. Work partially supported by N.S.F. Grant MCS 82-02115

considerations ν is supposed to be independent of λ for $0<\lambda\leqq 1$ and $\nu\cong 0.75$, 0.58 in dimensions 2 and 3 respectively [1]. In four or more dimensions ν is expected to be $\frac{1}{2}$ with logarithmic corrections in four dimensions [1].

On a mathematical level almost nothing has been proved about ν, except when $d=1$ where $\nu=1$. For example it would be very interesting to obtain a lower bound on ν such as $\nu\geqq\frac{1}{2}$ or an upper bound $\nu\leqq 1-\varepsilon$, $\varepsilon>0$.

In this paper we examine the case $d\geqq 5$ for small positive values of λ. We prove the following theorem.

Theorem. *For $d\geqq 5$, there is a $\lambda_0>0$ such that for each λ, $0\leqq\lambda\leqq\lambda_0$, there is a positive constant D and*

$$\langle\omega^2(T)\rangle_T(\lambda)=DT(1+O(T^{-1/16}))\,.$$

Furthermore the scaling limit of the endpoint is gaussian, i.e.

$$\lim_{s\to\infty}\langle e^{ik\omega(st)/\sqrt{s}}\rangle_{st}(\lambda)=e^{-Dtk^2/2}$$

uniformly on compact sets of k and t.

Dobrushin has informally announced results similar to ours but no proof has yet appeared.

A careful analysis of our proof shows that the coefficient of $\delta(\omega(s)-\omega(t))$ may be set equal to 1 for $|s-t|\geqq M$ large. The smallness in λ is only needed for $|s-t|\leqq M$.

The role of dimension can heuristically be seen from the fact that two independent random walks intersect only finitely often with probability one in dimension $d\geqq 5$.

Our result exemplifies the validity of mean field theory in high dimensions. Another such example are the remarkable results of Aizenman [2] and Fröhlich [3] on the gaussian nature of $\lambda\phi^4$ field models in five or more dimensions. [Formally, self avoiding walk may be regarded as the $N\to 0$ limit of $\lambda\phi^4$ theories, where N denotes the number of components of the field ϕ.] Another related result is the work of Lawler [4] who proved that the scaling limit of a loop reased random walk is Brownian motion when $d\geqq 5$.

The proof of Theorem 1 is based on an expansion in λ which uses the renormalization group philosophy. An outline of the proof appears in Sect. 2. We believe that our techniques will be useful for analysing a number of other problems such as random walk in a random environment. We refer the reader to [5] for other mathematical results which incorporate the renormalization group philosophy. We also recommend the viewpoint and results in [6], e.g. Sects. 1–4.

2.1. Outline of Proof

Let us consider walks which remember to self repel only for a time span τ, thus we define $V_{st}=\delta(\omega(s)-\omega(t))$, and

$$C_\tau(x,T)\equiv(2d)^{-T}\sum_{\substack{\omega,|\omega|=T\\ \omega(T)=x}}\prod_{\substack{0\leqq s<t\leqq T\\ |s-t|\leqq\tau}}(1-\lambda V_{st})\,. \tag{2.1}$$

We shall use the same letter C with different arguments to denote its Fourier transform,

$$C_\tau(k, T) \equiv \sum_x C_\tau(x, T) e^{ik \cdot x}, \tag{2.2}$$

and its Fourier-Laplace transform,

$$C_\tau(k, z) \equiv \sum_{T=0}^{\infty} C_\tau(k, T) z^T. \tag{2.3}$$

Here k lies in the torus dual to $\mathbb{Z}^d$. We are interested in the probability distribution of $\omega(T)$, the position after T steps, and in particular its variance given by

$$\langle \omega^2(T) \rangle_\tau \equiv \sum_x C_\tau(x, T) x^2 \Big/ \sum_x C_\tau(x, T).$$

The theory of generalized Markov processes shows that for some non-zero constant D_τ

$$\langle \omega^2(T) \rangle_\tau = D_\tau(T + o(T)) \quad \text{as} \quad T \to \infty.$$

[A proof of this, for λ small, is also contained in this paper.] D_τ is called the diffusion coefficient and our major task is to obtain bounds on D_τ which are uniform in τ as $\tau \to \infty$ when the dimension $d \geqq 5$. We do it by setting $\tau \equiv T_l \equiv 2^{\ell^\alpha} - 1$ and devising an inductive scheme of the form "true for $\ell \leqq m-1$"$\Rightarrow$"true for $\ell \leqq m$", $m = 1, 2, \ldots, \ell \in \mathbb{R}^+$. We will write C_ℓ for C_{T_ℓ} etc.

If $\lambda = 0$ or $\tau = 0$, then the product in (2.1) is one, and using the independence of nearest neighbor steps it is easy to show that

$$C_0(k, z) = (1 - z\, D(k))^{-1},$$

where

$$D(k) = \frac{\sum_{i=1}^{d} \cos k_i}{d}. \tag{2.4}$$

Let $r_\ell(k)$ denote the radius of convergence of (2.3). Note that $r_\ell(0)$ is increasing in ℓ because $(1 - \lambda V) \leqq 1$ and $r_\ell(k) \geqq r_\ell(0) \geqq 1$. For $\ell = 0$, $r_0(k) = |D(k)^{-1}|$.

Our strategy is first to show that $C_\ell(k, z)$ is close to $C_0(k, z)$ in dimension $d \geqq 5$. To make this idea precise let $\Pi_\ell(k, z)$ and $F_\ell(k, z)$ be implicitly defined by

$$C_\ell(k, z) = (1 - zD(k) - \Pi_\ell(k, z))^{-1} = F_\ell(k, z)^{-1}. \tag{2.5}$$

We shall prove that Π_ℓ is analytic and for $|u| \leqq 2$,

$$|\partial_k^u \Pi_\ell(k, z)|, \left| \frac{\partial}{\partial z} \Pi_\ell(k, z) \right| \leqq K\lambda, \tag{2.6}$$

for

$$z \in \mathcal{D}_\ell \equiv \{z : |z| \leqq r_\ell(0) + \tfrac{1}{16} T_\ell^{-1} \ln T_\ell\}. \tag{2.7}$$

Here K is a constant independent of ℓ.

We now explain how these estimates imply that the diffusion constant D_ℓ obtained using

$$\left.\frac{-\Delta_k C_\ell(k,T)}{C_\ell(k,T)}\right|_{k=0} = D_\ell T(1+o(T))$$

is uniformly founded. By (2.6) we see that $C_\ell(k,z)$ is bounded and meromorphic in $\mathscr{D}_\ell$ and has a unique real pole $r_\ell(k)$ for small values of k, which varies differentiably in k. See Sect. 6. By the Cauchy integral formula and (2.3),

$$\begin{aligned} r_\ell(0)^{T+1} C_\ell(k,T) &= \frac{1}{2\pi i}\oint C_\ell(k,z)\left(\frac{r_\ell(0)}{z}\right)^{T+1} dz \\ &= -\left(\frac{\partial F_\ell}{\partial z}\right)^{-1}\left(\frac{r_\ell(0)}{r_\ell(k)}\right)^{T+1} + \frac{1}{2\pi i}\oint_{\partial\mathscr{D}_\ell} C_\ell(k,z)\left(\frac{r_\ell(0)}{z}\right)^{T+1} dz \\ &\simeq \left(\frac{r_\ell(0)}{r_\ell(k)}\right)^{(T+1)} \simeq e^{-k^2 T D_\ell/2}\,. \end{aligned} \tag{2.8}$$

Here we have deformed a small circle about the origin to $\partial\mathscr{D}_\ell$ to pick up the contribution from the pole at $r_\ell(k)$ plus an error term which is shown to be small for large $T \geqq T_\ell$ in Sect. 7. Roughly speaking the error term is small because

$$\left|\frac{r_\ell(0)}{z}\right|^T = \left[r_\ell(0)\Big/\left(r_\ell(0)+\frac{1}{16}\frac{\ln T_\ell}{T_\ell}\right)\right]^T \approx e^{-\frac{1}{16}\frac{T}{T_\ell}\ln T_\ell}\,.$$

Now the diffusion constant is easily calculated in terms of the second derivative of $r_\ell(k)$ at $k=0$ (by symmetry $r'_\ell(0)=0$). Uniform bounds on $r''_\ell(0)$ are obtained using the implicit function theorem and (2.6).

In order to obtain bounds of the form (2.6) on $\Pi_\ell(k,z)$ we first develop a formula for Π_ℓ in terms of a convergent series. A detailed derivation of this formula in presented in Sect. 4. Section 5 is the core of this paper. It is there where our induction hypothesis is formulated and a bound on Π_ℓ and $\delta\Pi_\ell = \Pi_\ell - \Pi_{\ell-1}$ is obtained in terms of L_p norms of $C_{\ell-1}(x,T)$.

In the remainder of this section we shall sketch a formula for Π_ℓ and some of the ideas of Sect. 5. Let

$$U_{st} = -\lambda V_{st};\ \tau = T_\ell\,.$$

We begin by expanding the "interaction"

$$\prod_{|s-t|\leqq\tau}(1+U_{st}) = \sum_{\Gamma}\prod_{st\in\Gamma} U_{st}\,, \tag{2.9}$$

where Γ ranges over all collections of "lines" or pairs s,t such that

$$0 \leqq s < t \leqq T \quad \text{and} \quad |t-s| \leqq \tau\,.$$

Any such collection of lines is called a graph. A time a, $0<a\leqq T$, is said to be a *Markov time* for Γ if there is no line $st\in\Gamma$ such that $s<a<t$. Thus there is no interaction (or line) across a. A graph Γ on $[0,T]$ is said to be *primitive* if it has no Markov points in $(0,T)$ and 0 belongs to some line of Γ. Every graph decomposes uniquely into primitive subgraphs over the appropriate intervals and it is not

difficult to prove that this leads to the formula

$$\Pi_\ell(k,z)=\sum_T z^T \Pi_\ell(k,T), \tag{2.10}$$

where

$$\Pi_\ell(k,T)=\sum_{|\omega|=T}\ \sum_{\substack{\Gamma \text{ primitive} \\ \text{on } [0,T]}}\ \prod_{st\in\Gamma} U_{st}e^{ik\omega(T)} \tag{2.11}$$

(see Sect. 4). There are too many graphs in (2.10) in the sense that if absolute values are taken inside (2.11), (2.10) will diverge. For this reason we resum our family of primitive graphs. To each primitive graph Γ we associate a (unique) minimal subgraph $S(\Gamma)\subset\Gamma$ which is a primitive subgraph of Γ with as few lines as possible (see Fig. 1 below). We call such a graph $S(\Gamma)$ a *lace*. The precise definition of $S(\Gamma)$ is given in Sect. 4.

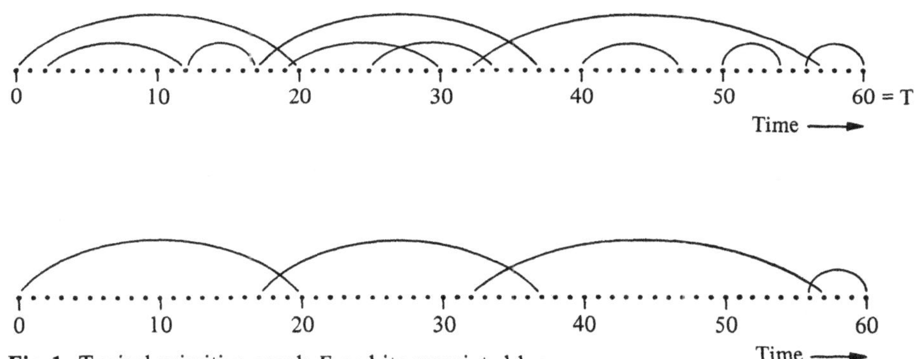

Fig. 1. Typical primitive graph Γ and its associated lace

We now sum over all primitive graphs Γ with $S(\Gamma)=L$ held fixed and then sum over L. The sum over Γ with $S(\Gamma)=L$ reconstructs part of the interaction, cf. (2.9), indeed

$$\sum_{\Gamma:S(\Gamma)=L}\prod U_{st}=\prod_{st\in\mathscr{C}_\tau(L)}(1+U_{st})\prod_{st\in L}U_{st}\,.$$

Here $\mathscr{C}_\tau(L)$ is defined in Sect. 4 to be the set of lines compatible with L. Our desired expression for Π_ℓ is

$$\Pi_\ell(k,T)=\sum_L\ \sum_{\substack{\omega \\ |\omega|=T}}\ \prod_{st\in\mathscr{C}_\tau(L)}(1+U_{st})\prod_{st\in L}U_{st}e^{ik\cdot\omega(T)}, \tag{2.12}$$

where $\sum_L$ is the sum over all lacings on $[0,T]$.

Let Π^N denote the contribution to the sum in (2.12) arising from all N lacings, i.e. lacings with precisely N lines. Then we have

$$\left|\frac{\partial}{\partial z}\Pi_\ell(k,z)\right|\leqq\sum_N\left|\frac{\partial}{\partial z}\Pi_\ell^N(k,z)\right|.$$

In Sect. 5 we prove that the right side is convergent for $z \in \mathscr{D}_\ell$ uniformly in ℓ and each term is bounded in terms of the L_p norms of $r_{\ell-1}(0)^T C_{\ell-1}(x, T)$, $p \geqq 2$, which by inductive assumption is less than $\|K \exp(-D_{\ell-1}k^2 T)\|$, so that e.g. in x space,

$$r_{\ell-1}(0)^T \|C_{\ell-1}(\cdot, T)\|_{L^\infty} \leqq \text{const}\, T^{-d/2}.$$

We illustrate this method by bounding

$$\left|\frac{\partial}{\partial z} \Pi_\ell^1\right|.$$

When $N=1$, $L=\{(0,T)\}$, and $\mathscr{C}(L) \equiv \mathscr{C}_T =$ set of all pairs st, $0 \leqq s < t \leqq T$ such that $|s-t| \leqq T_\ell$ and $|s-t| < T$. Thus

$$|\Pi_\ell^1(k,T)| = \lambda \left| \sum_{|\omega|=T} \prod_{st \in \mathscr{C}_T} (1-\lambda V_{st})\, V_{0T} e^{i\omega(T)\cdot k} \right|.$$

The factor V_{0T} above ensures that $\omega(T)=0$ so that

$$|\Pi_\ell^1(k,T)| = \lambda \sum_{\substack{|\omega|=T \\ \omega(T)=0}} \prod (1-\lambda V_{st}) \leqq \frac{\lambda}{1-\lambda} C_{\ell-1}(x=0,T).$$

In the last inequality we have used (2.1) and the fact that $0 \leqq (1-\lambda V) \leqq 1$. By inductive assumption [see (2.8) and Sect. 5 for details]

$$\begin{aligned} r_{\ell-1}(0)^T C_{\ell-1}(x=0,T) &= r_{\ell-1}(0)^T \int C_{\ell-1}(k,T)\, dk/(2\pi)^d \\ &< K T^{-d/2}. \end{aligned}$$

When $T > T_\ell$, $\Pi^1 = 0$, thus, by (2.10),

$$\begin{aligned} \left|\frac{\partial \Pi^1}{\partial z}\right| &\leqq \lambda \sum_{1 \leqq T \leqq T_\ell} T|z|^{(T-1)} \Pi^1(k,T) \\ &\leqq \lambda K \sum_{1 \leqq T \leqq T_\ell} T^{(1-d/2)} \left|\frac{z}{r_{\ell-1}(0)}\right|^T. \end{aligned}$$

Now for $d \geqq 5$ the right side converges uniformly in T_ℓ and $z \in \mathscr{D}_\ell$, so we have obtained part of (2.6) inductively[1]. In Sect. 7 it is proved, along the lines of our discussion of (2.8), that the bound on Π, (2.6), allows us to advance the inductive assumptions on C.

3. Notation and Conventions

In this paper we use K to denote a constant independent of all parameters, except the dimension d. In particular K is independent of the induction step, labelled by $m = 1, 2, \ldots$. For constants which do not enjoy these properties we will indicate dependences as arguments, e.g. $K(\delta)$. E, as in error, is used to denote a function of k, z, λ which is bounded in absolute value by some K. K and E are not necessarily

---

1 We are omitting an argument that $r_{\ell-1}(0) \simeq r_\ell(0)$, when ℓ is large or λ small. In addition we have simplified the estimate of the sum over T which in fact involves disection into subranges according to values of $T_{\ell'}$, $\ell' \leqq \ell$

the same in different locations including strings of inequalities for the same quantity. The rest of this section is a compendium of our notation.

For each $\ell \in \mathbb{R}$ we define a memory T_ℓ by

$$T_\ell \equiv \max(2^{\ell^\alpha}-1,0) \quad \text{with} \quad \alpha \equiv 9/8 . \tag{3.1}$$

We also define $\ell(s)$ by $T_{\ell(s)}=s$. To each ℓ is associated a random walk, self-avoiding (with strength λ) with memory T_ℓ,

$$C_\ell(x,T) \equiv \sum_{\omega:0\to x,|\omega|=T} (2d)^{-T} \prod_{\substack{st,|t-s|<T_\ell \\ 0\leqq s<t\leqq T_\ell}} (1-\lambda V_{st}), \tag{3.2}$$

$$V_{st} \equiv \delta(\omega(s)-\omega(t)) \equiv \begin{cases} 1 & \text{if} \quad \omega(s)=\omega(t) \\ 0 & \text{else} \end{cases}, \tag{3.3}$$

$0 \leqq \lambda \leqq 1$, $T=1,2,\ldots$.

$$C_\ell(k,T) \equiv \sum_{\omega:|\omega|=T} (2d)^{-T} \prod_{\substack{st,|t-s|\leqq T_\ell \\ 0\leqq s<t\leqq T}} (1-\lambda V_{st}) e^{ik\omega(T)} . \tag{3.4}$$

This is the fourier transform of $C_\ell(x,T)$,

$$\langle e^{ik\omega(T)} \rangle_{\ell,T} \equiv C_\ell(k,T)/C_\ell(0,T) ,$$

$$C_\ell(k,z) \equiv \sum_{T=0}^{\infty} z^T C_\ell(k,T) .$$

By convention $C_\ell(k,T=0)\equiv 1$. The arguments are to be used to distinguish the different transforms being denoted by C's,

$$F_\ell(k,z) \equiv [C_\ell(k,z)]^{-1} .$$

r_ℓ is the radius of convergence of $C_\ell(k=0,z)$. It will be seen that $F_\ell(k=0,z)$ has a simple zero at $z=r_\ell$ and the trajectory in $\mathbb{C}$ of this zero as k varies will be denoted by $r_\ell(k)$,

$$D(k) \equiv \frac{1}{d} \sum_{i=1}^{d} \cos k_i .$$

$k_i, \ldots, k_d$ are the components at k. $\Pi_\ell(k,z)$ is defined implicitly by

$$C_\ell(k,z) \equiv [1-zD(k)-\Pi_\ell(k,z)]^{-1} ,$$

$$\delta\Pi_\ell(k,z) \equiv \Pi_\ell(k,z)-\Pi_{\ell-1}(k,z) ,$$

$$\varepsilon_\ell \equiv \frac{1}{16} \ln[2+T_\ell]/[2+T_\ell] ,$$

$$d(u,v) \equiv \frac{d}{2} - \alpha\left(\frac{|u|}{2}+v\right) - \frac{9}{8} .$$

u is a multi-index which we use for k derivatives denoted by ∂_k^u. Likewise ∂_z^v denotes z derivatives,

$$\mathscr{D}_\ell(a) \equiv \{z \in \mathbb{C} : |z| \leqq r_\ell + a\varepsilon_\ell\} ; \; a>0 .$$

Throughout this paper $d \geqq 5$ and λ is assumed non-negative and small.

4. A Formula for Π

The interaction of a path with itself over a time interval $[0, T]$ is given by

$$P[0,T] \equiv \prod_{0 \leqq s < t,\, t-s \leqq \tau} (1+U_{st}),$$

where

$$U_{st} \equiv -\lambda V_{st} \equiv -\lambda\delta(\omega(s)-\omega(t)).$$

$\tau(=T_\ell)$ is the memory over which the random walk ω self-avoids. Following the outline given in Sect. 2 we will develop an expansion for P and deduce from it a formula for Π which appears in Theorem 4.1 below.

We expand P as a sum of all graphs as explained in Sect. 2. For each graph Γ we let $a(\Gamma)$, $0 < a(\Gamma) \leqq T$, denote the first Markovian time of Γ, as defined in Sect. 2. Then

$$P[0,T] = P[1,T] + \sum_{a=1}^{T} K[0,a]\, P_0[a,T],$$

where

$$K[0,a] = \sum_{\substack{\Gamma \text{ primitive} \\ \text{on } [0,a]}}{}' \prod_{st\in\Gamma} U_{st}. \tag{4.1}$$

The $P[1, T]$ is there because we insist $a(\Gamma) > 0$. It is the sum of all graphs with no bond joining 0 to some other time. The second equality follows by noticing that we can split Γ into two disjoint subsets corresponding to bonds on $[0, a]$, $[a, T]$, and sum over them independently.

The last equality implies

$$\begin{aligned} C_\tau(k,z) &\equiv \sum_{T=0}^{\infty} \left(\frac{z}{2d}\right)^T \sum_{\omega, |\omega|=T} P[0,T]\, e^{ik\omega(T)} \\ &= 1 + zD(k)\, C_\tau(k,z) + \sum_{T=1}^{\infty} \left(\frac{z}{2d}\right)^T \sum_{a=1}^{T} \sum_{\omega,|\omega|=T} \\ &\quad \cdot K[0,a]\, e^{ik\omega(a)}\, P[a,T]\, e^{ik(\omega(T)-\omega(a))}. \end{aligned}$$

(By convention $P[0, 0] = 1$.) We can factor the sum over ω across the time a and continue with

$$= 1 + zD(k)\, C_\tau(k,z) + \left\{ \sum_{a=1}^{\infty} \left(\frac{z}{2d}\right)^a \sum_{\omega,|\omega|=a} K[0,a]\, e^{ik\omega(a)} \right\} C_\tau(k,z).$$

On recalling that Π_τ is defined implicitly by

$$C_\tau(k,z) \equiv [1 - zD(k) - \Pi_\tau(k,z)]^{-1},$$

it follows that

$$\Pi_\tau(k,z) = \sum_{a=1}^{\infty} \left(\frac{z}{2d}\right)^a \sum_{\omega,|\omega|=a} K[0,a]\, e^{ik\omega(a)}. \tag{4.2}$$

The next step is to find a better expression for Π, which unlike the one above will be absolutely convergent, i.e. absolute values can be taken inside all sums including the sum over graphs inside $K[0,1]$ provided λ is small. Given a graph Γ on $[0,a]$ we define a subgraph $S(\Gamma)\equiv L$ consisting of bonds $s_i t_i$, $i=1,2,\ldots,N_\Gamma$ obtained inductively by

$$t_1\equiv\max\{t:0t\in\Gamma\},$$

$$t_j\equiv\max\{t:st\in\Gamma, s<t_{j-1}\};\ j\geqq 2,$$

$$N_\Gamma\equiv\text{first value of } j \text{ for which } t_j=a.$$

Having obtained t_j, s_j is defined by

$$s_j\equiv\min\{s:st_j\in\Gamma\}.$$

The set of all L that are obtained by this process applied to Γ as Γ runs over all graphs which occur in $K[0,a]$, (primitive graphs), is denoted $\mathscr{L}_\tau[0,a]$. If $L\in\mathscr{L}_\tau[0,a]$ we call it a *lace*. If it has N bonds it is called an *N-lace*. It is not difficult to see that $L=\{s_i t_i: i=1,2,\ldots,N\}\in\mathscr{L}_\tau[0,a]$ iff

$$\begin{aligned}&\text{a) } s_1=0,\ t_N=a,\ |t_j-s_j|\leqq\tau \quad\text{for}\quad j=1,2,\ldots,N,\\ &\text{b) for } j=2,3,\ldots,N-1\ s_j t_j \text{ overlaps with}\\ &\qquad s_{j-1}t_{j-1} \text{ and } s_{j+1}t_{j+1}. \text{ No other overlaps occur.}\end{aligned}\tag{4.3}$$

We say that st overlaps with $s't'$ iff the open intervals (s,t), (s',t') intersect.

The discussion so far shows that

$$K[0,a]=\sum_{L\in\mathscr{L}_\tau[0,a]}\Bigg(\sum_{\Gamma,S(\Gamma)=L}\ \prod_{st\in\Gamma/L}U_{st}\Bigg)\prod_{st\in L}U_{st}.\tag{4.4}$$

In order to simplify the sum in round brackets we observe that Γ satisfies $S(\Gamma)=L=\{s_i t_i: i=1,2,\ldots\}$ iff Γ/L is a subset of $\mathscr{C}_\tau(L)$ which, by definition, is all bonds st satisfying

$$\begin{aligned}&\text{a) } 0\leqq s<t\leqq a,\ |t-s|\leqq\tau,\ st\notin L.\\ &\text{b) If } s=0,\ t<t_1. \text{ If } s<t_i \text{ for some } i, \text{ then } t\leqq t_{i+1}.\\ &\text{c) If } t=t_i, \text{ then } s>s_i.\end{aligned}\tag{4.5}$$

This assertion is an easy consequence of the definition of L. It follows, cf. (2.9), that

$$\sum_{\Gamma,S(\Gamma)=L}\ \prod_{st\in\Gamma/L}U_{st}=\prod_{st\in\mathscr{C}_\tau(L)}(1+U_{st}).\tag{4.6}$$

When we collect up (4.1), (4.2), (4.4), (4.6) we obtain

Theorem 4.1.

$$\Pi_\tau(k,z)=\sum_{T\geqq 1}\ \sum_{\omega:|\omega|=T}\left(\frac{z}{2d}\right)^T e^{ik\omega(T)}\sum_{L\in\mathscr{L}_\tau[0,T]}\ \prod_{st\in L}(-\lambda V_{st})\prod_{st\in\mathscr{C}_\tau(L)}(1-\lambda V_{st}),\tag{4.7}$$

where $V_{st}=\delta(\omega(s)-\omega(t))$, $\mathscr{L}_\tau[0,T]$ *is the set of laces defined in* (4.3) *(with* $a=T$*) and* $\mathscr{C}_\tau(L)$ *is the set of bonds compatible with* L *defined in* (4.5).

We will use this theorem in Sect. 5 to get good bounds on Π. The following features of $\mathscr{C}_\tau(L)$ will be needed: if L is an N-lace, $N>1$, on $[0,T]$, then its bonds $s_i t_i$ define a partition of $[0,T]$ into $2N-1$ subintervals $I_1,\dots,I_{2N-1}$ with disjoint interiors, obtained by putting the times s_i, $i=1,2,\dots,N$, t_j, $j=1,2,\dots,N$ in order and letting I_k, $k=1,2,\dots,2N-1$, be the intervals between (see Fig. 2). The properties (4.5) imply that

$$st\in\mathscr{C}_\tau(L) \quad \text{if} \quad |t-s|\leqq\tau \quad \text{and} \quad s,t\in I_j \quad \text{for some } j. \tag{4.8}$$

This means that the random walk ω restricted to any I_j is still self-avoiding.

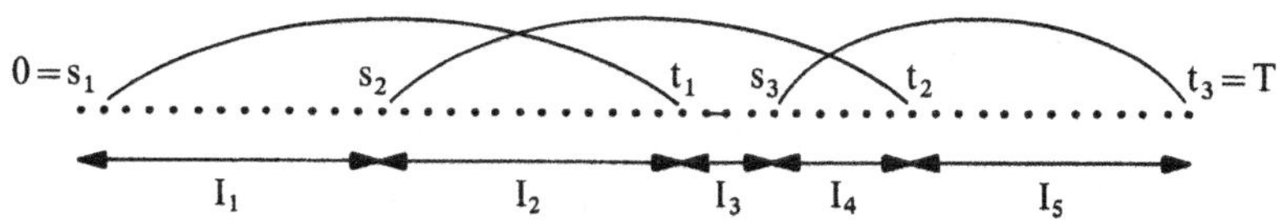

Fig. 2

The other feature of $\mathscr{C}_\tau(L)$ which we will use in Sect. 5 is that if all the bonds $st\in L$ satisfy $|t-s|\leqq\tau/2$, then

$$\mathscr{C}_{\tau'}(L)=\mathscr{C}_\tau(L) \quad \text{for} \quad \tau'\geqq\tau, \tag{4.9}$$

because of condition (b) in (4.5).

5. Inductive Assumptions on C, Consequences for $\delta\Pi$

The induction is on $m=1,2,\dots$. The inductive assumptions are: there exist $K^{(\mathrm{I1})}$, $K^{(\mathrm{I2})}$ fixed throughout our induction such that

(I1) $0\leqq r_\ell-r_{\ell-1}\leqq K^{(\mathrm{I1})}\sqrt{\lambda}(2+T_{\ell-1})^{9/8-d/2}$ for $\ell\leqq m-1$,

(I2) $\|\partial_k^u C_\ell(\cdot,T)\|_p\leqq K^{(\mathrm{I2})} r_\ell^{-T} T^{|u|/2-d/(2p)}$

for $\ell\leqq m-1$, $T\geqq T_{\ell+1}$, $T\geqq 1$, $|u|\leqq 2$.

(I2) says that for large T $r_\ell^T\partial_k^u C(k,T)\cong\partial_k^u e^{-Kk^2T}$ in L_p norm. At the beginning of the induction $m=1$ so $\ell\leqq 0$. The definitions of r_ℓ, C_ℓ for negative ℓ imply that $r_\ell=r_0$, $C_\ell=C_0$ so (I1) is true and (I2) holds by direct calculation on $C_0(k,T)=D(k)^T$, provided we take $K^{(\mathrm{I2})}$ sufficiently large.

In this section we will first prove a bound on $\delta\Pi_\ell=\Pi_\ell-\Pi_{\ell-1}$ in terms of x *space* L_p norms of $C_{\ell-1}$. This is Proposition 5.2 given below. x-space L_p norms are convenient since we can use x-space bounds such as $1-\lambda V\leqq 1$ on them. They are converted to k-space L_p bounds by the Hausdorff-Young inequality, and then the inductive assumptions are applied to obtain our main result for this section which is:

Proposition 5.1. (I1) *and* (I2) *imply that if λ is smaller than a universal constant, then*

$$|\partial_k^u\partial_z^v\,\delta\Pi_\ell(k,z)|\leqq K\lambda[2+T_{\ell-1}]^{-d(u,v)}, \tag{5.1}$$

for $\ell\leqq m$, $|u|\leqq 2$, and z in the disc

$$\{z:|z|\leqq r_{\ell-1}+2\varepsilon_\ell\}. \tag{5.2}$$

$\partial_k^u \partial_z^v \delta\Pi_\ell(k, z)$ *is analytic on this disc and continuous in* k. *The constants* $d(u, v)$, ε_ℓ *are given by*

$$d(u,v) \equiv \frac{d}{2} - \alpha\left(\frac{|u|}{2} + v\right) - \frac{9}{8},$$

$$\varepsilon_\ell \equiv \frac{1}{16}\ln[2+T_\ell]/[2+T_\ell].$$

Notice that we have started our inductive step because in Proposition 5.1 $\ell \leqq m$ as opposed to the $m-1$ in the inductive assumptions. In Sect. 6 we will use Proposition 5.1 to advance (I1) to $\ell \leqq m$. In Sect. 7 we will use Proposition 5.1 and results from Sect. 6 to advance I2.

Estimates on $\delta\Pi$. To illustrate the basic strategy we will first obtain a weak version of the estimate we will ultimately need.

We start with the expression of Theorem 4.1 for Π_ℓ and estimate $|\Pi_\ell|$ by taking absolute values inside the sums over T, L, and ω, and use

$$|1-\lambda\delta(\omega(s)-\omega(t))| \leqq 1 \tag{5.3}$$

on all $st \in \mathscr{C}_\ell(L)$. The result is

$$|\Pi_\ell(k,z)| \leqq \sum_{T=1}^{\infty} \sum_{\omega:|\omega|=T} \left(\frac{|z|}{2d}\right)^T \sum_{L\in\mathscr{L}_\ell[0,T]}{}' \prod_{st\in L} \lambda\delta(\omega(s)-\omega(t)). \tag{5.4}$$

The contribution of each L can be expressed in terms of $C_0(x, z)$; consider the first three laces

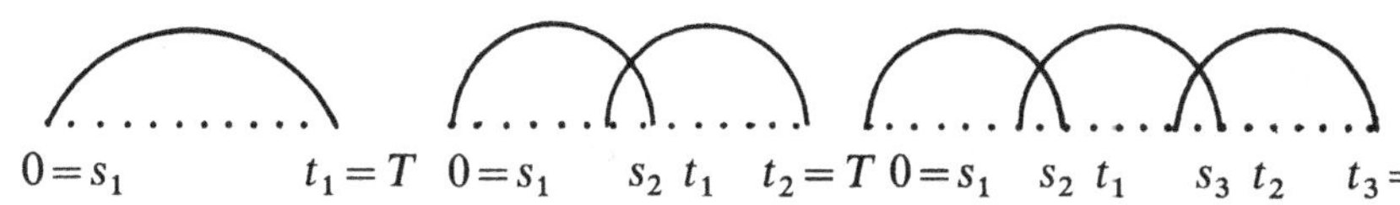

(5.5)

These diagrams, according to the product over $st \in L$ in (5.4) correspond to the sum of random walks that intersect themselves as indicated in the figure below

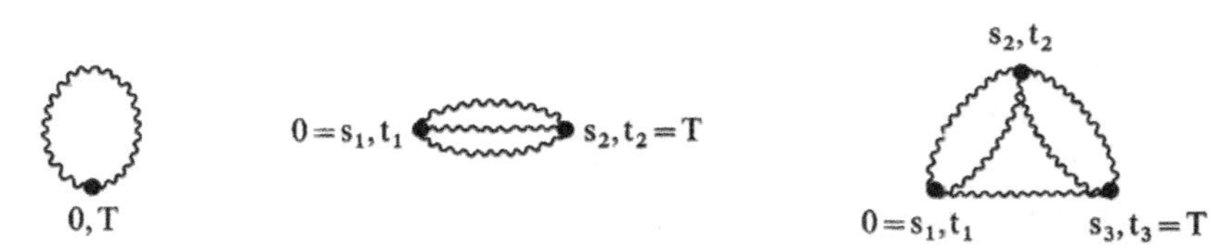

where the times opposite each intersection indicate the times at which the intersection takes place. With L and positions of intersection held fixed, we can rewrite the sum over ω in (5.4) as separate sums over $\omega_1, \omega_2, \ldots$, where ω_i is the restriction of ω to the time interval I_i in Fig. 2. We can also introduce times $\sigma_i \geqq 1$ for each ω_i:

$$\sigma_1 = s_2 - s_1,\ \sigma_2 = t_1 - s_2,\ \sigma_3 = s_3 - t_1, \ldots,$$

and replace the sum over T by sums over $\sigma_1, \sigma_2, \ldots \leqq T_\ell$ separately. Define

$$\tilde{C}_0(x,z) = \sum_{\sigma \leqq T_\ell} z^\sigma C_0(x,\sigma). \tag{5.6}$$

Now we can write down the result of doing the sums over $\sigma_1, \sigma_2, \ldots$, and positions of intersection. The contributions of the three diagrams (5.5′) to the righ-hand side of our upper bound on $|\Pi_\ell|$ are less than

$$\lambda \tilde{C}(0,|z|),\ \lambda^2 \sum_x \tilde{C}_0^3(x,|z|),\ \lambda^3 \sum_x ((\tilde{C}_0^2 * \tilde{C}_0^2)\tilde{C}_0)(x,|z|),$$

where $*$ denotes convolution. Let $\mathbf{M}$ denote the operation of multiplying by $\tilde{C}_0$, i.e.

$$\mathbf{M}: f(x) \to \tilde{C}_0(x,|z|)\, f(x),$$

and let $\mathbf{C}$ denote convolution by $\tilde{C}_0$:

$$\mathbf{C}: f(x) \to \sum_y \tilde{C}_0(x-y) f(y).$$

Then these three contributions may be rewritten in the form

$$\lambda \tilde{C}_0(0,|z|),\ \lambda^2(\mathbf{CM}\tilde{C})(0,|z|),\ \lambda^3(\mathbf{CMCM}\tilde{C}_0)(0,|z|).$$

This pattern continues; the next diagrams are

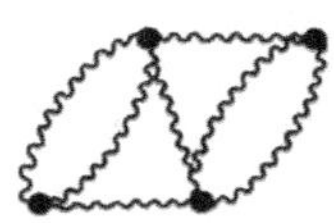 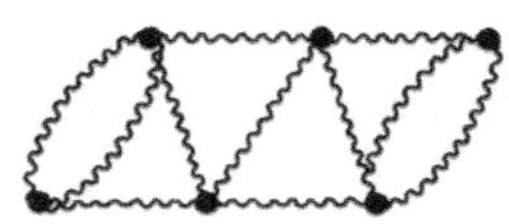

which correspond to

$$\lambda^4((\mathbf{CM})^3 \tilde{C}_0)(0,|z|),\ \lambda^5((\mathbf{CM})^4 \tilde{C}_0)(0,|z|), \ldots,$$

and an easy induction shows that this pattern persists so that the right-hand side of (5.4) equals

$$\sum_{i=0}^{\infty} ((\mathbf{CM})^i \tilde{C}_0)(0,|z|)\lambda^{i+1}. \tag{5.7}$$

We now introduce a bound for each term in this sum whose content is that each term may be bounded by a product of L^2 norms of $\tilde{C}_0$ with one factor of $\tilde{C}_0$ omitted and instead bounded by an L_∞ norm. Let, for $j=1,\ldots,i$, $\mathbf{C}_{2j}$ be convolution by a function f_{2j} and $\mathbf{M}_{2j}$ be multiplication by f_{2j-1}, then we claim that

$$\left\| \prod_{j=1}^{i} (\mathbf{C}_{2j}\mathbf{M}_{2j}) f_0 \right\|_\infty \leqq \left(\prod_{j=1, j\neq k}^{2i} \|f_j\|_2 \right) \|f_k\|_\infty, \tag{5.8}$$

for any $k=1,\ldots,2i$.

Proof. We use Young and Hölder's inequalities in the forms

$$\mathbf{M}_f: L_2 \to L_1, \quad \|\mathbf{M}_f\| \leqq \|f\|_2,$$

$$\mathbf{C}_g: L_1 \to L_2, \quad \|\mathbf{C}_g\| \leqq \|g\|_2,$$

so that

$$\|\mathbf{C}_g\mathbf{M}_f\|_{L_2\to L_2} \leqq \|f\|_2 \|g\|_2 .$$

Also using $\mathbf{M}_f : L_\infty \to L_2, \mathbf{C}_g : L_2 \to L_\infty$, we obtain

$$\|\mathbf{C}_g\mathbf{M}_f\|_{L_\infty\to L_\infty} \leqq \|f\|_2 \|g\|_2 .$$

Finally, using $\mathbf{M}_f : L_2 \to L_2$ or L_1 and $\mathbf{C}_g : L_2$ or $L_1 \to L_\infty$, we obtain

$$\|\mathbf{C}_g\mathbf{M}_f\|_{L_2\to L_\infty} \leqq \begin{cases} \|f\|_\infty \|g\|_2 \\ \|f\|_2 \|g\|_\infty \end{cases} .$$

The claim is a combination of these bounds.

We return to (5.7) and apply this bound (5.8) to obtain

$$|\Pi_m(k, z)| \leqq \sum_{i=1}^{\infty} \prod_{j=1}^{2i-1} (\|\tilde{C}_0(\cdot, |z|)\|_*)\lambda^i , \tag{5.9}$$

where $* = 2$ or ∞, with one $\| \; \|_\infty$.

Now this bound is insufficient for our needs but without destroying the structure of our argument it can easily be improved: in (5.4) it is not necessary to "throw away" all factors $1-\lambda V_{st}$ by the upper bound (5.3). Instead we can remove only those factors $1-\lambda V_{st}$ which couple walks ω_i, ω_j, $i \neq j$ in our argument above. These are the bonds which couple the parts of ω in a way which is inconsistent with the factorization of the sum over ω (lace L fixed) into independent sums over subwalks $\omega_1, \omega_2, \ldots$. With these bonds removed by a bound, the previous argument may be repeated yielding a bound in terms of L_2, L_∞ norms of

$$\sum_{s \leqq T_\ell} |z|^s C_\ell(x, s) .$$

See (4.8) and the remark after it. Since $s \leqq T_\ell$ this equals

$$\tilde{C}_\ell(x, |z|) \equiv \sum_{s \leqq T_\ell} |z|^s C_{\ell(s)}(x, s) , \tag{5.10}$$

where $\ell(s)$ is defined by $T_{\ell(s)} = s$, so that

$$|\Pi_\ell(k, z)| \leqq \frac{1}{1-\lambda} \sum_{i=1}^{\infty} \prod_{i=1}^{2i-1} (\|\tilde{C}_\ell(\cdot, |z|)\|_*)\lambda^i . \tag{5.11}$$

The $(1-\lambda)^{-1}$ restores the $0T$ bond in the first diagram in (5.5).

Next we consider $\delta\Pi_\ell$: $\delta\Pi_\ell \equiv \Pi_\ell - \Pi_{\ell-1}$. The expansion in Theorem 4.2 is applied to each Π on the right-hand side and we see that laces L in which every bond $st \in L$ is *short*, i.e. $|s-t| \leqq \frac{1}{2} T_{\ell-1}$, make identical contributions and cancel in $\delta\Pi$ [see (4.9)]. Therefore we may restrict the sums over L by requiring that each L contain at least one *long* line $st \in L$ such that $|s-t| \geqq \frac{1}{2} T_{\ell-1}$. We write $\delta\Pi_\ell = \Pi'_\ell - \Pi'_{\ell-1}$, where the primes mean that the sums of laces, L, in Theorem 4.2 are constrained so that each lace contains at least one long bond $st : |t-s| \geqq T_{\ell-1}/2$. The method of bounding Π applies also to $\delta\Pi'$, after using $|\delta\Pi_\ell| \leqq |\Pi'_\ell| + |\Pi'_{\ell-1}|$. A long bond implies a long walk, so when we apply the arguments leading to (5.11) to each Π' we can assume that one $\tilde{C}$ norm is replaced by a norm on

$$\delta\tilde{C}_\ell(x,|z|) \equiv \sum_{T_\ell \geqq s \geqq T_\ell/6} |z|^s \tilde{C}_\ell(\cdot, s).$$

Define

$$\|\delta^*\tilde{C}_\ell(\cdot,|z|)\|_* = \begin{cases} \|\tilde{C}_\ell(\cdot,|z|)\|_2 & \text{or} \\ \|\delta\tilde{C}_\ell(\cdot,|z|)\|_\infty \end{cases}.$$

The bound is

$$|\delta\Pi_\ell(k,z)| \leqq \frac{2}{1-\lambda} \sum_{i=1}^{\infty} \sum_{*} \prod_{j=1}^{2i-1} (\|\delta^*\tilde{C}_\ell(\cdot,|z|)\|_*)\lambda^i, \tag{5.12}$$

where the sum over $*$ is a sum over all ways of allocating the L_∞ norm. We have made use of the flexibility in L_∞ norm in (5.8) to coordinate the L_∞ norm with δ.

All these arguments can be extended to derivatives with respect to z and k of $\delta\Pi$. We differentiate under the sum over laces, L, in Theorem 4.2. The z derivatives act on the factor z^T which we write as

$$z^T = \prod_{j=1}^{2i-1} z^{s_i}.$$

Each z^s corresponds to a subwalk ω_i prescribed by lace L We distribute the derivatives over the factors z^s by Leibniz rule and proceed as before. k-derivatives act on the $\exp(ik\omega(T))$ in (4.8) which we can also factor along the walk $\omega = \omega_1 \cup \omega_2 \ldots \cup \omega_{2i-1}$, as prescribed by a lace L Again we apply Leibniz rule, take absolute values inside the sum ways of distributing derivatives and proceed as above. We obtain a bound on $\delta\Pi$ in terms of x space norms

Proposition 5.2.

$$|\partial_k^u \partial_z^v \delta\Pi_\ell(k,z)| \leqq \frac{2}{1-\lambda} \sum_{i=1}^{\infty} \lambda^i \sum \prod_{j=1}^{2i-1} \|(\cdot)^{u_j} \partial_{|z|}^{v_j} \delta^*\tilde{C}_\ell(\cdot,|z|)\|_*, \tag{5.13}$$

$v=0,1,\ldots,|u|=0,1,2,3,\ldots$, the unlabelled sum is over: non-negative multi-indices u_j such that $\sum|u_j|=|u|$, non-negative v_j's such that $\sum v_j = v$ and choosing one $\|\ \|_*$ norm to be $\|\ \|_\infty$, the rest are $\|\ \|_2$ norms. In each L_2 norm $\delta^*\tilde{C}_\ell = \tilde{C}_\ell$. The L_∞ norm has $\delta^*\tilde{C}_\ell = \delta\tilde{C}_\ell$ where

$$\begin{aligned} \tilde{C}_\ell(x,|z|) &\equiv \sum_{0 \leqq s \leqq T_\ell} |z|^s C_{\ell(s)}(x,s), \\ \delta\tilde{C}_\ell(x,|z|) &\equiv \sum_{T_\ell/6 \leqq s \leqq T_\ell} |z|^s C_{\ell(s)}(x,s), \end{aligned} \tag{5.14}$$

and $\ell(s)$ is such that $T_{\ell(s)} = s$.

Note that the numbers of terms in the unlabelled sum is less than

$$(2i-1)^{1+|u|+v},$$

which will be dominated by λ powers if λ is small.

Now we use Proposition 5.2 to obtain Proposition 5.1, our main result for this section.

For $\ell \leqq m$, $T_\ell \geqq 1$,

$$\left\| \sum_{s \leqq T_\ell} \partial^v_{|z|} |z|^s (\cdot)^u \tilde{C}_{\ell(s)}(\cdot, s) \right\|_{2 \text{ or } \infty} \leqq \sum |z|^{s-v} s^v \| (\cdot)^u \tilde{C}_{\ell(s)-1}(\cdot, s) \|^{(x \text{ space})}_{2 \text{ or } \infty}$$

$$\leqq \sum |z|^{s-v} s^v \| \partial^u_k C_{\ell(s)-1}(\cdot, s) \|^{(k \text{ space})}_{2 \text{ or } 1} \tag{5.15}$$

by the Hausdorff Young inequality, u and v are arbitrary. By the inductive assumption (I2), which we have for $\ell \leqq m-1$, and can use, because $\ell(s)-1 \leqq \ell-1$, $\leqq m-1$,

$$\leqq K \sum_{s \leqq T_\ell} |z|^{s-v} s^v s^{|u|/2 - d/(2p)} r^{-s}_{\ell(s)-1},$$

with $p=2$ or 1,

$$\leqq K T_\ell^{v+|u|/2} \sum_{s \leqq T_\ell} (|z|/r_{\ell(s)-1})^{s-v} s^{-d/2p}, \tag{5.16}$$

because $s \leqq T_\ell$ and for all ℓ, $r_\ell \geqq 1$. The $|z|$ factor has to be bounded carefully, using (5.2)

$$|z|/r_{\ell(s)-1} \leqq [r_{\ell-1} + 2\varepsilon_\ell]/r_{\ell(s)-1} \leqq 1 + [r_{\ell-1} - r_{\ell(s)-1}] + 2\varepsilon_\ell .$$

By summing (I1) over ℓ from $\ell-1$ down to $\ell \in [\ell(s)-1, \ell(s)-2]$,

$$|z|/r_{\ell(s)-1} \leqq 1 + K\sqrt{\lambda}\, T^{9/8-d/2}_{\ell(s)-2} + 2\varepsilon_\ell$$

$$\leqq 1 + K(\delta)\sqrt{\lambda}\, T^{(9/8-d/2)\delta}_{\ell(s)} + 2\varepsilon_\ell$$

for any $\delta < 1$. Since $T_{\ell(s)} = s$ and δ can be chosen so that $(9/8-d/2)\delta \leqq -1$,

$$|z|/r_{\ell(s)-1} \leqq \exp[K(\delta)\sqrt{\lambda}\, s^{(9/8-d/2)\delta} + 2\varepsilon_\ell] \leqq \exp(K(\delta)\sqrt{\lambda}\, s^{-1} + 2\varepsilon_\ell).$$

We substitute this upper bound for $|z|/r_{\ell(s)-1}$ into (5.16) to find that the norm (5.15) is less than

$$K T_\ell^{v+u/2} \sum_{s \leqq T_\ell} s^{-d/2p} \exp(2s\varepsilon_\ell).$$

By observing that the exponent is less than or equal to $\frac{1}{8}\log s$ for $s \geqq e$ we bound this by

$$K T_\ell^{v+u/2} \sum_s s^{-d/2p+\frac{1}{8}}.$$

Thus taking $p=2$ and $p=1$,

$$\left\| \sum_{s=0}^{T_\ell} \partial^v_{|z|} |z|^s (\cdot)^u \tilde{C}_{\ell(s)}(\cdot, s) \right\|_2 \leqq K T_\ell^{v+u/2},$$

$$\left\| \sum_{T_\ell \geqq s \geqq \frac{1}{6} T_{\ell-1}} \partial^v_{|z|} |z|^s (\cdot)^u \tilde{C}_{\ell(s)}(\cdot, s) \right\|_\infty \leqq K T_\ell^{v+u/2} T_{\ell-1}^{9/8-d/2}.$$

We substitute these bounds into Proposition 5.2 and find, using $\sum v_i = v$, $\sum |u_i| = |u|$ that $|\partial_k^u \partial_z^v \delta\Pi_\ell(k,z)| \leqq \lambda K T_\ell^{v+u/2} T_{\ell-1}^{9/8-d/2}$ for $\ell \leqq m$, provided λ is smaller than a universal constant. This is the desired bound (5.1), since $T_\ell \leqq K T_{\ell-1}^a$ for all ℓ. The analyticity of $\delta\Pi$ in z is immediate, the continuity in k follows from the expansion (4.8) and the bounds we have proven. Proposition 5.1 is proved.

6. Poles of $C_\ell(k,z)$

In this section we shall obtain information about the poles of $C_\ell(k,z)$ using (5.1) and the inductive assumption I1. More precisely let

$$\mathscr{D}_\ell(a) = \{z : |z| \leqq r_\ell + a\varepsilon_\ell\}. \tag{6.1}$$

We shall prove that for $\ell \leqq m$, $C_\ell(k,z)$ is meromorphic in the disc $\mathscr{D}_\ell(3/2)$ and has a unique real pole $r_\ell(k)$ which varies differentiably with k provided $k^2/2d \leqq \varepsilon_\ell$. These results will follow from implicit function type arguments.

Remark. By I1 note that

$$0 \leqq r_\ell - r_{\ell-1} \leqq a(\varepsilon_{\ell-1} - \varepsilon_\ell); \qquad \ell \leqq m-1,$$

for $a \geqq 1/4$ and λ smaller than a universal constant, hence the discs $\mathscr{D}_\ell(a)$ are decreasing in ℓ, for $\ell \leqq m-1$.

Let us observe that $C_\ell(k,z)$ has the following symmetries:

i) $C_\ell(k,z)$ is even in k_i, $i = 1, \ldots, d$,

ii) $C_\ell(k,\bar{z}) = \overline{C_\ell(k,z)}$,

iii) $C_\ell(k+\boldsymbol{\pi}, z) = C_\ell(k,-z)$,

where $\boldsymbol{\pi} = (\pi, \ldots, \pi)$. Identities i) and ii) follow easily from using the invariance of the sum (3.4) under $\omega^i \to -\omega^i$, and iii) follows from the observation that $\omega^1(T) + \ldots + \omega^d(T)$, T are either both even or odd. The same symmetries hold with C_ℓ replaced by Π_ℓ (by definition of Π_ℓ). From ii) it follows that the poles of $C_\ell(k,z)$ are real or come in complex conjugate pairs.

Lemma 6.1. *Assume that* I1 *holds for* $\ell \leqq m-1$ *and* (5.1) *holds for* $0 \leqq \ell \leqq m$. *If* λ *is taken small, independently of* m, *then*

(a) I1 *holds for* $\ell \leqq m$,

(b) *for* $\ell \leqq m$, *for all* k, $\Pi_\ell(k,z)$ *is analytic in* z, $z \in \mathscr{D}_\ell(3/2)$ *and*

$$\begin{aligned} |\partial_k^u \partial_z^v \Pi_\ell(k,z)| &\leqq K(\delta)\lambda \quad \text{if} \quad d(u,v) \geqq \delta \\ &\leqq K\lambda(2+T_\ell)^{-d(u,v)} \quad \text{if} \quad d(u,v) \leqq 0 \end{aligned} \tag{6.2}$$

for each $\delta > 0$, $z \in \mathscr{D}_\ell(3/2)$,

(c) *for* $\ell \leqq m$, $C_\ell(k,z)$ *is meromorphic in* z *for* $z \in \mathscr{D}_\ell(3/2)$.

In particular note that we have advanced one of our inductive hypotheses, I1.

For the next lemma, recall that

$$F_\ell(k,z) \equiv [1 - zD(k) - \Pi_\ell(k,z)]^{-1} \equiv C_\ell^{-1}(k,z).$$

We will also need $r(k,\lambda=0) = D^{-1}(k)$, which is the unperturbed zero solving $1 - r(k)D(k) = 0$.

Lemma 6.2. *Assume the conclusions of Lemma 6.1. For λ sufficiently small, independently of m, and all $\ell \leqq m$,*

(a) *When k satisfies either $k^2 \leqq 2d\varepsilon_\ell$ or $(k-\pi)^2 \leqq 2d\varepsilon_\ell$, then $C_\ell(k,z)$ has one simple pole, $r_\ell(k)$, in $\mathscr{D}_\ell(3/2)$ whose derivatives $\partial_k^u r_\ell(k)$, $0 \leqq |u| \leqq 2$, are continuous, uniformly bounded and obey*

$$\partial_k^u(r_\ell(k) - r(k, \lambda=0)) = \lambda E. \tag{6.3}$$

(b) *With the same condition on k as in part* (a), $|u|$, $v \leqq 2$:

$$\int_{\partial\mathscr{D}_\ell(3/2)} |\partial_k^u C^{v+1}(k,z)|\,|dz| \leqq K(2+T_\ell)^{|u|+v} \ln(2+T_\ell). \tag{6.4}$$

(c) *If the condition on k in part* (a) *is violated, then $F_\ell(k,z)$ has no zero in $\mathscr{D}_\ell(\frac{1}{2})$ and*

$$\int_{\partial\mathscr{D}_\ell(1/2)} |\partial_k^u C_\ell^{1+v}(k,z)|\,|dz| \leqq K(2+T_\ell)^{|u|+v} \ln(2+T_\ell). \tag{6.5}$$

Proof of Lemma 6.1. First note that (6.2) holds for $\ell \leqq m-1$ by writing Π_ℓ as a telescopic sum of $\delta\Pi$'s and bounding it with (5.1). We must check that the domains of z are compatible, i.e., for $\ell' \leqq \ell \leqq m-1$

$$r_{\ell'-1} + 2\varepsilon_{\ell'} \geqq r_{\ell'} + \tfrac{3}{2}\varepsilon_{\ell'} \geqq r_{\ell-1} + \tfrac{3}{2}\varepsilon_{\ell-1}$$

for $\ell \leqq m-1$. The first inequality follows from I1 and the second one follows from the remark after (6.1), provided λ is taken smaller than a universal constant.

Part (a). To advance the induction step I1 we introduce an interpolating function: for $\ell \in (m-1, m)$, $H(z,\sigma) = F_{\ell-1}(0,z) + \sigma\delta\Pi_\ell(0,z)$. Using (6.2) for $\ell \leqq m-1$ and (5.1) with $\ell \leqq m$, we see that

$$\frac{\partial H}{\partial z} = -1 + \lambda E \quad \text{for} \quad |z| \leqq r_{\ell-1} + 3/2\varepsilon_\ell.$$

By taking λ small (independently of m) and using the implicit function theorem $H(z,\sigma)$ has a locally unique simple zero $r(\sigma)$ as long as $\partial H/\partial z(r(\sigma),\sigma) \neq 0$. Local uniqueness forces $r(\sigma)$ to be real because zeros of H are either real or occur in complex pairs.

Let $\bar{\sigma}$ be the largest value of σ in $[0,1]$ such that $|r(\sigma)| \leqq r_{\ell-1} + \varepsilon_\ell$. For $\sigma \leqq \bar{\sigma}$ and λ small

$$\frac{d}{d\sigma} r(\sigma) \leqq \left| \frac{\dfrac{\partial H}{\partial \sigma}(r(\sigma),\sigma)}{\dfrac{\partial H}{\partial z}(r(\sigma),\sigma)} \right| \leqq |\delta\Pi_\ell(0, r(\sigma))| \cdot |(1+\lambda E)| \leqq K\lambda(2+T_{\ell-1})^{-d/2+9/8} \ll \varepsilon_\ell,$$

using (5.1). By integrating over σ from 0 to $\bar{\sigma}$ we find that $\bar{\sigma} = 1$ and

$$r(\sigma=1) - r_{\ell-1} \leqq K\lambda(2+T_{\ell-1})^{-d/2+9/8}.$$

Since r_ℓ lies on the positive real axis and is the zero of F_ℓ closest to the origin, local uniqueness forces $r(\sigma=1) = r_\ell$, hence I1 holds for $\ell \leqq m$, provided we take λ small so that $K\lambda \leqq K^{(\mathrm{I1})}\sqrt{\lambda}$.

Part (b). (6.2) is proved by writing Π as a telescopic sum of $\delta\Pi$'s and bounding each term using (5.1) for $\ell \leqq m$. By part (a) we now have

$$r_{\ell'-1} + 2\varepsilon_{\ell'} \geqq r_\ell + \tfrac{3}{2}\varepsilon_\ell \quad \text{for} \quad \ell' \leqq \ell \leqq m,$$

so that the z domains are compatible.

Part (c). Immediate consequence of part (b).

Proof of Lemma 6.2. Since the value of ℓ does not change in the course of this proof we suppress ℓ subscripts on C_ℓ, F_ℓ, Π_ℓ, T_ℓ, ε_ℓ, r_ℓ, and $\mathscr{D}_\ell(a)$.

From the definition of F and (6.2),

$$\frac{\partial F}{\partial z} = -D(k) - \frac{\partial \Pi}{\partial z} = -D(k) + \lambda E(k,z). \tag{6.7}$$

Suppose k is near 0 or π as in part (a) of the lemma, then $D(k)$ is bounded away from zero uniformly in ℓ. Furthermore, if F has two zeros $z(k)$ and $z'(k)$ both in $\mathscr{D}(3/2)$, then the integral of $\partial F/\partial z$ along a contour between $z(k)$ and $z'(k)$ must vanish. For λ small this is in contradiction with (6.7). Thus for λ small (uniformly in ℓ) $F(z,k)$ has at most one zero $r(k)$ in $\mathscr{D}(3/2)$ for each k in the region specified. For small k, $r(k)$ is in $\mathscr{D}(3/2)$ and $r(k=0)=r$ by the implicit function theorem. Zeros are either real or occur in complex conjugates so by local uniqueness $r(k)$ lies on the real axis. Its k derivatives are

$$\frac{\partial r(k)}{\partial k_i} = -\frac{\partial F}{\partial k_i}\Big/\frac{\partial F}{\partial z} = \left(\frac{1}{d}\sin(k_i)\, r(k) - \frac{\partial \Pi}{\partial k_i}(k, r(k))\right)\Big/\frac{\partial F}{\partial z}. \tag{6.8}$$

Since Π is even in k_i, (6.2) implies

$$\left|\frac{\partial \Pi}{\partial k_i}(k, r(k))\right| \leqq K\lambda |\sin k_i| \tag{6.9}$$

for $k \simeq 0$ as in part (a). Thus using (6.2) again,

$$\frac{\partial r(k)}{\partial k_i} = \frac{1}{d}\sin(k_i)\, r(k)\,(1 + \lambda E(k, r(k)))/D(k). \tag{6.10}$$

We take λ small (uniformly in ℓ) and conclude that $r(k)$ moves monotonically away from $r(k=0)=r$ towards the boundary of $\mathscr{D}(3/2)$ along the real axis and (6.10) holds provided $r(k) \in \mathscr{D}(3/2)$ and $k^2 \leqq 2d\varepsilon$. By symmetry there is another trajectory of solutions starting at $-r$ when $k=\pi$ for which analogous considerations hold.

The second derivatives of $r(k)$ are obtained by twice differentiating $F(k, r(k)) = 0$:

$$0 = \frac{\partial^2 F}{\partial k_i \partial k_j} + \frac{\partial^2 F}{\partial k_i \partial z}\frac{\partial r}{\partial k_j} + \frac{\partial^2 F}{\partial k_j \partial z}\frac{\partial r}{\partial k_i} + \frac{\partial F}{\partial z}\frac{\partial^2 r}{\partial k_i \partial k_j} + \frac{\partial^2 F}{\partial z^2}\frac{\partial r}{\partial k_i}\frac{\partial r}{\partial k_j}. \tag{6.11}$$

By (6.10) and (6.2) every term in (6.11) except possibly the fourth is equal to its $\lambda=0$ value $+\lambda E$. For $k \cong 0$ or $k \cong \pi$ as in part (a) $\partial F/\partial z$ is bounded away from zero according to (6.2), so we obtain (6.3) with $|u|=2$ by solving (6.11) for the double derivative of r. At present (6.3) is known provided k^2 or $(k-\pi)^2 \leqq 2d\varepsilon$ and

$r(k) \in \mathscr{D}(3/2)$. By integrating (6.3) with $|u|=2$ from zero to k using the fact that $\partial r/\partial k_i = 0$ at $k=0$ (by symmetry) we find that $k^2 \leqq 2d\varepsilon \Rightarrow r(k) \in \mathscr{D}(3/2)$ provided λ is smaller than a universal constant, consequently (6.3) is fully established for $|u|=2$. For $|u|=1$ it is immediate from (6.10). For $|u|=0$ we note that at $k=0$, $|r(k=0)-r(k=0, \lambda=0)| \leqq K\lambda$ by the implicit function theorem and (6.2), so we obtain (6.3) for $|u|=0$ by integrating (6.3) for $|u|=1$. Continuity of derivatives is evident.

Part (b). We will consider only the case $|u|=2$, $v=1$. The others are similar.

$$\partial_k^u C(k,z) = 2C^3(k,z)(\partial_{k_i}F)(\partial_{k_j}F) - C^2(k,z)\partial_{k_i}\partial_{k_j}F. \tag{6.12}$$

We bound derivatives of F using (6.2). By integrating (6.3) with $|u|=2$ from zero to k, $k^2 \leqq 2d\varepsilon$, we find

$$r(k) - r = r(k, \lambda=0) - 1 + \lambda E k^2. \tag{6.13}$$

From this we conclude that if λ is small (uniformly in ℓ) and $k \cong 0$ or $k \cong \pi$ as required in parts (a) and (b) then $r(k)$ is strictly inside $\mathscr{D}(3/2)$, indeed

$$|z - r(k)| \geqq \varepsilon/4 \quad \text{for} \quad z \in \partial\mathscr{D}(3/2). \tag{6.14}$$

Therefore by the fundamental theorem of calculus

$$|F(k,z)| = |z - r(k)| \left| \int_0^1 d\sigma\, \partial_z F(k, r(k) + \sigma(z - r(k))) \right| \geqq K|z - r(k)|, \tag{6.15}$$

because by (6.7) $\partial_z F \simeq 1$, when $z \in \mathscr{D}(3/2)$ and $k \simeq 0$ or π. Thus by integrating the inverse of (6.15),

$$\int_{\partial\mathscr{D}(3/2)} |C(k,z)|^{n+1}\, |dz| \leqq (K\varepsilon)^{-n} |\ln K\varepsilon|. \tag{6.16}$$

The logarithm is actually only necessary if $n=0$. We combine this inequality with (6.12) and obtain the inequality of part (b) for the case $|u|=2$, $v=1$.

Part (c). First let us suppose $2d\varepsilon \leqq k^2 \leqq 1$. To bound the contour integral we need a lower bound on F:

$$F(k,z) = F(0,z) + (1 - D(k))z - \int_0^k \partial_k \Pi(\tilde{k}, z)\, d\tilde{k} = \int_r^z \frac{\partial F}{\partial z}(0, \tilde{z})\, d\tilde{z} + (1 - D(k))z + \lambda k^2 E$$

[using (6.9)]

$$= r - D(k)z - \int_r^z \frac{\partial \Pi}{\partial z}(0, \tilde{z})\, d\tilde{z} + \lambda k^2 E = (r - D(k)z) + \lambda(z - r)E + \lambda k^2 E \tag{6.17}$$

[using (6.2)]. To prove that the error terms are relatively small we use $|a+ib| \geqq ||a| + |b||/\sqrt{2}$ to show that for $z \in \partial\mathscr{D}(\frac{1}{2})$, i.e., $z = (r + \varepsilon/2)e^{i\theta}$,

$$|r - D(k)z| \geqq \frac{r}{\sqrt{2}} \left| 1 - D(k)\left(1 + \frac{\varepsilon}{2r}\right)\cos\theta \right| + \frac{1}{\sqrt{2}} |D(k)|\, |\sin\theta|.$$

For $2d\varepsilon \leqq k^2 \leqq 1$, $D(k)$ is bounded away from 1 in such a way that this inequality implies $|r - D(k)z| \geqq K(k^2 + |D(k)\sin\theta|)$ with $K > 0$. We now see that the last error term in (6.17) is relatively small when λ is chosen small (uniformly in ℓ). For the second term in (6.17) we use

$$|r-z| \leqq |r-D(k)z| + |z|\,|1-D(k)| \leqq |r-D(k)z| + \left(r+\frac{\varepsilon}{2}\right)\frac{1}{2d}k^2,$$

and again we see that the first term dominates the second in (6.17). Thus for λ small (uniformly in ℓ) $|F(k,z)| \geqq K|r-D(k)z| \geqq K(\varepsilon+|\theta|)$ for some K's>0. The bound (6.5) on the contour integral now follows with the same arguments as in (b). By symmetry the same arguments are available for k such that $2d\varepsilon \leqq (k-\pi)^2 \leqq 1$. For the remaining values of k, $|1-zD(k)| \geqq K$, some $K>0$, and by (6.2) with λ small $|C(k,z)| \leqq K$. Part (c) is proved.

7. The Induction Completed. Proof of Main Theorems

In this section we complete our induction by proving that I2 holds for $\ell \leqq m$, i.e.

$$\|r_\ell^T \partial_k^u C_\ell(\cdot, T)\|_{L^p}^{(k\,\text{space})} \leqq K^{(\mathrm{I}2)} T^{|u|/2-d/(2p)} \tag{7.1}$$

for $T \geqq T_{\ell+1}$ and $\ell \leqq m$. We will prove (7.1) with $K^{(\mathrm{I}2)}$ equal to 4 times (the best constant for $\lambda=0$).

We will also prove that there is a constant $D>0$ such that

$$\langle \omega^2(T)\rangle_{\ell(T)} = -\sum_i \frac{\partial^2}{\partial k_i^2} C_{\ell(T)}(k,T)/C_{\ell(T)}(0,T)|_{k=0} = DT(1+T^{-1/16}E) \tag{7.2}$$

for T large. Finally we shall establish the gaussian nature of the scaling limit of the end point, i.e. for fixed t

$$\langle e^{ik\cdot\omega(st)/\sqrt{s}}\rangle(\lambda) \equiv \frac{C_{\ell(st)}\left(\dfrac{k}{\sqrt{s}}, [st]\right)}{C_{\ell(st)}(0,[st])} \Rightarrow e^{-Dk^2t/2}, \tag{7.3}$$

as s tends to infinity.

Proof of 7.1. It suffices to establish (7.1) for $m-1 \leqq \ell \leqq m$. We may also assume that either T or T_ℓ is large because we can achieve (7.1) on any bounded set of T's and T_ℓ's by taking λ small.

By Cauchy's formula

$$C_\ell(k,T) = \frac{1}{2\pi i}\oint C_\ell(k,z) z^{-T-1}\,dz = \frac{1}{2\pi i}\oint \frac{z^{-T-1}}{F_\ell(k,z)}\,dz, \tag{7.4}$$

where the contour is a small circle centered at the origin. If $k^2 \geqq 2d\varepsilon_\ell$ and $(k-\pi)^2 \geqq 2d\varepsilon_\ell$, then by Lemma 6.2, $C_\ell(k,z)$ is analytic inside the disc $\mathscr{D}(\frac{1}{2})$. We now deform the contour in (7.4) to $\partial\mathscr{D}_\ell(\frac{1}{2})$. By (6.5) and (7.4) we see that

$$\begin{aligned} r_\ell^T|\partial_k^u C_\ell(k,T)| &< \sup_{z\in\partial\mathscr{D}_\ell(1/2)} KT_\ell^{|u|}\ln T_\ell |r_\ell/z|^T \\ &= KT_\ell^{|u|}\ln T_\ell |r_\ell/(r_\ell+\varepsilon_\ell/2)|^T \\ &\leqq KT_\ell^{|u|}\ln T_\ell\, e^{-T\varepsilon_\ell/4} \leqq \tfrac{1}{4}K^{(\mathrm{I}2)}T^{-d/2}(2\pi)^{-d} \end{aligned} \tag{7.5}$$

for $T \geqq T_{\ell+1}$ and either T or T_ℓ large.

For the case $k^2 \leqq 2d\varepsilon_\ell$ we deform the small contour about the origin to $\mathscr{C} = \partial\mathscr{D}_\ell(3/2)$, which by Lemma 6.2 contains exactly one pole $r_\ell(k)$. Thus we have a contribution from the pole plus an error term:

$$r_\ell^T C_\ell(k, T) = -\left[\frac{\partial F_\ell}{\partial z}(k, r_\ell(k))\right]^{-1} r_\ell(k)^{-T-1} r_\ell^T + r_\ell^T \oint_{\mathscr{C}} C_\ell(k, z) z^{-(T+1)} dz. \quad (7.6)$$

As in our analysis of (7.5) we apply (6.4) to show that the second term on the right side of (7.6) is bounded by $\frac{1}{4}K^{(12)}T^{-d/2}$ for $T \geqq T_{\ell+1}$. To bound the first term on the right side of (7.6) we integrate (6.3) with $|u| = 2$ twice with respect to k, remembering that at $k = 0$ first derivatives vanish so that

$$r_\ell(k) - r_\ell = r(k, \lambda = 0) - 1 + \lambda k^2 E = D^{-1}(k) - 1 + \lambda k^2 E.$$

Thus

$$\int_{k^2 \leqq 2d\varepsilon_\ell} dk |(\partial_z F_\ell)^{-1} r_\ell(k)^{-T-1} r_\ell^T|^p$$
$$\sim (\partial_z F_\ell)^{-1}(k=0) r_\ell^{-1} \int_{-\infty}^{\infty} dk \exp\left(-\frac{p}{2}K(\ell, \lambda) T k^2\right),$$

where $\sim$ means asymptotic as $T \to \infty$, uniformly in ℓ and $K(\lambda, \ell) \to 1$ uniformly in ℓ as $\lambda \to 0$. From this we immediately obtain (7.1) for $u = 0$ provided λ is taken smaller than a universal constant.

Next we consider the case $|u| = 2$, $k^2 \leqq 2d\varepsilon_\ell$. We take the second derivative of both sides of (7.6). The second term on the right is again bounded by $T^{-d/2}$ using (6.4) as before. To analyze the first term let us set $r = r_\ell(k)$ and

$$H = \left[\frac{\partial F_\ell}{\partial z}(k, r_\ell(k))\right]^{-1}.$$

Then we have

$$\frac{d^2}{dk_i^2}[H r^{-(T+1)}] = \left\{\frac{d^2}{dk_i^2} H\right\} r^{-(T+1)} - 2\left(\frac{d}{dk_i} H\right)(T+1) r^{-(T+2)} \frac{\partial r}{\partial k_i}$$
$$+ H\left(\frac{\partial}{\partial k_i} r\right)^2 (T+1)(T+2) r^{-(T+3)} - H\left(\frac{d^2}{dk_i^2} r\right)(T+1) r^{-(T+2)}. \quad (7.7)$$

After multiplying by $r_\ell(0)^T$ we shall show that the last two terms of (7.7) are dominant in L^p norm for large T. Moreover if $k = 0$, the second and third terms on the right side of (7.7) vanish because $r_\ell(k)$ is even. To show that the first two terms are relatively small we shall use the following lemma.

Lemma 7.1. *For $k^2 \leqq 2d\varepsilon_\ell$ we have*

$$\left|\frac{d}{dk_1} H\right| < K T^\beta (1 + T^{1/2}|k_1|),$$

$$\left|\frac{d^2}{dk_1^2} H\right| < K[T^{2\beta}(1 + T^{1/2}|k_1|)^2 + T^{1/2+\beta} + T^{1+\beta}|k_1| + T^{3/2+\beta}k_1^2 + T^{1/2+\beta}],$$

where $\beta = \frac{1}{8} + 2(\alpha - 1)$.

Proof. The total derivatives with ℓ subscripts suppressed are given by

$$\frac{d}{dk_1}H = -H^2\left[\frac{\partial^2 F}{\partial k_1\,\partial z} + \frac{\partial^2 F}{\partial z^2}\frac{\partial r}{\partial k_1}\right],$$

and

$$\frac{d^2}{dk_1^2}H = 2H^3\left[\frac{\partial^2 F}{\partial k_1\,\partial z} + \frac{\partial^2 F}{\partial z^2}\frac{\partial r}{\partial k_1}\right]^2$$
$$-H^2\left[\frac{\partial^3}{\partial^2 k_1\,\partial z}F + 2\frac{\partial^3 F}{\partial k\,\partial z^2}\frac{\partial r}{\partial k} + \frac{\partial^3 F}{\partial z^3}\left(\frac{\partial r}{\partial k_1}\right)^2 + \frac{\partial^2 F}{\partial z^2}\frac{\partial^2 r(k)}{\partial^2 k_1}\right].$$

The lemma now follows directly by applying (6.2) and (6.3) to each of the terms above together with the bound $|\partial r/\partial k_1| \leqq K|k_1|$ (see 6.10). We multiply (7.7) by $r_\ell^T(0)$ and take the L^p norm of both sides remembering $k^2 \leqq 2d\varepsilon_\ell$. We analyse $[r_\ell(k)/r_\ell]^T$ as in the $u=0$ case, and we use Lemma 6.2 part (a) on k derivatives of $r_\ell(k)$. The above lemma shows that the first two terms of (7.7) give contributions which are down by $T^{1/8}$ relative to the second two terms, hence they can be dropped. The rest of the argument is as in the $|u|=0$ case. Mixed partials and the case $|u|=1$ are analogous. End of proof of (7.1).

Proof of (7.2). We begin by showing that for $k^2 \leqq Kd\varepsilon_\ell$, with K sufficiently small,

$$C_\ell(k,T) = -\left[\frac{\partial F}{\partial z}(k, r_\ell(k))\right]^{-1} r_\ell^{-T-1}(k)\,[1 + T^{-1/16}E] \tag{7.8}$$

for $T \leqq T_\ell$. Here $\ell \equiv \ell(T)$ is defined so that $T_\ell = T$. If $\ell_1 \geqq \ell(T) = \ell$, then recall that $C_\ell(k,T) = C_{\ell_1}(k,T)$. By (7.6)

$$C_\ell(k,T) = -\left[\frac{\partial F_\ell}{\partial z}(k, r_\ell(k)\right]^{-1} r_\ell(k)^{-T-1} + \frac{1}{2\pi i}\oint C_\ell(k,z)z^{-T-1}\,dz\,, \tag{7.9}$$

where $\mathscr{C} \equiv \partial\mathscr{D}_\ell(3/2)$ is the integration contour. By integrating with respect to k (6.3) with $|u|=2$, we find that $r_\ell(k) = r_\ell D^{-1}(k) + \lambda E k^2$, so that for λ small $r_\ell(k) \in \mathscr{D}_\ell(1/16)$. For $z \in \mathscr{C}$ we have

$$|r_\ell(k)/z|^{T+1} \leqq [(r_\ell + \tfrac{1}{16}\varepsilon_\ell)/(r_\ell + \tfrac{3}{2}\varepsilon_\ell)]^{T+1} \leqq KT^{-5/64}.$$

By (6.7), (6.4) and the above estimate the final term of (7.9) is bounded and (7.8) follows.

The second derivative of $C(k,T) = C_{\ell(T)}(k,T)$ at $k=0$ is again calculated using (7.6) with $\mathscr{C}$ as above. By (7.6), (7.7), and Lemma 7.1, we have

$$r^{T+1}(0)\left.\frac{\partial^2 C(k,T)}{\partial^2 k_i}\right|_{k=0} = \left[\frac{\partial F}{\partial z}(0, r(0))\right]^{-1}\frac{\partial^2 r}{\partial k_i^2}(T+1)\frac{1}{r(0)} + T^{(1/2+\beta)}E$$
$$+\left.\frac{1}{2\pi i}\oint_{\mathscr{C}} C^2(k,z)\frac{\partial^2 F}{\partial k_i^2}\left(\frac{r}{z}\right)^{T+1} dz\right|_{k=0}\,. \tag{7.10}$$

For notational simplicity we have set $r = r_{\ell(T)}(k)$, $F_{\ell(T)} = F$ etc. The term $T^{(1/2+\beta)}E$ represents contributions arising from $d^2H/dk_i^2|_{k=0}$. Note the $\partial H/\partial k$ and $\partial r/\partial k$

vanish at $k=0$. The final term on the right side of (7.10) is bounded by $KT^{15/16}$, using (6.4) and the fact that for $z\in\mathscr{C}|r(0)/z|^T\leqq KT^{-5/64}$. Thus (7.8) and (7.10) imply that

$$-\frac{\partial^2}{\partial k_i^2}C(k,T)/C(0,T)\bigg|_{k=0}=\frac{1}{r}\left(\frac{\partial^2}{\partial k_i^2}r\right)\bigg|_{k=0}T+T^{15/16}E\,.$$

To complete our proof of (7.2) we need only show that the T dependence of $r=r_{\ell(T)}$ and its second derivative in k are of lower order so that the diffusion constant D in (7.2) may be chosen independent of T. First note that by I1, $|r_\infty-r_{\ell(T)}|\leqq KT^{-5/4}$. We need to obtain a similar bound for the second derivative of $r_{\ell(T)}(k)$. Let $\ell=\ell(T)$. By (6.11) and (6.2),

$$\frac{\partial^2 r_\ell}{\partial k_i^2}(k)\bigg|_{k=0}=-\frac{\partial^2 F_\ell}{\partial k_i^2}(k,r_\ell(0))\left(\frac{\partial F}{\partial z}(0,r_\ell)\right)^{-1}\bigg|_{k=0}. \tag{7.11}$$

The ℓ dependence of the first factor on the right side of (7.11) can be analyzed as follows:

$$\begin{aligned}\frac{\partial^2 F_\ell}{\partial k_i^2}(k,r_\ell(0))-\frac{\partial^2 F_\ell}{\partial k_i^2}(k,r_\infty)\bigg|_{k=0}&\leqq\sup_{z\in\mathscr{D}_\ell(1/2)}\left|\frac{\partial^3}{\partial z\,\partial k_i^2}F_\ell(k,z)\right|_{k=0}|r_\ell-r_\infty|\\&\leqq KT_\ell^{1/2+\beta}T_\ell^{-5/4}\leqq KT_\ell^{-3/4+\beta}\leqq KT_\ell^{-1/2}\,.\end{aligned} \tag{7.12}$$

Recall that $r_\infty(0)$ belongs to $\mathscr{D}_\ell(\tfrac{1}{2})$. Also

$$\begin{aligned}\sum_{\ell_1\leqq\ell}\frac{\partial^2}{\partial k_i^2}(F_{\ell_1}(k,r_\infty(0))-F_{\ell_1+1}(k,r_\infty(0)))&=\sum_{\ell_1\geqq\ell}\frac{\partial^2}{\partial k_i^2}\Pi_{\ell_1+1}=\sum_{\ell_1\geqq\ell}T_{\ell_1+1}^{-1/4}E\\&=T_\ell^{-1/4}E\,.\end{aligned}$$

Hence $\partial_k^2F_\ell$ may be replaced by $\partial_k^2F_\infty$. The second factor of (7.11) may be analysed in a similar fashion, hence (7.2) follows.

Proof of the Scaling Limit (7.3). Let $st=T$. For any given values of k and t and $K>0$ we have $k^2/s\leqq Kd\varepsilon_{\ell(T)}$ as $s\to\infty$. Thus by (7.8) with $\ell=\ell(T)$ we have

$$C_\ell\left(\frac{k}{\sqrt{s}},T\right)=-\left[\frac{\partial F_\ell}{\partial z}\left(\frac{k}{\sqrt{s}},r_\ell(k/\sqrt{s})\right)\right]^{-1}r_\ell\left(\frac{k}{\sqrt{s}}\right)^{-T-1}(1+T^{-1/16}E)\,.$$

Lemma 7.1 implies that

$$\left|\frac{d}{dk_i}\left[\frac{\partial F_\ell}{\partial z}\left(\frac{k}{\sqrt{s}},r_\ell(k/\sqrt{s})\right)\right]^{-1}\right|\leqq K(t)\,T^{-1/8}\,,$$

so we can conclude that $C_\ell(k/\sqrt{s},T)\cong C_\ell(0,T)$ in that

$$\lim_{s\to\infty}C_{\ell(T)}(k/\sqrt{s},T)/C_{\ell(T)}(0,T)=\lim_{s\to\infty}\left[\frac{r_{\ell(T)}(0)}{r_{\ell(T)}(k/\sqrt{s})}\right]^T.$$

By Taylor's theorem with remainder

$$r_{\ell(T)}(k/\sqrt{s})=r_{\ell(T)}(0)+\sum_{|u|=2}\partial^u r_{\ell(T)}(\xi)\frac{k^u}{su!},$$

where ξ lies on the line segment joining 0 to $k/\sqrt{s}$. Note that $\partial^u r_{\ell(T)}(0)$ vanishes for $|u|=1$ and for mixed partials with $|u|=2$. The proof of the scaling limit follows once we show that the following limit exists for all k and t

$$\lim_{s\to\infty}\frac{1}{r_{\ell(T)}}\partial^u r_{\ell(T)}(\xi)=D^u,\ \xi\in\{\sigma k/\sqrt{s}:0\leqq\sigma\leqq 1\},$$

where D^u is a constant independent of k and t, which vanishes for mixed partials. Note that $D^{ii}=D$, the diffusion constant.

To establish the limit we return to (6.11) evaluated at $k=\xi$. The second, third and fifth terms in (6.11) go to zero as $st\equiv T\to\infty$ by (6.2) and (6.3) with $|u|=2$ integrated from $k=0$ to ξ. Existence of D^u follows as soon as we establish limits for the coefficient $\partial F/\partial z$ and the first term in (6.11).

Since $\partial^u\delta\Pi_j(k,z)$ is continuous in k for $|u|\leqq 2$ and is exponentially small for large j, it follows by summing over j that

$$\lim_{s\to\infty}|\partial^u F_\ell[\xi(s),r_\ell(\xi(s))]-\partial^u F_\ell[0,r_\ell(\xi(s))]|=0.$$

As in (7.12) one can show that

$$|\partial^u F_\ell(0,r_\ell(\xi(s))-\partial^u F_\ell(0,r_\infty(0))|\Rightarrow 0.$$

Finally $\partial^u F_\ell(0,r_\infty(0))$ is Cauchy by (5.1) together with the fact that $r_\infty(0)\in\mathscr{D}_\ell(\frac{1}{2})$ for all ℓ. The argument for $\partial F/\partial z$ in (6.11) is similar.

Acknowledgements. The authors would like to thank Mark Kac, Jürg Fröhlich, Alan Sokal, and Eugene Dynkin for teaching us about this and related problems.

References

1. Domb, C.: Self-avoiding random walk on lattices. In: Stochastic processes in chemical physics. Shuler, K. E. (ed.), pp. 229–260. New York: Wiley 1969
2. Aizenman, M.: Geometric analysis of ϕ^4 fields and Ising models. Parts I and II. Commun. Math. Phys. **86**, 1–48 (1982)
3. Fröhlich, J.: On the triviality of $\lambda\phi_d^4$ theories and the approach to the critical point in $d\geqq 4$ dimensions. Nucl. Phys. B **200**, 281–296 (1982)
4. Lawler, G.: A self-avoiding random walk. Duke Math. J. **47**, 655–693 (1980)
5. Fröhlich, J. (ed.): Scaling and self-similarity in physics – Renormalization in statistical mechanics and dynamics. Boston, Basel, Stuttgart: Birkhäuser 1983
6. Fröhlich, J.: Cargèse lectures

Communicated by A. Jaffe

Received April 2, 1984; in revised form June 14, 1984

Indecomposable Representations with Invariant Inner Product

A Theory of the Gupta-Bleuler Triplet

Huzihiro Araki

Research Institute for Mathematical Sciences, Kyoto University, Kyoto 606, Japan

Abstract. Consequences of the existence of an invariant (necessarily indefinite) non-degenerate inner product for an indecomposable representation π of a group G on a space $\mathfrak{H}$ are studied. If π has an irreducible subrepresentation π_1 on a subspace $\mathfrak{H}_1$, it is shown that there exists an invariant subspace $\mathfrak{H}_2$ of $\mathfrak{H}$ containing $\mathfrak{H}_1$ and satisfying the following conditions: (1) the representation $\pi_1^{\#} = \pi \bmod \mathfrak{H}_2$ on $\mathfrak{H} \bmod \mathfrak{H}_2$ is conjugate to the representation $(\pi_1, \mathfrak{H}_1)$, (2) $\mathfrak{H}_1$ is a null space for the inner product, and (3) the induced inner product on $\mathfrak{H}_2 \bmod \mathfrak{H}_1$ is non-degenerate and invariant for the representation

$$\pi_2 = (\pi|_{\mathfrak{H}_2}) \bmod \mathfrak{H}_1,$$

a special example being the Gupta-Bleuler triplet for the one-particle space of the free classical electromagnetic field with $\mathfrak{H}_1$ = space of longitudinal photons and $\mathfrak{H}_2$ = the space defined by the subsidiary condition.

1. Introduction

In the study of massless particles, one meets [1–6, 9–11, 15, 17–19] indecomposable representations π of a Lie group G on a space $\mathfrak{H}$ (with an invariant indefinite inner product), which take, for example, the following form:

$$\left.\begin{matrix} \pi_n \to \pi_{n-1} \to \dots \to \pi_1 \\ \mathfrak{H} = \mathfrak{H}_n \supset \mathfrak{H}_{n-1} \supset \dots \supset \mathfrak{H}_1 \end{matrix}\right\}. \tag{1.1}$$

Here $\mathfrak{H}_j$ is a $\pi(G)$-invariant subspace of $\mathfrak{H}_{j+1}$ without any $\pi(G)$-invariant complement, namely, there are no subspaces $\mathfrak{H}_j'$ such that $\mathfrak{H}_j + \mathfrak{H}_j' = \mathfrak{H}_{j+1}$, $\mathfrak{H}_j \cap \mathfrak{H}_j' = \mathbf{0}$, $\pi(G)\mathfrak{H}_j' \subset \mathfrak{H}_j'$. The representation π_j of G on $\mathfrak{H}_j/\mathfrak{H}_{j-1}$ is obtained by first restricting π to the subspace $\mathfrak{H}_j$ of $\mathfrak{H}$ and then considering it modulo $\mathfrak{H}_{j-1}$: $\pi_j(g)[\xi] = [\pi(g)\xi]$ for $\xi \in \mathfrak{H}_j$, where $[\xi] = \xi + \mathfrak{H}_{j-1}$ is a vector in $\mathfrak{H}_j/\mathfrak{H}_{j-1}$. (We note that the construction of representations on a quotient space, especially with respect to a null space, is now a popular game [14].)

It has been noticed in examples [1–3, 5, 6, 15] that $\pi_1 = \pi_n$, and it has been suggested that this is a general phenomenon [7]. The purpose of this note is to formulate this into a mathematical theorem.

A representation $\pi_1^\#$ of G on $\mathfrak{H}_1^\#$ will be called conjugate to a representation π_1 of G on $\mathfrak{H}_1$ if there is a sesquilinear form (ξ, n), $\xi \in \mathfrak{H}_1^\#$, $\eta \in \mathfrak{H}_1$ such that $\mathfrak{H}_1$ and $\mathfrak{H}_1^\#$ separate each other [i.e. $(\xi, \eta) = 0$ for all $\eta \in \mathfrak{H}_1$ implies $\xi = \mathbf{0}$ and $(\xi, \eta) = 0$ for all $\xi \in \mathfrak{H}_1^\#$ implies $\eta = \mathbf{0}$] and the following equality holds for all $\xi \in \mathfrak{H}_1^\#$, $\eta \in \mathfrak{H}_1$, and $g \in G$:

$$(\pi_1(g^{-1})\xi, \eta) = (\xi, \pi_1(g)\eta). \tag{1.2}$$

Our conclusion: if π_1 is an irreducible subrepresentation of an indecomposable representation π of G on a space $\mathfrak{H}$ with a G-invariant non-degenerate (indefinite) inner product, then π is of the form (1.1) ($n \geqq 2$) with π_n on $\mathfrak{H}_n/\mathfrak{H}_{n-1}$ dual to π_1 on $\mathfrak{H}_1$ (Theorem 1). (In examples quoted above, π_1 is selfdual.)

If $\mathfrak{H}_j$ has a complement $\mathfrak{K}_{j+1}$ in $\mathfrak{H}_{j+1}$ (i.e. $\mathfrak{H}_j + \mathfrak{K}_{j+1} = \mathfrak{H}_{j+1}$, $\mathfrak{H}_j \cap \mathfrak{K}_j = \mathbf{0}$) for $j = 1, \ldots, n-1$, then $\mathfrak{H} = \mathfrak{K}_1 + \mathfrak{K}_2 + \ldots + \mathfrak{K}_n$ with $\mathfrak{K}_1 \equiv \mathfrak{H}_1$ and $\pi(g)$ can be written in the following matrix form:

$$\pi(g) = \begin{pmatrix} \pi_1(g) & c_{12}(g) & \ldots & c_{1n}(g) \\ 0 & \pi_2(g) & \ldots & c_{2n}(g) \\ \ldots & \ldots & \ldots & \ldots \\ 0 & 0 & \ldots & \pi_n(g) \end{pmatrix}. \tag{1.3}$$

The off-diagonal elements $c_{ij}(g)$ $(i<j)$ can be studied in terms of cohomologies [8, 12, 13, 16]. For $n = 2$ and 3, we give a necessary and sufficient condition on representations π_j for (1.1) to have G-invariant non-degenerate inner product in terms of cohomologies related to π_j (Theorem 2).

All mathematical arguments are algebraic, topological complications being avoided in the following way: (1) We call a subspace $\mathfrak{E}$ of $\mathfrak{H}$ closed if $\mathfrak{E} = (\mathfrak{E}^\perp)^\perp$, where $\perp$ indicates the polar of $\mathfrak{E}$ in $\mathfrak{H}$ with respect to the (G-invariant) non-degenerate inner product in $\mathfrak{H}$:

$$\mathfrak{E}^\perp = \{\xi \in \mathfrak{H} \colon (\xi, \eta) = 0 \text{ for all } \eta \in \mathfrak{E}\}. \tag{1.4}$$

Irreducibility and indecomposability are defined in terms of G-invariant closed subspaces. Then Theorem 1 holds without any further assumption. (2) For discussing $c_{ij}(g)$, we introduce the assumption that subspaces $\mathfrak{H}_j$ have closed complements in $\mathfrak{H}_{j+1}$ (used in their definition). With this assumption, Theorem 2 holds.

2. Invariant Hermitian Form

Let π be a representation of a group G on a complex vector space $\mathfrak{H}$. Let (ξ, η), $\xi \in \mathfrak{H}$, $\eta \in \mathfrak{H}$ be a hermitian form on $\mathfrak{H}$, linear in ξ and conjugate linear in η [$\overline{(\xi, \eta)} = (\eta, \xi)$], $\pi(G)$-invariant [i.e. $(\pi(g)\xi, \pi(g)\eta) = (\xi, \eta)$ for all $g \in G$ and all $\xi \in \mathfrak{H}$, $\eta \in \mathfrak{H}$] and non-degenerate [i.e. $(\xi, \eta) = 0$ for all $\eta \in \mathfrak{H}$ implies $\xi = \mathbf{0}$].

We recall (cf. Sect. 1) that a subspace $\mathfrak{E}$ is defined to be closed if $\mathfrak{E} = (\mathfrak{E}^\perp)^\perp$.

Lemma 2.1. *$\mathfrak{E}$ is a closed subspace if and only if there is a subset Δ of $\mathfrak{H}$ such that $\mathfrak{E}=\Delta^{\perp}$.*

Proof. If $\mathfrak{E}$ is a closed subspace, we may choose $\Delta=\mathfrak{E}^{\perp}$. To prove the converse, we note (i) $\mathfrak{K}_1\subset\mathfrak{K}_2$, then $\mathfrak{K}_1^{\perp}\supset\mathfrak{K}_2^{\perp}$; (ii) from definition, $\mathfrak{K}\subset(\mathfrak{K}^{\perp})^{\perp}$. By taking $\mathfrak{K}=\mathfrak{E}$, we have $\mathfrak{E}\subset(\mathfrak{E}^{\perp})^{\perp}$ by (ii). By taking $\mathfrak{K}=\Delta$, we obtain $\Delta\subset(\Delta^{\perp})^{\perp}$ by (ii) and hence $\Delta^{\perp}\supset((\Delta^{\perp})^{\perp})^{\perp}$ by (i). Therefore, $\mathfrak{E}=(\mathfrak{E}^{\perp})^{\perp}$ *for* $\mathfrak{E}=\Delta^{\perp}$. Q.E.D.

We note that any finite dimensional subspace is closed by the non-degeneracy of the inner product.

Remark 2.2. The collection of seminorms $p_{\xi}(\eta)=|(\xi,\eta)|$ on $\eta\in\mathfrak{H}$ with ξ varying over $\mathfrak{H}$ defines the weak topology $\sigma(\mathfrak{H},\mathfrak{H})$ on $\mathfrak{H}$. A subspace $\mathfrak{E}$ of $\mathfrak{H}$ is closed in the above sense [i.e. $\mathfrak{E}=(\mathfrak{E}^{\perp})^{\perp}$] if and only if $\mathfrak{E}$ is $\sigma(\mathfrak{H},\mathfrak{H})$-closed due to the bipolar theorem.

We introduce the following definitions:

Definition 2.3. A representation π of G on $\mathfrak{H}$ is irreducible if there is no closed subspace $\mathfrak{E}$ of $\mathfrak{H}$ such that $\mathfrak{E}\neq\mathfrak{H}$, $\mathfrak{E}\neq\mathbf{0}$ and $\pi(g)\mathfrak{E}\subset\mathfrak{E}$ (i.e. $\mathfrak{E}$ is G-invariant) for all $g\in G$. A representation π of G on $\mathfrak{H}$ is topologically indecomposable if there are no non-zero closed G-invariant subspaces $\mathfrak{H}_1$ and $\mathfrak{H}_2$ of $\mathfrak{H}$ such that $((\mathfrak{H}_1+\mathfrak{H}_2)^{\perp})^{\perp}=\mathfrak{H}$ and $\mathfrak{H}_1\cap\mathfrak{H}_2=\mathbf{0}$ (i.e. $\mathfrak{H}_1$ and $\mathfrak{H}_2$ are topologically complementary).

In terms of the notion of conjugate representations introduced in Sect. 1, we can now present one of our main results:

Theorem 1. *Let π be a representation of a group G on a complex vector space $\mathfrak{H}$ with a G-invariant non-degenerate hermitian form. Let $\mathfrak{H}_1$ be a G-invariant closed subspace of $\mathfrak{H}$ such that the restriction π_1 of π to $\mathfrak{H}_1$ is irreducible and there are no G-invariant, closed topological complements of $\mathfrak{H}_1$ in $\mathfrak{H}$. Then* (1) *$\mathfrak{H}_1$ is a null space,* (2) *π is of the form* (1.1) *with $n=2$ or 3 such that π_n on $\mathfrak{H}_n/\mathfrak{H}_{n-1}$ is conjugate to π_1 on $\mathfrak{H}_1$, and* (3) *$\mathfrak{H}_2/\mathfrak{H}_1$ has a $\pi_2(G)$-invariant non-degenerate hermitian form.*

Proof. Consider $\mathfrak{K}=\mathfrak{H}_1\cap\mathfrak{H}_1^{\perp}$. Due to G-invariance of the hermitian form, the G-invariance of $\mathfrak{H}_1$ implies that of $\mathfrak{H}_1^{\perp}$ and hence that of $\mathfrak{K}$. As an intersection of closed subspaces, $\mathfrak{K}$ is closed. More explicitly, $(\mathfrak{H}_1+\mathfrak{H}_1^{\perp})^{\perp}=\mathfrak{H}_1^{\perp}\cap(\mathfrak{H}_1^{\perp})^{\perp}=\mathfrak{H}_1^{\perp}\cap\mathfrak{H}_1=\mathfrak{K}$. By the irreducibility assumption for π_1, the G-invariant closed subspace $\mathfrak{K}$ of $\mathfrak{H}_1$ must be either $\mathfrak{H}_1$ or $\mathbf{0}$.

If $\mathfrak{K}=\mathbf{0}$, then $(\mathfrak{H}_1+\mathfrak{H}_1^{\perp})^{\perp}=\mathfrak{K}=\mathbf{0}$ while $\mathfrak{H}_1\cap\mathfrak{H}_1^{\perp}=\mathbf{0}$. Hence $\mathfrak{H}_1^{\perp}$ is a G-invariant, closed, topological complement of $\mathfrak{H}_1$. This contradicts the assumption and $\mathfrak{K}=\mathbf{0}$ is excluded. [The assumption excludes the case of the zero complement (i.e. $\mathfrak{H}_1^{\perp}=\mathbf{0}$, $\mathfrak{H}_1=\mathfrak{H}$, π irreducible) as well.]

We now have $\mathfrak{H}_1\cap\mathfrak{H}_1^{\perp}=\mathfrak{K}=\mathfrak{H}_1$ and hence $\mathfrak{H}_1\subset\mathfrak{H}_1^{\perp}$. Therefore, $\mathfrak{H}_1$ is a null space [i.e. $\xi\in\mathfrak{H}$ and $\eta\in\mathfrak{H}$ imply $(\xi,\eta)=0$].

Let $\mathfrak{H}_2=\mathfrak{H}_1^{\perp}$ and $\mathfrak{H}_1^{\#}\equiv\mathfrak{H}/\mathfrak{H}_2$. For $\xi\in\mathfrak{H}$, denote the class $\xi+\mathfrak{H}_2$ of ξ in $\mathfrak{H}/\mathfrak{H}_2$ by $\xi^{\#}$. Denote $\pi^{\#}(g)\xi^{\#}\equiv(\pi(g)\xi)^{\#}$ which does not depend on the choice of ξ in the equivalence class $\xi^{\#}$ due to the G-invariance of $\mathfrak{H}_2$. Define $\langle\xi^{\#},\eta\rangle=(\xi,\eta)$ for $\xi\in\mathfrak{H}$ and $\eta\in\mathfrak{H}_1$. Again this does not depend on the choice of $\xi\in\xi^{\#}$ due to $\mathfrak{H}_2=\mathfrak{H}_1^{\perp}$. Due to the non-degeneracy of the hermitian form on $\mathfrak{H}$, $\mathfrak{H}_1^{\#}$ separates $\mathfrak{H}_1$, i.e. $\langle\xi^{\#},\eta\rangle=0$ for all $\xi^{\#}\in\mathfrak{H}_1^{\#}$ implies $\eta=0$. On the other hand, $\langle\xi^{\#},\eta\rangle=0$ for all $\eta\in\mathfrak{H}_1$ implies $\xi\in\mathfrak{H}_1^{\perp}=\mathfrak{H}_2$, and hence $\xi^{\#}=0$. Namely, $\mathfrak{H}_1$ separates $\mathfrak{H}_1^{\#}$. By G-invariance of the hermitian form, we have

$$\langle \pi_1^\#(g^{-1})\xi^\#, \eta\rangle = (\pi_1(g^{-1})\xi, \pi_1(g^{-1})\pi_1(g)\eta) = (\xi, \pi_1(g)\eta) = \langle \xi^\#, \pi_1(g)\eta\rangle .$$

Therefore, $\pi_1^\#$ on $\mathfrak{H}_1^\#$ is conjugate to π_1 on $\mathfrak{H}_1$.

If $\mathfrak{H}_2 = \mathfrak{H}_1^\perp$ coincides with $\mathfrak{H}_1$, we have the form (1.1) with $n=2$. If $\mathfrak{H}_2 \neq \mathfrak{H}_1$, restrict $\pi(g)$ to $\mathfrak{H}_2$, and then consider it modulo $\mathfrak{H}_1$. Denote the representation so obtained on $\mathfrak{H}_2/\mathfrak{H}_1$ by π_2. We then have the form with $n=3$. In either case $\pi_n = \pi_1^\#$ is conjugate to π_1 and (2) is proved.

For $\xi \in \mathfrak{H}_2$, denote $[\xi] = \xi + \mathfrak{H}_1 \in \mathfrak{H}_2/\mathfrak{H}_1$, and define $([\xi],[\eta]) = (\xi, \eta)$. Due to $\mathfrak{H}_2 = \mathfrak{H}_1^\perp$, this does not depend on representative vectors $\xi \in [\xi]$ and $\eta \in [\eta]$. Furthermore, $([\xi],[\eta]) = 0$ for all $[\eta] \in \mathfrak{H}_2/\mathfrak{H}_1$ implies $\xi \in \mathfrak{H}_2^\perp = (\mathfrak{H}_1^\perp)^\perp = \mathfrak{H}_1$, and hence $[\xi] = 0$. Since the hermitian form is $\pi_2(G)$-invariant, we have (3).

Remark 2.4. If the representation π is topologically indecomposable and has a proper irreducible subrepresentation π_1, then the assumption of Theorem 1 is satisfied.

3. Cohomology

In order to discuss $c_{ij}(g)$, we introduce a standard notation for a cohomology.

Definition 3.1. The set $C^n(\pi_1, \pi_2)$ of n-cochains for representations $\pi_i^\#$ on $\mathfrak{H}_i^\#(i=1,2)$ is the set of all functions $c(g_1, \dots, g_n)$ of $g_k \in G$ $(k=1, \dots, n)$, taking values in the set $L(\mathfrak{H}_1, \mathfrak{H}_2)$ of all everywhere-defined linear maps from $\mathfrak{H}_1$ into $\mathfrak{H}_2$.

In our application, there are representations π_j on $\mathfrak{H}_j$ $(j=1,2)$ conjugate to π_j on $\mathfrak{H}_j$, and we need the following restriction:

Definition 3.2. The set $C_b^n(\pi_1, \pi_2)$ of bounded n-cochains is the set of $c \in C^n(\pi_1, \pi_2)$ such that there exists $c^* \in C^n(\pi_2^\#, \pi_1^\#)$ satisfying

$$\langle \xi, c(g_1, \dots, g_n)\eta\rangle = \langle c^*(g_n^{-1}, \dots, g_n^{-1})\xi, \eta\rangle \tag{3.1}$$

for all $\eta \in \mathfrak{H}_1$ and $\xi \in \mathfrak{H}_2^\#$. The set $C_s^n(\pi_1\pi_1^\#)$ of self-adjoint cochains is the set of $c \in C^n(\pi_1, \pi_1^\#)$ such that (3.1) is satisfied for $c^* = c$ and $\xi, \eta \in \mathfrak{H}_1$.

The coboundary operation δ is defined by the following and satisfies $\delta^2 = 0$.

Definition 3.3.

$$\begin{aligned}(\delta c_n)(g_1, \dots, g_{n+1}) = {} & \pi_2(g_1)c_n(g_2, \dots, g_n) \\ & + \sum_{i=1}^{n} (-1)^i c(g_1, \dots, g_i g_{i+1}, \dots, g_{n+1}) \\ & + (-1)^{n+1} c_n(g_1, \dots, g_n)\pi_1(g_{n+1}) .\end{aligned} \tag{3.2}$$

Cocycles, coboundaries, and cohomologies are defined as usual.

Definition 3.4.

$$Z^n(\pi_1, \pi_2) = \{c_n \in C^n;\ \delta c_n = 0\} ,$$

$$B^n(\pi_1, \pi_2) = \delta C^{n-1}(\pi_1, \pi_2) \quad (=0 \text{ for } n=0) ,$$

$$H^n(\pi_1, \pi_2) = Z^n(\pi_1, \pi_2)/B^n(\pi_1, \pi_2) .$$

Similarly, Z^n, B^n, H^n with suffices b and s are defined. Note that $(\delta c_n)^* = (-1)^{n+1}\delta(c_n^*)$, as is seen by a straightforward calculation. Thus $B_s^n \equiv i^n \delta C_s^{n-1} = B_b^n \cap C_s^n$.

Definition 3.5. For $c_{32} \in C^m(\pi_2, \pi_3)$ and $c_{21} \in C^n(\pi_1, \pi_2)$, define $c_{32} \times c_{21} \in C^{m+n}(\pi_1, \pi_2)$ by

$$(c_{32} \times c_{21})(g_1, \ldots, g_{m+n}) = c_{32}(g_1, \ldots, g_m) c_{21}(g_{m+1}, \ldots, g_{m+n}). \tag{3.3}$$

Lemma 3.6.

$$Z^m(\pi_2, \pi_3) \times Z^n(\pi_1, \pi_2) \subset Z^{m+n}(\pi_1, \pi_3),$$

$$B^m(\pi_2, \pi_3) \times Z^n(\pi_1, \pi_2) \subset B^{m+n}(\pi_1, \pi_3),$$

$$Z^m(\pi_2, \pi_3) \times B^n(\pi_1, \pi_2) \subset B^{m+n}(\pi_1, \pi_3),$$

$$H^m(\pi_2, \pi_3) \times H^n(\pi_1, \pi_2) \subset H^{m+n}(\pi_1, \pi_3).$$

Proof. Straightforward consequences of the following formula:

$$\delta(c_{32} \times c_{21}) = (\delta c_{32}) \times c_{21} + (-1)^m c_{32} \times (\delta c_{21}). \tag{3.4}$$

4. Cohomological Conditions on $c_{ij}(g)$

In the rest of the paper, we assume that each $\mathfrak{H}_j$ has a complement so that $\pi(g)$ can be written in the form of (1.3) and discuss the off-diagonal elements $c_{ij}(g)$.

Lemma 4.1. *In order that* (1.2) *be a representation of a group G, it is necessary and sufficient that* (1) *each* π_i *is a representation and* (2) c_{ij}, $1 \leqq i < j \leqq n$, *satisfy*

$$\delta c_{kl} = -\sum_{j=k+1}^{l-1} c_{kj} \times c_{jl} \quad (1 \leqq k < l \leqq n), \tag{4.1}$$

where the right-hand side is understood to be 0 *if* $l = k+1$.

Proof. Follows from a straightforward computation of $(\pi(g_1)\pi(g_2))_{kl} = \pi(g_1 g_2)_{kl}$.

Remark 4.2. The right-hand side of (4.1) belongs to $Z^2(\pi_l, \pi_k)$ already due to other constraints involving only c_{ij} with $i+j < k+l$. [By (3.4), we have

$$\delta(\sum c_{kj} \times c_{jl}) = \sum (\delta c_{kj}) \times c_{jl} - \sum c_{kj} \times \delta c_{jl}.$$

By substituting (4.1) into δc_{kj} and δc_{jl}, we see that this vanishes by cancellation.] Equation (4.1) requires it to belong to $B^2(\pi_l, \pi_k)$.

For given π and an increasing sequence of G-invariant subspaces $\mathfrak{H}_1, \mathfrak{H}_2, \ldots, \mathfrak{H}_n$, $c_{ij}(g)$ depends on the choice of a complement $\mathfrak{K}_{k+1}$ of $\mathfrak{H}_k$ in $\mathfrak{H}_{k+1}$. Change of $\mathfrak{K}$'s can be described in terms of a triangular matrix S ($S_{kl} = 0$ for $k > l$) which leaves every $\mathfrak{H}_j$ invariant. A general triangular matrix S is generated by the following specific ones

$$S_i = \mathbb{1} + u_{ii}(R_i - \mathbb{1}_{\mathfrak{H}_i}), \qquad S_{ij} = \mathbb{1} + u_{ij} R_{ij}, \tag{4.2}$$

where u_{ij} is a matrix unit [i.e. $(u_{ij})_{kl} = \delta_{ik}\delta_{jl}$], R_i is an operator on $\mathfrak{H}_i$ and R_{ij} is a linear map from $\mathfrak{H}_j$ into $\mathfrak{H}_i$. Changes by transformation $\pi(s) \to S_*^{-1} \pi(g) S_*$ are as follows, where π's and c's not mentioned are not changed at all:

$$S_i : \pi_i(g) \to R_i^{-1}\pi_i(g)R_i, \quad c_{ij}(g) \to R_i^{-1}c_{ij}(g), \quad c_{ki}(g) \to c_{ki}(g)R_i,$$

$$S_{ij} : c_{ij}(g) \to c_{ij}(g) + (\delta R_{ij})(g), \quad c_{ik}(g) \to c_{ik}(g) - R_{ij}c_{jk}(g),$$

$$c_{kj}(g) \to c_{kj}(g) + c_{ki}(g)R_{ij}.$$

We now consider a specific case of $n=2$ and assume that the G-invariant subspace $\mathfrak{H}_1$ $(=\mathfrak{K}_1)$ has a complement $\mathfrak{K}_2$ so that π is of the form (1.3) with $n=2$. By Lemma 4.1, c_{12} belongs to $Z^1(\pi_2, \pi_1)$.

Proposition 4.3. *$\mathfrak{H}_1$ does not have a G-invariant complement if and only if $c_{12} \notin B^1(\pi_2, \pi_1)$.*

Proof. First assume that there exists a G-invariant complement $\mathfrak{K}_2'$. Then $\mathfrak{H} = \mathfrak{H}_1 + \mathfrak{K}_2 = \mathfrak{H}_1 + \mathfrak{K}_2'$. Hence any $\xi \in \mathfrak{H}$ has two decompositions

$$\xi = \xi_1 + \xi_2 = \xi_1' + \xi_2', \quad (\xi_1, \xi_1' \in \mathfrak{H}_1, \ \xi_2 \in \mathfrak{K}_2, \ \xi_2' \in \mathfrak{K}_2'). \tag{4.3}$$

Define $W = (W_{ij})_{i,j=1,2}$, $W_{ij}\xi_j = \xi_i'$, $(W^{-1})_{ij}\xi_j' = \xi_i$. Then $W_{21} = (W^{-1})_{21} = 0$, $W_{11} = (W^{-1})_{11} = 1$, $(W_{22})^{-1} = (W^{-1})_{22}$, and $(W^{-1})_{12} = -W_{12}W_{22}^{-1}$. Since $\pi(g)$ has a diagonal form $\pi_1(g) \oplus \pi_2'(g)$ for the decomposition $\mathfrak{H} = \mathfrak{H}_1 + \mathfrak{K}_2'$, we have

$$\pi(g) = W^{-1}\begin{pmatrix} \pi_1(g) & 0 \\ 0 & \pi_2'(g) \end{pmatrix} W \tag{4.4}$$

for the decomposition $\mathfrak{H} = \mathfrak{H}_1 + \mathfrak{K}_2$. Hence $\pi_2(g) = W_{22}^{-1}\pi_2'(g)W_{22}$ and

$$c_{12} = \delta W_{12}. \tag{4.5}$$

Conversely, assume that $c_{12} = \delta W_{12}$ for some $W_{12} \in C^0(\pi_2, \pi_1)$. Then define $\mathfrak{K}_2' = (1 - W_{12})\mathfrak{K}_2$. If $(1 - W_{12})\xi_2 \in \mathfrak{H}_1$ for $\xi_2 \in \mathfrak{K}_2$, then $\xi_2 = (1 - W_{12})\xi_2 + W_{12}\xi_2 \in \mathfrak{H}_1$ (due to $W_{12}\xi_2 \in \mathfrak{H}_1$) and $\xi_2 \in \mathfrak{H}_1 \cap \mathfrak{K}_2 = \mathbf{0}$. Hence $\mathfrak{H}_1 \cap \mathfrak{K}_2' = \mathbf{0}$. Furthermore,

$$W = \begin{pmatrix} 1 & W_{12} \\ 0 & 1 - W_{12} \end{pmatrix} \tag{4.6}$$

obviously intertwines $\mathfrak{H} = \mathfrak{H}_1 + \mathfrak{K}_2$ into $\mathfrak{H}_1 + \mathfrak{K}_2'$, so that $\mathfrak{K}_2'$ is a complement of $\mathfrak{H}$. Furthermore, since $W_{12}\xi_2 \in \mathfrak{H}_1$ for $\xi_2 \in \mathfrak{K}_2$,

$$\begin{aligned} \pi(g)(1 - W_{12})\xi_2 &= \pi_2(g)\xi_2 - \pi_1(g)W_{12}\xi_2 + c_{12}(g)\xi_2 \\ &= (1 - W_{12})\pi_2(g)\xi_2 \in \mathfrak{K}_2', \end{aligned}$$

due to $c_{12} = \delta W_{12}$. Therefore, $\mathfrak{K}_2'$ is a $\pi(G)$-invariant complement of $\mathfrak{H}_1$. Q.E.D.

For our application, we would be interested in closed subspaces. Let $\mathfrak{H}^{\#}$ be a subspace of the algebraic dual of $\mathfrak{H}$ and assume that $\mathfrak{H}^{\#}$ separates $\mathfrak{H}$. We consider the weak topology $\sigma(\mathfrak{H}, \mathfrak{H}^{\#})$ on $\mathfrak{H}$ given by seminorms $\xi \in \mathfrak{H} \to |\zeta(\xi)| \equiv p_\zeta(\xi)$, $\zeta \in \mathfrak{H}^{\#}$.

Assumption. The $\pi(G)$-invariant subspace $\mathfrak{H}_1$ of $\mathfrak{H}$ has a closed complement $\mathfrak{K}_2 \equiv \mathfrak{H}_2$ such that $\mathfrak{H}_2^{\perp} + \mathfrak{H}_1^{\perp} = \mathfrak{H}^{\#}$, where $\mathfrak{H}_1^{\perp}$ and $\mathfrak{H}_2^{\perp}$ are the polar of $\mathfrak{H}_1$ and $\mathfrak{H}_2$ in $\mathfrak{H}^{\#}$.

Since $\mathfrak{H}$ separates its dual and hence $\mathfrak{H}^{\#}$ by definition, $\mathfrak{H}_1^{\perp}\cap\mathfrak{H}_2^{\perp}=0$. Then $\sigma(\mathfrak{H},\mathfrak{H}^{\#})$ topology on $\mathfrak{H}_i$ is the $\sigma(\mathfrak{H}_i,\mathfrak{H}_i^{\#})$ topology, where $\mathfrak{H}_1^{\#}=\mathfrak{H}_2^{\perp}$, $\mathfrak{H}_2^{\#}\equiv\mathfrak{H}_1^{\perp}$. Let $\pi(g)$ be of the form (1.3) with $n=2$ for the decomposition $\mathfrak{H}=\mathfrak{H}_1+\mathfrak{H}_2$.

Proposition 4.3′. *Under the assumption above, $\mathfrak{H}_1$ does not have a G-invariant closed complement $\mathfrak{K}_2'$ satisfying $\mathfrak{H}_1^{\perp}+(\mathfrak{K}_2')^{\perp}=\mathfrak{H}^{\#}$ if and only if $c_{12}\notin B_b^1(\pi_2,\pi_1)$.*

Proof. First assume that there exists a G-invariant closed complement $\mathfrak{K}_2'$ of $\mathfrak{H}_1$ satisfying $\mathfrak{H}_1^{\perp}+(\mathfrak{K}_2')^{\perp}=\mathfrak{H}^{\#}$. By the same procedure as before we have (4.5), and we have only to show that W_{12}^* exists so that $W_{12}\in C_b^0(\pi_2,\pi_1)$. Let $\xi_\alpha\in\mathfrak{H}$ be a net tending to 0 in $\sigma(\mathfrak{H},\mathfrak{H}^{\#})$ topology. Let $\xi_\alpha=\xi_{\alpha1}'+\xi_{\alpha2}'$, $\xi_{\alpha1}'\in\mathfrak{H}_1$, $\xi_{\alpha2}'\in\mathfrak{K}_2'$. For any $\zeta\in\mathfrak{H}^{\#}$, we have $\zeta=\zeta_1'+\zeta_2'$ with $\zeta_1'\in(\mathfrak{K}_2')^{\perp}$, $\zeta_2'\in\mathfrak{H}_1^{\perp}$, and $(\zeta,\xi_{\alpha1}')=(\zeta_1',\xi_{\alpha1}')=(\zeta_1',\xi_\alpha)\to0$. Hence W_{12} is $\sigma(\mathfrak{H},\mathfrak{H}^{\#})$ continuous. For any $\zeta\in\mathfrak{H}^{\#}$, we have $\zeta=\zeta_1+\zeta_2$ with $\zeta_1\in\mathfrak{K}_2^{\perp}$, $\zeta_2\in\mathfrak{H}_1$, and $(\zeta,\xi)=(\zeta_2,\xi)$ for $\xi\in\mathfrak{K}_2$. Thus $\sigma(\mathfrak{H},\mathfrak{H}^{\#})$ topology on $\mathfrak{K}_2$ is the same as $\sigma(\mathfrak{K}_2,\mathfrak{H}_1^{\perp})$ topology. Similarly, $\sigma(\mathfrak{H},\mathfrak{H}^{\#})$ topology on $\mathfrak{H}_1$ is the same as $\sigma(\mathfrak{H}_1,\mathfrak{H}_1^{\#})$ topology with either $\mathfrak{H}_1^{\#}=\mathfrak{K}_2^{\perp}$ or $\mathfrak{H}_1^{\#}=(\mathfrak{K}_2')^{\perp}$. For $\zeta\in\mathfrak{H}_1^{\#}$, $(\zeta,W_{12}\xi)$ is $\sigma(\mathfrak{K}_2,\mathfrak{H}_1^{\perp})$ continuous (due to the continuity of W_{12} shown above), and hence there exists a unique $W_{12}^*\zeta\in\mathfrak{H}_1^{\perp}=\mathfrak{K}_2^{\#}$ satisfying $(W_{12}^*\zeta,\xi)=(\zeta,W_{12}\xi)$. Here the uniqueness follows from the separating property of $\mathfrak{K}_2$ on $\mathfrak{K}_2^{\#}$ ($\mathfrak{H}_1^{\perp}\cap\mathfrak{K}_2^{\perp}=0$) and implies the linearity of W_{12}^*. Therefore, $W_{12}\in C_b^0(\pi_2,\pi_1)$. Conversely, assume that $c_{12}=\delta W_{12}$ for some $W_{12}\in C_b^0(\pi_2,\pi_1)$. We then prove that $\mathfrak{K}_2'=(1-W_{12})\mathfrak{K}_2$ is closed and $\mathfrak{H}^{\#}=\mathfrak{H}_1^{\perp}+(\mathfrak{K}_2')^{\perp}$. Let $(1-W_{12})\xi_\alpha$ ($\xi_\alpha\in\mathfrak{K}_2$) be a net in $\mathfrak{K}_2'$ tending to $\bar{\xi}\in\mathfrak{H}$. Let $\bar{\xi}=\bar{\xi}_1+\bar{\xi}_2$, $\bar{\xi}_1\in\mathfrak{H}_1$, $\bar{\xi}_2\in\mathfrak{K}_2$. For any $\zeta\in\mathfrak{H}_1^{\perp}=\mathfrak{K}_2^{\#}$, we have $(\zeta,\xi_\alpha-\bar{\xi}_2)=(\zeta,(1-W_{12})\xi_\alpha-\bar{\xi})\to0$, and hence $\xi_\alpha\to\bar{\xi}_2$ in $\sigma(\mathfrak{K}_2,\mathfrak{K}_2^{\#})$ topology, which is the same as $\sigma(\mathfrak{H},\mathfrak{H}^{\#})$ topology on $\mathfrak{K}_2$. For any $\zeta\in\mathfrak{H}_1^{\#}$, we have $(\zeta,W_{12}(\xi_\alpha-\bar{\xi}))=(W_{12}^*\zeta,\xi_\alpha-\bar{\xi}_2)\to0$. Hence $W_{12}(\xi_\alpha-\bar{\xi}_2)\to0$ in $\sigma(\mathfrak{H}_1,\mathfrak{H}_1^{\#})$ topology which is the same as $\sigma(\mathfrak{H},\mathfrak{H}^{\#})$ topology on $\mathfrak{H}_1$. Hence $(1-W_{12})\xi_\alpha\to(1-W_{12})\bar{\xi}_2\in\mathfrak{K}_2'$ and $\mathfrak{K}_2'$ is $\sigma(\mathfrak{H},\mathfrak{H}^{\#})$ closed.

Let $(1-W_{12})\xi_\alpha$ ($\xi_\alpha\in\mathfrak{K}_2$) be a net in $\mathfrak{K}_2'$ tending to 0 in $\sigma(\mathfrak{K}_2',\mathfrak{H}_1^{\perp})$ topology. Then, for any $\zeta\in\mathfrak{H}_1^{\perp}$, $(\zeta,\xi_\alpha)=(\zeta,(1-W_{12})\xi_\alpha)\to0$. Since $\sigma(\mathfrak{H},\mathfrak{H}^{\#})$ topology on $\mathfrak{K}_2$ is the same as $\sigma(\mathfrak{K}_2,\mathfrak{H}_1^{\perp})$ topology due to the assumption, we have $\xi_\alpha\to0$. Hence $(\zeta,W_{12}\xi_\alpha)=(W_{12}^*\zeta,\xi_\alpha)\to0$ for all $\zeta\in\mathfrak{K}_2^{\perp}$, which implies $W_{12}\xi_\alpha\to0$ in $\sigma(\mathfrak{H}_1,\mathfrak{K}_2^{\perp})$ topology or, equivalently, in $\sigma(\mathfrak{H},\mathfrak{H}^{\#})$ topology. Therefore, we have $(1-W_{12})\xi_\alpha\to0$ in $\sigma(\mathfrak{H},\mathfrak{H}^{\#})$ topology. This means that $\sigma(\mathfrak{K}_2',\mathfrak{H}_1^{\perp})$ and $\sigma(\mathfrak{H},\mathfrak{H}^{\#})$ topologies are the same on $\mathfrak{K}_2'$. Hence, for any $\zeta\in\mathfrak{H}^{\#}$, there exists $\zeta_2'\in\mathfrak{H}_1^{\perp}$ such that $(\zeta,\xi)=(\zeta_2',\xi)$ for all $\xi\in\mathfrak{K}_2'$, i.e. $\zeta_1'\equiv\zeta-\zeta_2'\in(\mathfrak{K}_2')^{\perp}$. Therefore, $\mathfrak{H}^{\#}=\mathfrak{H}_1^{\perp}+(\mathfrak{K}_2')^{\perp}$. Q.E.D.

Proposition 4.3 is generalized to the case of $\mathfrak{H}=\mathfrak{K}_1+\ldots+\mathfrak{K}_n$ with $\mathfrak{K}_i\cap\mathfrak{K}_j=0$ for $i\neq j$ and $\mathfrak{H}_j=\mathfrak{K}_1+\ldots+\mathfrak{K}_j$.

Corollary 4.4. *For a representation π of the form (1.3), $\mathfrak{H}_j$ does not have any G-invariant complement in $\mathfrak{H}_{j+1}$ for $j=1,\ldots,n-1$ if and only if*

$$c_{k,k+1}\notin B^1(\pi_{k+1},\pi_k)\,,\qquad k=1,\ldots,n-1\,,\tag{4.7}$$

$$\delta c_{k,l}=-\sum_{j=k+1}^{l-1}c_{kj}\times c_{jl}\,,\qquad 1\leqq k<l+1\leqq n+1\,.\tag{4.8}$$

Proof. Equation (4.8) is necessary and sufficient for π of the form (1.3) to be a representation by Lemma 4.1. Then (4.7) is necessary and sufficient by Proposition 4.3.

Corollary 4.5. *Let π_1 on $\mathfrak{H}_1$ and $\pi_1^\#$ on $\mathfrak{H}_1^\#$ be irreducible representations of a group G which are mutually conjugate with respect to a mutually separating sesquilinear form $\langle\xi,\eta\rangle$, $\eta\in\mathfrak{H}_1$, $\xi\in\mathfrak{H}_1^\#$. (For irreducibility, the closedness of invariant subspaces is relative to the weak topologies on $\mathfrak{H}_1$ by $\mathfrak{H}_1^\#$ and on $\mathfrak{H}_1^\#$ by $\mathfrak{H}_1$.) In order that there exists a representation of the form $\pi=\pi_1^\#\to\pi_1$ on $\mathfrak{H}=\mathfrak{H}_1+\mathfrak{H}_1^\#$ with a $\pi(G)$-invariant hermitian form*

$$(\eta_1+\xi_1,\eta_2+\xi_2)=\langle\xi_1,\eta_2\rangle+\overline{\langle\xi_2\eta_1\rangle} \tag{4.9}$$

for $\eta_j\in\mathfrak{H}_1$, $\xi_j\in\mathfrak{H}_1^\#$ $(j=1,2)$ and with no $\pi(G)$-invariant closed complement $\mathfrak{K}_2'$ of $\mathfrak{H}_1$ satisfying $(\mathfrak{K}_2')^\perp+\mathfrak{K}_1^\perp=\mathfrak{H}$, it is necessary and sufficient that $H_s^1(\pi_1^\#,\pi_1)\neq 0$.

Proof. If such a representation exists, we have a nonzero $c_{12}\in H_b^1(\pi_1^\#,\pi_1)$, which satisfies for $\xi_j\in\mathfrak{H}_1^\#$ $(j=1,2)$

$$\begin{aligned}\langle\xi_1,c_{12}(g)\xi_2\rangle&=(\xi_1,\pi(g)\xi_2)=(\pi(g^{-1})\xi_1,\xi_2)\\&=\overline{\langle\xi_2,c_{12}(g^{-1})\xi_1\rangle}\equiv\langle c_{12}(g^{-1})\xi_1,\xi_2\rangle,\end{aligned}$$

where the second equality is due to the $\pi(G)$-invariance of the inner product and the last equality is the definition of the sesquilinear form between $\xi_2\in\mathfrak{H}_1^\#$ and $c_{12}(g^{-1})\xi_1\in(\mathfrak{H}_1^\#)^\#\equiv\mathfrak{H}_1$. Hence $c_{12}=c_{12}^*\in Z_s^1(\pi_1^\#,\pi_1)$ and $c_{12}\notin B_s^1=B_b^1\cap C_s^1$.

Conversely, if there exists a nonzero $c_{12}\in\mathfrak{H}_s^1(\pi_1^\#,\pi_1)$, then π given by (1.3) on $\mathfrak{H}=\mathfrak{H}_1+\mathfrak{H}_1^\#$ is a representation of G, under which (4.9) is invariant, as is easily checked, and $\mathfrak{H}_1$ does not have an invariant complement by Proposition 4.3. (Note that c_{12} is nonzero in H_b^1 due to $B_b^1\cap C_s^1=B_s^1$.)

5. A Standard Form of the Gupta-Bleuler Triplet

The triplet of the form (1.1) with $n=3$ given in the preceding section (i.e. $\mathfrak{H}_j$ does not have any G-invariant closed complement in $\mathfrak{H}_{j+1}$, $j=1,2$, and $\mathfrak{H}_2=\mathfrak{H}_1^\perp$) will be called the Gupta-Bleuler triplet. We assume that $\mathfrak{H}_i$ has a closed complement $\mathfrak{K}_{i+1}$ in $\mathfrak{H}_{i+1}$ so that the representation is of the form (1.3).

Definition 5.1. A Gupta-Bleuler triplet

$$\pi=\pi_1^\#\to\pi_2\to\pi_1 \quad \text{on} \quad \mathfrak{H}=\mathfrak{K}_1\oplus\mathfrak{K}_2\oplus\mathfrak{K}_3 \tag{5.1}$$

with $\mathfrak{K}_1=\mathfrak{H}_1$, $\mathfrak{K}_3=\mathfrak{H}_1^\#$,

$$\pi(g)=\begin{pmatrix}\pi_1(g) & c_{12}(g) & c_{13}(g)\\ 0 & \pi_2(g) & c_{23}(g)\\ 0 & 0 & \pi_1^\#(g)\end{pmatrix},$$

and the inner product

$$(\xi_1+\xi_2+\xi_3,\xi_1'+\xi_2'+\xi_3')=\langle\xi_3,\xi_1'\rangle+\overline{\langle\xi_3',\xi_1\rangle}+(\xi_2,\xi_2')_2 \tag{5.2}$$

for $\xi_j,\xi_j'\in\mathfrak{K}_j$ $(j=1,2,3)$ is said to be in a standard form, where $(\xi_2,\xi_2')_2$ is a $\pi_2(G)$-invariant non-degenerate hermitian form on $\mathfrak{K}_2$.

Theorem 2. *Let conjugate irreducible representations π_1 on $\mathfrak{H}_1$ and $\pi_1^\#$ on $\mathfrak{H}_1^\#$ as in Corollary 4.5 and a representation π_2 on $\mathfrak{K}_2$ with an invariant non-degenerate hermitian form be given. In order that a Gupta-Bleuler triplet $\pi=\pi_1^\#\to\pi_2\to\pi_1$ on $\mathfrak{H}=\mathfrak{H}_1+\mathfrak{K}_2+\mathfrak{H}_1^\#$ in a standard form to exist, it is necessary and sufficient that there exists a $c_{12}\in H_b^1(\pi_2,\pi_1)$ ($\pi_2^\#$ on $\mathfrak{H}_2^\#$ taken to be π_2 on $\mathfrak{H}_2$ with the duality given by the G-invariant hermitian form on $\mathfrak{H}_2$) such that $c_{12}\times c_{12}^*\in B_s^2(\mathfrak{H}_1^\#,\mathfrak{H}_1)$ and c_{12} and c_{12}^* are nonzero in $H^1(\pi_2,\pi_1)$ and $H^1(\pi_1^\#,\pi_2)$.*

Proof. Assume the existence of the triplet $\pi=\pi_1^\#\to\pi_2\to\pi_1$ on $\mathfrak{H}=\mathfrak{H}_1+\mathfrak{K}_2+\mathfrak{H}_1^\#$ in a standard form. By Corollary 4.5, $c_{12}\in H^1(\pi_2,\pi_1)$ and $c_{23}\in H^1(\pi_1^\#,\pi_2)$ are nonzero and $c_{12}\times c_{23}=\delta\psi$ with $\psi=-c_{13}\in C^1(\pi_1^\#,\pi_1)$. Furthermore, the $\pi(G)$-invariance of (5.2) implies $c_{23}=c_{12}^*$ and $c_{13}=c_{13}^*$. This proves the necessity.

For the converse, we take the given c_{12} and define $c_{23}\equiv c_{12}^*$ and $c_{13}=-\psi$. Then $\pi(g)$ given by (1.3) is a representation of G on $\mathfrak{H}$ and (5.2) is $\pi(G)$-invariant. Furthermore, the non-existence of $\pi(G)$-invariant closed complement follows from Proposition 4.3 due to $c_{12}\neq 0$ and $c_{12}^*\neq 0$.

Proposition 5.3. *A Gupta-Bleuler triplet $\pi=\pi_1^\#\to\pi_2\to\pi_1$ on $\mathfrak{H}=\mathfrak{H}_1+\mathfrak{K}_2+\mathfrak{K}_3$ can be transformed into a standard form by a proper choice of the closed subspaces $\mathfrak{K}_2$ and $\mathfrak{K}_3$ (with $\mathfrak{H}_2=\mathfrak{H}_1+\mathfrak{K}_2=\mathfrak{H}_1^\perp$ fixed) if the weak topology given by $\mathfrak{H}_1$ on $\mathfrak{H}_1^\#=\mathfrak{H}/\mathfrak{H}_2=\mathfrak{K}_3$ coincides with the weak topology given by $\mathfrak{H}$ on $\mathfrak{K}_3$.*

Remark. The assumption on the weak topology is equivalent to $\mathfrak{K}_3^\perp+\mathfrak{H}_1=\mathfrak{H}$.

Proof. Assume the coincidence of two topologies. For each $\xi\in\mathfrak{K}_2$ and $\eta\in\mathfrak{K}_3$, there exist $\sigma_{12}\xi$ and $\sigma_{13}\eta$ in $\mathfrak{H}_1$ satisfying

$$(\xi,\eta')=(\sigma_{12}\xi,\eta'),\qquad (\eta,\eta')=(\sigma_{13}\eta,\eta') \tag{5.3}$$

for all $\eta'\in\mathfrak{K}_3$. Define

$$\mathfrak{K}_2'=(1-\sigma_{12})\mathfrak{K}_2,\qquad \mathfrak{K}_3'=(1-\sigma_{13}/2)\mathfrak{K}_3\,. \tag{5.4}$$

We have $\mathfrak{H}_1+\mathfrak{K}_2'=\mathfrak{H}_1+\mathfrak{K}_2=\mathfrak{H}_2$, $\mathfrak{H}_2+\mathfrak{K}_3'=\mathfrak{H}_2+\mathfrak{K}_3=H$. Furthermore, $\mathfrak{K}_2'$ is orthogonal to $\mathfrak{K}_3$, hence to $\mathfrak{K}_3'$ (due to $\mathfrak{H}_1\perp\mathfrak{H}_2\supset\mathfrak{K}_2'$). Further,

$$((1-\sigma_{13}/2)\eta,(1-\sigma_{13}/2)\eta')=(\eta,\eta')-((\sigma_{13}\eta,\eta')+(\eta,\sigma_{13}\eta'))/2=0\,,$$

i.e. $\mathfrak{K}_3'$ is a null space. Therefore, the inner product in $\mathfrak{H}$ takes the form of (5.2) relative to the decomposition $\mathfrak{H}=\mathfrak{H}_1+\mathfrak{K}_2'+\mathfrak{K}_3'$ and the Gupta-Bleuler triplet is now in a standard form.

Remark 5.4. Let $R_{12}\in C_b^0(\pi_2,\pi_1)$, $R_{13}=-R_{13}^*\in L(\mathfrak{H}_1^\#,\mathfrak{H}_1)$, and set

$$W=\begin{pmatrix}1 & -R_{12} & -R_{12}R_{12}^*/2-R_{13}\\ 0 & 1 & R_{12}^*\\ 0 & 0 & 1\end{pmatrix}. \tag{5.5}$$

Then W preserves the inner product (5.2) of the normal form and

$$W^{-1}=\begin{pmatrix}1 & R_{12} & -R_{12}R_{12}^*/2+R_{13}\\ 0 & 1 & -R_{12}^*\\ 0 & 0 & 1\end{pmatrix}. \tag{5.6}$$

The coordinate change $\pi(g)\to W\pi(g)W^{-1}$ changes c_{12} by a coboundary:

$$c_{12}\to c_{12}+\delta R_{12}, \tag{5.7}$$

$$c_{13}\to c_{13}-\{(\delta R_{12})R_{12}^{*}-R_{12}\delta(R_{12}^{*})\}/2-c_{12}R_{12}^{*}-R_{12}c_{12}^{*}+\delta R_{13}. \tag{5.8}$$

Note that $(\delta R)^{*}=-\delta(R^{*})$, so that the new c_{13} is again selfadjoint. The relation $\delta c_{13}=-c_{12}c_{12}^{*}$ is also preserved.

6. Discussion

Our results suggest the following method of analysis for a representation π of a group G on a vector space $\mathfrak{H}$ with a $\pi(G)$-invariant, nondegenerate indefinite inner product. The first step is to decompose the representation as a sum of indecomposable representations. Obviously, there would be cases, where this is not possible. (An example is a direct integral of non-equivalent indecomposable representations.)

Suppose we succeed in this first step and suppose $(\pi, \mathfrak{H})$ is now topologically indecomposable. The second step is then to find an irreducible subrepresentation $(\pi_1, \mathfrak{H}_1)$ unless $(\pi, \mathfrak{H})$ is already irreducible.

The third step is then to apply our results. Especially, if $\mathfrak{H}_1$ has a closed complement $\mathfrak{K}_2$ in $\mathfrak{H}_1^{\perp}$ ($\mathfrak{H}_1^{\perp}\supset\mathfrak{H}_1$ by Theorem 1) and if $\mathfrak{H}_1^{\perp}$ has a closed complement $\mathfrak{K}_3$ such that $\mathfrak{H}_1+\mathfrak{K}_3^{\perp}=\mathfrak{H}$, then π is of the standard form of Definition 5.1 due to Proposition 5.3, and the analysis of Theorem 2 is applicable. (It may happen that $\mathfrak{H}_1=\mathfrak{H}_1^{\perp}$, in which case we have $\mathfrak{K}_2=0$.) Now we are left with the analysis of π_2 on $\mathfrak{K}_2$ with a non-degenerate invariant inner product, and we can follow the same procedure as before.

While there are several points in the above procedure for possible obstructions, there seem to be many examples for which the above procedure works fine.

Acknowledgement. The author is indebted to Prof. C. Fronsdal for drawing his attention to the problem treated in this paper.

References

1. Binegar, B.T.: On the state space of the dipole ghost. Lett. Math. Phys. **8**, 149–158 (1984)
2. Binegar, B., Fronsdal, C., Heidenreich, W.: Linear conformal quantum gravity. Phys. Rev. D **27**, 2249–2261 (1983)
3. Binegar, B., Fronsdal, C., Heidenreich, W.: Conformal QED. J. Math. Phys. **24**, 2828–2846 (1983)
4. Bleuler, K.: Eine neue Methode zur Behandlung der longitudinalen und skalaren Photonen. Helv. Phys. Acta **23**, 567–586 (1950)
5. Flato, M., Fronsdal, C.: Quantum field theory of singletons (preprint)
6. Fronsdal, C.: Generalized gauge theories and the constituents of massless particles. In: Lecture Notes in Physics, Vol. 153, pp. 329–347. Berlin, Heidelberg, New York: Springer 1982
7. Fronsdal, C.: Private communication
8. Guichardet, A.: Extensions des représentations induites des produits semi-directs. J. Reine Angew. Math. **310**, 7–32 (1979)
9. Gupta, S.N.: Theory of longitudinal photons in quantum electrodynamics. Proc. Phys. Soc. **63**, 681–691 (1950)

10. Gupta, S.N.: On the supplementary condition in quantum electrodynamics. Proc. Phys. Soc. **64**, 850–852 (1951)
11. Parvizi, D.: Indecomposable representations of Poincaré group and quantization of weak gravitational and electromagnetic fields. Rep. Math. Phys. **8**, 401–420 (1975)
12. Pinczon, G., Simon, J.: On the 1-cohomology of Lie groups. Lett. Math. Phys. **1**, 83–91 (1975)
13. Pinczon, G., Simon, J. Extensions of representations and cohomology. Rep. Math. Phys. **16**, 49–77 (1979)
14. Rawnsley, J., Schmid, W., Wolf, J.A.: Singular unitary representations and indefinite harmonic theory. J. Funct. Anal. **51**, 1–114 (1983)
15. Rideau, G.: Noncompletely reducible representations of the Poincaré group associated with the generalized Lorentz gauge. J. Math. Phys. **19**, 1627–1634 (1978)
16. Rideau, G.: On the extensions of mass-zero representations of the Poincaré group. Rep. Math. Phys. **16**, 251–263 (1979)
17. Rideau, G.: Extensions of unitary irreducible representations of Poincaré group. Application to quantum field theory. In: Lecture Notes in Physics, Vol. 153, pp. 302–304. Berlin, Heidelberg, New York: Springer 1982
18. Strocchi, F., Wightman, A.S.: Proof of the charge superselection rule in local relativistic quantum field theory. J. Math. Phys. **15**, 2198–2224 (1974)
19. Todorov, I.T., Mintchev, M.C., Petkova, V.B.: Conformal invariance in quantum field theory. Scuola Normale Superiore, Pisa, 1978

Communicated by H. Araki

Received April 1, 1984

Classical Models of Confinement II

Harry Lehmann* and Tai Tsun Wu**

1 II. Institut für Theoretische Physik der Universität Hamburg, D-2000 Hamburg, Federal Republic of Germany
2 Deutsches Elektronen-Synchrotron DESY, D-2000 Hamburg, Federal Republic of Germany, and Gordon McKay Laboratory, Harvard University, Cambridge, MA 02138, USA

Dedicated to the memory of Kurt Symanzik

Abstract. We generalize the class of Abelian models of paper I which lead to a linear potential between opposite charges. The electric field E is here taken as a power series in the electric displacement D raised to the σth power. We solve to first order the case of two opposite static point charges asymptotically for large separations.

1. Introduction

In a recent paper [1] (hereafter referred to as I), we investigated classical models of confinement. These models are essentially those previously studied by Pagels and Tomboulis [2] and more extensively by Adler and Piran [3]. They can be viewed as ordinary electrostatics with a field-dependent dielectric constant.

In I we considered the case

$$E = f(D) = E_0 + b_1 D + b_2 D^2 + \dots, \tag{1.1}$$

with $b_1 > 0$. Without loss of generality we can set $E_0 = b_1 = 1$. For two unit static charges of opposite sign, separated by a large distance $2R$, it was found that the flux is confined within an ellipsoid of revolution with semi-minor axis $(2R/\sqrt{\pi})^{1/2}$, leading to a linear potential with a correction of order $\ln R$. We also studied briefly the more general case,

$$E = f(D) = 1 + D^\sigma + \dots, \tag{1.2}$$

with $\sigma > 0$ and we showed that the transverse dimension of the confinement domain increases as $R^{1/(1+\sigma)}$ for large R.

It is the purpose of the present note to examine further the case (1.2). This more general case is of interest in view of the uncertainty in the shape of the confinement domain for a realistic theory such as QCD.

---

* Work supported in part by NATO Research Grant No. 432/84
** Work supported in part by the U.S. Department of Energy under Grant No. DE-FG02-84ER40158 with Harvard University

2. Zeroth-Order Solution

With the charges located at $\pm R$ along the z-axis of a cylindrical coordinate system, we can express D in terms of a flux function $\Psi(z, \varrho)$ as [3],

$$\varrho D_z = \frac{1}{2\pi} \partial_\varrho \Psi, \quad \varrho D_\varrho = -\frac{1}{2\pi} \partial_z \Psi. \tag{2.1}$$

This Ψ satisfies the partial differential equation [Eq. (2.11) of I]

$$\partial_z \left[\frac{\partial_z \Psi f(D)}{[(\partial_z \Psi)^2 + (\partial_\varrho \Psi)^2]^{1/2}} \right] + \partial_\varrho \left[\frac{\partial_\varrho \Psi f(D)}{[(\partial_z \Psi)^2 + (\partial_\varrho \Psi)^2]^{1/2}} \right] = 0, \tag{2.2}$$

with the boundary conditions given by (2.10) of I. If we express ϱ as a function of z and Ψ, viz.,

$$\varrho = \varrho(z, \Psi), \tag{2.3}$$

(2.2) becomes

$$\frac{-g(D)\partial_{zz}\varrho \cdot \mathrm{sgn}(\partial_\Psi \varrho)}{[1 + (\partial_z \varrho)^2]^{3/2}} + \frac{1}{2\pi} \left[\frac{1}{\partial_\Psi \varrho} \partial_\Psi \frac{1}{\varrho \partial_\Psi \varrho} + \frac{\partial_z \varrho}{\partial_\Psi \varrho} \partial_\Psi \frac{\partial_z \varrho}{\varrho \partial_\Psi \varrho} - \partial_z \frac{\partial_z \varrho}{\varrho \partial_\Psi \varrho} \right] = 0, \tag{2.4}$$

where

$$g(D) = \frac{f(D)}{f'(D)} - D. \tag{2.5}$$

In the limit of large R, the weak field approximation $f(D) = 1 + D^\sigma$ and, hence,

$$g(D) = \sigma^{-1} D^{1-\sigma} \tag{2.6}$$

can be applied far away from the charges. In this region, $\partial_\Psi \varrho < 0$ and (2.4) simplifies to zeroth-order in R:

$$\partial_{zz}\varrho - \frac{1}{(2\pi)^\sigma} \cdot \frac{1}{\partial_\Psi \varrho} \partial_\Psi \left(\frac{-1}{\varrho \partial_\Psi \varrho} \right)^\sigma = 0. \tag{2.7}$$

The separation of variables

$$\varrho(z, \Psi) = Z(z) G(\Psi), \tag{2.8}$$

leads to the results

$$z = \frac{2R}{\sqrt{\pi}} \frac{\Gamma\left(\frac{1}{2\sigma} + 1\right)}{\Gamma\left(\frac{1}{2\sigma} + \frac{1}{2}\right)} \sqrt{1 - Z^{2\sigma}}\, F\left(-\frac{1}{2\sigma} + \frac{1}{2}, \frac{1}{2}; \frac{3}{2}; 1 - Z^{2\sigma} \right), \tag{2.9}$$

and

$$G(\Psi) = (\pi\sigma)^{-1/2} \left(\sqrt{2} R (\sigma+1)^{\sigma/2} \frac{\Gamma\left(\frac{1}{2\sigma}\right)}{\Gamma\left(\frac{1}{2\sigma} + \frac{1}{2}\right)} \right)^{\frac{1}{\sigma+1}} \left[1 - \Psi^{\frac{\sigma}{\sigma+1}} \right]^{1/2}, \tag{2.10}$$

where F is the hypergeometric function. For $\sigma=1/n$ $(n=1,2,3,\ldots)$ this hypergeometric function is elementary.

It follows from this explicit solution that

$$\Psi(z,\varrho)=\left(1-\frac{\bar{\varrho}^2}{Z^2(z)}\right)^{\frac{\sigma+1}{\sigma}}, \tag{2.11}$$

where $\bar{\varrho}$ is related to ϱ by

$$\varrho=(\pi\sigma)^{-1/2}\left(\sqrt{2}R(\sigma+1)^{\sigma/2}\frac{\Gamma\left(\frac{1}{2\sigma}\right)}{\Gamma\left(\frac{1}{2\sigma}+\frac{1}{2}\right)}\right)^{\frac{1}{\sigma+1}}\bar{\varrho}\,. \tag{2.12}$$

The boundary B of the confinement domain is accordingly given by

$$\bar{\varrho}_b=Z(z)\,. \tag{2.13}$$

Note in particular that, when $z=0$, $Z=1$ and hence the maximal radius of the confinement domain is given by (2.12) with $\bar{\varrho}=1$.

3. Electrostatic Energy

The total electrostatic energy $V(R)$ is given by

$$V(R)=\int d^3x\int_0^D dD'f(D')\sim\int d^3x\left(D+\frac{1}{\sigma+1}D^{\sigma+1}\right). \tag{3.1}$$

The substitution of the zeroth-order solution (2.11) into (3.1) leads to

$$V(R)=2R+\frac{2(1+\sigma)}{(1+2\sigma)(1-\sigma)}\left(\frac{1+\sigma}{2}\right)^{\frac{\sigma}{\sigma+1}}\left(\frac{\Gamma\left(\frac{1}{2\sigma}+\frac{1}{2}\right)}{\Gamma\left(\frac{1}{2\sigma}\right)}\right)^{\frac{2\sigma}{1+\sigma}}R^{\frac{1-\sigma}{1+\sigma}}. \tag{3.2}$$

When $\sigma\to 1$, this agrees with (4.17) of I. When $\sigma>1$, a divergent integral is encountered which has been interpreted in the sense of distributions [4]. With a suitable choice of an additive constant in $V(R)$, (3.2) gives the leading correction to the linear potential.

4. Reformulation of the Problem

In (2.11) we have exhibited the zeroth-order of (2.2) for large R. To go beyond this zeroth-order in the present case of general σ, the choice of variables is important. The underlying reason is that the zeroth-order solution already involves a transcendental function, as seen from (2.9). In the remainder of this paper, we shall consider the case $f(D)=1+D^\sigma+b_2D^{2\sigma}+\ldots$.

We generalize the procedure of Sect. 5.2 of I. First we transform from z to a variable ζ by

$$z=\frac{\zeta}{\beta}F\left(-\frac{1}{2\sigma}+\frac{1}{2},\frac{1}{2};\frac{3}{2};\zeta^2\right), \tag{4.1}$$

where

$$\beta = \frac{\sqrt{\pi}}{2R'} \frac{\Gamma\left(\frac{1}{2\sigma} + \frac{1}{2}\right)}{\Gamma\left(\frac{1}{2\sigma} + 1\right)}, \tag{4.2}$$

and R', which by (2.9) is equal to R in zeroth-order, will be chosen later. Next, ϱ is scaled by

$$\varrho = \frac{\bar{\varrho}}{\alpha}, \tag{4.3}$$

where

$$\alpha = \sqrt{\pi\sigma} \left(\sqrt{2}(\sigma+1)^{\sigma/2} \frac{\Gamma\left(\frac{1}{2\sigma}\right)}{\Gamma\left(\frac{1}{2\sigma} + \frac{1}{2}\right)} R'' \right)^{\frac{-1}{\sigma+1}}. \tag{4.4}$$

Here R'' is defined by the condition that the boundary B intersects the plane $z=0$ at $\varrho = 1/\alpha$. It follows from the last remark in Sect. 2 that R'' is also equal to R in zeroth-order. Instead of $\bar{\varrho}$ it is more convenient to use the variable τ [see (4.22) of I] defined by

$$\tau = \frac{\bar{\varrho}}{r(\zeta)}, \tag{4.5}$$

where the boundary B is given by $\bar{\varrho}_b = r(\zeta)$. We now define the expansion parameter Λ in terms of α and β by

$$\Lambda = \frac{8\alpha^2}{\beta^2} = \frac{4}{\sigma}\left(\frac{\sigma+1}{2}\right)^{\frac{-\sigma}{\sigma+1}} \left(\frac{\Gamma\left(\frac{1}{2\sigma}\right)}{\Gamma\left(\frac{1}{2\sigma}+\frac{1}{2}\right)} \right)^{\frac{2\sigma}{\sigma+1}} R'^2 R''^{\frac{-2}{\sigma+1}}. \tag{4.6}$$

Finally, we rescale Ψ by

$$\Psi = \left(\frac{R''}{R'}\right)^{\frac{2}{\sigma}} \psi. \tag{4.7}$$

[For the special case $\sigma=1$ treated in I, R' and R'' are respectively $\Lambda'/4\sqrt{\pi}$ and $\Lambda''/4\sqrt{\pi}$, while the Λ of (4.6) reduces to that of (5.9) of I.]

For the purpose of obtaining the first-order solution, we begin with (2.2) written as:

$$\frac{g(D)}{D}[(\partial_\varrho \Psi)^2 \partial_{zz}\Psi + (\partial_z\Psi)^2 \partial_{\varrho\varrho}\Psi - 2(\partial_z\Psi)(\partial_\varrho\Psi)(\partial_{z\varrho}\Psi)]$$
$$+ \left[(\partial_z\Psi)^2 + (\partial_\varrho\Psi)^2][\partial_{zz}\Psi + \partial_{\varrho\varrho}\Psi - \frac{1}{\varrho}\partial_\varrho\Psi\right] = 0. \tag{4.8}$$

To exhibit the dependence of $g(D)/D$ on Λ, we define $\bar{D}$ by

$$\bar{D}=\frac{\pi\sigma}{(\sigma+1)\alpha^2}\left(\frac{R''}{R'}\right)^{\frac{2}{\sigma}}D. \tag{4.9}$$

In terms of this $\bar{D}$,

$$\frac{g(D)}{D}=\frac{\Lambda}{4}\bar{D}^{-\sigma}h\left(\frac{4}{\sigma\Lambda}\bar{D}^{\sigma}\right), \tag{4.10}$$

where

$$h(x)=1+(1-\sigma-2b_2)x+\dots. \tag{4.11}$$

Carrying out the change of variables, we obtain from (4.8) and (4.9)

$$\begin{aligned}
&2\bar{D}^{-\sigma}h\left(\frac{4}{\sigma\Lambda}\bar{D}^{\sigma}\right)(1-\zeta^2)^{1-\frac{1}{\sigma}}\Big\{(\partial_\zeta\psi)^2\,\partial_{\tau\tau}\psi+(\partial_\tau\psi)^2\Big[2\frac{\partial_\zeta r}{r}\,\partial_\zeta\psi-\tau\frac{\partial_{\zeta\zeta}r}{r}\,\partial_\tau\psi\\
&+\partial_{\zeta\zeta}\psi-\frac{(\sigma-1)\zeta}{\sigma(1-\zeta^2)}\partial_\zeta\psi+\frac{(\sigma-1)\zeta}{\sigma(1-\zeta^2)}\tau\frac{\partial_\zeta r}{r}\,\partial_\tau\psi\Big]-2(\partial_\zeta\psi)(\partial_\tau\psi)(\partial_{\zeta\tau}\psi)\Big\}\\
&+(\partial_\tau\psi)^2\left[1+\frac{8}{\Lambda}(1-\zeta^2)^{1-\frac{1}{\sigma}}\left(\tau\partial_\zeta r-r\frac{\partial_\zeta\psi}{\partial_\tau\psi}\right)^2\right]\Big\{\frac{1}{r^2}\tau\partial_\tau\left(\frac{\partial_\tau\psi}{\tau}\right)\\
&+\frac{8}{\Lambda}(1-\zeta^2)^{1-\frac{1}{\sigma}}\Big[\left(\frac{\partial_\zeta r}{r}\right)^2\tau\partial_\tau(\tau\partial_\tau\psi)-2\tau\frac{\partial_\zeta r}{r}\partial_{\zeta\tau}\psi-\tau\partial_\zeta\left(\frac{\partial_\zeta r}{r}\right)\partial_\tau\psi\\
&+\partial_{\zeta\zeta}\psi-\frac{(\sigma-1)\zeta}{\sigma(1-\zeta^2)}\partial_\zeta\psi+\frac{(\sigma-1)\zeta}{\sigma(1-\zeta^2)}\tau\frac{\partial_\zeta r}{r}\,\partial_\tau\psi\Big]\Big\}=0,
\end{aligned} \tag{4.12}$$

with

$$\bar{D}=\frac{-\sigma}{2(\sigma+1)}\frac{\partial_\tau\psi}{\tau r^2}\left[1+\frac{8}{\Lambda}(1-\zeta^2)^{1-\frac{1}{\sigma}}\left(\tau\partial_\zeta r-r\frac{\partial_\zeta\psi}{\partial_\tau\psi}\right)^2\right]^{1/2}. \tag{4.13}$$

The partial differential Eq. (4.12), which is the generalization of (5.10) of I, is exact.

5. First-Order Solution

To first order in Λ^{-1},

$$r(\zeta)=(1-\zeta^2)^{\frac{1}{2\sigma}}\left[1+\frac{1}{\Lambda}w(\zeta)\right]. \tag{5.1}$$

The fact that (4.12) is homogeneous in ψ simplifies the perturbation series of $\psi(\zeta,\tau)$. To first order the implication is that

$$\psi(\zeta,\tau)=(1-\tau^2)^{\frac{\sigma+1}{\sigma}}\left\{1+\frac{1}{\Lambda}[c+\tau^2k(\zeta)]\right\}. \tag{5.2}$$

In (5.1) and (5.2) w and k are even functions of ζ. The zeroth-order solution (2.11) has been used. Because of the boundary condition at $\tau=0$,

$$\psi(\zeta,0)=\left(\frac{R'}{R''}\right)^{\frac{2}{\sigma}}=1+\frac{c}{\Lambda}, \tag{5.3}$$

c does not depend on ζ.

The substitution of (5.1) and (5.2) into (4.12) yields two ordinary differential equations for $w(\zeta)$ and $k(\zeta)$:

$$\sigma(1-\zeta^2)^2 w''-\zeta(\sigma+1)(1-\zeta^2)w'-2(\sigma+1)w+\sigma(c+k)+\frac{4(\sigma+4)\zeta^2}{\sigma^2(1-\zeta^2)}=0, \tag{5.4}$$

and

$$\sigma(1-\zeta^2)^2 k''-\zeta(\sigma+1)(1-\zeta^2)k'-4(2\sigma+1)k-\frac{8(\sigma+1)(1-2b_2)}{\sigma^2(1-\zeta^2)} -\frac{32(\sigma+1)^2\zeta^2}{\sigma^3(1-\zeta^2)}=0, \tag{5.5}$$

with the boundary condition $w(0)=0$.

The solution of (5.5) is

$$k(\zeta)=\frac{2(\sigma+1)}{\sigma^2(\sigma+3)}\left\{\frac{2\sigma-1}{2\sigma+1}(1-2b_2)+\frac{4(\sigma+1)(3\sigma+2)}{\sigma(2\sigma+1)} -2\left[1-2b_2+\frac{4(\sigma+1)}{\sigma}\right]\frac{1}{(1-\zeta^2)}\right\}. \tag{5.6}$$

The solution of (5.4) is more subtle. The following two points are relevant. First, this equation involves the constant c which we are at liberty to choose by varying R' as seen from (5.3). Secondly, the even solution of the homogeneous equation is transcendental. Therefore, the solution of (5.4) can be elementary only for a special value of c. This particular value of c can be found with the ansatz

$$w(\zeta)=\frac{K\zeta^2}{1-\zeta^2}. \tag{5.7}$$

The substitution of (5.6) and (5.7) into (5.4) leads to a linear equation in ζ^2. This determines both K and c:

$$K=\frac{2}{\sigma^2(\sigma-1)(\sigma+3)}[-\sigma(\sigma+1)b_2+2\sigma^2+\sigma-4], \tag{5.8}$$

$$c=\frac{2}{\sigma^2(\sigma-1)(\sigma+3)(2\sigma+1)}[6(\sigma+1)b_2-2\sigma^3-\sigma^2+8\sigma+1]. \tag{5.9}$$

This first-order solution is given by (5.1) and (5.2) together with (5.6)–(5.9). Unfortunately, this elementary solution is not completely satisfactory because it fails to reduce to our previous solution given in I for $\sigma\to 1$. Indeed, in this limit both K and c approach infinity.

We proceed to give an alternative choice of c which avoids this difficulty. First, the even solution of the homogeneous equation corresponding to (5.4) is

$$w_0(\zeta)=1+\frac{\sigma+1}{\sigma}\zeta^2\cdot(1-\zeta^2)^{-\frac{\sigma+1}{2\sigma}}F\left(-\frac{1}{2\sigma}+\frac{1}{2},\frac{1}{2};\frac{3}{2};\zeta^2\right). \tag{5.10}$$

In the limit $\sigma\to 1$ this w_0 is simply

$$w_0\to\frac{1+\zeta^2}{1-\zeta^2}. \tag{5.11}$$

Since $w_0(0)=1$ and $w(0)=0$ for all σ, we look for a solution of the form

$$w(\zeta)=\text{const}\left[\frac{1+\tilde{\alpha}\zeta^2}{1-\zeta^2}-w_0(\zeta)\right], \tag{5.12}$$

where $\tilde{\alpha}=\tilde{\alpha}(\sigma)$ must satisfy $\tilde{\alpha}(1)=1$. Aside from this condition, $\tilde{\alpha}(\sigma)$ is arbitrary. Therefore, there are many solutions. We have found that a convenient choice is

$$\tilde{\alpha}(\sigma)=\frac{1}{\sigma}. \tag{5.13}$$

With this choice, it follows from (5.4) that w and c are given by

$$w=\frac{2}{\sigma^2(\sigma-1)(\sigma+3)}[-\sigma(\sigma+1)b_2+2\sigma^2+\sigma-4]\frac{\zeta^2}{1-\zeta^2} \cdot\left[1-(1-\zeta^2)^{\frac{\sigma-1}{2\sigma}}F\left(-\frac{1}{2\sigma}+\frac{1}{2},\frac{1}{2};\frac{3}{2};\zeta^2\right)\right], \tag{5.14}$$

$$c=\frac{2(\sigma+1)}{\sigma^2(\sigma+3)(2\sigma+1)}[-2(2\sigma+3)b_2+6\sigma+7]. \tag{5.15}$$

Here (5.14) and (5.15) replace (5.7)–(5.9).

It is seen from either (5.8) or (5.14) that the boundary B remains unchanged to first order when

$$b_2=\frac{2\sigma^2+\sigma-4}{\sigma(\sigma+1)}. \tag{5.16}$$

This is the generalization of the phenomenon found in I for the case $b_2=-1/2$.

The two versions of the first-order solution correspond to different choices of R'. By (4.1) and (4.2) a different choice of R' implies a redefinition of the variable ζ which amounts to a non-elementary transformation between the old and the new ζ.

Acknowledgements. One of us (T.T.W.) wishes to thank Professor Fritz Gutbrod, Professor Hans Joos, Professor Paul Söding, and Professor Volker Soergel for their kind hospitality at DESY.

References

1. Lehmann, H., Wu, T.T.: Nucl. Phys. B **237**, 205 (1984)
2. Pagels, H., Tomboulis, E.: Vacuum of the quantum Yang-Mills theory and magnetostatics. Nucl. Phys. B **143**, 485 (1978)
3. Adler, S.L.: Effective-action approach to mean-field non-Abelian statics and a model for bag formation. Phys. Rev. D **23**, 2905 (1981)
   Adler, S.L., Piran, T.: Flux confinement in the leading logarithm model. Phys. Lett. **113** B, 405 (1982)
   Adler, S.L., Piran, T.: Relaxation methods for gauge field equilibrium equations. Rev. Mod. Phys. **56**, 1 (1984)
4. Gelfand, I.M., Shilov, G.E.: Generalized functions, Vol. 1. New York: Academic Press, 1964

Communicated by G. Mack

Received June 27, 1984

The Energy-Momentum Spectrum in Local Field Theories with Broken Lorentz-Symmetry

Hans-Jürgen Borchers[1] and Detlev Buchholz[2]

1 Institut für Theoretische Physik der Universität Göttingen, D-3400 Göttingen, Federal Republic of Germany

2 II. Institut für Theoretische Physik der Universität Hamburg, D-2000 Hamburg, Federal Republic of Germany

Abstract. Assuming locality of the observables and positivity of the energy it is shown that the joint spectrum of the energy-momentum operators has a Lorentz-invariant lower boundary in all superselection sectors. This result is of interest if the Lorentz-symmetry is (spontaneously) broken, such as in the charged sectors of quantum electrodynamics.

1. Introduction

The familiar theoretical explanation of the fact that the energy p_0 and the momentum $\mathbf{p}$ of an elementary particle are related by the fundamental equation $p_0 = (\mathbf{p}^2 + m^2)^{1/2}$, m being the mass of the particle, is based on the assumption that particle states can be described by vectors in some irreducible representation of the Poincaré group [1]. Alternatively, this form of the energy-momentum spectrum can be deduced from the hypothesis that the vectors describing particles can be generated from a vacuum vector with the help of local field operators [2]. The existence of Lorentz-transformations is not needed in the latter argument.

Both of these explanations are based on the (implicit) idea that, disregarding small "tails," particle states can be localized in bounded regions of space. Yet this picture is not correct in the case of particles carrying an electric charge. Since the electric charge is the source of the long-range Coulomb field, such states have poor localization properties which are incompatible with the existence of a continuous, unitary representation of the Lorentz group in the corresponding superselection sectors [3]. (Note that the geometrical action of a Lorentz transformation grows with the distance from the origin. So its effect on dislocalized states is large.) By the same token, there cannot exist local field operators generating these states from the vacuum [4]. Hence the conventional arguments establishing the Lorentz-invariance of the energy-momentum spectrum are not applicable in this case.

In the present article we reconsider the problem of the shape of the energy-momentum spectrum in the general framework of local quantum theory [5]. This setting is appropriate for the discussion of theories with long range forces, such as quantum electrodynamics. We recall that the basic object in local quantum theory is the algebra of all local observables which, in the present analysis, may be regarded as a concrete *-algebra $\mathfrak{A}$ of (bounded) operators acting on the physical Hilbert space $\mathscr{H}$. As usual, the space-time translations $x \in \mathbb{R}^4$ are assumed to act as automorphisms α_x on $\mathfrak{A}$, but the existence of Lorentz-transformations is not required here.

Since the vectors in $\mathscr{H}$ are to describe elementary physical systems we assume that energy-momentum operators can be defined on $\mathscr{H}$ and that the energy is positive in all Lorentz-frames. This means that on $\mathscr{H}$ there exists a continuous, unitary representation $x \rightarrow U(x)$ of the translations implementing the automorphisms α_x,

$$U(x)AU(x)^{-1} = \alpha_x(A) \quad \text{for} \quad A \in \mathfrak{A}, \tag{1.1}$$

and the joint spectrum of the generators of U is contained in the closed forward lightcone $\bar{V}$,

$$\operatorname{sp} U \subset \bar{V}. \tag{1.2}$$

It has been shown in [6] that under these conditions the operators $U(x)$ can be chosen from the weak closure of the observable algebra $\mathfrak{A}$,

$$U(x) \in \mathfrak{A}^{-}. \tag{1.3}$$

This fact is essential for the interpretation of the generators of U as energy-momentum operators. Yet since we do not rely on the existence of Lorentz-transformations or fields generating the physical states from a vacuum vector, there exists a variety of representations U of this kind: given any $p \in \bar{V}$ one can proceed from U to the representation,[1] $x \rightarrow U'(x) = U(x) \cdot e^{ip \cdot x}$, which also fulfills the conditions (1.1) to (1.3), but the spectra $\operatorname{sp} U$ and $\operatorname{sp} U'$ are displaced by p. If $\mathscr{H}$ consists of several superselection sectors (i.e. if $\mathfrak{A}^{-}$ has a non-trivial center) one can modify U in this manner on each coherent subspace of $\mathscr{H}$, and this will in general change the spectrum substantially. So there arises the question as to which choice of the representation U should be regarded as the most natural one.

We employ here the condition that a reasonable representation U of the translations has to give rise to a minimal energy-operator. More precisely: if (with respect to a fixed Lorentz-frame) H is the generator of the time-translations of U, and if H' is the corresponding generator of any other representation U' satisfying the conditions (1.1)–(1.3), then

$$H \leqq H'. \tag{1.4}$$

We will show in the subsequent section that such a minimal representation of the translations always exists. At the very end of our analysis it will also become clear that in local quantum theory this representation is unique.

Given these general constraints we will study in Sect. 3 which shape of the energy-momentum spectrum $\operatorname{sp} U$ is compatible with the principle of locality

1 If $a, b \in \mathbb{R}^4$ we denote by $a \cdot b$ the Lorentz-scalar product, which in proper coordinates is given by $a \cdot b = a_0 \cdot b_0 - \mathbf{a} \cdot \mathbf{b}$

(spacelike commutativity) of observables. It is essential for this part of our analysis that, according to this principle,

$$[A^*, \alpha_x(A)] = 0 \tag{1.5}$$

for all local operators $A \in \mathfrak{A}$ and space-like translations[2] $x \notin (\bar{V} - a) \cup (-\bar{V} + a)$, where a is a sufficiently large time-like vector allowing for the finite localization of A. This property of the observables ties the translations U to the causal structure of Minkowski space and thereby restricts the admissible shape of $\operatorname{sp} U$. Our main result can be stated as follows:

The lower boundary of $\operatorname{sp} U$ *consists of the hyperboloid* $\{p \in \mathbb{R}^4 : p^2 = m^2, p_0 \geqq 0\}$ *for some* $m \geqq 0$ *and, possibly, of the point* $\{0\}$.

So also in the absence of Lorentz-transformations it turns out that the states at the bottom of the energy-momentum spectrum (amongst which one expects the elementary particles) satisfy the familiar relation between energy and momentum. This fact is of particular interest in quantum electrodynamics, where it adds to the understanding of the infra-particle problem as discussed in [7] (cf. also [8]). Another aspect of our results will be pointed out in Sect. 4.

Let us finally remark that, using locality, one can establish continuity properties of the spectral resolution of the spatial translations [9], just as in the presence of Lorentz-transformations [10, 11]. So it seems plausible that the general invariance and continuity properties of the energy-momentum spectrum following from Lorentz-covariance are actually a consequence of the more fundamental principle of locality. Unfortunately, we have not yet been able to completely settle this point. But we believe that our methods can be improved so as to fill the remaining gaps. We hope to return to this problem in a future publication.

2. Minimal Translations

Let us now turn to the problem of exhibiting a representation U of the translations which is minimal in the sense outlined in the introduction. To this end we fix a Lorentz-frame Σ with time-direction given by a positive timelike vector e. This vector enters into the following

Definition. Let U be a representation of the translations with the properties (1.1)–(1.3). We say U is Σ-minimal if the generator H of the time-translations $\mathbb{R} \ni t \to U(t \cdot e)$ is minimal, i.e. if the generator H' corresponding to any other choice U' of the translations satisfies $H' \geqq H$. [Note that the generators of U and U' commute because of condition (1.3). So there exists a joint core for H and H'.]

In the analysis of the subsequent section we will make use of another characterization of Σ-minimal translations which can be taken from the following

Lemma 2.1. *Let U be a representation of the translations satisfying the conditions (1.1)–(1.3). Then the following two statements are equivalent.*

2 If S, S' are subsets of $\mathbb{R}^4$ we define $S \pm S' = \{s \pm s' : s \in S, s' \in S'\}$

i) *U is Σ-minimal*

ii) *There is no non-trivial subspace $\mathscr{H}' \subset \mathscr{H}$ which is stable under the action of $\mathfrak{A}$ such that*[3]

$$\mathrm{sp}\, U \upharpoonright \mathscr{H}' \subset p + \bar{V}$$

for some $p \in \mathbb{R}^4$ with $p \cdot e > 0$.

We omit the simple proof of this equivalence.

For the proof that Σ-minimal translations always exist it is instructive to study first the case where the algebra $\mathfrak{A}$ acts irreducibly on $\mathscr{H}$. In this case it follows from condition (1.1) that any two representations U and U' of the translations are related by

$$U'(x) = U(x) \cdot e^{-ip \cdot x} \tag{2.1}$$

for some $p \in \mathbb{R}^4$. So given U we must specify a p such that U' is Σ-minimal. From the spectrum condition $\mathrm{sp}\, U' \subset \bar{V}$ it follows that p has to satisfy

$$\mathrm{sp}\, U - p \subset \bar{V}. \tag{2.2}$$

The requirement that the generator of the time-translations $t \to U'(t \cdot e)$ is minimal amounts to the condition that $p \cdot e$ has the maximal value which is compatible with relation (2.2). That such vectors p exist follows from the fact that the set of vectors p' satisfying (2.2) and $p' \cdot e \geqq 0$ is compact and non-empty (since $\mathrm{sp}\, U \subset \bar{V}$). It should, however, be noticed that p need not be unique since $\bar{V}$ is not a simplicial cone. This fact causes some problem of choice in the analysis of the general case, where $\mathfrak{A}$ is reducible[4].

Turning now to this case, let U be a fixed representation of the translations satisfying the conditions (1.1)–(1.3). We denote by $\mathfrak{W}$ the set of all continuous, unitary representations $x \to W(x)$ of $\mathbb{R}^4$ with values in the center of $\mathfrak{A}^-$ for which the translations U' given by

$$U'(x) = U(x) \cdot W(x)^{-1} \tag{2.3}$$

fulfill again the spectrum condition. It is obvious that any U' satisfying the conditions (1.1)–(1.3) can be represented in this way. What we want to show is that there is always a choice of $W \in \mathfrak{W}$ such that the resulting translations U' are Σ-minimal. To this end we divide $\mathfrak{W}$ into classes $\mathfrak{W}(K)$ of representations W having the same generator K for the subgroup $\mathbb{R} \cdot e \subset \mathbb{R}^4$. Between these classes we establish the following partial ordering:

$$\mathfrak{W}(K') \succ \mathfrak{W}(K'') \quad \text{if} \quad K' \geqq K''. \tag{2.4}$$

We will show, using Zorn's lemma [12], that there exists a unique maximal element with respect to this ordering.

---

3 If $x \to V(x)$ is a representation of $\mathbb{R}^4$ on $\mathscr{H}$ and if $\mathscr{K} \subset \mathscr{H}$ is a subspace which is invariant under V we denote by $V \upharpoonright \mathscr{K}$ the restriction of V to $\mathscr{K}$

4 The reader who is not interested in these technicalities may proceed directly to the subsequent section

To begin with we note that every pair of classes $\mathfrak{W}(K')$, $\mathfrak{W}(K'')$ has a joint upper bound. Namely, if $W' \in \mathfrak{W}(K')$ and $W'' \in \mathfrak{W}(K'')$, we can construct a representation W setting

$$W(x) = W'(x) \cdot F + W''(x) \cdot (1-F), \tag{2.5}$$

where F is the projection onto the non-negative part of $K' - K''$. It is obvious from relation (2.5) that $W \in \mathfrak{W}$. Moreover, the generator $K = K' \cdot F + K'' \cdot (1-F)$ of $t \to W(t \cdot e)$ is larger than K' and K'', respectively. Hence $\mathfrak{W}(K)$ is an upper bound for both, $\mathfrak{W}(K')$ and $\mathfrak{W}(K'')$. For the purpose of finding a maximal element we may therefore restrict our attention to the classes $\mathfrak{W}(K) \succ \mathfrak{W}(0)$.

Next, we analyse the spectrum of the representations $W \in \mathfrak{W}(K)$ on certain subspaces of $\mathscr{H}$. Let

$$U(x) = \int e^{ip \cdot x} dE(p) \tag{2.6}$$

be the spectral representation of the given translations and let $E(\Delta)$, where $\Delta \subset \mathbb{R}^4$ is any Borel set, be the corresponding spectral projections. We denote by F_n, $n \in \mathbb{N}$ the central support projection[5] of $E_n = E(n \cdot e - \bar{V})$. Since $E_n \leqq E_{n'}$ if $n \leqq n'$ and $\operatorname{s\,lim}_{n \to \infty} E_n = 1$ we have clearly $F_n \leqq F_{n'}$ for $n \leqq n'$ and $\operatorname{s-lim}_{n \to \infty} F_n = 1$. It is essential for our argument that the restrictions of the representations $W \in \mathfrak{W}$ to the subspaces $F_n \cdot \mathscr{H}$ have their spectrum in $n \cdot e - \bar{V}$. This follows immediately from the inclusion

$$\operatorname{sp} U \restriction F\mathscr{H} \subset \bar{V} + \operatorname{sp} W \restriction F\mathscr{H} \tag{2.7}$$

(which holds for every non-zero central projection F) and the fact that $(n \cdot e - \bar{V}) \cap \operatorname{sp} U \restriction F_n\mathscr{H} \neq \emptyset$. On the other hand, every representation $W \in \mathfrak{W}(K) \succ \mathfrak{W}(0)$ has its spectrum in the half space $\Sigma_+ = \{k \in \mathbb{R}^4 : k \cdot e \geqq 0\}$, hence

$$\operatorname{sp} W \restriction F_n\mathscr{H} \subset (n \cdot e - \bar{V}) \cap \Sigma_+, \tag{2.8}$$

and this set is compact.

We are now in a position to show that every linearly ordered set of classes $\mathfrak{W}(K_\iota) \succ \mathfrak{W}(0)$, $\iota \in \mathbb{I}$ ($\mathbb{I}$ being some index set) has an upper bound. We pick from each class $\mathfrak{W}(K_\iota)$ a representation W_ι and consider the corresponding (four) generators Q_ι. Since, according to relation (2.8), each representation $W_\iota \restriction F_n\mathscr{H}$ has its spectrum in the compact set $(n \cdot e - \bar{V}) \cap \Sigma_+$, it follows that the operators $Q_\iota \cdot F_n$ are uniformly bounded if n is kept fixed. Hence, because of the weak compactness of the unit ball in a von Neumann algebra, the sequence $Q_\iota \cdot F_n$, $\iota \in \mathbb{I}$ contains a weakly convergent subnet. Its limit, which we denote by $Q^{(n)}$, is selfadjoint and bounded.

Now, according to the spectrum condition entering into the definition of $\mathfrak{W}$ we have for each $f \in \bar{V}$ the inequalities $P \cdot f \geqq Q_\iota \cdot f$ and $P \cdot f \geqq 0$, where P is the generator of the translations U. Consequently $P \cdot f \geqq Q_\iota F_n \cdot f$ which gives in the above limit $P \cdot f \geqq Q^{(n)} \cdot f$. This means that the representation

$$x \to W_n(x) = e^{iQ^{(n)}x} \tag{2.9}$$

5 The central support projection F of a projection $E \in \mathfrak{A}^-$ is the smallest projection F in the center of $\mathfrak{A}^-$ such that $E \leqq F$

is an element of $\mathfrak{W}$. Setting $K^{(n)} = Q^{(n)} \cdot e$ and taking into account that the set $\mathfrak{W}(K_\iota)$, $\iota \in \mathbb{I}$ is linearly ordered, we have furthermore

$$K^{(n)} \geqq K_\iota \cdot F_n . \tag{2.10}$$

To complete our argument we must proceed to the limit n going to infinity. We do this as follows: let $G_n = F_n - F_{n-1}$, $n \in \mathbb{N}$ with the convention that $F_0 = 0$. Since $F_n \geqq F_{n-1}$ and $\operatorname{s-lim}_{n \to \infty} F_n = 1$, it is obvious that the operators G_n are mutually orthogonal projections which sum up to 1. So we can define a continuous, unitary representation $\bar{W}$ of $\mathbb{R}^4$, setting

$$\bar{W}(x) = \sum_{n=1}^{\infty} W_n(x) \cdot G_n . \tag{2.11}$$

Since $W_n \in \mathfrak{W}$ for each $n \in \mathbb{N}$ it is clear that $\bar{W} \in \mathfrak{W}$.

It remains to show that $\mathfrak{W}(\bar{K})$, where $\bar{K}$ is the generator of $t \to \bar{W}(t \cdot e)$, is an upper bound of $\mathfrak{W}(K_\iota)$, $\iota \in \mathbb{I}$. According to the definition (2.11) we have $\bar{W} \upharpoonright G_n \mathscr{H} = W_n \upharpoonright G_n \mathscr{H}$, so from relation (2.11) and the fact that $G_n \leqq F_n$ we obtain $\bar{K} G_n \geqq K_\iota G_n$. This holds for any $n \in \mathbb{N}$, and since $\sum_{n=1}^{\infty} G_n = 1$ we arrive at the desired result $\bar{K} \geqq K_\iota$.

From the fact that every linearly ordered set of classes $\mathfrak{W}(K)$ has an upper bound it follows by Zorn's lemma that there is some maximal element $\mathfrak{W}(K_0)$. Moreover, since each pair of classes has a joint upper bound, it is also clear that $\mathfrak{W}(K_0) \succ \mathfrak{W}(K)$ for all $\mathfrak{W}(K)$, hence $\mathfrak{W}(K_0)$ is unique.

Now we pick any $W_0 \in \mathfrak{W}(K_0)$ and define the translations

$$x \to U_0(x) = U(x) \cdot W_0(x)^{-1} . \tag{2.12}$$

It is then an immediate consequence of the definition of $\mathfrak{W}(K_0)$ that these translations are Σ-minimal. So we arrive at

Theorem 2.2. *There exists a representation $x \to U_0(x)$ of the translations which is Σ-minimal.*

As we already mentioned, the representation $x \to U_0(x)$ need not be unique since $\bar{V}$ is not a simplicial cone. This situation is, however, better in the cases where $\mathfrak{A}$ consists of local operators [in the sense of relation (1.5)]. Then one can proceed from condition (1.2) to the weaker assumption that $\operatorname{sp} U \subset \bar{V}_s$, where $\bar{V}_s$ is some simplicial cone whose faces are made up of characteristic planes (i.e. planes containing some lightlike vector). Since $\bar{V}_s$ is simplicial it follows from the spectral theory for automorphism groups (cf. for example [13]) that there exist unique minimal translations U_0 with $\operatorname{sp} U_0 \subset \bar{V}_s$. Using locality and the special geometrical features of $\bar{V}_s$ one can argue then as in the subsequent section and show that actually $\operatorname{sp} U_0 \subset \bar{V}$. So $\bar{V}$ is an admissible cone (cf. [13]) if $\mathfrak{A}$ consists of local operators.

The analysis of the present section, however, shows that no specific information on $\mathfrak{A}$ is needed if one only wants to select representation U of the translations with a minimal energy-operator.

3. Locality and the Shape of the Spectrum

Assuming from now on that the algebra $\mathfrak{A}$ consists of local operators [in the sense of relation (1.5)] we will analyze in this section which shape of the energy-momentum spectrum is compatible with the principle of locality. To this end we fix a Σ-minimal representation U of the translations and study the family of functions given in the following

Definition. Let $A \in \mathfrak{A}$ be any local operator and let $\Phi \in \mathscr{H}$. We consider the functions of $x \in \mathbb{R}^4$

$$\begin{aligned} \Delta^+_{A,\Phi}(x) &= (\Phi, A^* U(x) A \Phi)\,, \\ \Delta^-_{A,\Phi}(x) &= (\Phi, \alpha_x(A) \cdot A^* U(x) \Phi)\,, \end{aligned} \tag{3.1}$$

and the commutator function

$$\Delta_{A,\Phi}(x) = (\Phi, [A^*, \alpha_x(A)]\, U(x) \Phi)\,, \tag{3.2}$$

which is the difference between $\Delta^+_{A,\Phi}$ and $\Delta^-_{A,\Phi}$.

The functions $\Delta^+_{A,\Phi}$ and $\Delta^-_{A,\Phi}$ are continuous and uniformly bounded, so their Fourier-transforms $\tilde{\Delta}^+_{A,\Phi}$ and $\tilde{\Delta}^-_{A,\Phi}$, respectively, are defined as temperate distributions. Because of the spectrum condition (1.2) we have the following information on the supports of these distributions[6]:

$$\begin{aligned} \operatorname{supp} \tilde{\Delta}^+_{A,\Phi} &\subset \operatorname{sp} U \subset \bar{V}\,, \\ \operatorname{supp} \tilde{\Delta}^-_{A,\Phi} &\subset (2 \cdot \operatorname{supp} \Phi - \operatorname{sp} U) \subset (2 \cdot \operatorname{supp} \Phi - \bar{V})\,, \end{aligned} \tag{3.3}$$

where $\operatorname{supp} \Phi$ denotes the support of the Fourier-transform of $x \to U(x)\Phi$. Consequently

$$\operatorname{supp} \tilde{\Delta}_{A,\Phi} \subset \operatorname{sp} U \cup (2 \cdot \operatorname{supp} \Phi - \operatorname{sp} U) \subset \bar{V} \cup (2 \cdot \operatorname{supp} \Phi - \bar{V})\,. \tag{3.4}$$

The latter region has a non-trivial, open complement if $\operatorname{supp} \Phi$ is compact, which we will assume henceforth.

Using locality, it follows from the "edge of the wedge theorem" that $\tilde{\Delta}_{A,\Phi}$ is the discontinuity on the reals of some analytic function. This fact imposes restrictions on the possible shape of $\operatorname{supp} \tilde{\Delta}_{A,\Phi}$ which we will exploit later in our analysis of $\operatorname{sp} U$. For the convenience of the reader we recall here the essential steps leading to this result: the commutator function $\Delta_{A,\Phi}$ has, according to relation (1.5), support in the region $(\bar{V} - a) \cup (-\bar{V} + a)$, where $a \in \mathbb{R}^4$ depends on the localization properties of $A \in \mathfrak{A}$. Therefore one can decompose $\Delta_{A,\Phi}$ into the difference of two continuous functions G_+ and G_- which have support in $(\bar{V} - a)$ and $(-\bar{V} + a)$, respectively. It follows from these support properties that the Fourier-transforms of $G_\pm$ are the boundary values (in the sense of distributions) of functions $H_\pm$ which are analytic in the tubes $T_\pm = \{z \in \mathbb{C}^4 : \operatorname{Im} z \in \pm V\}$, respectively. These boundary values coincide [because of relation (3.4)] in some open set $C \subset \mathbb{R}^4$. Thus H_+ and H_- are, by the edge of the wedge theorem [14, Sect. 27], the restrictions to T_+ and T_-, respectively, of a function H which is analytic in $T_+ \cup T_- \cup C^\sim$, where $C^\sim$ is

6 If S is a subset of $\mathbb{R}^4$ we set $2 \cdot S = S + S$; cf. footnote 2

some complex neighbourhood of C. Hence one obtains the basic relation

$$\tilde{\Delta}_{A,\Phi}(k) = \lim_{\varepsilon} (H(k+i\varepsilon) - H(k-i\varepsilon)) \tag{3.5}$$

which holds on $\mathbb{R}^4$ (in the sense of distributions) for suitable sequences $\varepsilon \in V$ tending to 0.

If one knows that $\tilde{\Delta}_{A,\Phi}$ vanishes in some region $C \subset \mathbb{R}^4$ it follows from relation (3.5) that $\tilde{\Delta}_{A,\Phi}$ also vanishes in the (in general larger) region $\hat{C}$ consisting of all real points of the envelope of holomorphy of the domain $T_+ \cup T_- \cup C^\sim$. We will make use of this mechanism for various types of coincidence regions C, for which the completions $\hat{C}$ are explicitly known.

The first result, which is of a local character, is the "double cone theorem" [14, Sect. 28]. This theorem tells us that if the coincidence region C contains an open neighbourhood of the timelike line $[0,1] \ni \lambda \to (1-\lambda) \cdot p_1 + \lambda \cdot p_2$, where $p_2 \in p_1 + \bar{V}$, then the completion $\hat{C}$ contains the double cone $(p_1 + \bar{V}) \cap (p_2 - \bar{V})$. Actually, one can relax in this theorem the assumption that the points p_1 and p_2 are connected by a straight line. It suffices that C contains a neighbourhood of some continuous curve $[0,1] \ni \lambda \to p(\lambda)$ with $p(0) = p_1$, $p(1) = p_2$, and $p(\lambda) \in p_1 + \bar{V}$ for all λ. This generalized form of the double cone theorem follows by repeated application of the restricted version.

The second result, which is of a global nature, applies to coincidence regions C which are the complement of regions of the form $\Gamma = (\Gamma_1 + \bar{V}) \cup (\Gamma_2 - \bar{V})$, where Γ_1 and Γ_2 are closed subsets of $\bar{V}$. For such regions the envelope of holomorphy of $T_+ \cup T_- \cup C^\sim$ has been computed in [15]. It coincides with the one given by the Jost, Lehmann, Dyson representation [14, Sect. 32] cf. also [16]. Again it is possible to find the completion $\hat{C}$ of C by elementary geometrical considerations: denoting by $H(a,\mu)$ the hyperboloids with apex $a \in \mathbb{R}^4$ and mass $\mu \geqq 0$,

$$H(a,\mu) = \{k \in \mathbb{R}^4 : (k-a)^2 = \mu^2\}, \tag{3.6}$$

one must determine all sets $H(a,\mu)$ which do not intersect with C and take their union

$$\hat{\Gamma} = \bigcup \{H(a,\mu) : H(a,\mu) \cap C = \emptyset\}. \tag{3.7}$$

The complement of the closed set $\hat{\Gamma} \subset \mathbb{R}^4$ is then the desired completion $\hat{C}$. We will refer to this method of finding $\hat{C}$ as the "JLD-technique".

With this information on the support properties of $\tilde{\Delta}_{A,\Phi}$ we can now turn to the analysis of $\operatorname{sp} U$. It follows from relation (3.3) that the support of $\tilde{\Delta}^+_{A,\Phi}$ is a subset of $\operatorname{sp} U$ and that the distributions $\tilde{\Delta}^+_{A,\Phi}(k)$ and $\tilde{\Delta}_{A,\Phi}(k)$ coincide in the (open) complement of the region $(2 \cdot \operatorname{supp} \Phi - \bar{V})$. So the commutator functions $\tilde{\Delta}_{A,\Phi}$, $A \in \mathfrak{A}$ can be used to study that part of the spectrum $\operatorname{sp} U$ which does not lie in the region $(2 \cdot \operatorname{supp} \Phi - \bar{V})$. Thus there arises the question for which points $q \in \mathbb{R}^4$ one can find vectors $\Phi \neq 0$ such that $q \notin (2 \cdot \operatorname{supp} \Phi - \bar{V})$, i.e. $2 \cdot \operatorname{supp} \Phi \cap (q + \bar{V}) = \emptyset$.

Definition. Let $q \in \mathbb{R}^4$. A vector $\Phi \in \mathscr{H}$, $\Phi \neq 0$ is said to be q-admissible if $\operatorname{supp} \Phi$ is compact and $2 \cdot \operatorname{supp} \Phi \cap (q + \bar{V}) = \emptyset$. (Note that a q-admissible vector Φ is also q'-admissible for any $q' \in q + \bar{V}$.)

Using locality and the fact that U is Σ-minimal, we will now show that there is a large supply of q-admissible vectors if $q \neq 0$ and $q \in \Sigma_+ = \{k \in \mathbb{R}^4 : k \cdot e \geqq 0\}$. In view of the above remark it suffices to establish this result for $q \in \Sigma_0 = \{k \in \mathbb{R}^4 : k \cdot e = 0\}$.

Lemma 3.1. *For every $q \in \Sigma_+$, $q \neq 0$ there exist q-admissible vectors Φ. Moreover, if for a spectral projection $E(\Delta)$ (cf. relation (2.6)) corresponding to some Borel set $\Delta \subset \mathbb{R}^4$, $(\Phi, A^* E(\Delta) A \Phi) = 0$ for all q-admissible vectors Φ and local operators A, then $E(\Delta) = 0$.*

Proof. The proof of the existence of q-admissible vectors amounts to showing that there exists a $p \in \operatorname{sp} U$ such that $p \notin (\frac{1}{2}q + \bar{V})$. Since $(\frac{1}{2}q + \bar{V})$ is closed there exists then also a neighbourhood $\mathcal{N}_p$ of p such that $2 \cdot \mathcal{N}_p$ is disjoint from $(q + \bar{V})$, hence any vector Φ with $\operatorname{supp} \Phi \subset \mathcal{N}_p$ is q-admissible.

Now if such a p does not exist, then $\operatorname{sp} U \subset \bar{V} \cap (\frac{1}{2}q + \bar{V})$, and it suffices to exclude this possibility for $q \in \Sigma_0$, $q \neq 0$. To this end we fix a compact, non-negligible set $\Gamma_0 \subset \operatorname{sp} U$ and pick a vector Ψ with $\operatorname{supp} \Psi \subset \Gamma_0$. Then we consider the commutator function $\tilde{\Delta}_{A,\Psi}$ which, for any local operator A, vanishes according to relation (3.4) in the complement of $(\bar{V} \cap (\frac{1}{2}q + \bar{V})) \cup (2\Gamma_0 - \bar{V})$. To this coincidence region we can apply the JLD-technique, the conclusion being that $\tilde{\Delta}_{A,\Psi}$ also vanishes in the complement of $\Gamma_1 + \bar{V} \cup (-\bar{V})$, where $\Gamma_1 = \bar{V} \cap (\frac{1}{2}q + \bar{V}) \cap (2\Gamma_0 - \bar{V})$. Consequently, the positive part $\tilde{\Delta}^+_{A,\Psi}$ of the commutator function vanishes in the complement of $(\Gamma_1 + \bar{V}) \cup \Gamma_2$, where $\Gamma_2 = (2\Gamma_0 - \bar{V}) \cap \bar{V}$ is a compact set which swallows the overlap region of the supports of $\tilde{\Delta}^+_{A,\Psi}$ and $\tilde{\Delta}^-_{A,\Psi}$. If one rewrites this result in terms of matrix elements of the spectral projections $E(\Delta)$ one gets

$$(1 - E(\Gamma)) \cdot A\Psi = 0\,, \tag{3.8}$$

if $\Gamma \supset ((\Gamma_1 + \bar{V}) \cup \Gamma_2) \cap \operatorname{sp} U$. Now Γ_1 and $\Gamma_2 \cap \operatorname{sp} U$ are compact subsets of $\bar{V} \cap (\frac{1}{2}q + \bar{V})$, and since for any spacelike vector $q \neq 0$ the region $\bar{V} \cap (\frac{1}{2}q + \bar{V})$ is contained in the open cone $(\frac{1}{4}q + V)$, there exists an open neighbourhood $\mathcal{N}$ of 0 such that $\Gamma_1 \cup (\Gamma_2 \cap \operatorname{sp} U) + \mathcal{N} \subset (\frac{1}{4}q + \bar{V})$. Hence relation (3.8) holds in particular if Γ is such that $\Gamma + \mathcal{N} \supset \frac{1}{4}q + \bar{V}$.

If $\mathfrak{A}$ is irreducible, then the set of vectors $A\Psi$, $A \in \mathfrak{A}$ is dense in $\mathcal{H}$. So it follows from (3.8) that $E(\Gamma) = 1$ if $\Gamma + \mathcal{N} \supset \frac{1}{4}q + \bar{V}$, which implies that $\operatorname{sp} U + \mathcal{N} \subset \frac{1}{4}q + \bar{V}$. Since $q \in \Sigma_0$ and $\mathcal{N}$ contains negative timelike vectors this is a contradiction to the assumption that U is Σ-minimal. So in this case there exists a q-admissible vector Φ, and using the irreducibility of $\mathfrak{A}$ once more it is also clear that $E(\Delta) = 0$ if $(\Phi, A^* E(\Delta) A \Phi) = 0$ for all $A \in \mathfrak{A}$.

In the general case we conclude from relation (3.8) (because it holds for all Ψ with $\operatorname{supp} \Psi \subset \Gamma_0$) that $(1 - E(\Gamma))\mathfrak{A}^- E(\Gamma_0) = 0$. So the central support projection Z of $E(\Gamma_0)$ is orthogonal to $(1 - E(\Gamma))$. But the relation $E(\Gamma) \cdot Z = Z$ for $\Gamma + \mathcal{N} \subset \frac{1}{4}q + \bar{V}$ implies that $\operatorname{sp} U \upharpoonright Z\mathcal{H} + \mathcal{N} \subset \frac{1}{4}q + \bar{V}$, so again we are led to a contradiction. This establishes the existence of q-admissible vectors Φ in the general case. Finally, if for all q-admissible vectors Φ and local operators A $(\Phi, A^* E(\Delta) A \Phi) = 0$, then $E(\Delta)\mathfrak{A}^-(1 - E(\frac{1}{2}q + \bar{V})) = 0$, hence $E(\frac{1}{2}q + \bar{V}) \cdot Z' = Z'$, where Z' is the central support projection of $E(\Delta)$. Now if $Z' \neq 0$ this would mean that $\operatorname{sp} U \upharpoonright Z'\mathcal{H} \subset \bar{V} \cap (\frac{1}{2}q + \bar{V})$, and applying the previous arguments it would follow that $\operatorname{sp} U \upharpoonright Z'\mathcal{H} + \mathcal{N}$

$\complement \frac{1}{4}q+\bar{V}$. This is not possible if U is Σ-minimal. So $Z'=0$ and consequently $E(\Delta)=0$. □

The above result shows that for any $q\in\Sigma_0$, $q\neq 0$ there exist points in $\operatorname{sp} U$ which lie in[7] $\bar{V}\cap\complement(q+\bar{V})$. If the Euclidean length of q is small, this region is a thin layer about a part of the surface $\partial\bar{V}$ of $\bar{V}$, so there exist (almost) lightlike points in $\operatorname{sp} U$. Next, we explore the structure of $\operatorname{sp} U$ in the neighbourhood of points which can be connected with the complement of $\bar{V}$ by a curve which does not intersect with $\operatorname{sp} U$.

Lemma 3.2. *Let $q\in V$ be a point which can be connected with the complement of $\bar{V}$ by a continuous curve which does not intersect with* $\operatorname{sp} U$. *Then, either $(q-V)\cap\operatorname{sp} U=\emptyset$ or $(q-V)\cap\operatorname{sp} U=\{0\}$. In the latter case* 0 *is an isolated (atomic) point of* $\operatorname{sp} U$.

Proof. Let $\mathscr{C}$ be the curve connecting q with the complement of $\bar{V}$. Since $\mathscr{C}\cap\operatorname{sp} U=\emptyset$ and $\operatorname{sp} U$ is closed, there is a whole neighbourhood of $\mathscr{C}$ which does not intersect with $\operatorname{sp} U$. So by slightly modifying $\mathscr{C}$ (if necessary) we can always assume that $\mathscr{C}$ is smooth. Moreover, we can assume that the (first) point l where $\mathscr{C}$ penetrates $\partial\bar{V}$ is different from 0. Let l_Σ be the projection of l onto the spacelike plane Σ_0. Then we continue the curve from l to the point $\varepsilon\cdot l_\Sigma$, $\varepsilon>0$ along a straight line. This section lies outside of $\bar{V}$, so it is clearly disjoint from $\operatorname{sp} U$. It is also obvious that the whole curve connecting q and $\varepsilon\cdot l_\Sigma$ is contained in the cone $\varepsilon\cdot l_\Sigma+\bar{V}$ if ε is sufficiently small.

Next we fix such an ε and pick a vector Φ which is $\varepsilon\cdot l_\Sigma$-admissible; such vectors exist according to the previous lemma since $\varepsilon\cdot l_\Sigma\in\Sigma_0$ and $\varepsilon\cdot l_\Sigma\neq 0$. The above geometrical preparations guarantee that the commutator functions $\tilde{\Delta}_{A,\Phi}$ vanish in a neighbourhood of the curve connecting q and $\varepsilon\cdot l_\Sigma$. Moreover, this curve is contained in the cone $\varepsilon\cdot l_\Sigma+\bar{V}$. So we can apply the (generalized) double cone theorem, which says that $\tilde{\Delta}_{A,\Phi}$ vanishes in the double cone $(q-\bar{V})\cap(\varepsilon\cdot l_\Sigma+\bar{V})$. Since the vector Φ is $\varepsilon\cdot l_\Sigma$-admissible, $\tilde{\Delta}^+_{A,\Phi}$ vanishes too in this region, i.e.

$$(\Phi, A^*E((q-V)\cap(\varepsilon\cdot l_\Sigma+V))A\Phi)=0\,. \tag{3.9}$$

This result is independent of the choice of the local operator A and the $\varepsilon\cdot l_\Sigma$-admissible vector Φ. So applying once more the previous lemma it follows that $E((q-V)\cap(\varepsilon\cdot l_\Sigma+V))=0$, which means that there is no energy-momentum spectrum in the open set $(q-V)\cap(\varepsilon\cdot l_\Sigma+V)$. This shows, since ε can be made arbitrarily small, that $\operatorname{sp} U$ is disjoint from $(q-V)\cap V$ as well as from the part of $(q-V)\cap\partial\bar{V}$ lying in the half-space $\{k\in\mathbb{R}^4: k\cdot l_\Sigma<0\}$.

For the analysis of the remaining part we note that one can replace in this statement l by any $l'\in(q-V)\cap\partial\bar{V}\cap\{k\in\mathbb{R}^4: k\cdot l_\Sigma<0\}$ because, as a consequence of the above result, any of these points can be connected with q by a continuous curve in the complement of $\operatorname{sp} U$. Repeating this procedure, one obtains sufficiently many points l' so that the corresponding half-spaces $\{k\in\mathbb{R}^4: k\cdot l'_\Sigma<0\}$ generate a covering of $\mathbb{R}^4\backslash\{0\}$[8]. It is then clear that $\operatorname{sp} U$ and $(q-V)\cap\bar{V}$ can only have the point 0 in common, if any. If $(q-V)\cap\operatorname{sp} U=\{0\}$, then 0 is an isolated point of $\operatorname{sp} U$

7 If $S\subset\mathbb{R}^4$ we denote by $\complement S$ the complement of S in $\mathbb{R}^4$

8 Note that this statement would not hold in two space-time dimensions

since $(q-V)$ contains an open neighbourhood of 0. So in this case 0 belongs to the atomic part of $\mathrm{sp}\,U$. □

We are now in a position to determine the shape of the lower boundary of the energy-momentum spectrum. We begin with a formal

Definition. Let $S \subset \bar{V}$ be a closed set. A point $p \in S$ is said to be a lower boundary point of S if in any neighbourhood of p there exists a q which can be connected with the complement of $\bar{V}$ by a continuous curve which does not intersect with S. The set of all lower boundary points of S is called the lower boundary of S.

The preceding lemma implies that the lower boundary of $\mathrm{sp}\,U$ is a continuous surface (disregarding the point 0 in the cases where 0 is an isolated point of $\mathrm{sp}\,U$). To make this obvious let us parametrize the lower boundary of $\mathrm{sp}\,U$ by the points in a spacelike plane, given in proper coordinates by $\{(q_0, \mathbf{q}): q_0 = 0\}$. For any such point $(0, \mathbf{p})$ it follows from Lemma 3.2 that the timelike line $(\lambda, \mathbf{p})$, $\lambda \in \mathbb{R}$ has a non-trivial intersection with $\mathrm{sp}\,U$ (excluding the trivial case $\mathrm{sp}\,U = \{0\}$). On the other hand, $(\lambda, \mathbf{p}) \notin \mathrm{sp}\,U$ for negative λ, so we can define a mapping $\sigma: (0, \mathbf{p}) \to (p_0, \mathbf{p})$ of the spacelike plane into the lower boundary of $\mathrm{sp}\,U$ setting

$$p_0 = \inf\{\lambda > 0 : (\lambda, \mathbf{p}) \in \mathrm{sp}\,U\}. \tag{3.10}$$

This mapping is also onto (apart from the cases where 0 is an isolated point of $\mathrm{sp}\,U$ which then is not contained on the range of σ) since by Lemma 3.2 any two lower boundary points $p' \neq 0$ and $p'' \neq 0$ are spacelike or lightlike separated. From the latter fact it follows also that σ is continuous and that its range is connected.

It is a simple but important consequence of this discussion that $\mathrm{sp}'\,U$ (where the prime indicates that 0 has to be removed from the set $\mathrm{sp}\,U$ if it is an isolated point) and $\mathrm{sp}'\,U + \bar{V}$ have the same lower boundary. We will exploit this fact in the proof of our main result.

Theorem 3.3. *The lower boundary of* $\mathrm{sp}\,U$ *is a Lorentz invariant set. If* $\mathfrak{A}$ *is reducible, this statement holds also for* $\mathrm{sp}\,U \upharpoonright \mathscr{H}'$ *whenever the subspace* $\mathscr{H}' \subset \mathscr{H}$ *is stable under the action of* $\mathfrak{A}$. *Moreover,* $x \to U(x)$ *is the only representation of the translations with this property (amongst the representations satisfying the conditions (1.1)–(1.3)).*

Proof. We distinguish 3 cases corresponding to the possible shape of $\mathrm{sp}\,U$ at 0.

i) $0 \notin \mathrm{sp}\,U$. Then there is a whole neighbourhood of 0 which does not belong to $\mathrm{sp}\,U$. Consequently, there exists some timelike vector $f \in V$ such that $\mathrm{sp}\,U \subset f + \Sigma_+$ and $\mathrm{sp}\,U \cap (f + \Sigma_0) \neq \emptyset$, i.e. the spacelike plane $f + \Sigma_0$ is tangent to $\mathrm{sp}\,U$. Now let $H(0, m)$, $m \neq 0$ be the unique hyperboloid with apex 0 for which $f + \Sigma_0$ is a tangent plane, and let p be the point where the hyperboloid and the plane intersect. We want to show that this hyperboloid is the lower boundary of $\mathrm{sp}\,U$. To this end we choose any $q \in \Sigma_0$, $q \neq 0$ and $\varepsilon > 0$. Then we pick a vector Φ which is $\varepsilon \cdot q$-admissible and consider the commutator functions $\tilde{\Delta}_{A,\Phi}$. These functions vanish in the region

$$\complement((\mathrm{sp}\,U + \bar{V}) \cup (2 \cdot \mathrm{supp}\,\Phi - \bar{V})), \tag{3.11}$$

so that we can apply the JLD-technique. Now the hyperboloids $H(a, \mu)$ which do not enter this region are centered at points

$$a \in \bar{V} \cap \complement(\varepsilon \cdot q + V), \tag{3.12}$$

and satisfy

$$H_+(a,\mu)\cap(\varepsilon\cdot q+\bar{V})\subset \operatorname{sp} U+\bar{V}, \tag{3.13}$$

where $H_+(a,\mu)$ denotes the upper shell of $H(a,\mu)$. In both of these statements we have used the fact that Φ is $\varepsilon\cdot q$-admissible. Note, however, that in the conditions (3.12) and (3.13) the dependence of the hyperboloids on the specific choice of the support of Φ has been removed. So, irrespective of the choice of the local operator A and the $\varepsilon\cdot q$-admissible vector Φ, all commutator functions $\tilde{\Delta}_{A,\Phi}$ vanish in a neighbourhood of any point $k\in(\varepsilon\cdot q+\bar{V})$ which does not lie on one of the hyperboloids satisfying (3.12) and (3.13). Since for $k\in(\varepsilon\cdot q+V)$ we have $\tilde{\Delta}_{A,\Phi}(k) =\tilde{\Delta}^+_{A,\Phi}(k)$, it follows then from Lemma 3.1 that $\operatorname{sp} U\cap(\varepsilon\cdot q+V)$ is contained in the union of these hyperboloids.

In order to simplify the subsequent geometrical discussion we proceed at this point to the limit $\varepsilon\to 0$. In this limit, the information obtained so far can be summarized in the inclusion

$$\begin{aligned}\operatorname{sp} U\cap V\subset \bigcup\{H_+(l,\mu):H_+(l,\mu)\subset \operatorname{sp} U+\bar{V},\\ l\in\partial\bar{V}, l\cdot q\geqq 0,\mu\geqq 0\}\,.\end{aligned} \tag{3.14}$$

The apices l and masses μ of the hyperboloids in (3.14) satisfy $(p-l)^2\leqq\mu^2$, where p is the point of contact of $H(0,m)$ and $f+\Sigma_0$: namely, either l and p are spacelike or lightlike separated, i.e. $(p-l)^2\leqq 0$, or l lies below the plane $(f+\Sigma_0)$. In the latter case, the upper and lower shells of the hyperboloids in (3.14) are separated by $(f+\Sigma_0)$, and since p is a point in this plane we get $(p-l)^2\leqq\mu^2$. Let us consider now the points k on the hyperboloids in (3.14) which can be represented in the form

$$k=p+\delta\cdot q+f' \tag{3.15}$$

for some $f'\in\bar{V}$ and $\delta\geqq 0$. Since $(k-l)^2=\mu^2\geqq(p-l)^2$, $l\in\partial\bar{V}$, $l\cdot q\geqq 0$ and $p^2=m^2$ we get

$$k^2-m^2\geqq 2l\cdot(k-p)=2(\delta l\cdot q+l\cdot f')\geqq 0\,. \tag{3.16}$$

This shows that every point $k\in\operatorname{sp} U$ which can be represented in the form (3.15) satisfies $k^2\geqq m^2$. But the choice of $q\in\Sigma_0$ in our argument was completely arbitrary, and since any point in $\operatorname{sp} U$ can be represented in the form (3.15) with a suitable $q\in\Sigma_0$, it follows from (3.14) that $\operatorname{sp} U\cap V\subset H_+(0,m)+\bar{V}$. Consequently, there can be no spectrum on $\partial\bar{V}$ (otherwise the lower boundary of $\operatorname{sp} U$ would consist of disconnected parts, which is impossible), so we arrive at $\operatorname{sp} U\subset H_+(0,m)+\bar{V}$. Moreover, p must be an element of $\operatorname{sp} U$ because p is the only point in $H_+(0,m)+\bar{V}$ which is also an element of the spacelike plane $f+\Sigma_0$, and by assumption $\operatorname{sp} U\cap(f+\Sigma_0)\neq\emptyset$.

In the final step of our argument we feed this information into relation (3.14): since $p\in\operatorname{sp} U$ there must be a hyperboloid $H(l,\mu)$ which contains p and whose upper shell lies in $\operatorname{sp} U+\bar{V}\subset H_+(0,m)+\bar{V}$. It is easy to verify that the only hyperboloid with these properties is $H(0,m)$, so we have $H_+(0,m)\subset\operatorname{sp} U+\bar{V}$, and consequently $H_+(0,m)$ is the lower boundary of $\operatorname{sp} U+\bar{V}$. But the lower boundary of $\operatorname{sp} U+\bar{V}$ and $\operatorname{sp} U$ coincide, so the first part of the theorem follows if $0\notin\operatorname{sp} U$.

ii) $0 \in \mathrm{sp}\, U$ is an isolated point. Then one can argue as in the previous case, the only difference being that one must replace everywhere the set $\mathrm{sp}\, U$ by $\mathrm{sp}' U$. Actually, one has much stronger results in this case: it has been shown in [2] that if $\mathscr{H}$ contains a vector Ω (vacuum) which is cyclic for $\mathfrak{A}$ and invariant under $U(x)$, then the whole spectrum $\mathrm{sp}\, U$ is a Lorentz-invariant set.

iii) $0 \in \mathrm{sp}\, U$ is not isolated. Then, according to Lemma 3.2, there does not exist any lower boundary point in the interior of $\bar{V}$, and consequently the lower boundary of $\mathrm{sp}\, U$ consists of $\partial \bar{V}$.

So $\mathrm{sp}\, U$ has in all these cases a Lorentz-invariant lower boundary. This is also true for $\mathrm{sp}\, U \restriction \mathscr{H}'$ whenever the subspace $\mathscr{H}' \subset \mathscr{H}$ is stable under $\mathfrak{A}$, since one can apply the previous argument to any subrepresentation of $\mathfrak{A}$.

Finally, let $x \to U_1(x) \in \mathfrak{A}^-$ be another unitary representation inducing the translations on $\mathfrak{A}$ whose spectrum has a Lorentz-invariant lower boundary in every subrepresentation of $\mathfrak{A}$. From the fact that both, U and U_1 implement the translations on $\mathfrak{A}$, it follows that $W(x) = U_1(x) U(x)^{-1}$ is an element of the center of $\mathfrak{A}^-$. So $x \to W(x)$ is a continuous, unitary representation of $\mathbb{R}^4$. We denote the spectral projections corresponding to W by $F(\Delta)$ and consider the subrepresentation of $\mathfrak{A}$ on $\mathscr{H}' = F(\Delta)\mathscr{H}$. Since $U_1(x) = U(x) W(x)$ we get, provided $\mathscr{H}' \neq 0$,

$$\mathrm{sp}\, U_1 \restriction \mathscr{H}' \subset \Delta + \mathrm{sp}\, U \restriction \mathscr{H}' . \tag{3.17}$$

If Δ is a neighbourhood of any point $q \notin (-\bar{V})$ this leads immediately to a contradiction to the assumption that $\mathrm{sp}\, U \restriction \mathscr{H}'$ and $\mathrm{sp}\, U_1 \restriction \mathscr{H}'$ have a Lorentz-invariant lower boundary. So for such sets we get $F(\Delta) = 0$. Now we interchange the rôle of U and U_1, using $U(x) = U_1(x) W(x)^{-1}$. This leads to

$$\mathrm{sp}\, U \restriction \mathscr{H}' \subset -\Delta + \mathrm{sp}\, U_1 \restriction \mathscr{H}' . \tag{3.18}$$

So $F(\Delta)$ also vanishes if Δ is a neighbourhood of any point $p \notin \bar{V}$, and combining both results we arrive at $F(\Delta) = 0$ if $0 \notin \Delta$. This shows that $W(x) = 1$ and completes the proof of the theorem. □

In the derivation of the above results we profited from the explicit information on the envelope of holomorphy of domains of the type $T_+ \cup T_- \cup C^\sim$, where C is an order convex region which lies between the two shells of some hyperboloid $H(a, \mu)$ (the JLD-technique). As is clear from this discussion, an extension of our methods to the analysis of the full energy-momentum spectrum, i.e. the problem of the possible shape of "holes" in $\mathrm{sp}\, U$, would require a (partial) knowledge of the domain of holomorphy for coincidence regions C consisting of an order convex part C_1 of the form $\complement((p - \bar{V}) \cup (-p + \bar{V}))$ and a double cone C_2 which is disjoint (but not spacelike separated) from C_1. Similar geometrical situations have been discussed in [17] (cf. also the remark in [18, p. 170]), and it seems plausible that these results can be applied to the present problem so as to establish the Lorentz invariance of the full spectrum. However, the details have yet to be worked out.

Without any additional effort we can, however, say something about the structure of the spectrum above the lower boundary in the special cases, where a part of the lower boundary is isolated from the rest of the spectrum.

Definition. Let $S \subset \bar{V}$ be a closed set and let $p \in S$ be a lower boundary point. The lower boundary of S is said to be isolated at p if there exists a neighbourhood $\mathscr{N}_p$ of

p such that $\mathcal{N}_p \cap S$ contains only lower boundary points. The lower boundary of S is said to be completely isolated if it is isolated at all of its points.

Theorem 3.4. *If the lower boundary of* sp$'U$ *is isolated at some point p, then it is completely isolated. Moreover, if one removes from* sp$'U$ *the lower boundary, then the remaining set has again a Lorentz-invariant lower boundary.*

Proof. According to the previous theorem the lower boundary of sp$'U$ consists of the hyperboloid $H(0, m)$, $m^2 = p^2$, since p is a lower boundary point of sp$'U$. We remove from sp$'U$ its lower boundary and denote the closure of the resulting set by sp$''U$. Since the boundary of sp$'U$ is isolated at p, there exists then an open neighborhood $\mathcal{N}_p$ of p such that sp$''U \cap \mathcal{N}_p = \emptyset$.

Next we remove from the positive part $\tilde{\Delta}^+_{A,\Phi}$ of the commutator functions $\tilde{\Delta}_{A,\Phi}$ the contributions coming from "intermediate states" with support on the lower boundary of sp$'U$. To this end we multiply $\tilde{\Delta}^+_{A,\Phi}(k)$ by $k^2 - m^2$. Since $d^4k\tilde{\Delta}^+_{A,\Phi}(k)$ is a measure, it is clear that the resulting distributions $(k^2 - m^2)\tilde{\Delta}^+_{A,\Phi}(k)$ have support in sp$''U$. On the other hand, the supports of all these distributions are dense in sp$''U$. It is crucial now that multiplying the commutator functions $\tilde{\Delta}_{A,\Phi}(k)$ by the polynomial $(k^2 - m^2)$ does not destroy their analytic properties [in the sense of relation (3.5)]. So one can apply the same arguments as in Lemma 3.2 and Theorem 3.3 in order to show that the lower boundary of sp$''U$ is a Lorentz-invariant set. Since sp$''U \cap \mathcal{N}_p = \emptyset$, it follows then that $H(0, m)$ is disjoint from sp$''U$, so the lower boundary of sp$'U$ is completely isolated. □

It needs no explanation that in the situation described in this theorem one is dealing with a representation space $\mathcal{H}$ of $\mathfrak{A}$ containing single particle states. A detailed analysis of these representations can be found in [9].

Let us remark in conclusion that our results hold in any number of space-time dimensions. In two dimensions, however, one finds in the analysis of Lemma 3.2 the additional possibility that sp$\,U$ can have an isolated part, consisting of the right or left branch of $\partial \bar{V}$ (cf. footnote 8). This kind of a spectrum describes massless particles moving only in one direction. Using the above methods one can also handle these cases, so the statements of both theorems hold true without modifications in any number of space-time dimensions.

4. Concluding Remarks

We have seen in the present investigations that there exists a canonical choice of the representation $x \to U(x)$ of the translations which is distinguished by the fact that its spectrum has a Lorentz-invariant lower boundary in each superselection sector. This result is a consequence of the principle of locality and holds even if the Lorentz-transformations are broken. In view of the uniqueness of the representation $x \to U(x)$ it is natural to regard the corresponding generators P as the (global) observables of energy and momentum.

It should be noticed, however, that the energy momentum operators P can in general only be used to compare the energy-content of states in the same superselection sector (in spite of the fact that P is uniquely defined on the whole physical Hilbert space $\mathscr{H}$). The energy-difference of states in different superselection sectors of $\mathscr{H}$ is a priori not defined in terms of the spectrum of P (cf. the example given below). So there arises the question of how to compare energies in different sectors.

If all physical states are local excitations of a unique vacuum state, then the answer is well known: given any charged state one can add to it a compensating charge "behind the moon" such that the resulting state has total charge 0. Because of the large separation of the charges it then follows that the total energy of this state (in terms of P) is just the sum of the energies of the two constituents. This "additivity of the energy" (cf. for example [2]) makes it possible to relate the energies in different sectors so that the spectrum of P gains an absolute meaning. This result can also be established without any a priori assumptions on the localization properties of the physical states if the particle spectrum of the theory is completely massive [9].

The situation is, however, less clear in theories with long-range forces and massless particles. The states in these models have in general poor localization properties, and, as a matter of fact, one can often distinguish the various superselection sectors by the different behaviour of the corresponding states at spacelike infinity (cf. for example [8]). So the heuristic ideas which led to a proof of the additivity of the energy in massive theories cannot readily be applied to the massless case, and P need not always have an absolute meaning.

Yet since we are only interested in comparing energies in different sectors we could be content with less information. It would be sufficient to know how much energy is needed to proceed from one superselection sector to another. In our setting this quantity can be defined as follows: let $\mathscr{H}_1$ and $\mathscr{H}_2$ be arbitrary coherent subspaces of $\mathscr{H}$ (superselection sectors). It is then a very general fact (Fell's theorem[9], cf. [5]) that any state $\Psi \in \mathscr{H}_2$ can be approximated by states in $\mathscr{H}_1$ in the weak topology, i.e. there exists a sequence (or, more generally, a net) $\Phi_\iota \in \mathscr{H}_1$ such that

$$\lim_\iota (\Phi_\iota, A\Phi_\iota) = (\Psi, A\Psi) \quad \text{for} \quad A \in \mathfrak{A}. \tag{4.1}$$

Now let $\Psi \in \mathscr{H}_2$ be a vector with energy-momentum about the lowest lying point $(m_2, \mathbf{0})$ in $\operatorname{sp} U \restriction \mathscr{H}_2$. Then we can ask whether it is possible to choose a sequence Φ_ι in relation (4.1) with finite maximal energy m_1, i.e.

$$\Phi_\iota \in E(\Delta)\mathscr{H}_1 \quad \text{where} \quad \Delta = ((m_1, \mathbf{0}) - \bar{V}) \cap \bar{V}. \tag{4.2}$$

If such an m_1 exists we say the sector $\mathscr{H}_2$ is energy-connected with $\mathscr{H}_1$; the energy which is necessary in order to proceed from $\mathscr{H}_1$ to $\mathscr{H}_2$ is then given by $\inf(m_1 - m_2)$. It is a simple consequence of our results on the shape of the energy-momentum spectrum that this quantity is always non-negative.

In this context there arises the following interesting question: does the property of energy-connectedness define an equivalence relation between superselection sectors? And if so, are the energies which are needed in order to proceed from $\mathscr{H}_1$ to $\mathscr{H}_2$ and from $\mathscr{H}_2$ to $\mathscr{H}_1$, respectively, equal? Both questions

---

9 We refrain from giving the weakest conditions under which Fell's theorem holds. For our purposes it is sufficient (and physically reasonable) to assume that $\mathfrak{A}$ is a simple C^*-algebra

have clearly an affirmative answer in theories of massive particles. There the energy which is needed in order to proceed from one sector to another is the same for both directions, because to each (compensating) charge there exists, by the PCT-theorem, a conjugate charge with the same mass.

Let us finally discuss the significance of states lying in energy-disconnected sectors. An experimenter, who would try to prepare such a state in larger and larger space-time regions $\mathcal{O}$ would realize that this requires unlimited energy, although the energy inside the region $\mathcal{O}$ stays finite. So the energy which is needed for the preparation of this state is essentially a surface energy.

The simplest model in which these circumstances occur is the field-theory of a free, massless, scalar particle. In this model there exists a one-parameter family Ω_λ, $\lambda \in \mathbb{R}$ of decent vacuum states [19] and it is straightforward to verify that the various vacuum sectors are energy-disconnected[10]. Note, however, that all vacuum states Ω_λ are eigenstates of the energy-momentum operator P corresponding to the eigenvalue 0; so P does not have a global meaning in this model.

It is not known whether the existence of energy-disconnected sectors in a model is always related to the existence of several vacuum states.

Acknowledgements. We are grateful to J. Bros for communicating to us his results on the analytic completion problem mentioned at the end of Sect. 3. One of us (D.B.) would also like to thank S. Doplicher, Universitá di Roma and A. S. Wightman, Princeton University for the warm hospitality extended to him during the early and final stages of this work. He also acknowledges financial support from the CNR, DFG, and Princeton University.

References

1. Wigner, E.P.: On unitary representations of the inhomogeneous Lorentz group. Ann. Math. **40**, 149 (1939)
2. Borchers, H.J.: Local rings and the connection of spin with statistics. Commun. Math. Phys. **1**, 281 (1965)
3. Fröhlich, J., Morchio, G., Strocchi, F.: Infrared problem and spontaneous breaking of the Lorentz group in QED. Phys. Lett. **89** B, 61 (1979)
4. Strocchi, F., Wightman, A.S.: Proof of the charge superselection rule in local relativistic quantum field theory. J. Math. Phys. **15**, 2198 (1974)
5. Haag, R., Kastler, D.: An algebraic approach to quantum field theory. J. Math. Phys. **5**, 848 (1964)
6. Borchers, H.J.: Translation group and spectrum condition. Commun. Math. Phys. **96**, 1–13 (1984)
7. Fröhlich, J., Morchio, G., Strocchi, F.: Charged sectors and scattering states in quantum electrodynamics. Ann. Phys. **119**, 241 (1979)
8. Buchholz, D.: The physical state space of quantum electrodynamics. Commun. Math. Phys. **85**, 49 (1982)
9. Buchholz, D., Fredenhagen, K.: Locality and the structure of particle states. Commun. Math. Phys. **84**, 1 (1982)
10. Hepp, K., Jost, R.: Über die Matrixelemente des Translationsoperators. Helv. Phys. Acta **35**, 34 (1962)
11. Maison, D.: Eine Bemerkung zur Clustereigenschaft. Commun. Math. Phys. **10**, 48 (1968)
12. Yosida, K.: Functional analysis. Berlin, Heidelberg, New York: Springer 1966

10 This result holds in four space-time dimensions. It is noteworthy that in three dimensions the corresponding sectors are energy-connected

13. Pedersen, G.K.: C^*-algebras and their automorphism groups. London, New York, San Francisco: Academic Press 1979
14. Vladimirov, V.S.: Methods of the theory of functions of many complex variables. Cambridge, MA, London: MIT-Press 1966
15. Bros, J., Messiah, A., Stora, R.: A problem of analytic completion related to the Jost-Lehmann-Dyson formula. J. Math. Phys. **2**, 639 (1961)
16. Pflug, P.: Eine Bemerkung über die Konstruktion von Holomorphiehüllen. Zeszyty Nankowe Uniwersytetu Jagiellonskiego, Vol. 23
17. Borchers, H.J., Stora, R.: Unpublished notes (1963)
18. Stora, R.: Hyperfunctions and Theoretical Physics, Nice 1973. In: Lecture Notes in Mathematics, Vol. 449. Berlin, Heidelberg, New York: Springer 1975
19. Doplicher, S.: Fock representation and massless particles. Commun. Math. Phys. **3**, 228 (1966)

Communicated by G. Mack

Received May 18, 1984

Large Deviations for Stationary Gaussian Processes*

M. D. Donsker and S. R. S. Varadhan

Courant Institute of Mathematical Sciences, New York University, New York, NY 10012, USA

Abstract. In their previous work on large deviations the authors always assumed the base process to be Markovian whereas here they consider the base process to be stationary Gaussian. Similar large deviation results are obtained under natural hypotheses on the spectral density function of the base process. A rather explicit formula for the entropy involved is also obtained.

1. Introduction

Let $\{X_k\}$, $-\infty<k<\infty$, be a stationary Gaussian process with $E\{X_k\}=0$ and $E\{X_0X_j\}=\varrho_j=\frac{1}{2\pi}\int_0^{2\pi} e^{ij\theta}f(\theta)d\theta$. We assume that the spectral density function $f(\theta)$ is continuous on $[0,2\pi]$, $f(0)=f(2\pi)$, and

$$\int_0^{2\pi} \log f(\theta)d\theta > -\infty . \tag{1.1}$$

Let $\Omega=\prod_{j=-\infty}^{\infty} \mathbb{R}_j$ where, for each j, $\mathbb{R}_j$ is the real line, i.e., Ω is the space of doubly infinite sequences of real numbers. We specify a point $\omega\in\Omega$ by $\omega=\{x_k\}$, $-\infty<k<\infty$, and let $\omega(j)=x_j$ for $-\infty<j<\infty$. The process $\{X_k\}$ induces a probability measure P on Ω. We will denote integration over Ω with respect to P measure by $E^P\{\}$.

For each positive integer n and each $\omega\in\Omega$, let $\omega^{(n)}$ be the point in Ω obtained by the periodic extension in both directions of the elements $x_1, x_2, \ldots, x_n$ of ω, i.e., if $\omega=\{x_k\}$, $-\infty<k<\infty$, then $\omega^{(n)}$ is the point

$$\ldots, x_1, \ldots, x_{n-1}, x_n, x_1, x_2, \ldots, x_n, x_1, x_2, \ldots, x_n, \ldots .$$

* The research in this paper was supported by the National Science Foundation, Grant No. MCS-80-02568

Let T be the shift operator acting on the points of Ω. For each $\omega \in \Omega$ and each positive integer n, we define a probability measure on Ω, call it $\pi_n(\omega)$, as follows:

$$\pi_n(\omega) = \frac{1}{n}(\delta_{\omega^{(n)}} + \delta_{T\omega^{(n)}} + \ldots + \delta_{T^{n-1}\omega^{(n)}}). \tag{1.2}$$

In other words, given $\omega \in \Omega$ and $n \geqq 1$, $\pi_n(\omega)$ is the probability measure on Ω which assigns mass $1/n$ to $\omega^{(n)}$ and to each of its $(n-1)$ translates. We note that $\pi_n(\omega)$ is a translation invariant measure on Ω, i.e., a stationary measure on Ω. Let $\underline{M}_S$ be the space of all stationary measures on Ω and impose on $\underline{M}_S$ the topology of weak convergence. We can use the mapping $\pi_n : \Omega \to \underline{M}_S$ and the measure P on Ω to construct a probability measure on $\underline{M}_S$ by defining, for each n, $Q_n = \pi_n P^{-1}$, i.e., if A is a set of stationary measures in $\underline{M}_S$, then

$$Q_n(A) = P\{\omega \in \Omega : \pi_n(\omega) \in A\}. \tag{1.3}$$

The assumptions we made on the Gaussian process $\{X_k\}$ imply that it is an ergodic process, and it then follows from the ergodic theorem that, for almost all ω (P-measure), the measure $\pi_n(\omega)$ converges weakly to the measure P as $n \to \infty$, i.e.,

$$P\{\omega \in \Omega : \pi_n(\omega) \Rightarrow P\} = 1. \tag{1.4}$$

Thus, with reference to the Q_n measure on $\underline{M}_S$ given by (1.3), we expect that if the set A contains the measure P, then $Q_n(A) \to 1$ as $n \to \infty$, whereas if P is not in the closure of A, we expect $Q_n(A) \to 0$ as $n \to \infty$. In this paper we show that in this latter case $Q_n(A)$ approaches zero exponentially fast as $n \to \infty$ and we determine the constant in the exponential rate. To be more specific, let R denote a general element of $\underline{M}_S$, i.e., a stationary measure on Ω, and recall that f is the spectral density of our basic stationary Gaussian process $\{X_k\}$. We define a functional $H_f(R)$ which is, in fact, the entropy of the stationary process R with respect to the stationary Gaussian process $\{X_k\}$, and $H_f(R)$: $\underline{M}_S \to [0, \infty]$ is such that:

If $C \subset M_S$ is closed,

$$\overline{\lim_{n \to \infty}} \frac{1}{n} \log Q_n(C) \leqq -\inf_{R \in C} H_f(R), \tag{1.5}$$

and if $G \subset \underline{M}_S$ is open,

$$\underline{\lim_{n \to \infty}} \frac{1}{n} \log Q_n(G) \geqq -\inf_{R \in G} H_f(R). \tag{1.6}$$

Consistent with remarks made above it will indeed be true that $H_f(P) = 0$. We give an explicit formula for $H_f(R)$ in (1.9) below.

Let $\Phi : \underline{M}_S \to \mathbb{R}$ be bounded and continuous, and let $E^{Q_n}\{\,\}$ denote integration over $\underline{M}_S$ with respect to Q_n measure. From (1.5) and (1.6) it follows easily that

$$\lim_{n \to \infty} \frac{1}{n} \log E^{Q_n}\{e^{n\Phi(R)}\} = \sup_{R \in \underline{M}_S} [\Phi(R) - H_f(R)]. \tag{1.7}$$

Since $Q_n = \pi_n P^{-1}$, equivalent to (1.7) is

$$\lim_{n \to \infty} \frac{1}{n} \log E^P\{e^{n\Phi(\pi_n(\omega))}\} = \sup_{R \in M_S} [\Phi(R) - H_f(R)]. \tag{1.8}$$

The main results of this paper then are (1.5), (1.6), their implication (1.8), and the formula (1.9) for $H_f(R)$ which we discuss now.

Let $R \in \underline{M}_S$ and let, for $A \subset \mathbb{R}$, $R(A|\omega) = R(X_0 \in A | X_{-1}, X_{-2}, \ldots)$ be the regular conditional probability distribution of X_0 given the entire past. Denote by $r(y|\omega)$ the corresponding density. If $G(\theta)$ is the spectral measure of the stationary process R, then the formula for $H_f(R)$ is

$$H_f(R) = E^R\left\{\int_{-\infty}^{\infty} r(y|\omega)\log r(y|\omega)dy\right\} + \frac{1}{2}\log 2\pi + \frac{1}{4\pi}\int_0^{2\pi}\frac{dG(\theta)}{f(\theta)} + \frac{1}{4\pi}\int_0^{2\pi}\log f(\theta)d\theta, \tag{1.9}$$

where it is understood that $H_f(R)$ is defined to be $+\infty$ if for any reason we cannot define any of the ingredients in (1.9).

The authors have developed a theory of large deviations for Markov processes (see [1, 2], for theoretical results and e.g. [3–5] for some applications thereof). In theory, our methods should apply to rather general stationary processes, but it is difficult to see in any great generality what the natural hypotheses to impose on a stationary process are in order to obtain the analogues of (1.5), (1.6), and (1.8), much less to obtain the analogue of (1.9) in anything like explicit form. For recent work in this direction see Orey [6]. In this paper we consider the special case of a Gaussian stationary process because as seen above the hypotheses needed are natural and we obtain an explicit formula for the corresponding entropy.

In Sect. 2 of this paper we prove (1.5) and (1.6) but in doing so we must use properties of $H_f(R)$ and other entropies which occur in the proofs. So as not to interrupt the argument we prove these properties in Sect. 3.

2. Proofs of the Main Theorems

As in the introduction, let $\{X_k\}$ be a stationary Gaussian process and $E\{X_0X_j\} = \varrho_j = \frac{1}{2\pi}\int_0^{2\pi} e^{ij\theta}f(\theta)d\theta$, where we assume that the spectral density function $f(\theta)$ is continuous on $[0, 2\pi]$, $f(0) = f(2\pi)$, and $\int_0^{2\pi}\log f(\theta)d\theta > -\infty$.

Let $\{\xi_k\}$, $-\infty < k < \infty$, be a sequence of independent Gaussian random variables each with mean 0 and variance 1. If we let $\{a_n\}$ be the Fourier coefficients of $\sqrt{f(\theta)}$, i.e., $\sqrt{f(\theta)} = \sum_{n=-\infty}^{\infty} a_n e^{in\theta}$, then we can write

$$X_k = \sum_{n=-\infty}^{\infty} a_{n-k}\xi_k = \sum_{n=-\infty}^{\infty} a_n\xi_{n+k}. \tag{2.1}$$

Let N be a positive integer and define

$$h_N(\theta) = \sum_{|j|<N} a_j\left(1 - \frac{|j|}{N}\right)e^{ij\theta}, \tag{2.2}$$

where the a_j's are defined in the preceding paragraph. Since $f(\theta)$ is continuous and $f(0)=f(2\pi)$, we have by Fejer's theorem,

$$\lim_{N\to\infty} \sup_{0\leqq\theta\leqq 2\pi} |h_N(\theta)-\sqrt{f(\theta)}|=0. \tag{2.3}$$

For each N, define the process

$$X_k^N = \sum_{|j|<N} a_j\left(1-\frac{|j|}{N}\right)\xi_{j+k}. \tag{2.4}$$

If we let $f_N(\theta)=h_N^2(\theta)$, then $f_N(\theta)$ is the spectral density of the process $\{X_k^N\}$.

In the introduction we noted that the process $\{X_k\}$ induces a probability measure P on Ω. Similarly the process $\{X_k^N\}$ just defined induces a probability measure P^N on Ω. Indeed, the family $\{\xi_k\}$ of independent, Gaussian, mean 0, variance 1, random variables itself induces a probability measure on Ω which we will denote by μ. Each of these three stationary measures on Ω together with the mapping $\pi_n:\Omega\to\underline{M}_S$ of the introduction generates a probability measure on $\underline{M}_S$. We have already introduced $Q_n=P\pi_n^{-1}$. Analogously, we define $v_n=\mu\pi_n^{-1}$ and for each N, $Q_n^N=P^N\Pi_n^{-1}$.

For convenience let $b_k=a_k\left(1-\frac{|k|}{N}\right)$ and define the mapping $F_N:\Omega\to\Omega$ by

$$(F_N(\omega))(j)=\sum_{|k|<N} b_k x_{j+k}, \tag{2.5}$$

so that F_N is a moving average map of Ω into itself. The mapping $F_N:\Omega\to\Omega$ induces a corresponding map $\tilde{F}_N:\underline{M}_S\to\underline{M}_S$.

We also have the mapping $F:\Omega\to\Omega$ given by

$$(F(\omega))(j)=\sum_{k=-\infty}^{\infty} a_k x_{j+k}, \tag{2.6}$$

which exists for almost all ω (μ measure on Ω), since $\sum_{n=-\infty}^{\infty} a_n^2<\infty$. We note that for sets $A\subset\underline{M}_S$:

$$\begin{aligned}
Q_n(A)&=P\{\omega\in\Omega:\pi_n(\omega)\in A\}=\mu\{\omega\in\Omega:\pi_n\cdot F(\omega)\in A\},\\
Q_n^N(A)&=P^N\{\omega\in\Omega:\pi_n(\omega)\in A\}=\mu\{\omega\in\Omega:\pi_n\cdot F_N(\omega)\in A\},\\
v_n(A)&=\mu\{\omega\in\Omega:\pi_n(\omega)\in A\}.
\end{aligned} \tag{2.7}$$

It will be important to introduce also the probability measure $\tilde{Q}_n^N$ on $\underline{M}_S$ given by $\tilde{Q}_n^N=v_n\tilde{F}_n^{-1}$ so that

$$\tilde{Q}_n^N(A)=v_n\{R\in\underline{M}_S:\tilde{F}_N(R)\in A\}=\mu\{\omega\in\Omega:\tilde{F}_N\cdot\pi_n(\omega)\in A\}. \tag{2.8}$$

To deduce the asymptotic behavior of Q_n as $n\to\infty$, i.e., to prove (1.5) and (1.6), we first determine the asymptotic behavior of Q_n^N as $n\to\infty$ and then let $N\to\infty$. To obtain the asymptotics of Q_n^N we compare it with $\tilde{Q}_n^N$ the asymptotic behavior of which as $n\to\infty$ follows from our earlier work and which we state below as Theorem 2.2. Looking at (2.8) and (2.7) we see that to compare $\tilde{Q}_n^N$ and Q_n^N means comparing $\pi_n \cdot F_N(\omega)$ with $\tilde{F}_N \cdot \pi_n(\omega)$. Now,

$$\tilde{F}_N \cdot \pi_n(\omega) = \frac{1}{n}[\delta_{F_N(\omega^{(n)})} + \delta_{F_N(T\omega^{(n)})} + \ldots + \delta_{F_N(T^{n-1}\omega^{(n)})}], \tag{2.9}$$

and

$$\pi_n \cdot F_N(\omega) = \frac{1}{n}[\delta_{(F_N(\omega))^{(n)}} + \delta_{T(F_N(\omega))^{(n)}} + \ldots + \delta_{T^{n-1}(F_N(\omega))^{(n)}}]. \tag{2.10}$$

Thus, to compare the two measures $\tilde{F}_N \cdot \pi_n$ and $\pi_n \cdot F_N$, we need to determine when $T^{j-1}(F_N(\omega))^{(n)} = F_N(T^{j-1}\omega^{(n)})$ with $1 \leqq j \leqq n$. It is clear that the difference of the two measures $\pi_n \cdot F_N$ and $\tilde{F}_N \cdot \pi_n$ can be estimated in the variation norm over the σ-field generated by the first L components. We obtain

$$\|\pi_n \cdot F_N - \tilde{F}_N \cdot \pi_n\|_{F'_L} \leqq \frac{2N+L}{n}. \tag{2.11}$$

In the space $\underline{M}_S$ we have the topology of weak convergence. This topology can be induced by the following metric. Let R_1 and R_2 be in $\underline{M}_S$ and let L be a positive integer. Denote by R_1^L and R_2^L the respective L-dimensional measures one obtains by considering the joint distribution of L successive random variables in the processes R_1 and R_2. We then define the distance between R_1 and R_2 in $\underline{M}_S$ by

$$d(R_1, R_2) = \sum_{L=1}^{\infty} \frac{1}{2^L} d_L(R_1^L, R_2^L), \tag{2.12}$$

where d_L is the Prohorov distance for L-dimensional measures. This metric in $\underline{M}_S$ yields the topology of weak convergence in $\underline{M}_S$. We now can prove the following lemma.

Lemma 2.1. *For any $\delta > 0$ and large enough n,*

$$\mu\{\omega \in \Omega : d(\pi_n \cdot F_N, \tilde{F}_N \cdot \pi_n) \geqq \varepsilon\} = 0. \tag{2.13}$$

Proof. From (2.12), (2.11), and the fact that the variation norm dominates the Prohorov norm in L-dimensions, we have for $n > N$,

$$d(\pi_n \cdot F_N, \tilde{F}_n \cdot \pi_n) \leqq \sum_{L=1}^{\infty} \frac{1}{2^L} \|\pi_n \cdot F_N - \tilde{F}_N \cdot \pi_n\|_{\underline{F}'_L} \leqq \frac{1}{n} \sum_{L=1}^{\infty} \frac{2N+L}{2^L}.$$

Since the infinite series on the right converges, we can make the distance on the left as small as we please by taking n large enough which implies (2.13).

As noted in (2.8),

$$\tilde{Q}_n^N(A) = \nu_n\{R \in \underline{M}_S : \tilde{F}_N(R) \in A\} = \mu\{\omega \in \Omega : \tilde{F}_N \cdot \pi_n(\omega) \in A\}.$$

Since μ measure on Ω comes from the sequence $\{\xi_k\}$ of independent Gaussian variables with mean 0 and variance 1, the results of [2-IV] give us the asymptotics of ν_n measure on $\underline{M}_S$ and hence of $\tilde{Q}_n^N$ measure on $\underline{M}_S$.

In the introduction we defined for $R \in \underline{M}_S$ the density $r(y|\omega)$ of the regular condition probability distribution of X_0 given the entire past. For $R \in \underline{M}_S$ define

$$I(R) = E^R \left\{ \int_{-\infty}^{\infty} r(y|\omega) \log \left(\frac{r(y|\omega)}{\frac{1}{\sqrt{2\pi}} e^{-y^2/2}} \right) dy \right\}, \tag{2.14}$$

and define for $R \in \underline{M}_S$

$$H_{f_N}(R) = \inf_{R' \in \underline{M}_S : R' \tilde{F}_N^{-1} = R} I(R'). \tag{2.15}$$

Then in Sect. 3, formula (3.21), we show

$$H_{f_N}(R) = E^R \left\{ \int_{-\infty}^{\infty} r(y|\omega) \log r(y|\omega) dy \right\} + \frac{1}{2} \log 2\pi$$
$$+ \frac{1}{4\pi} \int_0^{2\pi} \frac{dG(\theta)}{f_N(\theta)} + \frac{1}{4\pi} \int_0^{2\pi} \log f_N(\theta) d\theta,$$

where $G(\theta)$ is the spectral measure of R and $f_N(\theta)$ (see just after (2.4)) is the spectral density of the $\{X_R^N\}$ process.

Thus, from [2-IV] and the contraction principle discussed in that paper, we have

Theorem 2.2. *If $C \subset \underline{M}_S$ is closed,*

$$\overline{\lim_{n \to \infty}} \frac{1}{n} \log \tilde{Q}_n^N(C) \leqq - \inf_{R \in C} H_{f_N}(R), \tag{2.16}$$

and if $G \subset \underline{M}_S$ is open,

$$\underline{\lim_{n \to \infty}} \frac{1}{n} \log \tilde{Q}_n^N(G) \geqq - \inf_{R \in G} H_{f_N}(R). \tag{2.17}$$

Using this and Lemma 2.1 we can prove the following theorem.

Theorem 2.3. *If $C \subset \underline{M}_S$ is closed,*

$$\overline{\lim_{n \to \infty}} \frac{1}{n} \log Q_n^N(C) \leqq - \inf_{R \in C} H_{f_N}(R), \tag{2.18}$$

and if $G \subset \underline{M}_S$ is open,

$$\underline{\lim_{n \to \infty}} \frac{1}{n} \log Q_n^N(G) \geqq - \inf_{R \in G} H_{f_N}(R). \tag{2.19}$$

Proof. Let $\delta > 0$ and $C^\delta = \bigcup_{R \in C} S(R, \delta)$ where $S(R, \delta)$ is the ball in $\underline{M}_S$ of radius δ centered at R. From (2.7) and (2.8) we see that

$$\begin{aligned} Q_n^N(C) &= \mu\{\omega \in \Omega : \pi_n \cdot F_N(\omega) \in C\} \\ &\leqq \mu\{\omega \in \Omega : \tilde{F}_N \cdot \pi_n(\omega) \in C^\delta\} + \mu\{\omega \in \Omega : d(\pi_n \cdot F_N, \tilde{F}_n \cdot \pi_n) \geqq \delta\} \\ &= \tilde{Q}_n^N(C^\delta) + \mu\{\omega \in \Omega : d(\pi_n \cdot F_N, \tilde{F}_N \cdot \pi_n) \geqq \delta\}. \end{aligned}$$

Let $\bar{C}^\delta$ be the closure of C^δ. Then, from Lemma 2.1 and Theorem 2.2,

$$\overline{\lim_{n\to\infty}} \frac{1}{n} \log Q_n^N(C) \leqq \overline{\lim_{n\to\infty}} \frac{1}{n} \log \tilde{Q}_n^N(\bar{C}^\delta) \leqq - \inf_{R\in\bar{C}^\delta} H_{\tilde{f}_N}(R). \tag{2.20}$$

Since C is closed, $H_{f_N}(R)$ is lower semicontinuous, and $\{R : H_{f_N}(R) \leqq K\}$ is tight, (cf. Lemma 3.10) we obtain (2.18) from (2.20) on letting $\delta \to 0$.

For the lower bound, let G be open in $\underline{M}_S$ and let $R \in G$. Pick $\delta > 0$ and an open set V such that $R \in V \subset V^\delta \subset G$. Then, again noting (2.7) and (2.8),

$$\begin{aligned} Q_n^N(G) &\geqq Q_n^N(V^\delta) = \mu\{\omega \in \Omega : \pi_n \cdot F_N(\omega) \in V^\delta\} \\ &\geqq \mu\{\omega \in \Omega : \tilde{F}_N \cdot \pi_n(\omega) \in V\} - \mu\{\omega \in \Omega : d(\pi_n \cdot F_N, \tilde{F}_N \cdot \pi_n) \geqq \delta\} \\ &= \tilde{Q}_n^N(V) - \mu\{\omega \in \Omega : d(\pi_n \cdot F_N, \tilde{F}_N \cdot \pi_n) \geqq \delta\}. \end{aligned}$$

Using again Lemma 2.1 and Theorem 2.2,

$$\underline{\lim_{n\to\infty}} \frac{1}{n} \log Q_n^N(G) \geqq \underline{\lim_{n\to\infty}} \frac{1}{n} \log \tilde{Q}_n^N(V) \geqq - \inf_{R'\in V} H_{f_N}(R') \geqq - H_{f_N}(R). \tag{2.21}$$

Since inequality (2.21) holds for any $R \in G$ we obtain (2.19).

Before proving the next theorem we need some preliminary lemmas.

Lemma 2.4. *Let* $\{X_k\}$ *and* $\{X_k^N\}$ *be defined respectively by* (2.1) *and* (2.4). *There exists a function* $\varepsilon(N) = o(1)$ *as* $N \to \infty$ *such that if* $0 < \lambda < \frac{1}{2\varepsilon(N)}$, *then for all n,*

$$\frac{1}{n} \log E^\mu \left\{ \exp\left\{ \lambda \sum_{k=1}^n (X_k - X_k^N)^2 \right\} \right\} \leqq -\frac{1}{2} \log(1 - 2\lambda\varepsilon(N)). \tag{2.22}$$

Proof. First of all, let $\{Y_k\}$ be any stationary Gaussian process with mean 0, covariance $\gamma_j = \frac{1}{2\pi} \int_0^{2\pi} e^{ij\theta} g(\theta) d\theta$, and strictly positive definite covariance matrix

$$A_n = \begin{pmatrix} \gamma_0 \; \gamma_1 \cdots \gamma_{n-1} \\ \gamma_1 \; \gamma_0 \cdots \gamma_{n-2} \\ \cdots \quad \cdots \quad \cdots \\ \gamma_{n-1} \; \cdots \quad \gamma_0 \end{pmatrix}.$$

Let $c = \sup_{0 \leqq \theta \leqq 2\pi} |g(\theta)|$. We show now that if $\lambda < \frac{1}{2c}$, then

$$E\left\{ \exp\left\{ \lambda \sum_{k=1}^n Y_k^2 \right\} \right\} \leqq (1 - 2\lambda c)^{-n/2}. \tag{2.23}$$

To see this, first observe that if c_n is the maximum eigenvalue of A_n,

$$\begin{aligned} E\left\{ \exp\left\{ \lambda \sum_{k=1}^n Y_k^2 \right\} \right\} &= \frac{1}{\sqrt{|A_n|}} \frac{1}{(2\pi)^{n/2}} \int_{-\infty}^{\infty} \cdots \int \exp\left\{ \lambda \sum_{k=1}^n y_k^2 - \frac{1}{2} Y' A_n^{-1} Y \right\} dy_1 \ldots dy_n \\ &= \frac{1}{\sqrt{|A_n|}} \frac{1}{\sqrt{|A_n^{-1} - 2\lambda I|}} = \frac{1}{\sqrt{|I - 2\lambda A_n|}} \leqq (1 - 2\lambda c_n)^{-n/2}, \end{aligned}$$

providing $\lambda < 1/2c_n$. To estimate c_n, consider

$$\sum_{j,k=1}^{n} \gamma_{k-j}\eta_k\bar{\eta}_j = \frac{1}{2\pi}\int_0^{2\pi}\left|\sum_{k=1}^{n}\eta_k e^{ik\theta}\right|^2 g(\theta)d\theta \leqq \sup_{0\leqq\theta\leqq 2\pi}|g(\theta)|\sum_{j=1}^{n}|\eta_j|^2,$$

which implies $c_n \leqq \sup_{0\leqq\theta\leqq 2\pi}|g(\theta)| = c$. Thus, $(1-2\lambda c_n)^{-n/2} \leqq (1-2\lambda c)^{-n/2}$ and $\lambda < 1/2c$ implies $\lambda < 1/2c_n$, so we have (2.23). Now, in particular let

$$y_k = X_k^N - X_k = \sum_{|j|<N} a_j\left(1-\frac{|j|}{N}\right)\xi_{j+k} - \sum_{j=-\infty}^{\infty} a_j\xi_{j+k}.$$

The spectral density $g(\theta)$ for the process $\{Y_k\}$ is now given by

$$g(\theta) = \left|\sum_{j=-\infty}^{\infty} a_j e^{ij\theta} - \sum_{|j|\leqq N} a_j\left(1-\frac{|j|}{N}\right)e^{ij\theta}\right|^2 = |\sqrt{f(\theta)} - h_N(\theta)|^2.$$

Let $\varepsilon(N) = \sup_{0\leqq\theta\leqq 2\pi}|g(\theta)| = \sup_{0\leqq\theta\leqq 2\pi}|\sqrt{f(\theta)} - h_N(\theta)|^2$. From (2.3) we see $\varepsilon(N) = o(1)$ as $N\to\infty$, and from (2.23) if $0 < \lambda < 1/2\varepsilon(N)$, then we have for all n,

$$\frac{1}{n}\log E\left\{\exp\left\{\lambda\sum_{k=1}^{n}(x_k - x_k^N)^2\right\}\right\} \leqq -\frac{1}{2}\log(1-2\lambda\varepsilon(N)),$$

which is the lemma.

Lemma 2.5. *Let $\omega\in\Omega$ be given by $\omega = \{x_k\}$, $-\infty < k < \infty$ and let $\omega'\in\Omega$ be given by $\omega' = \{x_k + \varepsilon_k\}$, $-\infty < k < \infty$. Then, there is a constant c such that*

$$d(\pi_n(\omega), \pi_n(\omega')) \leqq c\sqrt{\frac{\varepsilon_1^2 + \varepsilon_2^2 + \ldots + \varepsilon_n^2}{n}}. \tag{2.24}$$

Proof. By definition [cf. (2.12)],

$$d(\pi_n(\omega), \pi_n(\omega')) = \sum_{L=1}^{\infty}\frac{1}{2^L}d_L([\pi_n(\omega)]_L^1, [\pi_n(\omega')]_L^1). \tag{2.25}$$

Let α and β be two L-dimensional prob distributions and $\hat{\alpha}$ and $\hat{\beta}$ their characteristic functions. Since uniform convergence of characteristic functions on compact sets in L-dimensions is equivalent to weak convergence in L-dimensions which is equivalent to convergence in the Prohorov metric d_L in L-dimensions, we can write

$$d_L(\alpha, \beta) = \sum_{k=1}^{\infty}\frac{1}{2^k}\sup_{|\eta|\leqq k}|\hat{\alpha}(\eta) - \hat{\beta}(\eta)|. \tag{2.26}$$

In particular, if $\alpha = \delta_{(x_1, x_2, \ldots, x_L)}$ and $\beta = \delta_{(x_1+\varepsilon_1, \ldots, x_L+\varepsilon_L)}$, we have from (2.26),

$$\begin{aligned} d_L(\alpha, \beta) &= \sum_{k=1}^{\infty}\frac{1}{2^k}\sup_{|\eta|\leqq k}|e^{i\langle x, \eta\rangle} - e^{i\langle x+\varepsilon, \eta\rangle}| \\ &= \sum_{k=1}^{\infty}\frac{1}{2^k}\sup_{|\eta|\leqq k}|e^{i\langle \varepsilon, \eta\rangle} - 1| \\ &\leqq \|\varepsilon\|\sum_{k=1}^{\infty}\frac{k}{2^k} = 2\|\varepsilon\|. \end{aligned}$$

Thus, from the definition of π_n,

$$d_L([\pi_n(\omega)]_L^1, [\pi_n(\omega')]_L^1)$$

$$\leqq \frac{2}{n}\left(\sqrt{\varepsilon_1^2+\varepsilon_2^2+\ldots+\varepsilon_L^2}+\sqrt{\varepsilon_2^2+\ldots+\varepsilon_{1+L}^2}+\ldots+\sqrt{\varepsilon_n^2+\ldots+\varepsilon_{n+L-1}^2}\right),$$

where $\varepsilon_{n+j}=\varepsilon_j$. Using Minkowski's inequality on this last, we get

$$d_L([\pi_n(\omega)]_L^1, [\pi_n(\omega')]_L^1)$$

$$\leqq 2\sqrt{\frac{1}{n}[(\varepsilon_1^2+\ldots+\varepsilon_L^2)+(\varepsilon_2^2+\ldots+\varepsilon_{2+L}^2)+\ldots+(\varepsilon_n^2+\ldots+\varepsilon_{n+L}^2)]}$$

$$=2\sqrt{\frac{L}{n}(\varepsilon_1^2+\varepsilon_2^2+\ldots+\varepsilon_n^2)}\,.$$

Finally, from (2.25) and this last inequality,

$$d(\pi_n(\omega), \pi_n(\omega'))\leqq \sum_{L=1}^{\infty}\frac{1}{2^L}\cdot 2\sqrt{\frac{L(\varepsilon_1^2+\ldots+\varepsilon_n^2)}{n}}=c\sqrt{\frac{\varepsilon_1^2+\ldots+\varepsilon_n^2}{n}}\,.$$

Lemma 2.6. *Let F_N and F be defined by* (2.5) *and* (2.6) *respectively. For any $\delta>0$,*

$$\mu\{\omega\in\Omega: d(\pi_n\cdot F_N, \pi_n\cdot F)\geqq\delta\}\leqq\mu\left\{\omega\in\Omega:\sum_{k=1}^{n}\frac{(X_k^N-X_k)^2}{n}\geqq\frac{\delta^2}{c^2}\right\}, \quad (2.27)$$

where c is the constant in Lemma 2.5.

Proof. From (2.1) and (2.4), $X_k=\sum_{n=-\infty}^{\infty}a_n\xi_{n+k}$ and $x_k^N=\sum_{|j|<N}a_j\left(1-\frac{|j|}{N}\right)\xi_{j+k}$. If we let $\varepsilon_k=X_k-X_k^N$, then $X_k=X_k^N+\varepsilon_k$. Thus, from Lemma 2.5,

$$d(\pi_n\cdot F_N, \pi_n\cdot F)\leqq c\sqrt{\frac{1}{n}\sum_{k=1}^{n}(X_k-x_k^N)^2}\,,$$

and (2.27) follows.

Lemma 2.7. *For every $\delta>0$,*

$$\lim_{N\to\infty}\overline{\lim_{n\to\infty}}\frac{1}{n}\log\mu\{\omega\varepsilon\,\Omega: d(\pi_n\cdot F_N, \pi_n\cdot F)\geqq\delta\}=-\infty\,. \quad (2.28)$$

Proof. For fixed N, and $\lambda<1/2\varepsilon(N)$, where $\varepsilon(N)$ is defined above in Lemma 2.4, we have from (2.27) and (2.24) for any $\delta>0$,

$$\mu\{\omega\in\Omega: d(\pi_n\cdot F_N, \pi_n\cdot F)\geqq\delta\}\leqq\mu\left\{\omega\in\Omega:\frac{1}{n}\sum_{k=1}^{n}(X_k^N-X_k)^2\geqq\frac{\delta^2}{c^2}\right\}$$

$$\leqq E^{\mu}\left\{\exp n\lambda\left(\frac{1}{n}\sum_{k=1}^{n}(X_k^N-X_k)^2\right)\right\}e^{-n\delta^2\lambda/c^2}\leqq e^{-n\delta^2\lambda/c^2}e^{-\frac{n}{2}\log(1-2\lambda\varepsilon(N))}\,.$$

Therefore, for any $\delta>0$ and $\lambda<1/2\varepsilon(N)$ with N fixed,

$$\overline{\lim_{n\to\infty}}\frac{1}{n}\log\mu\{\omega\in\Omega: d(\pi_n\cdot F_N, \pi_n\cdot F)\geqq\delta\}\leqq-\frac{\delta^2\lambda}{c^2}-\frac{1}{2}\log(1-2\lambda\varepsilon(N)). \quad (2.29)$$

Since $\varepsilon(N)=o(1)$ as $N\to\infty$, we get from (2.29) that for all $\delta>0$ and all $\lambda>0$,

$$\varlimsup_{N\to\infty}\varlimsup_{n\to\infty}\frac{1}{n}\log\mu\{\omega\in\Omega: d(\pi_n\cdot F_N,\pi_n\cdot F)\geqq\delta\}\leqq-\frac{\delta^2\lambda}{c^2}.$$

In this last if we let $\lambda\to\infty$ we obtain (2.28).

We are now ready to prove (1.5) and (1.6) of the introduction.

Theorem 2.8. *If* $C\subset\underline{M}_S$ *is closed, then*

$$\varlimsup_{n\to\infty}\frac{1}{n}\log Q_n(C)\leqq-\inf_{R\in C}H_f(R), \tag{2.30}$$

and if $G\subset\underline{M}_S$ *is open, then*

$$\varliminf_{n\to\infty}\frac{1}{n}\log Q_n(G)\geqq-\inf_{R\in G}H_f(R), \tag{2.31}$$

where $H_f(R)$ *is defined by* (1.9).

Proof. Let C be closed in $\underline{M}_S$. Then,

$$Q_n(C)=P\{\omega\in\Omega:\pi_n(\omega)\in C\}=\mu\{\omega\in\Omega:\pi_n\cdot F(\omega)\in C\}.$$

For any $\delta>0$, we have

$$\begin{aligned}Q_n(C)&\leqq\mu\{\omega\in\Omega:\pi_n\cdot F_n\in C^\delta\}+\mu\{\omega\in\Omega:d(\pi_n\cdot F_N,\pi_n\cdot F)\geqq\delta\}\\&=P^N\{\omega\in\Omega:\pi_n(\omega)\in C^\delta\}+\mu\{\omega\in\Omega:d(\pi_n\cdot F_N,\pi_n\cdot F)\geqq\delta\}\\&=Q_n^N(C^\delta)+\mu\{\omega\in\Omega:d(\pi_n\cdot F_N,\pi_n\cdot F)\geqq\delta\}\\&\leqq Q_n^N(\bar C^\delta)+\mu\{\omega\in\Omega:d(\pi_n\cdot F_N,\pi_n\cdot F)\geqq\delta\}.\end{aligned}$$

From Theorem 2.3, this last inequality implies

$$\varlimsup_{n\to\infty}\frac{1}{n}\log Q_n(C)\leqq\max\Big(-\inf_{R\in\bar C^\delta}H_{f_N}(R),A_N(\delta)\Big),$$

where $A_N(\delta)=\varlimsup_{n\to\infty}\frac{1}{n}\log\mu\{\omega\in\Omega:d(\pi_n\cdot F_N,\pi_n\cdot F)\geqq\delta\}$. From Lemma 2.7 we then obtain,

$$\varlimsup_{n\to\infty}\frac{1}{n}\log Q_n(C)\leqq-\varliminf_{N\to\infty}\inf_{R\in\bar C^\delta}H_{f_N}(R). \tag{2.32}$$

From Lemma 3.10 below it follows that

$$\varliminf_{N\to\infty}\inf_{R\in\bar C^\delta}H_{f_N}(R)=\inf_{R\in\bar C^\delta}H_f(R), \tag{2.33}$$

where $H_f(R)$ is given by (1.9). Using (2.33) in (2.32) we get

$$\varlimsup_{n\to\infty}\frac{1}{n}\log Q_n(C)\leqq-\inf_{R\in\bar C^\delta}H_f(R). \tag{2.34}$$

Since C is closed and $H_f(R)$ is lower semicontinuous (2.34) implies (2.30) on letting $\delta\to0$.

For the lower bound, let G be open in $\underline{M}_S$ and let R be any element of G. Pick $\delta>0$ and an open set V such that $R\in V\subset V^\delta\subset G$. Then

$$\begin{aligned} Q_n(G) &\geqq Q_n(V^\delta)=\mu\{\omega\in\Omega:\pi_n\cdot F(\omega)\in V^\delta\}\\ &\geqq\mu\{\omega\in\Omega:\pi_n\cdot F_N\in V\}-\mu\{\omega\in\Omega:d(\pi_n\cdot F_N,\pi_n\cdot F)\geqq\delta\}\\ &=Q_n^N(V)-\mu\{\omega\in\Omega:d(\pi_n\cdot F_N,\pi_n\cdot F)\geqq\delta\}\,. \end{aligned} \tag{2.35}$$

Without loss of generality we can assume that $H_f(R)<\infty$ for the $R\in G$ we picked.

As before, let

$$A_N(\delta)=\overline{\lim_{n\to\infty}}\frac{1}{n}\log\mu\{\omega\in\Omega:d(\pi_n\cdot F_N,\pi_n\cdot F)\geqq\delta\}\,.$$

From Lemma 2.7 we can choose N so large that $A_N(\delta)\leqq-(H_f(R)+3)$, i.e., we can choose N so large that

$$\mu\{\omega\in\Omega:d(\pi_n\cdot F_N,\pi_n\cdot F)\geqq\delta\}\leqq e^{-n(H_f(R)+3)+o(n)}\,. \tag{2.36}$$

Also, since V is open we obtain from Theorem 2.3,

$$\underline{\lim_{n\to\infty}}\frac{1}{n}\log Q_n^N(V)\geqq-\inf_{R'\in V}H_{f_N}(R')\,. \tag{2.37}$$

In Lemma 3.9 below we prove that for an $R\in\underline{M}_S$ such that $H_f(R)<\infty$, there exists a sequence $\{R_N\}$ in $\underline{M}_S$ which converges weakly to R and for which $\lim_{N\to\infty}H_{f_N}(R_N)=H_f(R)$. If we let $0<\eta<1$ be given, then from (2.37) and Lemma 3, since V is a neighborhood of R, we can take N so large that

$$\underline{\lim_{n\to\infty}}\frac{1}{n}\log Q_n^N(V)\geqq-\inf_{R'\in V}H_{f_N}(R')\geqq-H_{f_N}(R_N)\geqq-H_f(R)-\eta\,.$$

In other words, for N large enough,

$$Q_n^N(V)\geqq e^{-n(H_f(R)+\eta)+o(n)}\,. \tag{2.38}$$

Using (2.38) and (2.36) in (2.35) we see that for N large enough

$$Q_n(G)\geqq e^{-n(H_f(R)+\eta)+o(n)}-e^{-n(H_f(R)+3)+o(n)}\,.$$

Since $\eta<1$ the first expression on the right is the controlling asymptotic term and we get,

$$\underline{\lim_{n\to\infty}}\frac{1}{n}\log Q_n(G)\geqq-H_f(R)-\eta\,. \tag{2.39}$$

Since η is arbitrary and R is an arbitrary element of G, we conclude finally that

$$\underline{\lim_{n\to\infty}}\frac{1}{n}\log Q_n(G)\geqq-\inf_{R\in G}H_f(R)\,,$$

which completes the proof of the theorem.

3. Properties of Entropy

In this section we derive the formula for $H_{f_N}(R)$ occurring in Theorem 2.2. We also prove those properties of $H_{f_N}(R)$ and $H_f(R)$ which we called upon in obtaining the upper and lower bounds in Theorem 2.8.

We will be somewhat more general at first and start with $I(R)$ as defined by (2.16). Note, that for $R \in \underline{M}_S$, and with $\phi(y) = \frac{1}{\sqrt{2\pi}} e^{-y^2/2}$,

$$I(R) = E^R\left\{\int_{-\infty}^{\infty} r(y|\omega) \log\left(\frac{r(y|\omega)}{\phi(y)}\right) dy\right\}$$
$$= E^R\left\{\int_{-\infty}^{\infty} r(y|\omega) \log r(y|\omega) dy\right\} + \frac{1}{2}\log 2\pi + \frac{1}{2} E^R\{\omega^2(0)\}. \tag{3.1}$$

Let $\{a_k\}$, $-\infty < k < \infty$ be any sequence of reals such that $\sum_{j=1}^{\infty} |a_j| < \infty$ and $|h(\theta)| = \left|\sum_{j=-\infty}^{\infty} a_j e^{ij\theta}\right| \neq 0$ for all $\theta \in [0, 2\pi]$. Consider the mapping $\tau: \Omega \to \Omega$ based on such a sequence $\{a_k\}$ given by $(\tau\omega)(k) = \sum_{j=-\infty}^{\infty} a_j \omega(j+k)$. We will call such mappings τ "admissible."

For τ such an admissible mapping, define for $R \in \underline{M}_S$

$$H^\tau(R) = \inf_{R' : R'\tau^{-1} = R} I(R'). \tag{3.2}$$

We prove below that if R has spectral measure $G(\theta)$, then

$$H^\tau(R) = I(R) + \frac{1}{4\pi}\int_0^{2\pi} \frac{dG(\theta)}{|h(\theta)|^2} - \frac{1}{4\pi}\int_0^{2\pi} dG(\theta) + \frac{1}{2\pi}\int_0^{2\pi} \log|h(\theta)d\theta. \tag{3.3}$$

Before proving (3.3) we need to examine how $I(R\tau^{-1})$ is related to $I(R)$ for an admissible map τ. We start with some preliminary lemmas. The first is easy and we omit the proof. The second is a classic result found in standard references on matrices.

Lemma 3.1. *Let $R \in \underline{M}_S$ and let $R_n(y_1, y_2, \ldots, y_n)$ be the n-dimensional joint density for the R process, i.e., $R_n(y_1, y_2, \ldots, y_n) dy_1 \ldots dy_n = R\{X_1 \in dy_1, \ldots, X_n \in dy_n\}$. Define, with $\phi(y) = \frac{1}{\sqrt{2\pi}} e^{-y^2/2}$,*

$$I_n(R) = \int_{-\infty}^{\infty} \cdots \int R_n(y_1, y_2, \ldots, y_n) \log\left(\frac{R_n(y_1, y_2, \ldots, y_n)}{\prod_{j=1}^{n} \phi(y_j)}\right) dy_1 \ldots dy_n.$$

Then, $I_n(R)-I_{n-1}(R)$ *is monotone increasing in* n, $\lim_{n\to\infty}(I_n(R)-I_{n-1}(R))=I(R)$ *so that also*

$$\lim_{n\to\infty}\frac{I_n(R)}{n}=I(R).$$

Lemma 3.2. *Let* D *be the determinant of a circulant matrix of order* n, *i.e.,*

$$D=\begin{vmatrix} d_1 & d_2 & \dots & d_n & \\ d_n & d_1 & d_2 \dots & d_{n-1} & \\ \vdots & & & & \\ d_2 & d_3 & \dots & d_n & d_1 \end{vmatrix}.$$

Then

$$D=\prod_{\xi\in Z_n}(d_1+d_2\xi+\dots+d_n\xi^{n-1}), \tag{3.4}$$

where Z_n *is the set of* n^{th} *roots of unity.*

Lemma 3.3. *Let* $f(x)$ *be a probability density in* d*-dimensions and let* $g(x)$ *be the* d*-dimensional Gaussian density,* $g(x)=\frac{1}{(2\pi)^{d/2}}e^{-(1/2)\|x\|^2}$. *Let* A *be a nonsingular matrix. Then,*

$$\int_{R^d}\left(\frac{f(A^{-1}x)}{|A|}\right)\log\left(\frac{f(A^{-1}x)/|A|}{g(x)}\right)dx=\int_{R^d}f(y)\log\frac{f(y)}{g(y)}dy$$
$$+\frac{1}{2}\int_{R^d}(\|Ay\|^2-\|y\|^2)f(y)dy-\log|A|. \tag{3.5}$$

Proof. We observe that if $\alpha(dx)=f(x)dx$, then $(\alpha A^{-1})(dx)=\frac{f(A^{-1}x)}{|A|}dx$. Now, letting $x=Ay$,

$$\int_{R^d}\left(\frac{f(A^{-1}x)}{|A|}\right)\log\left(\frac{f(A^{-1}x)/|A|}{g(x)}\right)dx=\int_{R^d}f(y)\log\left(\frac{f(y)}{g(Ay)}\right)dx-\log|A|$$
$$=\int_{R^d}f(y)\log\left(\frac{f(y)}{g(y)}\right)dy+\int_{R^d}f(y)\log\left(\frac{g(y)}{g(Ay)}\right)dy-\log|A|$$
$$=\int_{R^d}f(y)\log\frac{f(y)}{g(y)}dy+\frac{1}{2}\int_{R^d}(\|Ay\|^2-\|y\|^2)f(y)dy-\log|A|.$$

As a preparatory lemma we consider admissible maps of a special kind – where only a finite number of the elements in $\{a_k\}$ are different from 0.

Lemma 3.4. *For any* $R\in\underline{M}_S$, *let* $I(R)$ *be defined by* (3.1) *and let* $\tau:\Omega\to\Omega$ *be defined by* $(\tau\omega)(j)=\sum_{k=-\infty}^{\infty}a_k\omega(k+j)$, *where, for some* K, $a_k=0$ *if* $|k|>K$ *and* $|h(\theta)| = \left|\sum_{k=-\infty}^{\infty}a_ke^{ik\theta}\right| \neq 0$ *for all* $\theta\in[0,2\pi]$. *Then,*

$$I(R\tau^{-1})\leq I(R)+\frac{1}{2}[E^{R\tau^{-1}}\{\omega^2(0)\}-E^R\{\omega^2(0)\}-\frac{1}{2\pi}\int_0^{2\pi}\log|h(\theta)|d\theta. \tag{3.6}$$

Proof. Let $n>K$ where K is such that $a_k=0$ if $|k|>K$. Consider the $(2n+1)\times(2n+1)$ circulant matrix

$$A_{n+1}=\begin{vmatrix} a_0 & a_1\dots a_n & a_{-n}\dots a_{-1} \\ a_{-1} & a_0\dots a_{n-1} & \dots a_{-2} \\ a_1 & a_2 \quad a_{-n} & a_0 \end{vmatrix}.$$

Let E_{2n+1} be $(2n+1)$ dimensional space and define a mapping $E_{2n+1}\to E_{2n+1}$ by $x\to A_{2n+1}x$. If $\omega\in\Omega$, let $\Gamma_{2n+1}\omega=(\omega(j))_{|j|\leqq n}$, i.e., if $\omega=\{x_k\}$, $-\infty<k<\infty$, then Γ_{2n+1} is the section:

$$x_{-n}, x_{-n+1},\dots,x_{-1},x_0,x_1,\dots,x_{n-1},x_n.$$

It is easy to see that if $|i|\leqq n-K$, then

$$(\tau\omega)(i)=(A_{2n+1}\Gamma_{2n+1}\omega)(i). \tag{3.7}$$

Define the density $R'_{2n+1}(y_{-n},\dots,y_0,\dots,y_n)$ by

$$R'_{2n+1}(y_{-n},\dots,y_0,\dots,y_n)dy_{-n}\dots dy_0\dots dy_n=R\tau^{-1}\Gamma^{-1}_{2n+1},$$

and similarly define the density $\hat{R}'_{2n+1}$ by

$$\hat{R}'_{2n+1}(y_{-n},\dots,y_0,\dots,y_n)dy_{-n},\dots,dy_0\dots dy_n=R\Gamma^{-1}_{2n+1}A^{-1}_{2n+1}.$$

From the definition of $I_n(R)$ in Lemma 3.1,

$$\begin{aligned} I_{2k+1}(R\tau^{-1})=\int_{-\infty}^{\infty}\!\!\cdots\!\int R'_{2k+1}(y_{-k},\dots,y_0,\dots,y_k) \\ \cdot\log\left(\frac{R'_{2k+1}(y_{-k},\dots,y_0,\dots,y_k)}{\left(\frac{1}{\sqrt{2\pi}}\right)^{2k+1}\exp\left\{-1\frac{1}{2}\sum_{j=-k}^{k}y_j^2\right\}}\right)dy_{-k}dy_0\dots dy_k. \end{aligned}\tag{3.8}$$

From (3.7) we see that $R\tau^{-1}\Gamma^{-1}_{2n+1}=R\Gamma^{-1}_{2l+1}A^{-1}_{2l+1}\Gamma^{-1}_{2n+1}$ provided $l\geqq n+K$. In particular, if we choose $n=k$, $l=k+K$, we have

$$R\tau^{-1}\Gamma^{-1}_{2k+1}=R\Gamma^{-1}_{2(k+K)+1}A^{-1}_{2(k+K)+1}\Gamma^{-1}_{2k+1}.$$

Therefore, from (3.8) we obtain

$$\begin{aligned} I_{2k+1}(R\tau^{-1})\leqq\int_{-\infty}^{\infty}\!\!\cdots\!\int \hat{R}'_{2(k+K)+1}(y_{-k-k},\dots,y_0,\dots,y_{k+K}) \\ \cdot\log\frac{\hat{R}'_{2(k+K)+1}(y_{-K+k},\dots,y_0,\dots,y_{k+K})}{\left(\frac{1}{\sqrt{2\pi}}\right)^{2(k+K)+1}\exp\left\{-\frac{1}{2}\sum_{j=-k-K}^{k+K}y_j^2\right\}}\,dy_{-k-K}\dots dy_{k+K}. \end{aligned}\tag{3.9}$$

From Lemma 3.3 and (3.9) we get the estimate

$$\begin{aligned} I_{2k+1}(R\tau^{-1})\leqq I_{2(k+K)+1}(R)-\log|A_{2(k+K)+1}| \\ +\frac{1}{2}E^R\{\|A_{2(k+K)+1}\Gamma_{2(k+K)+1}\omega\|^2-\|\Gamma_{2(k+K)+1}\omega\|^2\}. \end{aligned}\tag{3.10}$$

Since R is stationary and $a_k=0$ if $|k|>K$,

$$\lim_{n\to\infty}\frac{1}{2n+1}E^R\{\|A_{2n+1}\Gamma_{2n+1}\omega\|^2\}=E^R\left\{\left|\sum_{j=-\infty}^{\infty}a_j\omega(j)\right|^2\right\}=E^{R\tau^{-1}}\{\omega^2(0)\},$$

and
$$\lim_{n\to\infty}\frac{1}{2n+1}E^R\{\|\Gamma_{2n+1}\omega\|^2\}=E^R\{\omega^2(0)\}.$$

Let $c=\lim\limits_{n\to\infty}\frac{1}{2n+1}\log|A_{2n+1}|$, divide both sides of (3.10) by $2k+1$, let $k\to\infty$, and use Lemma 3.1 to obtain

$$I(R\tau^{-1})\leqq I(R)-c+\frac{1}{2}[E^{R\tau^{-1}}\{\omega^2(0)\}-E^R\{\omega^2(0)\}]. \tag{3.11}$$

Thus, it remains to show that

$$c=\frac{1}{2\pi}\int_0^{2\pi}\log|h(\theta)|d\theta,$$

where $h(\theta)=\sum\limits_{k=-\infty}^{\infty}a_ke^{ik\theta}$. Since A_{2n+1} is a circulant matrix, Lemma 3.2 applies. We note that if $\xi\in Z_{2n+1}$ then $\xi^{-j}=\xi^{2n+1-j}$. Since $a_k=0$ if $|k|>K$, we get

$$\lim_{n\to\infty}\frac{1}{2n+1}\log|A_{2n+1}|=\lim_{n\to\infty}\frac{1}{2n+1}\log\prod_{\xi\in Z_{2n+1}}\left(\sum_{|j|\leqq n}a_j\xi^j\right)$$
$$=\lim_{n\to\infty}\frac{1}{2n+1}\sum_{\xi\in Z_{2n+1}}\log\sum_{j=-\infty}^{\infty}a_j\xi^j=\frac{1}{2\pi}\int_0^{2\pi}\log\left|\sum_{j=-\infty}^{\infty}a_je^{ij\theta}\right|d\theta,$$

where, of course, in the last equality we used the properties of roots of unity and that the last expression is finite since $|h(\theta)|\neq 0$ for all $\theta\in[0,2\pi]$ by hypothesis.

In the preceding lemma we showed inequality (3.6) under the hypothesis that the admissible mapping τ was in fact based on coefficients $\{a_k\}$ only a finite number of which were different from zero. Using that result we can now show that for any admissible mapping equality actually holds in (3.6).

Lemma 3.5. *Let τ be any admissible mapping. Then, for any $R\in\underline{M}_S$,*

$$I(R\tau^{-1})=I(R)+\frac{1}{2}[E^{R\tau^{-1}}\{\omega^2(0)\}-E^R\{\omega^2(0)\}]-\frac{1}{2\pi}\int_0^{2\pi}\log\left|\sum_{k=-\infty}^{\infty}a_ke^{ik\theta}\right|d\theta. \tag{3.12}$$

Proof. First define the map τ_n by $(\tau_n\omega)(j)=\sum\limits_{k=-n}^{n}a_k\omega(k+j)$ with the same a_k's as in the statement of the lemma. We apply Lemma 3.4 to τ_n so that for any $R\in\underline{M}_S$,

$$I(R\tau_n^{-1})\leqq I(R)+\frac{1}{2}[E^{R\tau_n^{-1}}\{\omega^2(0)\}-E^R\{\omega^2(0)\}]$$
$$-\frac{1}{2\pi}\int_0^{2\pi}\log\left|\sum_{k=-n}^{n}a_ke^{ik\theta}\right|d\theta. \tag{3.13}$$

Since $I(R)$ is lower semicontinuous on $\underline{M}_S$, $I(R\tau^{-1}) \leqq \varliminf_{n\to\infty} I(R\tau_n^{-1})$. Moreover, $\lim_{n\to\infty} E^{R\tau_n^{-1}}\{\omega^2(0)\} = E^{R\tau^{-1}}\{\omega^2(0)\}$, and

$$\lim_{n\to\infty} \frac{1}{2\pi}\int_0^{2\pi} \log\left|\sum_{k=-n}^{n} a_k e^{ik\theta}\right| d\theta = \frac{1}{2\pi}\int_0^{2\pi} \log\left|\sum_{k=-\infty}^{\infty} a_k e^{ik\theta}\right| d\theta,$$

since $\left|\sum_{k=-\infty}^{\infty} a_k e^{ik\theta}\right| \neq 0$ for all $\theta \in [0, 2\pi]$. Thus, from (3.13) we conclude, for $R \in \underline{M}_S$,

$$I(R\tau^{-1}) \leqq I(R) + \frac{1}{2}[E^{R\tau^{-1}}\{\omega^2(0)\} - E^R\{\omega^2(0)\}] - \frac{1}{2\pi}\int_0^{2\pi} \log\left|\sum_{k=-\infty}^{\infty} a_k e^{ik\theta}\right| d\theta \tag{3.14}$$

for any admissible mapping τ. Let $g(\theta) = \sum_{j=-\infty}^{\infty} a_j e^{ij\theta}$. By our hypothesis this is an absolutely convergent Fourier series and $g(\theta) \neq 0$ for any $\theta \in [0, 2\pi]$. Wiener's theorem says that the L^2-function $1/g(\theta)$ has a Fourier series $\frac{1}{g(\theta)} = \sum_{j=-\infty}^{\infty} b_j e^{ij\theta}$ and $\sum_{j=-\infty}^{\infty} |b_j| < \infty$. Of course, $\left|\sum_{j=-\infty}^{\infty} b_j e^{ij\theta}\right| \neq 0$ for all $\theta \in [0, 2\pi]$.

Consider then the admissible map σ given by $(\sigma\omega)(j) = \sum_{k=-\infty}^{\infty} b_k \omega(k+j)$ and note that $\sigma = \tau^{-1}$. Apply (3.14) to σ and we get

$$I(R\sigma^{-1}) \leqq I(R) + \frac{1}{2}[E^{R\sigma^{-1}}\{\omega^2(0)\} - E^R\{\omega^2(0)\}] + \frac{1}{2\pi}\int_0^{2\pi} \log\left|\sum_{k=-\infty}^{\infty} a_k e^{ik\theta}\right| d\theta, \tag{3.15}$$

where the sign in the last term on the right is now positive since $\left|\sum_{k=-\infty}^{\infty} a_k e^{ik\theta}\right| = \frac{1}{\left|\sum_{k=-\infty}^{\infty} b_k e^{ik\theta}\right|}$. But (3.15) is true for all $R \in \underline{M}_S$ and, therefore, true in particular if we replace R in (3.15) by $R\tau^{-1}$ yielding, since $\sigma = \tau^{-1}$,

$$I(R\tau^{-1}\sigma^{-1}) = I(R) \leqq I(R\tau^{-1}) + \frac{1}{2}[E^R\{\omega^2(0)\} - E^{R\tau^{-1}}\{\omega^2(0)\}] + \frac{1}{2\pi}\int_0^{2\pi} \log\left|\sum_{k=-\infty}^{\infty} a_k e^{ik\theta}\right| d\theta,$$

or

$$I(R\tau^{-1}) \geqq I(R) + \frac{1}{2}[E^{R\tau^{-1}}\{\omega^2(0)\} - E^R\{\omega^2(0)\}] - \frac{1}{2\pi}\int_0^{2\pi} \log\left|\sum_{k=-\infty}^{\infty} a_k e^{ik\theta}\right| d\theta. \tag{3.16}$$

Inequalities (3.14) and (3.16) give us (3.12) for any admissible map τ which is the lemma.

From the preceding lemma we see that for any admissible mapping τ, $I(R)<\infty$ if and only if $I(R\tau^{-1})<\infty$. Since an admissible mapping τ is $1-1$ we observe that for any τ in this class, and with $\sigma=\tau^{-1}$,

$$\begin{aligned}H^{\tau}(R) &= \inf_{R':R'\tau^{-1}=R} I(R')=I(R\sigma^{-1}) \\ &= I(R)+\frac{1}{2}[E^{R\sigma^{-1}}\{\omega^2(0)\}-E^R\{\omega^2(0)\}]+\frac{1}{2\pi}\int_0^{2\pi}\log|h(\theta)|d\theta\,, \qquad (3.17)\end{aligned}$$

where $h(\theta)=\sum_{j=-\infty}^{\infty} a_j e^{ij\theta}$, and τ is given by $(\tau\omega)(j)=\sum_{k=-\infty}^{\infty} a_k\omega(k+j)$. Note, that in (3.17) the sign before the last term on the right is + [cf. (3.12)] because $\sigma=\tau^{-1}$ and the sequence $\{a_k\}$ belongs to τ.

Moreover if $R\in \underline{M}_S$ has spectral measure $G(\theta)$, then $\frac{1}{2\pi}\int_0^{2\pi} dG(\theta)=E^R\{\omega^2(0)\}$ and also since $\sigma=\tau^{-1}$ we have $E^{R\sigma^{-1}}\{\omega^2(0)\}=\frac{1}{2\pi}\int_0^{2\pi}\frac{dG(\theta)}{|h(\theta)|^2}$. Hence, we can rewrite (3.17) in the form

$$\begin{aligned}H^{\tau}(R)=I(R)&+\frac{1}{4\pi}\int_0^{2\pi}\frac{dG(\theta)}{|h(\theta)|^2} \\ &-\frac{1}{4\pi}\int_0^{2\pi}dG(\theta)+\frac{1}{2\pi}\int_0^{2\pi}\log|h(\theta)|d\theta\,, \qquad (3.18)\end{aligned}$$

which is formula (3.3) at the beginning of this section.

In formula (3.18) which holds for any admissible τ, replace $I(R)$ by the last line in (3.1), using again that $E^R\{w^2(0)\}=\frac{1}{2\pi}\int_0^{2\pi}dG(\theta)$, so

$$\begin{aligned}H^{\tau}(R)=E^R&\left\{\int_{-\infty}^{\infty} r(y|\omega)\log r(y|\omega)dy\right\} \\ &+\frac{1}{2}\log 2\pi+\frac{1}{4\pi}\int_0^{2\pi}\frac{dG(\theta)}{|h(\theta)|^2}+\frac{1}{2\pi}\int_0^{2\pi}\log|h(\theta)|d\theta\,. \qquad (3.19)\end{aligned}$$

Now, in particular, consider the mapping $F_N:\Omega\to\Omega$ given by (2.5) or more precisely, its induced map $\tilde{F}_N$ in $\underline{M}_S$. This is an admissible map since

$$(F_N\omega)(j)=\sum_{|k|<N} a_k\left(1-\frac{|k|}{N}\right)\omega(k+j)\,,$$

and therefore at most a finite number of the coefficients defining this map are different from 0. Moreover, $h_N(\theta)=\sum_{|k|<N} a_k\left(1-\frac{|k|}{N}\right)e^{ik\theta}$, defined in (2.2), never vanishes since by Fejer's theorem,

$$h_N(\theta)=\int_0^{2\pi}\sqrt{f(\phi)}\,\psi_N(\theta-\phi)d\phi\,,$$

where

$$\psi_N(\theta)=\sum_{|k|<N}\left(1-\frac{|k|}{N}\right)e^{ik\theta},$$

is strictly positive. If as before we let $f_N(\theta)=|h_N(\theta)|^2$, then $f_N(\theta)$ is the spectral density of the process $\{X_k^N\}$, and so with $\tau=F_N$ and $H_{f_N}(R)=H^\tau(R)$, we get from (3.19)

$$H_{f_N}(R)=E^R\left\{\int_{-\infty}^{\infty} r(y|\omega)\log r(y|\omega)dy\right\}+\frac{1}{2}\log 2\pi$$
$$+\frac{1}{4\pi}\int_0^{2\pi}\frac{dG(\theta)}{f_N(\theta)}+\frac{1}{4\pi}\int_0^{2\pi}\log f_N(\theta)d\theta\,. \tag{3.20}$$

Formula (3.20) is the same as in Sect. 2 just after (2.17).

Finally, we must see in what sense $H_{f_N}(R)$ approximates $H_f(R)$ as defined by formula (1.9). We begin with the following well known lemma.

Lemma 3.6. *Let $\phi(x)$ be a probability density on $\mathbb{R}$ with two moments and let $\sigma^2(\phi)$ be the variance of ϕ. Then, for any $\alpha>0$,*

$$\int_{-\infty}^{\infty}\phi(y)\log\phi(y)dy\geqq-\frac{1}{2}\log 2\pi\alpha-\frac{1}{2\alpha}\sigma^2(\phi)\,. \tag{3.21}$$

Proof. Let $\phi_{\sigma,\mu}=\frac{1}{\sigma\sqrt{2\pi}}\exp\left\{-\frac{(y-\mu)^2}{2\sigma^2}\right\}$. We note

$$\int_{-\infty}^{\infty}\phi(y)\log\phi(y)dy=\int_{-\infty}^{\infty}\phi(y)\log\left(\frac{\phi(y)}{\phi_{\sigma,\mu}(y)}\right)dy+\int_{-\infty}^{\infty}\phi(y)\log\phi_{\sigma,\mu}(y)dy\,.$$

The first term on the right is nonnegative being the entropy of one probability density with respect to another. Now, $\log\phi_{\sigma,\mu}(y)=-\log(\sigma\sqrt{2\pi})-\frac{1}{2\sigma^2}(y-\mu)^2$, and therefore

$$\int_{-\infty}^{\infty}\phi(y)\log\phi(y)dy\geqq-\frac{1}{2}\log(2\pi\sigma^2)-\frac{1}{2\sigma^2}\int_{-\infty}^{\infty}(y-\mu)^2\phi(y)dy\,.$$

If, in particular, we choose $\mu=\int_{-\infty}^{\infty}y\phi(y)dy$ and $\sigma^2=\alpha$ we obtain (3.21).

Lemma 3.7. *Let τ be an admissible map and for $R\in\underline{M}_S$ let*

$$e(R)=\int_{-\infty}^{\infty}r(y|\omega)\log r(y|\omega)dy\,.$$

If $R'=R\tau^{-1}$, then

$$e(R')=e(R)-\frac{1}{2\pi}\int_0^{2\pi}\log|h(\theta)|d\theta\,, \tag{3.22}$$

where $h(\theta)=\sum_{j=-\infty}^{\infty} a_j e^{ij\theta}$ *and* $\{a_j\}$, $-\infty<j<\infty$ *are the coefficients defining the map* τ.

Proof. From (3.12) and (3.1)

$$\begin{aligned} I(R')&=I(R)+\frac{1}{2}[E^{R'}\{\omega^2(0)\}-E^R\{\omega^2(0)\}]-\frac{1}{2\pi}\int_0^{2\pi}\log|h(\theta)|d\theta \\ &=e(R)+\frac{1}{2}\log 2\pi+\frac{1}{2}E^{R'}\{\omega^2(0)\}-\frac{1}{2\pi}\int_0^{2\pi}\log|h(\theta)|d\theta. \end{aligned}$$

Using (3.1) on the left of this last equation we see that

$$\begin{aligned} &e(R')+\frac{1}{2}\log 2\pi+\frac{1}{2}E^{R'}\{\omega^2(0)\} \\ &\quad=e(R)+\frac{1}{2}\log 2\pi+\frac{1}{2}E^{R'}\{\omega^2(0)\}-\frac{1}{2\pi}\int_0^{2\pi}\log|h(\theta)|d\theta, \end{aligned}$$

from which (3.22) follows.

Lemma 3.8. *With* $f_N(\theta)$ *and* $f(\theta)$ *as in Sect. 2,*

$$\lim_{N\to\infty}\int_0^{2\pi}\log f_N(\theta)d\theta=\int_0^{2\pi}\log f(\theta)d\theta. \tag{3.23}$$

Proof. It suffices to show, since $f_N(\theta)=h_N^2(\theta)$, that

$$\lim_{N\to\infty}\int_0^{2\pi}\log h_N(\theta)d\theta=\int_0^{2\pi}\log\sqrt{f(\theta)}d\theta.$$

Now, $h_N(\theta)=\sqrt{f}*\psi_N$, where ψ_N is Fejer's kernel. Therefore, by Jensen's inequality,

$$\log h_N(\theta)=\log(\sqrt{f}*\psi_N)\geqq\log\sqrt{f}*\psi_N,$$

so that

$$\int_0^{2\pi}\log h_N(\theta)d\theta\geqq\int_0^{2\pi}(\log\sqrt{f}*\psi_N)d\theta=\int_0^{2\pi}\log\sqrt{f(\theta)}\,d\theta,$$

and hence

$$\lim_{N\to\infty}\int_0^{2\pi}\log h_N(\theta)d\theta\geqq\int_0^{2\pi}\log\sqrt{f(\theta)}d\theta.$$

On the other hand, from (2.3) we have that $\{h_N(\theta)\}$ are uniformly bounded and $\log h_N(\theta)\to\log\sqrt{f}$, so that from Fatou's lemma

$$\overline{\lim_{N\to\infty}}\int_0^{2\pi}\log h_N(\theta)d\theta\leqq\int_0^{2\pi}\log\sqrt{f(\theta)}d\theta.$$

Lemma 3.9. *Let $R \in \underline{M}_S$ such that $H_f(R) < \infty$; then there exists a sequence $\{R_N\} \in \underline{M}_S$ such that $\{R_N\}$ converges weakly to R and*

$$\lim_{N\to\infty} H_{f_N}(R_N) = H_f(R). \tag{3.24}$$

Proof. From its definition (1.9),

$$H_f(R) = e(R) + \frac{1}{2}\log 2\pi + \frac{1}{4\pi}\int_0^{2\pi} \frac{dG(\theta)}{f(\theta)} + \frac{1}{4\pi}\int_0^{2\pi} \log f(\theta)d\theta, \tag{3.25}$$

where $G(\theta)$ is the spectral measure of R and $e(R)$ is as defined in Lemma 3.7. From (3.20), for any $R' \in \underline{M}_S$,

$$H_{f_N}(R') = e(R') + \frac{1}{2}\log 2\pi + \frac{1}{4\pi}\int_0^{2\pi} \frac{dG'(\theta)}{f_N(\theta)} + \frac{1}{4\pi}\int_0^{2\pi} \log f_N(\theta)d\theta, \tag{3.26}$$

where $G'(\theta)$ is the spectral measure of R'. The way we construct the desired sequence $\{R_N\}$ is to define $R_N = R\tau_N^{-1}$, where $\{\tau_N\}$ is a suitably chosen sequence of admissible mappings which first of all will approach the identity map as $N\to\infty$, so that R_N will converge weaky to R. Secondly, $\{\tau_N\}$ will have to be selected so that (3.24) holds. Choosing τ_N means choosing a suitable sequence of coefficients. Let $g_N(\theta) = \sum_{j=-\infty}^{\infty} c_k^{(N)} e^{ik\theta}$, where $\{c_k^{(N)}\}$ are the defining coefficients of the admissible map τ_N. Taking into account Lemma 3.7 and 3.8, we see that for τ_N to satisfy the necessary requirements, $g_N(\theta)$ must satisfy:

(1) $$\lim_{N\to\infty} \int_0^{2\pi} \log|g_N(\theta)|d\theta = 0,$$

(2) $$\lim_{N\to\infty} \int_0^{2\pi} \frac{|g_N(\theta)|^2}{f_N(\theta)} dG(\theta) = \int_0^{2\pi} \frac{dG(\theta)}{f(\theta)},$$

(3) $$\lim_{N\to\infty} \int_0^{2\pi} |g_N(\theta) - 1|^2 dG(\theta) = 0.$$

Condition (1), because of (3.22), guarantees $e(R_N) \to e(R)$. Since the spectral measure of R_N, call it $dG_N(\theta) = |g_N(\theta)|^2 dG(\theta)$, we see that (2), (1) and Lemma 3.8 insure that $H_{f_N}(R_N) \to H_f(R)$. Condition (3) insures that τ_N approaches the identity because it is the same as requiring $E^R\{|\omega(0) - (\tau_N\omega)(0)|^2\} \to 0$ as $N\to\infty$.

Let $g_{N,\varepsilon}(\theta) = \dfrac{\sqrt{f_N(\theta)}}{\sqrt{f_N(\theta)+\varepsilon}}$. We will show

(1)′ $$\lim_{\varepsilon\to 0} \lim_{N\to\infty} \left| \int_0^{2\pi} \log|g_{N,\varepsilon}(\theta)|d\theta \right| = 0,$$

(2)′ $$\lim_{\varepsilon\to 0} \lim_{N\to\infty} \left| \int_0^{2\pi} \frac{|g_{N,\varepsilon}(\theta)|^2 dG(\theta)}{f_N(\theta)} - \int_0^{2\pi} \frac{dG(\theta)}{f(\theta)} \right| = 0,$$

and

(3)′ $$\lim_{\varepsilon\to 0} \lim_{N\to\infty} \int_0^{2\pi} |g_{N,\varepsilon}(\theta) - 1|^2 dG(\theta) = 0.$$

This will mean there exists a sequence $\varepsilon_N \to 0$ as $N \to \infty$ such that if we define $g_N(\theta) = g_{N,\varepsilon_N}(\theta)$, then (1)′, (2)′, and (3)′ imply (1), (2), (3) and hence the lemma. Because of (2.3), $g_{N,\varepsilon}(\theta) \to \sqrt{f(\theta)}/(\sqrt{f(\theta)}+\varepsilon)$ uniformly for $\theta \in [0,2\pi]$. Also $\frac{|g_{N,\varepsilon}(\theta)|^2}{f_N(\theta)} \to \frac{1}{(\sqrt{f(\theta)}+\varepsilon)^2}$ uniformly for $\theta \in [0,2\pi]$. Thus,

$$\lim_{N\to\infty} \int_0^{2\pi} \log|g_{N,\varepsilon}(\theta)d\theta = \int_0^{2\pi} \log\left(\frac{\sqrt{f(\theta)}}{\sqrt{f(\theta)}+\varepsilon}\right)d\theta, \tag{3.27}$$

$$\lim_{N\to\infty} \int_0^{2\pi} \frac{|g_{N,\varepsilon}(\theta)|^2 dG(\theta)}{f_N(\theta)} = \int_0^{2\pi} \frac{dG(\theta)}{(\sqrt{f(\theta)}+\varepsilon)^2}, \tag{3.28}$$

and

$$\lim_{N\to\infty} \int_0^{2\pi} |g_{N,\varepsilon}(\theta)-1|^2 dG(\theta) = \int_0^{2\pi} \left|\frac{\sqrt{f(\theta)}}{\sqrt{f(\theta)}+\varepsilon}-1\right|^2 dG(\theta). \tag{3.29}$$

In (3.27), since by hypothesis $\int_0^{2\pi} \log f(\theta)d\theta > -\infty$, we see $f(\theta)>0$ almost everywhere on $[0,2\pi]$. Thus

$$\lim_{\varepsilon\to 0} \int_0^{2\pi} \log\left(\frac{\sqrt{f(\theta)}}{(\sqrt{f(\theta)})+\varepsilon}\right)d\theta = 0,$$

which shows (1)′. By the monotone convergence theorem

$$\lim_{\varepsilon\to 0} \int_0^{2\pi} \frac{dG(\theta)}{((\sqrt{f(\theta)})+\varepsilon)^2} = \int_0^{2\pi} \frac{dG(\theta)}{f(\theta)},$$

which shows (2)′. Finally by the bounded convergence theorem

$$\lim_{\varepsilon\to 0} \int_0^{2\pi} \left|\frac{\sqrt{f(\theta)}}{\sqrt{f(\theta)}+\varepsilon}-1\right|^2 dG(\theta) = 0,$$

which shows (3)′ and completes the proof.

The preceding lemma was needed to complete the proof of the lower bound in Theorem 2.8. We now prove a lemma providing what was needed for the upper bound.

Lemma 3.10. *Let $H_{f_N}(R)$ be given by (3.20) and $H_f(R)$ be given by (1.9). Then for any $l>0$,*

$$\bigcup_{N=1}^{\infty} \{R \in \underline{M}_S : H_{f_N}(R) \leqq l\} \tag{3.30}$$

is tight in $\underline{M}_S$.

If $\{R_N\}$ is any sequence in $\underline{M}_S$ converging weakly to $R \in \underline{M}_S$, then

$$\varliminf_{N\to\infty} H_{f_N}(R_N) \geqq H_f(R). \tag{3.31}$$

Proof. From (2.3) we have that $f_N(\theta)\to f(\theta)$ uniformly in $[0,2\pi]$. From Lemma 3.8 we have that $\lim_{N\to\infty}\int_0^{2\pi}\log f_N(\theta)d\theta=\int_0^{2\pi}\log f(\theta)d\theta$. Since

$$H_{f_N}(R)=E^R\left\{\int_{-\infty}^{\infty} r(y|\omega)\log r(y|\omega)dy\right\}$$
$$+\frac{1}{2}\log 2\pi+\frac{1}{4\pi}\int_0^{2\pi}\frac{dG(\theta)}{f_N(\theta)}+\frac{1}{4\pi}\int_0^{2\pi}\log f_N(\theta)d\theta\,,$$

and

$$H_f(R)=E^R\left\{\int_{-\infty}^{\infty} r(y|\omega)\log r(y|\omega)dy\right\}$$
$$+\frac{1}{2}\log 2\pi+\frac{1}{4\pi}\int_0^{2\pi}\frac{dG(\theta)}{f(\theta)}+\frac{1}{4\pi}\int_0^{2\pi}\log f(\theta)d\theta\,,$$

it suffices, in order to show (3.30) and (3.31), to show

$$\bigcup_{N=1}^{\infty}\{R\in \underline{M}_S:\Phi_N(R)\leqq l\} \tag{3.32}$$

is tight in $\underline{M}_S$, and, for any sequence $\{R_N\}$ converging to R,

$$\varliminf_{N\to\infty}\Phi_N(R_N)\geqq\Phi(R)\,, \tag{3.33}$$

where

$$\Phi_N(R)=E^R\left\{\int_{-\infty}^{\infty} r(y|\omega)\log r(y|\omega)dy\right\}+\frac{1}{4\pi}\int_0^{2\pi}\frac{dG(\theta)}{f_N(\theta)}\,,$$

and

$$\Phi(R)=E^R\left\{\int_{-\infty}^{\infty} r(y|\omega)\log r(y|\omega)dy\right\}+\frac{1}{4\pi}\int_0^{2\pi}\frac{dG(\theta)}{f(\theta)}\,.$$

We start with Lemma 3.6 where the ϕ of that lemma is $r(y|\omega)$, giving us for any $\alpha>0$,

$$E^R\left\{\int_{-\infty}^{\infty} r(y|\omega)\log r(y|\omega)dy\right\}\geqq-\frac{1}{2}\log 2\pi\alpha-\frac{1}{2\alpha}\cdot\frac{1}{2\pi}\int_0^{2\pi}dG(\theta)\,. \tag{3.34}$$

Since $f_N(\theta)\to f(\theta)$, there exists a c such that $f_N(\theta)\leqq c$ on $[0,2\pi]$. Thus, from (3.34)

$$E^R\left\{\int_{-\infty}^{\infty} r(y|\omega)\log r(y|\omega)dy\right\}\geqq-\frac{1}{2}\log 2\pi\alpha-\frac{1}{2\alpha}\cdot\frac{1}{2\pi}\int_0^{2\pi}\frac{f_N(\theta)dG(\theta)}{f_N(\theta)}$$
$$\geqq-\frac{1}{2}\log 2\pi\alpha-\frac{1}{2\alpha}\frac{c}{2\pi}\int_0^{2\pi}\frac{dG(\theta)}{f_N(\theta)}\,.$$

In this last if we let $\alpha=c/\varepsilon$ for $0<\varepsilon<1$, we get

$$E^R\left\{\int_{-\infty}^{\infty} r(y|\omega)\log r(y|\omega)dy\right\}\geqq-\frac{1}{2}\log\frac{2\pi c}{\varepsilon}-\frac{\varepsilon}{4\pi}\int_0^{2\pi}\frac{dG(\theta)}{f_N(\theta)}\,,$$

so that if $\Phi_N(R) \leqq l$, then

$$\frac{1}{4\pi}\int_0^{2\pi}\frac{dG(\theta)}{f_N(\theta)} \leqq \frac{l+\frac{1}{2}\log\frac{2\pi c}{\varepsilon}}{1-\varepsilon}.$$

Since $f_N(\theta)$ is bounded by c on $[0, 2\pi]$, this last implies

$$\frac{1}{2\pi}\int_0^{2\pi} dG(\theta) \leqq c\left(\frac{l+\frac{1}{2}\log\frac{2\pi c}{\varepsilon}}{1-\varepsilon}\right).$$

This means

$$E^R\{\omega^2(0)\} \leqq c\left(\frac{l+\frac{1}{2}\log\frac{2\pi c}{\varepsilon}}{1-\varepsilon}\right),$$

and since the constant on the right depends only on l and not on N or R, we see that $\bigcup_{N=1}^{\infty} \{R \in \underline{M}_S : \Phi_N(R) \leqq l\}$ is tight, which is (3.22).

To prove (3.23) we want to show that if $\{R_N\}$ converges weakly to R and $\Phi_N(R_N) \leqq M$, then $\Phi(R) \leqq M$.

We just proved that if $\Phi_N(R) \leqq l$ then $E^R\{\omega^2(0)\}$ is bounded by a constant independent of N. Thus if $\Phi_N(R_N) \leqq M$, we conclude that the second moments of $\{R_N\}$ are uniformly bounded which in turn implies the total mass of the spectral measure of R_N is uniformly bounded, i.e., $\int_0^{2\pi} dG_N(\theta)$ is uniformly bounded. Hence, by Helly's theorem (choosing a subsequence if necessary) $\{G_N\}$ converges weakly to $G' = G + G_1$, where G is the spectral measure of R and G_1 is nonnegative.

By hypothesis

$$\Phi_N(R_N) = E^{R_N}\left\{\int_{-\infty}^{\infty} r_N(y|\omega)\log r_N(y|\omega)dy\right\} + \frac{1}{4\pi}\int_0^{2\pi}\frac{dG_N(\theta)}{f_N(\theta)} \leqq M. \tag{3.35}$$

In (3.1) if, instead of entropy of R with respect to the stationary process generated by independent Gaussian, mean 0 variance 1 variables, we let $I_\lambda(R)$ be entropy of R with respect to the same stationary Gaussian process but with variance λ for $\lambda > 0$, then, of course,

$$I_\lambda(R) = E^R\left\{\int_{-\infty}^{\infty} r(y|\omega)\log r(y|\omega)dy\right\} + \frac{1}{2}\log 2\pi\lambda + \frac{1}{2\lambda}E^R\{\omega^2(0)\}.$$

From (3.35), we see that for any $\lambda > 0$,

$$I_\lambda(R_N) + \frac{1}{4\pi}\int_0^{2\pi}\frac{dG_N(\theta)}{f_N(\theta)} \leqq M + \frac{1}{2}\log 2\pi\lambda + \frac{1}{2\lambda}E^{R_N}\{\omega^2(0)\}. \tag{3.36}$$

Now, $I_\lambda(R)$ is lower semicontinuous on $\underline{M}_S$, and so since $R_N \Rightarrow R$, $f_N(\theta) \to f(\theta)$, $G_N \Rightarrow G' = G + G_1$, we conclude that

$$I_\lambda(R) + \frac{1}{4\pi}\int_0^{2\pi}\frac{dG'(\theta)}{f(\theta)} \leqq M + \frac{1}{2}\log 2\pi\lambda + \frac{1}{4\pi\lambda}\int_0^{2\pi} dG'(\theta). \tag{3.37}$$

Therefore, using the expression for $I_\lambda(R)$,

$$E^R\left\{\int_{-\infty}^{\infty} r(y|\omega)\log r(y|\omega)\right\}$$

$$+\frac{1}{2}\log 2\pi\lambda+\frac{1}{4\pi\lambda}\int_0^{2\pi} dG(\theta)+\frac{1}{4\pi}\int_0^{2\pi}\frac{dG'(\theta)}{f(\theta)}$$

$$\leqq M+\frac{1}{2}\log 2\pi\lambda+\frac{1}{4\pi\lambda}\int_0^{2\pi} dG'(\theta).$$

Since this last is true for any $\lambda>0$, we let $\lambda\to\infty$, after cancelling $\frac{1}{2}\log 2\pi\lambda$ from both sides, to obtain

$$E^R\left\{\int_{-\infty}^{\infty} r(y|\omega)\log r(y|\omega)dy\right\}+\frac{1}{4\pi}\int_0^{2\pi}\frac{dG'(\theta)}{f(\theta)}\leqq M.$$

Since $G'(\theta)=G(\theta)+G_1(\theta)$ and $G_1(\theta)$ is nonnegative, we conclude that in fact

$$\Phi(R)=E^R\left\{\int_{-\infty}^{\infty} r(y|\omega)\log r(y|\omega)dy\right\}+\frac{1}{4\pi}\int_0^{2\pi}\frac{dG(\theta)}{f(\theta)}\leqq M,$$

which completes the proof.

References

1. Donsker, M.D., Varadhan, S.R.S.: Asymptotic evaluation of certain Wiener integrals for large time. Functional Integration and its Applications, Proceedings of the International Conference held at the Cumberland Lodge, Windsor Great Park, London, in April 1974, A. M. Arthurs, (ed.). Oxford: Clarendon 1975
2. Donsker, M.D., Varadhan, S.R.S.: Asymptotic evaluation of certain Markov process expectations for large time, I. Commun. Pure Appl. Math. **28**, 1–47 (1976); II. Commun. Pure Appl. Math. **28**, 279–301 (1975); III. Commun. Pure Appl. Math. **29**, 389–461 (1976); IV. Commun. Pure Appl. Math. **36**, 183–212 (1983)
3. Donsker, M.D., Varadhan, S.R.S.: Asymptotics for the Wiener Sausage. Commun. Pure Appl. Math. **28**, 525–565 (1975)
4. Donsker, M.D., Varadhan, S.R.S.: On laws of the iterated logarithm for local times. Commun. Pure Appl. Math. **30**, 707–753 (1977)
5. Donsker, M.D., Varadhan, S.R.S.: Asymptotics for the Polaron. Commun. Pure Appl. Math. **36**, 505–528 (1983)
6. Orey, S.: Large deviations and Shanon-McMillan theorems (preprint)

Communicated by A. Jaffe

Received July 6, 1984

Reduction in the Number of Coupling Parameters

W. Zimmermann

Max-Planck-Institut für Physik und Astrophysik, Werner-Heisenberg-Institut für Physik, D-8000 München 40, Federal Republic of Germany

Dedicated to the memory of Kurt Symanzik

Abstract. A method is developed for reducing the formulation of massless models with several independent couplings to a description in terms of a single coupling parameter. The original as well as the reduced system are supposed to be renormalizable and invariant under the renormalization group. For most models there are, if any, only a finite number of reductions possible including those which correspond to symmetries of the system. The reduction method leads to a consistent formulation of the reduced model in any order of perturbation theory even in cases where it is difficult to establish a symmetry in higher orders. An example where no symmetry seems to be involved is the interaction of a spinor field with a pseudoscalar field. For this model the reduction method determines the quartic coupling constant uniquely as a function of the Yukawa coupling constant. The Wess-Zumino model is an exceptional case for which the reduction method admits an infinite number of solutions besides the supersymmetric one.

1. Introduction

Symmetry considerations provide a natural method of reducing the number of independent parameters in models of quantum field theory. If a symmetry is imposed, otherwise unconstrained coupling parameters become related among each other so that the number of independent parameters is decreased. Renormalizability of the model is maintained provided anomalies are absent and the symmetry can be implemented in all orders of perturbation theory.

In this paper a more general approach for reducing the number of coupling parameters is taken which is based on the principles of renormalizability and invariance under the renormalization group. It turns out that these requirements severely limit the possibilities of constraining the coupling parameters to a single independent one. The method is developed for the reduction of massless models from $n+1$ coupling parameters $\lambda_0, \lambda_1, \ldots, \lambda_n$ to a description in terms of λ_0 only. Any symmetry requirement leading to a renormalizable formulation is certainly included by this treatment. In fact, hidden symmetries could be detected in this way. On the other hand there are cases where the general reduction is possible and unique, but no symmetry is known to be involved. It is also conceivable that a symmetry can only be implemented in low orders while the general reduction method leads to a unique prescription in all orders of perturbation theory. In such a case a renormalizable formulation of the reduced model is obtained for which, however, the relevant symmetry is only realized in low orders.

In Sect. 2 the general conditions are studied under which a reduction is possible. For the coupling parameters λ_j as functions of λ_0 the ordinary differential equations

$$\beta_0 \frac{d\lambda_j}{d\lambda_0} = \beta_j \tag{1.1}$$

with

$$\lim_{\lambda_0 \to 0} \lambda_j = 0 \tag{1.2}$$

are found. β_j denotes the β-function corresponding to λ_j. Equation (1.1) can be derived either from the Callan-Symanzik equations [1, 2] or the evolution equations of the effective couplings. An interesting possibility is the special case that the β-function of the reduced system vanishes identically [1]. Then, after inserting the functions $\lambda_j(\lambda_0)$ the β-functions of the original system also vanish identically

$$\beta_i \equiv 0, \quad i = 0, 1, \ldots, n, \tag{1.3}$$

and the system (1.1) is trivially satisfied.

Renormalizability for the original as well as the reduced system implies that the functions $\lambda_j(\lambda_0)$ allow for power series expansions in λ_0. In lowest order one finds a system of quadratic equations for the constant lowest order approximations $\varrho_0^{(j)}$ of the ratios

$$\frac{\lambda_j}{\lambda_0} = \varrho_0^{(j)} + o(\lambda_0). \tag{1.4}$$

These are the eigenvalue conditions proposed by Chang for the ratios of coupling constants [4] [2]. They form necessary conditions for the possibility of reducing the system. But without further restrictions they are not sufficient. For sometimes higher order effects prevent the extension of (1.4) to power series solutions of (1.1).

In Sect. 3 the case of two coupling parameters g^2 and λ is treated in detail by applying results from [6] and [7] [3]. The β-functions are assumed to be of the form

$$\beta_0 = b_0 g^4 + \ldots, \qquad \beta_1 = c_1 \lambda^2 + c_2 \lambda g^2 + c_3 g^4 + \ldots. \tag{1.5}$$

1 For some models arguments have been given indicating that the β-function vanishes to all orders of perturbation theory. See for instance [3]

2 Chang et al. applied the eigenvalue conditions to grand unification in order to build asymptotically free models with only one coupling constant. Unfortunately this program turned out to be too ambiguous due to the freedom in introducing heavy particles. See [5] which contains further references

3 The purpose of [6] was to find all asymptotically free solutions of the evolution equations with two coupling parameters. In this context the solutions of (1.1–2) were constructed by asymptotic expansions. Among the solutions found only the power series solutions are relevant for the present paper

A reduction to a renormalizable description in terms of g^2 is only possible if the quadratic equation

$$c_1\varrho_0^2+(c_2-b_0)\varrho_0+c_3=0 \tag{1.6}$$

has real roots, i.e. if the discriminant

$$\Delta=(c_2-b_0)^2-4c_1c_3\geqq 0 \tag{1.7}$$

is non-negative. For asymptotically free gauge theories with a Higgs coupling (1.7) coincides with the condition for asymptotic freedom found by Gross and Wilczek [8]. It is always satisfied for supersymmetric gauge theories where $\lambda=h^2$ with h describing a matter or Higgs interaction. In lowest order the ratio of the coupling parameters is given by one of the roots $\varrho_\pm$ of (1.6)

$$\frac{\lambda}{g^2}=\varrho_\pm+o(g^2)\,. \tag{1.8}$$

Unless

$$\xi=-\frac{c_1}{b_0}(\varrho_+-\varrho_-)\qquad(\varrho_+\geqq\varrho_-) \tag{1.9}$$

is an integer the lowest order term (1.8) can be completed to a power series expansion in g^2. The precise conditions under which an expansion for integral ξ is possible are stated in Sect. 4. It is further shown that by a reparametrization it can be arranged that the lowest order of a power series (1.8) becomes exact

$$\lambda'=\varrho_\pm g^2\quad\text{if}\quad\varrho_\pm\neq 0 \tag{1.10}$$

and

$$\lambda'=\varrho_n g^{2n+2}\quad\text{if}\quad\lambda=o(g^{2n+2}) \tag{1.11}$$

with a suitably defined new coupling parameter λ'.

In the remainder of the paper the reduction method is applied to two models of special interest: Sect. 4 concerns the interaction of a spinor field with a pseudoscalar field. For a consistent formulation of the renormalization it is necessary to introduce a quartic selfinteraction of a scalar field since the Yukawa interaction alone would not render the four pseudoscalar vertex part convergent. The model thus involves two independent coupling constants, g for the Yukawa coupling and λ for the quartic interaction. No symmetry is known which would relate the two coupling constants. While the bare scalar coupling constant cannot be dropped, one might think of setting the renormalized coupling constant λ equal to zero in order to eliminate the additional parameter. However, formulations with different normalization points would then be inequivalent. On the other hand, the general reduction method leads to a unique power series expansion

$$\lambda=\varrho_\pm g^2+\varrho_1 g^4+\dots \tag{1.12}$$

of λ, thus providing a consistent renormalizable description with one coupling constant g only. The two values ϱ_+ and ϱ_- correspond to different signs of λ.

Finally the reduction method is discussed for models which become supersymmetric by imposing relations among the coupling constants. Special problems may occur for models which are not asymptotically free. In Sect. 5 the massless Wess-

Zumino model[4] [9] is treated with independent coupling constants g for the Yukawa and λ for the quartic coupling. A peculiar situation is found due to the fact that ξ is a negative integer

$$\xi = -3. \tag{1.13}$$

This leads to an asymptotic expansion of λ in the form[5]

$$\lambda = \varrho_+ g^2 + \varrho_1 g^4 + \varrho_2 g^6 + \varrho_3 g^8 + d g^8 \ln g^2 + \dots. \tag{1.14}$$

The coefficient ϱ_3 is arbitrary and d is determined uniquely by lower orders including the order g^6. For $d=0$ logarithms are absent and (1.14) represents a power series with arbitrary ϱ_3. Without using supersymmetry, calculations of order g^6 would be required to check whether or not $d=0$. But the existence of a renormalized supersymmetric formulation excludes the occurrence of logarithms so that $d=0$. With suitable supersymmetric normalization conditions one has

$$\lambda = \varrho_+ g^2 \tag{1.15}$$

for the supersymmetric solution and

$$\lambda = \varrho_+ g^2 + \varrho_3 g^8 + \sum_{n=4}^{\infty} \varrho_n g^{2n+2} \tag{1.16}$$

with arbitrary ϱ_3. Thus the general reduction method is not unique in this special case, but also admits infinitely many asymmetric reduced systems[6]. Even the relation (1.15) is not characteristic for the supersymmetric case since by an asymmetric redefinition of λ the relation (1.15) can always be restored.

No such problems seem to occur for supersymmetric models where the primary β-function is negative or vanishes in lowest order. For the $N=2$ and $N=4$ super Yang-Mills theories it was found that the relevant lowest order solutions can indeed be uniquely extended to power series expansions in the primary coupling constant [11]. General statements can be made about two-parametric models with β-functions of the form (1.5) and $b_0 \leqq 0$. If $b_0 = 0$ and $\Delta > 0$ two power series can be constructed for λ with uniquely determined coefficients [7]. One of the expansions corresponds to the supersymmetric case. This includes a variety of models, in particular those which may have vanishing β-functions in any order of perturbation theory. If $b_0 < 0$ and $\Delta > 0$ the model is asymptotically free. Usually supersymmetric models with asymptotic freedom are unstable against perturbations of the symmetry [7, 12]. In the unstable case a unique power series for λ can be constructed [6, 7]. Thus in all these cases the general reduction method provides a unique formulation of the reduced model in every order of perturbation theory even though the symmetry may have been established for low orders only.

---

4 For the renormalization of the massless model see [10]

5 There is also a power series for λ which is not related to supersymmetry

6 Recently it has been shown by O. Piguet and K. Sibold that there is only one realization of supersymmetry in the perturbative treatment of the massless Wess-Zumino model [17]. Therefore, the additional reduced systems do not seem to be supersymmetric

2. General Method of Reduction

We consider a massless model of quantum field theory described by $n+1$ dimensionless coupling parameters $\lambda_0, \lambda_1, \ldots, \lambda_n$ and a normalization mass κ. The model is supposed to be invariant under the renormalization group. Our aim is to express $\lambda_1, \ldots, \lambda_n$ as functions of λ_0 so that a model involving a single coupling parameter λ_0 is obtained which is again invariant under the renormalization group. Accordingly we write each λ_j as a function of

$$\lambda_j = \lambda_j(\lambda_0), \tag{2.1}$$

independent of the normalization mass κ. The functions $\lambda_j(\lambda_0)$ should be differentiable in the domain of λ_0 considered and vanish in the weak coupling limit[7]

$$\lim_{\lambda_0 \to 0} \lambda_j(\lambda_0) = 0. \tag{2.2}$$

For the Green's functions of the original system the invariance under the renormalization group implies the Callan-Symanzik equations

$$\left(\kappa^2 \frac{\partial}{\partial \kappa^2} + \sum \beta_j \frac{\partial}{\partial \lambda_j} + \gamma\right) \tau = 0, \tag{2.3}$$

while for the Green's functions of the reduced system the equations

$$\left(\kappa^2 \frac{\partial}{\partial \kappa^2} + \beta' \frac{\partial}{\partial \lambda_0} + \gamma'\right) \tau' = 0 \tag{2.4}$$

follow. The β- and γ-functions depend on the coupling parameters only. β' and γ' are functions of the single variable λ_0. γ and γ' are additive in the contributions from the field operators occurring in the Green's functions. τ is a function of the momenta, the coupling parameters and the normalization mass κ. τ' is obtained from τ by substituting the functions (2.1) for the parameters λ_j. Accordingly,

$$\frac{\partial \tau'}{\partial \lambda_0} = \frac{\partial \tau}{\partial \lambda_0} + \sum_{j=1}^{n} \frac{\partial \tau}{\partial \lambda_j} \frac{d\lambda_j}{d\lambda_0}.$$

Linear independence of the Green's functions and their derivatives leads to the relations

$$\beta' = \beta_0, \gamma' = \gamma, \beta' \frac{d\lambda_j}{d\lambda_0} = \beta_j.$$

Hence the functions (2.1) must satisfy the following system of ordinary differential equations

$$\beta_0 \frac{d\lambda_j}{d\lambda_0} = \beta_j. \tag{2.5}$$

On the other hand, if the functions (2.1) satisfy (2.5), the reduced form (2.4) of the Callan-Symanzik equations follows. Thus the system (2.5) forms a necessary and sufficient condition for reducing the original system by the functions $\lambda_j(\lambda_0)$.

---

7 The condition of renormalizability requiring that the functions λ_j can be expanded with respect to powers of λ_0 will not be used for the time being

It is instructive to use an alternative method for the derivation of (2.5) by eliminating the scale variable from the evolution equations of the effective couplings. At a normalization mass κ'^2 we impose on the coupling parameters that the values λ_j' of λ_j are given functions of the value λ_0' of λ_0,

$$\lambda_j' = \lambda_j(\lambda_0') \quad \text{at} \quad \kappa^2 = \kappa'^2 . \tag{2.6}$$

We want to investigate under which restrictions on the functions the same dependence holds at other normalization points:

$$\lambda_j = \lambda_j(\lambda_0) \quad \text{at} \quad \kappa^2 . \tag{2.7}$$

If the normalization mass is changed from κ'^2 to κ^2 the field operators φ_i of the system undergo a transformation of the renormalization group,

$$\varphi_j(x, \lambda_0, \ldots, \lambda_n, \kappa^2) = z_j^{1/2} \varphi_j(x, \lambda_0', \ldots, \lambda_n', \kappa'^2)$$

with positive z_j[8]. The new values of the coupling parameters are given by

$$\lambda_0 = \bar{\lambda}_0(u, \lambda_0', \lambda_1', \ldots, \lambda_n') , \tag{2.8}$$

$$\lambda_j = \bar{\lambda}_j(u, \lambda_0', \lambda_1', \ldots, \lambda_n') , \tag{2.9}$$

$$u = \kappa^2/\kappa'^2, \qquad j = 1, \ldots, n .$$

The functions $\bar{\lambda}_0$, $\bar{\lambda}_j$ denote effective couplings suitably defined as analytic functions of u which are regular at any positive value of u[9].

In order to determine the constraints on the functions λ_j we take a fixed initial value $\lambda_0' \neq 0$, and first discuss the case where

$$\beta_0(\lambda_0', \lambda_1', \ldots, \lambda_n') \neq 0 . \tag{2.10}$$

Expression (2.10) equals the value of $\partial\bar{\lambda}_0/\partial u$ at $u=1$, $\lambda_0 = \lambda_0'$, $\lambda_j = \lambda_j'$. Since $\bar{\lambda}_0$ is regular analytic at $u=1$ the derivative $\partial\bar{\lambda}_0/\partial u$ is continuous near $u=1$ so that

$$\frac{\partial\bar{\lambda}_0}{\partial u}(u, \lambda_0', \lambda_1', \ldots, \lambda_n') \neq 0 \tag{2.11}$$

in a neighborhood of $u=1$. Therefore, Eq. (2.8) can be inverted with respect to u. Inserting the inversion

$$u = u(\lambda_0; \lambda_0', \lambda_1', \ldots, \lambda_n')$$

into (2.9) we find that the λ_j necessarily become functions of λ_0 which are independent of the normalization mass κ. By definition they represent the functions λ_j in (2.6–7):

$$\lambda_j(\lambda_0) = \bar{\lambda}_j(u(\lambda_0; \lambda_0', \ldots, \lambda_n'), \lambda_0', \ldots, \lambda_n') . \tag{2.12}$$

With the help of the evolution equations

$$u\frac{\partial\bar{\lambda}_i}{\partial u} = \beta_i(\bar{\lambda}_0, \lambda_1(\bar{\lambda}_0), \ldots, \lambda_n(\bar{\lambda}_0)), \qquad i = 0, 1, \ldots, n , \tag{2.13}$$

---

8 For the concept of the renormalization group used here see [13]

9 For the possibility of defining effective couplings as analytic functions see [14]

the scale variable u can be eliminated near $u=1$. We thus obtain (2.5) in the form

$$\frac{d\lambda_j}{d\lambda_0}=\frac{\beta_j(\lambda_0,\lambda_1,\ldots,\lambda_n)}{\beta_0(\lambda_0,\lambda_1,\ldots,\lambda_n)},\quad j=1,\ldots,n,$$

valid in the neighborhood of $\lambda_0=\lambda_0'$.

We next discuss the case where

$$\beta_0(\lambda_0',\lambda_1',\ldots,\lambda_n')=0. \tag{2.14}$$

In this case the function $\beta_0(\lambda_0,\lambda_1(\lambda_0),\ldots,\lambda_n(\lambda_0))$ has a zero at $\lambda_0=\lambda_0'$. Then (2.13) implies

$$\frac{\partial\bar{\lambda}_0}{\partial u}=0\quad\text{at}\quad u=1 \tag{2.15}$$

for the function

$$\bar{\lambda}_0=\bar{\lambda}_0(u,\lambda_0',\lambda_1',\ldots,\lambda_n'). \tag{2.16}$$

Since $\bar{\lambda}_0$ is regular analytic in u at $u=1$ it is

$$\frac{\partial\bar{\lambda}_0}{\partial u}\neq 0\quad\text{for}\quad u\neq 1\quad\text{near}\quad u=1 \tag{2.17}$$

or

$$\frac{\partial\bar{\lambda}_0}{\partial u}\equiv 0. \tag{2.18}$$

Hence the function (2.16) is either variable in u and stationary at $u=1$ or it is constant[10].

In case (2.17) of variable $\bar{\lambda}_0$ we may invert (2.8) for $u<1$ as well as $u>1$, obtaining

$$u=u_-(\lambda,\lambda_0',\lambda_1',\ldots,\lambda_n')\quad\text{for}\quad u<1, \tag{2.19}$$

$$u=u_+(\lambda,\lambda_0',\lambda_1',\ldots,\lambda_n')\quad\text{for}\quad u>1. \tag{2.20}$$

If $\bar{\lambda}_0$ has an extremal value at $u=1$ the inversions u_- and u_+ denote different branches of u both defined for $\lambda_0<\lambda_0'$ in case of a maximum or $\lambda_0>\lambda_0'$ in case of a minimum. Inserting (2.19) and (2.20) into (2.9) we find two sets of functions λ_j of λ_0 which must be identical to (2.6) and thus to each other. For $\lambda_0\neq\lambda_0'$ again (2.5) follows. Equation (2.5) can be extended to $\lambda_0=\lambda_0'$ by taking the limit $\lambda_0\to\lambda_0'$.

We now turn to the case (2.18) of constant $\bar{\lambda}_0$. Equations (2.7) and (2.8) imply that λ_j does not depend on u either, so that by (2.13) also the other β-functions vanish. The system (2.5) is then trivially satisfied.

We summarize the results as follows: In all cases the functions satisfy the system (2.5) of ordinary differential equations in agreement with the derivation given in the first part of this section. If λ_0 is a zero of one of the β-functions – with the other coupling parameters expressed as functions (2.6) of λ_0' – the system (2.5) implies that λ_0' is a zero of all β-functions. It follows that the effective couplings are either variable in u and stationary at $u=1$ or they are all independent of u.

We provide some further information on the zeroes of the β-functions considered. Zeroes of the first type with variable effective couplings are always

10 For a discussion of zeroes of the β-function which correspond to stationary values of the effective coupling see [14]

isolated. For it is

$$\beta(\bar{\lambda}_0, \lambda_1(\bar{\lambda}_0), \ldots, \lambda_n(\bar{\lambda}_0)) = \frac{\partial \bar{\lambda}_0}{\partial u} \neq 0 \quad \text{for} \quad \bar{\lambda}_0 \neq \lambda'_0$$

in a neighborhood of λ'_0. We further observe that a zero must necessarily be of the second type with constant effective couplings if all derivatives

$$\frac{d^m}{d\lambda_0^m} \beta(\lambda_0, \lambda_1(\lambda_0), \ldots, \lambda_n(\lambda_0)) \tag{2.21}$$

exist at $\lambda_0 = \lambda'_0$. For then all derivatives of $\bar{\lambda}_0$ with respect to u vanish at $u=1$ as follows by differentiating the evolution equation (2.13) of $\bar{\lambda}_0$. Sufficient for the existence of the derivatives (2.21) is the existence of all partial derivatives of β[11]. The functions $\lambda_j(\lambda_0)$ were assumed to be differentiable and the existence of their higher derivatives follows by differentiating the system (2.5).

Zeroes of the β-functions with the effective couplings independent of the scale variable need not be isolated. In fact, arguments have been given for some supersymmetric models that reduced forms exist with β-functions vanishing in any order of perturbation theory. If this should prevail independent of perturbation theory the relations

$$\beta_i(\lambda_0, \lambda_1(\lambda_0), \ldots, \lambda_n(\lambda_0)) = 0, \qquad i = 0, 1, \ldots, n, \tag{2.22}$$

would vanish identically in λ_0 for some functions $\lambda_j(\lambda_0)$.

The reducibility condition (2.5) allows for a large class of solutions unless further restrictions are imposed. In a region of non-vanishing β_0 the Lipschitz condition can be verified for the ratios β_j/β_0 provided certain differentiability assumptions on the β-functions are made. With this the Picard-Lindelöf theorem applies according to which exactly one solution $\lambda_j(\lambda_0)$ of (2.5) passes through any point $\lambda'_0, \lambda'_1, \ldots, \lambda'_n$. Due to the singular nature of the system (2.5) at $\lambda_0 = \lambda_j = 0$, the standard existence theorems cannot be applied there. On the other hand it is difficult to gain control over the asymptotic behavior in the weak coupling limit for solutions with prescribed non-vanishing initial values $\lambda'_0, \ldots, \lambda'_n$. In general, uniqueness properties do not hold for solutions passing through the origin $\lambda_0 = \lambda_j = 0$: For some systems there are no solutions of (2.5) which satisfy (2.2). For others there are infinitely many such solutions.

Further constraints are imposed if we require renormalizability for the original as well as the reduced system. Then the Green's functions of the original system have power series expansions in $\lambda_0, \lambda_1, \ldots, \lambda_n$ and the Green's functions of the reduced system can be expanded with respect to powers of λ_0[12]. This leads to the requirement that the solutions $\lambda_j(\lambda_0)$ of (2.5) possess power series expansions in λ_0.

It is easy to work out the conditions necessary for the renormalizable reduction of a system in lowest order of the primary coupling constant. As example we

---

11 A stationary value of the effective coupling indeed leads to a singular behavior for the derivatives of the β-function (see [14])

12 We do not consider here expansions with respect to fractional powers or logarithms of coupling constants which may arise due to infrared singularities of conventional perturbation theories (see for instance [15, 16])

consider β-functions with the expansions

$$\beta_0 = b_0\lambda_0^2 + \sum_{n=3}^{\infty} \sum_{m=0}^{n-1} \sum_{j_1 \dots j_m} b_{n-m, j_1 \dots j_m} \lambda_0^{n-m} \lambda_{j_1} \dots \lambda_{j_m}, \tag{2.23}$$

$$\begin{aligned} \beta_j = & \sum_{ik} c_{ik}^{(j)} \lambda_i \lambda_k + \sum_i c_i^{(j)} \lambda_i \lambda_0 + c^{(j)} \lambda_0^2 \\ & + \sum_{n=3}^{\infty} \sum_{m=0}^{n} \sum_{j_1 \dots j_m} c_{n-m, j_1 \dots j_m} \lambda_0^{n-m} \lambda_{j_1} \dots \lambda_{j_m} \end{aligned} \tag{2.24}$$

where λ_0 is the square $\lambda_0 = g^2$ of the primary coupling parameter g. Since all β-functions are even functions of g it is natural to require that the coupling parameters λ_j of the reduced system are also even in g. Renormalizability combined with the condition (2.2) implies that the coupling parameters of the reduced system have power series expansions

$$\lambda_j = \varrho_0^{(j)} g^2 + \sum_{n=2}^{\infty} \varrho_n^{(j)} g^{2n+2}. \tag{2.25}$$

Comparing the coefficients of g^4 in (2.5) we find the quadratic equations [4]

$$\sum_{ik} c_{ik}^{(j)} \varrho_0^{(i)} \varrho_0^{(k)} + \sum_i (c_i^{(j)} - \delta_{ij} b_0) \varrho_0^{(i)} + c^{(j)} = 0 \tag{2.26}$$

Its solutions $\varrho_0^{(j)}$ represent the lowest order values of the ratios λ_j/g^2. As such they should be real and – if required by the model – satisfy constraints like the positivity of coupling parameters. The equations (2.26) are necessary for the renormalizability of the reduced system, but not always sufficient. For in some cases the lowest order approximation based on a solution of (2.26) cannot be extended to power series expansions. Examples for that will be found in the following section.

3. Two Coupling Parameters

We are going to discuss in some detail the reduction of systems involving two coupling constants. The notation used is

$$\left(\kappa^2 \frac{\partial}{\partial \kappa^2} + \beta_0(g^2, \lambda) \frac{\partial}{\partial g^2} + \beta_1(g^2, \lambda) \frac{\partial}{\partial \lambda} + \gamma \right) \tau = 0 \tag{3.1}$$

for the Callan-Symanzik equations. The β-functions are assumed to have expansions of the form

$$\beta_0 = b_0 g^4 + \sum_{n=3}^{\infty} \sum_{m=0}^{n-1} b_{n-m, m} g^{2(n-m)} \lambda^m, \tag{3.2}$$

$$\begin{aligned} \beta_1 = & c_1 \lambda^2 + c_2 \lambda g^2 + c_3 g^4 \\ & + \sum_{n=3}^{\infty} \sum_{m=0}^{n} c_{n-m, m} g^{2(n-m)} \lambda^m, \end{aligned} \tag{3.3}$$

which cover a large variety of models. We want to investigate under which conditions the model can be reduced by

$$\lambda = \lambda(g^2) \tag{3.4}$$

to a renormalizable system involving a single coupling constant g. The reducibility condition (2.5) takes the form

$$\beta_0 \frac{d\lambda}{dg^2} = \beta_1 . \tag{3.5}$$

Renormalizability and condition (2.2) impose on the solutions that they can be expanded in the form

$$\lambda = \varrho_0 g^2 + \sum_{j=1}^{\infty} \varrho_j g^{2j+2} . \tag{3.6}$$

The first coefficient ϱ_0 is determined to be a root of the quadratic equation

$$c_1 \varrho_0^2 + (c_2 - b_0)\varrho_0 + c_3 = 0 . \tag{3.7}$$

ϱ_0 is only real if the discriminant

$$\Delta = (c_2 - b_0)^2 - 4c_1 c_2 \tag{3.8}$$

is non-negative

$$\Delta \geqq 0 . \tag{3.9}$$

This requirement already precludes the reduction for a large number of models. In the work that follows (3.9) will be assumed. There may be further restrictions on the values of the first coefficients ϱ_0. For instance, in some models λ is the square of a coupling parameter and cannot be negative for that reason. In this case only non-negative values of $\varrho_\pm$ are admissible.

For the case

$$b_0 \neq 0, \qquad c_1 \neq 0 , \tag{3.10}$$

we may take over the results obtained in [6] concerning power series solutions of (3.5). The following notations will be used. $\varrho_\pm$ denotes the roots of (3.7) with ϱ_+ being the larger value,

$$\varrho_+ \geqq \varrho_- . \tag{3.11}$$

A number ξ is defined by

$$\xi = -\frac{c_1}{b_0}(\varrho_+ - \varrho_-), \qquad (b_0 \neq 0) . \tag{3.12}$$

Since usually $c_1 > 0$ positive ξ implies asymptotic freedom.

If $\xi < 0$ a power series solution

$$\lambda_- = \varrho_- g^2 + \sum_{j=1}^{\infty} \varrho_{-j} g^{2j+2} \tag{3.13}$$

of (3.5) exists with uniquely determined coefficients. Further the solution

$$\lambda_+ = \varrho_+ g^2 + \sum_{j=1}^{\infty} \varrho_{+j} g^{2j+2} \tag{3.14}$$

exists if $\xi<0$ is not integral. If ξ is a negative integer one finds the general solution of (3.5) as an expansion involving logarithms

$$\lambda_+=\varrho_+g^2+\sum_{j=1}^{|\xi|-1}\varrho_{+j}g^{2j+2}+\varrho_{+|\xi|}g^{2\xi+2}+dg^{2|\xi|+2}\ln g^2+\dots . \tag{3.15}$$

The coefficient $\varrho_{+|\xi|}$ is arbitrary, the coefficient d of the first logarithmic term is uniquely determined by lower orders. If $d=0$ no power series solution λ_+ of (3.5) exists. In that case a solution with asymptotic behavior ϱ_+g^2 for $g\to 0$ can only be formed by including logarithmic terms which do not correspond to a renormalizable Lagrangian. If $d=0$ the power series solution (3.15) exists with arbitrary coefficient $\varrho_{+|\xi|}$ and represents the general solution. Thus for negative integral ξ either no power series solution λ_+ exists or λ_+ represents the general solution of (3.5) with an arbitrary parameter.

If $\xi>0$ a power series (3.14) always exists for λ_+. The power series (3.13) for λ_- exists provided ξ is not integral. If ξ is a positive integer either no power series solution (3.13) exists or it represents the general solution with arbitrary coefficient $\varrho_{-(\xi+1)}$.

For $\xi=0$ both expansions (3.13–14) coincide and exist with uniquely determined coefficients.

The case $b_0=0$, $c_1\neq 0$ was treated in [7]. If $\Delta>0$, there exist two distinct power series solutions λ_+ and λ_- of the form (3.13–14). Although their coefficients are unique, they may include the general solution. For the difference of two solutions with the same weak coupling behavior ϱ_+g^2 (or ϱ_-g^2 respectively) is exponentially decreasing for $g\to 0$. If $b_0=0$ and $\Delta=0$ no power series solution of (3.5) exists unless all coefficients of terms g^{2n} in β_1 vanish. In the latter case $\lambda\equiv 0$ is the only power series solution.

We now discuss the simplifications which occur for supersymmetric gauge theories with $\lambda=h^2$, where h describes a matter or Higgs interaction. In that case all coefficients of the terms g^{2n} in β_1 vanish. As the square of a coupling parameter λ is non-negative. We further assume $c_1>0$ which is usually the case.

The absence of a g^4-term in β_1 implies that (3.9) is always satisfied. This eliminates a major obstacle in constructing renormalizable reduced models. The roots of (3.7) become

$$\varrho_0=0 \quad\text{and}\quad \varrho_0=\frac{b_0-c_2}{c_1}. \tag{3.16}$$

Since β_1 vanishes at $\lambda=0$, Eq. (3.5) has the solution $\lambda\equiv 0$. Apart from this trivial solution we list the following power series solutions of (3.5) under the positivity constraint $\lambda\geqq 0$:

(1) $b_0<0$, $c_2<b_0$, or equivalently $\xi>0$, $\varrho_+>0$, $\varrho_-=0$.

There is the expansion (3.14) of λ_+ with unique coefficients. If ξ is an integer there is further an expansion of λ_-,

$$\lambda_-=dg^{2\xi+2}+\sum_{j=\xi+1}^{\infty}\varrho_{-j}g^{2j+2}, \qquad \xi=1,2,\dots,d>0, \tag{3.17}$$

with arbitrary positive coefficient d.

(2a) $b_0>0$, $c_2<b_0$, or equivalently $\xi<0$, $\varrho_+>0$, $\varrho_-=0$.

If ξ is not an integer there is the expansion (3.14) of λ_+ with unique coefficients. If ξ is a negative integer either λ_+ does not exist or the coefficient $d_{+|\xi|}$ is arbitrary.

(2b) $b_0>0$, $c_2>b_0$, or equivalently $\xi<0$, $\varrho_+=0$, $\varrho_-<0$.

If ξ is an integer there is the expansion

$$\lambda_+=dg^{2|\xi|+2}+\sum_{j=|\xi|+1}^{\infty}\varrho_{+j}g^{2j+2},\qquad \xi=-1,-2,\ldots,d>0\,. \tag{3.18}$$

(3) $b_0=0$, $c_2<0$ implying $\varrho_+>0$, $\varrho_-=0$.

There is the expansion (3.14) of λ_+ with unique coefficients.

In all other cases, namely $b_0=c_2$ or $b_0\leqq 0$ with $c_2\geqq b_0$, there are no power series solutions except $\lambda\equiv 0$, which have $\lambda\geqq 0$ for sufficiently small g^2.

Finally we remark that the lowest order form $\lambda\approx\varrho_0 g^2$ of a power series expansion can be made exact by reparametrizing λ provided $\varrho_0\neq 0$. For the coefficients of

$$\lambda'=\lambda+a_1\lambda^2+a_2\lambda^3+\ldots \tag{3.19}$$

can be chosen such that

$$\lambda'\equiv\varrho_0 g^2\,. \tag{3.20}$$

If $\varrho_0=0$ the transformation (3.19) in general does not even lead to a polynomial form of λ'. But

$$\lambda'=\lambda+b_1\lambda g^2+b_2\lambda g^4+\ldots$$

can be used to transform a power series solution

$$\lambda=\varrho_n g^{2n+2}+o(g^{2n+4})$$

into

$$\lambda'\equiv\varrho_n g^{2n+2}\,.$$

4. Model of a Spinor and Pseudoscalar Field

We consider the massless renormalizable model of a single spinor field ψ interacting with a pseudoscalar field A. The interaction terms are

$$ig\bar{\psi}\gamma_5 A\psi-\frac{\lambda}{4!}A^4\,.$$

The model contains two independent parameters g and λ. We try to reduce the system to a renormalizable description in terms of g only. In lowest order the β-functions are

$$\begin{aligned}\beta_0&=\frac{1}{16\pi^2}5g^4+\ldots,\\ \beta_1&=\frac{1}{16\pi^2}\left(\frac{3}{2}\lambda^2+4\lambda g^2-24g^4\right)+\ldots.\end{aligned} \tag{4.1}$$

From this the values

$$\begin{aligned}\varrho_+&=\tfrac{1}{3}+\tfrac{1}{3}\sqrt{145}>0\,,\\ \varrho_-&=\tfrac{1}{3}-\tfrac{1}{3}\sqrt{145}<0\,,\\ \xi&=-\tfrac{1}{5}\sqrt{145}<0\end{aligned} \tag{4.2}$$

follow. If λ is positive for sufficiently small g there is only one power series solution,

$$\lambda=\tfrac{1}{3}(1+\sqrt{145})g^2+\varrho_{+1}g^4+\varrho_{+2}g^6+\dots, \tag{4.3}$$

uniquely determining λ as a function of g.* By a redefinition of λ, the lowest order can be made exact

$$\lambda=\tfrac{1}{3}(1+\sqrt{145})g^2. \tag{4.4}$$

For completeness we quote the generalization of (3.15) from [6]

$$\begin{aligned}\lambda=&\tfrac{1}{3}(1+\sqrt{145})g^2+\varrho_{+1}g^4+\varrho_{+2}g^6\\ &+d_{11}g^{\frac{2}{5}\sqrt{145}+2}+\varrho_{+3}g^8+d_{12}g^{\frac{3}{5}\sqrt{145}+4}\\ &+\varrho_{+4}g^{10}+d_{21}g^{\frac{4}{5}\sqrt{145}+2}+\dots.\end{aligned} \tag{4.5}$$

The terms are ordered according to decreasing magnitude for $g\to 0$. ϱ_{+1} and ϱ_{+2} are unique. d_{11} is arbitrary, all other coefficients are determined for given d_{11}. The power series solution (4.3) is stable since the general solution (4.5) has the same asymptotic behavior for $g\to 0$. If λ is negative for sufficiently small g there is the power series

$$\lambda=\tfrac{1}{3}(1-\sqrt{145})g^2+\varrho_{-1}g^4+\varrho_{-2}g^6+\dots \tag{4.6}$$

which is an unstable solution.

5. Wess-Zumino Model

We study the massless Wess-Zumino model with the coupling constants g and λ of the interaction terms

$$g\bar{\psi}(A+i\gamma_5 B)\psi-\frac{\lambda}{2}(A^2+B^2)^2$$

treated as independent parameters. In lowest order the β-functions are

$$\begin{aligned}\beta_0&=\frac{1}{16\pi^2}12g^4+\dots\\ \beta_1&=\frac{1}{16\pi^2}(20\lambda^2+8\lambda g^2-16g^4)+\dots.\end{aligned} \tag{5.1}$$

From this the values

$$\varrho_+=1,\qquad \varrho_-=-\tfrac{4}{5},\qquad \xi=-3 \tag{5.2}$$

follow. The solutions corresponding to the supersymmetric ratio $\lambda/g^2\approx\varrho_+$ have the asymptotic expansion (3.15)

$$\lambda=g^2+\varrho_{+1}g^4+\varrho_{+2}g^6+\varrho_{+3}g^8+dg^8\ln g^2+\dots. \tag{5.3}$$

ϱ_{+1}, ϱ_{+2}, and d are uniquely determined. ϱ_{+3} is arbitrary. The higher order coefficients are determined for given ϱ_{+3}. The existence of a renormalized version of the supersymmetric model implies that a power series solution of λ exists.

---

* See Note added in proof on p. 225

Therefore $d=0$, so that (5.3) takes the form

$$\lambda=g^2+\sum_{j=1}^{\infty}\varrho_{+j}g^{2j+2} \tag{5.4}$$

with arbitrary ϱ_{+3}. Only one of those corresponds to the supersymmetric case. With suitable supersymmetric normalization conditions it is

$$\lambda=g^2 \tag{5.5}$$

for the supersymmetric system and

$$\lambda=g^2+\varrho_{+3}g^8+\sum_{j=4}^{\infty}\varrho_{+j}g^{2j+2} \tag{5.6}$$

for the asymmetric reduced systems with arbitrary $\varrho_{+3}\neq 0$. The solution (5.3) is stable since its asymptotic behavior is the same as for the general solution (5.6). In addition there is the power series solution starting with $\varrho_{-}g^2$,

$$\lambda=-\frac{4}{5}g^2+\sum_{j=1}^{\infty}\varrho_{-j}g^{2j+2},$$

which is unstable and not related to supersymmetry.

Acknowledgements. I am grateful to Drs. Oehme, Piguet, and Sibold for many helpful discussions. Further I wish to thank Dr. Kuiper for a stay at the Institut des Hautes Études Scientifiques, Bures-sur-Yvette, in the spring of 1982 where part of this work was done.

References

1. Callan, C.: Broken scale invariance in scalar field theory. Phys. Rev. D **2**, 1541–1547 (1970)
2. Symanzik, K.: Small distance behavior in field theory and power counting. Commun. Math. Phys. **18**, 227–246 (1970)
3. Mandelstam, S.: Light-cone superspace and the ultraviolet finiteness of the $N=4$ model. Nucl. Phys. B **213**, 149–168 (1983)
4. Chang, N.P.: Eigenvalue conditions and asymptotic freedom for Higgs scalar gauge theories. Phys. Rev. D **10**, 2706–2709 (1974)
5. Chang, N.P., Das, A., Perez-Mercader, J.: Asymptotically free SU(5) model with three generations. Phys. Rev. D **22**, 1829–1832 (1980)
6. Oehme, R., Zimmermann, W.: Relation between effective couplings for asymptotically free models. Commun. Math. Phys. (to appear)
7. Oehme, R., Sibold, K., Zimmermann, W.: Renormalization group equations with vanishing lowest order of the primary β-function (to be published in Phys. Lett.)
8. Gross, D., Wilczek, F.: Asymptotically free gauge theories. I. Phys. Rev. D **8**, 3633–3652 (1973)
9. Wess, J., Zumino, B.: A Lagrangian model invariant under supergauge transformations. Phys. Lett. **49** B, 52–54 (1974)
10. Clark, T., Piguet, O., Sibold, K.: Supercurrents, renormalization and anomalies. Nucl. Phys. B **143**, 445–484 (1978)
11. Maison, D.: Determination of correct β-functions for super Yang-Mills theories using a supersymmetry violating renormalization scheme. Preprint Werner-Heisenberg-Institut für Physik MPI/PTh 43/84 (to be published)

12. Suzuki, M.: On instability of asymptotic freedom of supergauge Yang-Mills theories. Nucl. Phys. B **83**, 269–275 (1974)
13. Zimmermann, W.: The renormalization group of the model of A^4-coupling in the abstract approach of quantum field theory. Commun. Math. Phys. **76**, 39–64 (1980)
14. Oehme, R., Zimmermann, W.: Analyticity of effective coupling and propagators in massless models of quantum field theory. Commun. Math. Phys. **85**, 363–379 (1982)
15. Symanzik, K.: On some massless superrenormalizable and non-renormalizable theories. In: Lecture Notes in Physics, Vol. 39, pp. 101–106, H. Araki (ed.). Berlin, Heidelberg, New York: Springer 1975
16. Bandelloni, G., Becchi, C., Blasi, A., Collina, R.: Renormalization of models with radiative mass generation. Commun. Math. Phys. **67**, 147–178 (1978)
17. Piguet, O., Sibold, K.: Renormalizing supersymmetry without auxiliary fields. Preprint Université de Genève, UGVA-DPT 1984/10–443 (to be published)

Communicated by G. Mack

Received July 23, 1984

Note added in proof. This relation between the quartic and the Yukawa coupling constant was found independently by E. de Rafael and R. Stora (private communication).

Conformal Gauges and Renormalized Equations of Motion in Massless Quantum Electrodynamics

V. B. Petkova, G. M. Sotkov, and I. T. Todorov

Institute for Nuclear Research and Nuclear Energy,
Bulgarian Academy of Sciences, BG-1184 Sofia, Bulgaria

To the memory of Kurt Symanzik

Abstract. A formulation of massless QED is studied with a non-singular Lagrangian and conformal invariant equations of motion. It makes use of non-decomposable representations of the conformal group G and involves two dimensionless scalar fields (in addition to the conventional charged field and electromagnetic potential) but gauge invariant Green functions are shown to coincide with those of standard (massless) QED. Assuming that the (non-elementary) representation of G for the 5-potential which leaves the equations of motion invariant and leads to the free photon propagator of Johnson-Baker-Adler (JBA) conformal QED remains unaltered by renormalization, we prove that consistency requirements for conformal invariant 2-, 3-, and 4-point Green functions satisfying (renormalized) equations of motion and standard Ward identities lead to either a trivial solution (with $e\psi=0$) or to a subcanonical dimension $d=\frac{1}{2}$ for the charged field.

1. Introduction

The search for a conformal invariant quantum field theory (QFT) is one way to look for a (critical) renormalization group fixed point (see, e.g., [S2] where the essential equivalence between the two problems has been spelled out). It is, therefore, intimately related to the existence problem for a local relativistic QFT (see [A3, F5, M5]).

The study of conformal quantum electrodynamics (QED) [J1, A1, 2, E1, M4, F6, B1] (see also Chap. VII to [T1]) differs in at least two points from a parallel investigation of a nongauge, Yukawa-type QFT (see [M2, 3, D3, 4, F4, T1] and references therein). First, current conservation and the Maxwell equations imply that the dimension of one of the basic fields, the 4-potential $A_\mu(x)$, is canonical (while the dimension of the charged field $\Psi(x)$ is gauge dependent). Secondly, although conformal invariance of the classical (vacuum) Maxwell equations has been known since the time when application of group theory to physics was a novelty (see [C1, B2]), the problem of finding a conformal invariant gauge fixing term has been treated only recently (starting with [M7, B3]. The non-invariance of conventional (local) gauges poses problems in exploiting conformal symmetry for gauge dependent Green functions like the photon propagator.

It was gradually realized that the electromagnetic potential should transform under a non-elementary representation of the conformal group $G=SU(2,2)$ (extended by space reflections) or its Euclidean counterpart Spin(5,1) (otherwise the invariant photon propagator comes out purely longitudinal). The most attractive approach, in our view, uses non-decomposable representations of G and leads to a 5-potential (see the evolution of these ideas in[1] [S3, B4, Z1, F7]). Alternatively, direct sums of irreducible subrepresentations and factor representations of exceptional elementary representations for A_μ have also been tried ([F3, P2]).

The present paper studies possible conformal invariant solutions of the model of massless QED, put forward in [F7]. We prove that if renormalization does not change the transformation law for the 5-potential (which leaves the classical equations of motion invariant and leads to a free photon propagator as in the JBA approach), then the only conformal solution of the model corresponds to the vanishing of a product of the type $eC_\psi Z(d)\frac{\Gamma(d+\frac{1}{2})}{\Gamma(\frac{5}{2}-d)}$, where e is the electric charge, C_ψ is a normalization constant of the (renormalized) electron propagator, and $Z(d)$ is the finite part of the electron field renormalization constant Z_2, d being the scale dimension of the (interacting) electron field. From the analysis of the (renormalized) Schwinger-Dyson form of the Dirac equation and the small distance behaviour of the 3-point Schwinger function $\langle A_\mu(x_1)\Psi(x_2)\Psi(x_3)\rangle_E$ we deduce (in Sect. 4) that this consistency condition amounts to

$$e(\tfrac{7}{2}-d)(\tfrac{5}{2}-d)(d-\tfrac{1}{2})=0\,. \tag{1.1}$$

It will be made clear in Sect. 3 that if we do not allow a logarithmic behaviour for the Euclidean 2-point function of ψ, then its conformal invariant Wightman function vanishes for $d-\frac{5}{2}=0, 1, \ldots$. Thus Eq. (1.1) means that either $e=0$ (a result asserted in [K1] for the model of [F3, P2]) or $\psi=0$, or else the (gauge dependent!) scale dimension d takes the subcanonical value $d=\frac{1}{2}$. Thus finite conformal electrodynamics with a free (JBA) photon propagator is either trivial or only exists for a particular class of gauges that imply $d=\frac{1}{2}$ for $e\neq 0$. Strictly speaking, we only verify that the Schwinger-Dyson equations and Ward identities involving 2-, 3-, and some 4-point functions are consistent with the conformal postulate provided that Eq. (1.1) is satisfied. The existence proof for a non-trivial model for $d=\frac{1}{2}$ is a challenging open problem. In any case, it is already clear that a more careful argument is needed to justify recent claims [K1, M6] that finite charge renormalization in QED is impossible. We also point out that conformal invariance does not necessarily lead to free photon propagators. The possibility of using more complicated operator transformation laws (including non-decomposable representations of the subgroup of dilations [D1, F1] that lead, in

1 The last three papers contain essentially equivalent sets of equations for a 5-potential interacting with a conserved (external) current. However, it is only in [F7] that a non-singular conformal action is proposed which also includes charged (spinor) fields

particular, to a logarithmic term in the transverse part of the photon propagator) has not been fully explored.

We start in Sect. 2 with a new concise derivation of the equations of motion (and the Lagrangian) of the model within the classical framework. We give a formal argument in Sect. 2C (based on Feynman path integration) that in spite of the presence of non-conventional terms in the equations involving two dimensionless scalar fields the gauge invariant Green functions coincide (at least within the framework of perturbation theory) with those of standard massless QED. Section 3 is devoted to the "conformal quantization" of the model (in which some of the canonical commutation relations are traded for the assumption that renormalized Green functions are conformal invariant). Ward identities are studied in detail. It is demonstrated that they imply a deformation of the charged field transformation law (with deformation parameter e). Section 4 is devoted to the renormalization program. The discussion is based on combined use of conformal invariance and Wilson (operator product) expansions (together with a systematic application of Euclidean symmetric integration) see [S6, P1, D5].

Notational Conventions

Minkowski space metric: $xp = x^\mu \eta_{\mu\nu} p^\nu = \mathbf{xp} - x^0 p^0$; $[\gamma_\mu, \gamma_\nu]_+ = 2\eta_{\mu\nu}$ (the physical time variable is $t = x^0$, the energy is $E = -p_0$).

Euclidean (Schwinger) functions and their Fourier transforms. Scalar field of dimension δ:

$$\Delta_\delta(x) = \frac{1}{(4\pi)^2} \frac{\Gamma(\delta)}{\Gamma(2-\delta)} \left(\frac{4}{x^2}\right)^\delta = \int \frac{e^{ipx}}{(p^2)^{2-\delta}} d_4 p, \tag{1.2}$$

Spinor field ψ of dimension d:

$$\begin{aligned} S_d &= \frac{1}{C_\psi} \langle \psi(x) \bar{\psi}(0) \rangle_E = \frac{1}{(4\pi)^2} \frac{\Gamma(d+\frac{1}{2})}{\Gamma(\frac{5}{2}-d)} \left(\frac{4}{x^2}\right)^{d+1/2} \frac{\not{x}}{2} \\ &= -i \int \frac{\not{p} e^{ipx}}{(p^2)^{5/2-d}} d_4 p, \, (S_\psi(x) = C_\psi S_d(x)), \not{p} = \gamma^\mu p_\mu, \end{aligned} \tag{1.3}$$

$$\begin{aligned} d_4 p &= \frac{d^4 p}{(2\pi)^4}, \quad x^2 = \mathbf{x}^2 + x_4^2, \quad x_4 = -ix^0, \\ \gamma_4 &= i\gamma_0, \quad (C_\psi = C_\psi(d), C_\psi(\tfrac{3}{2}) = 1). \end{aligned} \tag{1.4}$$

2. Conformal Electrodynamics Involving a Non-Decomposable 5-Potential

A) Nonsingular Conformal Lagrangian for the Electromagnetic Potential from a Manifestly Covariant Connection Form

The most natural way to introduce a non-trivial conformal 5-potential is to use Dirac's manifestly covariant formalism [D2, M1].

Compactified Minkowski space $\bar{M}$ is realized as the projective quadric:

$$\bar{M}=Q/\mathbb{R}^*, \quad \mathbb{R}^*=\mathbb{R}\backslash\{0\},$$
$$Q=\{\xi=(\xi^0,\boldsymbol{\xi},\xi^5,\xi^6)\in\mathbb{R}^6;$$
$$\xi^2=\xi^a\eta_{ab}\xi^b=-\xi_0^2+\boldsymbol{\xi}^2+\xi_5^2-\xi_6^2=0, \xi\neq 0\}.$$

Identifying the points at infinity with the projective 3-cone

$$I_\infty=Q_\infty/\mathbb{R}^*, \quad Q_\infty=\{\xi\in Q; \kappa\equiv\xi^5+\xi^6=0, (=\xi_\mu\xi^\mu)\},$$

we imbed (real) Minkowski space M into $\bar{M}$ by setting

$$(M\ni)x^\mu=\frac{1}{\kappa}\xi^\mu, \quad \mu=0,1,2,3, \quad \text{for} \quad \kappa(=\xi^5+\xi^6)\neq 0;$$
$$\text{thus} \quad M\simeq\bar{M}\backslash I_\infty.$$

The 5-potential is defined as a homogeneous (in ξ) real valued 1-form on the quadric Q:

$$\mathscr{A}_a(\xi)d\xi^a(=\mathscr{A}_a(\varrho\xi)d\varrho\xi^a) \quad \text{for} \quad \xi_a d\xi^a=0(=\xi_a\xi^a).$$

In the local coordinates (x,κ) on $Q\backslash Q_\infty$ we can write

$$\mathscr{A}_a(\xi)d\xi^a=A_\mu(x)dx^\mu+A_-(x)\frac{d\kappa}{\kappa}, \tag{2.1}$$

where $\mathscr{A}_\mu(x)=\kappa\mathscr{A}_\mu(\xi)+(\mathscr{A}_6(\xi)-\mathscr{A}_5(\xi))\xi_\mu$ and $A_-(x)=\xi^a\mathscr{A}_a(\xi)$ are κ-independent Poincaré covariant vector and scalar fields with the following transformation properties under dilations $x\to\varrho x$, $\varrho>0$ and (infinitesimal) special conformal transformations

$$x^\mu\to{}'x^\mu=\omega(x^\mu+x^2c^\mu)\,(\approx x^\mu(1-2c\cdot x)+x^2c^\mu \quad \text{for} \quad c^\mu\to 0),$$
$$\omega^{-1}=1+2cx+c^2x^2, \tag{2.2}$$
$$A_-(x)\to A_-(\varrho x), \quad A_\nu(x)\to\varrho A_\nu(\varrho x); \tag{2.3}$$
$$A_N(x)\to A_N(x)+c^\mu[C_\mu, A_N(x)],$$

where $N=\nu$, $-$ and

$$[A_-(x), C_\mu]=(2x_\mu x\partial-x^2\partial_\mu)A_-(x), \tag{2.4a}$$
$$[A_\nu(x), C_\mu]=[2x_\mu(1+x\partial)-x^2\partial_\mu]A_\nu(x)$$
$$+2(\eta_{\mu\nu}xA(x)-x_\nu A_\mu(x)-\eta_{\mu\nu}A_-(x)). \tag{2.4b}$$

The crucial property is the appearance of A_- in the right-hand side of (2.4b), which is responsible for the non-vanishing of $[A_\nu(x), C_\mu]|_{x=0}$. As a consequence the Maxwell tensor $F_{\mu\nu}=\partial_\mu A_\nu-\partial_\nu A_\mu$ is also transformed under a non-elementary non-decomposable representation of the (local) conformal group, together with

$\partial_\nu A_-$; we have

$$[\partial_\mu A_-(x), C_\lambda] = [2x_\lambda(1+x\partial) - x^2\partial_\lambda]\partial_\mu A_-(x) + 2(\eta_{\lambda\mu}x\partial - x_\mu\partial_\lambda)A_-(x), \tag{2.5}$$

$$[F_{\mu\nu}(x), C_\lambda] = [2x_\lambda(2+x\partial) - x^2\partial_\lambda]F_{\mu\nu}(x) + 2(x^\kappa\eta_{\lambda[\mu}F_{\kappa\nu]} - x_{[\mu}F_{\lambda\nu]} + \eta_{\lambda[\mu}\partial_{\nu]}A_-), \tag{2.6}$$

where $\eta_{\lambda[\mu}F_{\kappa\nu]} \equiv \eta_{\lambda\mu}F_{\kappa\nu} - \eta_{\lambda\nu}F_{\kappa\mu}$, etc.

Taking the divergence ∂^ν of both sides of (2.6) and the d'Alembertian of (2.5), we deduce that only the combination

$$J^\mu = \partial_\nu F^{\mu\nu} - \tfrac{1}{2}\Box\partial^\mu A_- \tag{2.7}$$

transforms as a conformal 4-vector. Therefore, if J^μ is identified with the electromagnetic current of a charged field, setting, e.g.,

$$J^\mu = ie\bar{\psi}\gamma^\mu\psi, \tag{2.8}$$

then Eq. (2.7) can be regarded as a conformal extension of the standard Maxwell equation. On the other hand, applying $\Box\partial^\nu$ to both sides of (2.4b), we obtain

$$[\Box\partial A, C_\mu] = [2x_\mu(3+x\partial) - x^2\partial_\mu]\Box\partial A + 2\partial^\nu F_{\mu\nu} - \Box\partial_\mu A_- . \tag{2.9}$$

Hence, the equation

$$\tfrac{1}{2}\Box\partial A = J_- \tag{2.10}$$

can be regarded as a covariant complement of the conformal Maxwell equation (2.7) provided that (J^μ, J_+) transforms as a 5-current; in particular,

$$\left[\begin{pmatrix} J_+ \\ J^\mu \end{pmatrix}, C_\lambda\right]_{x=0} = \begin{pmatrix} 2J_\lambda(0) \\ 0 \end{pmatrix}, \tag{2.11}$$

so that we can not use the simple conformal gauge condition $\Box\partial A = 0$ (of refs. [M2, B3]) in the interacting case.

It follows from the properties of conformal representation spaces (see [P3]) that the integral

$$\int J_+(x)d^4x \tag{2.12}$$

is convergent and, according to (2.10) and Gauss theorem, it is expressed in terms of the asymptotic values of $\Box A_\mu$ at infinity. Current conservation

$$\partial^\mu J_\mu = 0 \tag{2.13}$$

guarantees its conformal invariance. As we shall see (in Sect. 2.C below), it is also gauge invariant.

Assuming that $(J^N) = (J^\nu, J_+)$ is a given conserved external 5-current, we can write the following canonical Lagrangian for the 5-potential A_N which yields the conformal invariant equations of motion (2.7), (2.10), and (2.13):

$$L(A, J) = \tfrac{1}{2}F^{\mu\nu}(\tfrac{1}{2}F_{\mu\nu} - \partial_\mu A_\nu + \partial_\nu A_\mu) - \tfrac{1}{2}A_-\Box\partial A + \frac{\beta}{8}(\Box A_-)^2 + J_+A_- + J^\mu A_\mu . \tag{2.14}$$

It is non-singular; in particular, each independent field variable A_μ and A_- has a non-vanishing conjugate momentum:

$$\pi^\mu = \frac{\partial L}{\partial \partial_0 A_\mu} = -F^{0\mu} + \frac{1}{2}\delta_0^\mu \Box A_- , \qquad \pi_- = -\frac{1}{2}\partial_0 \partial A + \frac{\beta}{8}\partial_0 \Box A_- .$$

The results of this section can be summarized as follows.

Proposition 2.1. *If the electromagnetic 5-potential $A_N = (A_\nu, A_-)$ obeys the non-decomposable conformal transformation law* (2.3) (2.4) *(derived from the invariance of the homogeneous connection form* (2.1) *on the 5-dimensional quadric Q), then the Lagrangian* (2.14) *gives rise to the conformal invariant field equations*

$$\mathcal{M}(\partial)(A^N) \equiv \begin{pmatrix} \partial^\mu \partial_\nu \Box & -\frac{1}{2}\partial^\mu \Box \\ \frac{1}{2}\partial_\nu \Box & -\frac{1}{4}\beta \Box^2 \end{pmatrix} \begin{pmatrix} A^\nu(x) \\ A_-(x) \end{pmatrix} = \begin{pmatrix} J^\mu(x) \\ J_+(x) \end{pmatrix} \tag{2.15a}$$

that describe the interaction of a conserved (external) (5-)current with A_N. Current conservation (2.13) *implies the generalized free field equation*

$$\Box^2 A_- = 0 , \tag{2.15b}$$

which yields β-independence of the equations. The Lagrangian is non-singular, the (Euclidean) p-space matrix $\mathcal{M}(ip)$ being invertible for $p \neq 0$ whatever the value of the gauge parameter β. ($\mathcal{M}^{-1}(ip)$ will be identified with the propagator matrix in Sect. 3B.)

B) Construction of J_+. Conformal Dirac Equation

The fifth component J_+ of the 5-current cannot be expressed in terms of a conventional (charged) Dirac field Ψ only. An economic way to construct J_+, proposed in [S3, F7] uses a longitudinal 5-potential coming from the closed form $dS(\xi) = \partial_\mu S(x) dx^\mu + q\frac{d\kappa}{\kappa}$, where q is a constant. $(\partial_\mu S, q)$ spans an invariant subspace of the non-decomposable space of 5-potentials (A_μ, A_-) (obeying the same transformation law (2.3), (2.4) which could be also deduced from the following law for the dimensionless scalar field $S(x)$:

$$\begin{aligned} &S(x) \to S(\varrho x) - q \ln \varrho ; \\ &[S(x), C_\mu] = 2x_\mu(x\partial S(x) - q) - x^2 \partial_\mu S(x) \end{aligned} \tag{2.16}$$

see [S1, D1, F1]).

Proposition 2.2. *If j^μ is a conserved conformal vector current and S obeys* (2.16), *then*

$$J^\mu = q j^\mu \quad \text{and} \quad J_+ = -j^\nu \partial_\nu S + \alpha_1 j^\nu \partial_\nu A_- + \alpha_2 \Box^2 S \tag{2.17}$$

transform as a 5-current for any values of the (real) parameters α_1 and α_2.

Proof. Equation (2.16) and the infinitesimal conformal transformation law for j^ν imply $[J_+(x), C_\mu] = (2x_\mu(4 + x\partial) - x^2\partial_\mu) J_+ + 2J_\mu(x)$ in accord with (2.11).

In the classical theory we could set $q = 1$. We shall see, however, that quantization forces us to regard q as a (constant) dynamical variable. It seems, nevertheless, consistent to assume at both the classical and the quantum level that

$$J^{\nu}(x)=ie:\bar{\psi}(x)\gamma^{\nu}\psi(x):=qJ^{\nu}(x)\,. \tag{2.18a}$$

We shall also restrict our attention to the choice $\alpha_1=0=\alpha_2$ in Eq. (2.17), thus having

$$J_{+}(x)=-J^{\nu}(x)\partial_{\nu}S(x)\,. \tag{2.18b}$$

Combining (2.14) with (2.8) and (2.18) and adding (massless) conformal invariant kinetic terms for ψ and S, we end up with the following Lagrangian for conformal electrodynamics:

$$L(A,S,\psi,\bar{\psi})=L(A,J)-\frac{1}{2}\bar{\psi}\overleftrightarrow{\partial\!\!\!/}\psi+\frac{\upsilon}{2}(\Box S)^2\,,\qquad \partial\!\!\!/=\partial_{\mu}\gamma^{\mu}\,. \tag{2.19}$$

It reproduces Eqs. (2.7) and (2.10) (with J^N expressed in terms of ψ and S) and implies in addition the (modified) Dirac equation

$$[\partial\!\!\!/-ie(A\!\!\!/-A_{-}\partial\!\!\!/S)]\Psi=0 \tag{2.20}$$

and a 4th order equation for S

$$\upsilon\Box^2 S=\partial_{\mu}(A_{-}J^{\mu})=ie\bar{\psi}(\partial\!\!\!/A_{-})\psi\,. \tag{2.21}$$

In order to verify the conformal invariance of Eq. (2.20), we use the infinitesimal laws (2.4) (2.16) and

$$[\psi(x),C_{\mu}]=\{2x_{\mu}(\tfrac{3}{2}+x\partial)-x^2\partial_{\mu}+\tfrac{1}{2}[\gamma_{\mu},x\!\!\!/]\}\psi(x)\,, \tag{2.22a}$$

$$[\partial\psi(x),C_{\mu}]=\{2x_{\mu}(\tfrac{5}{2}+x\partial)-x^2\partial_{\mu}+\tfrac{1}{2}[\gamma_{\mu},x\!\!\!/]\}\partial\!\!\!/\psi(x)\,. \tag{2.22b}$$

[They imply, in particular, that $(A\!\!\!/-A_{-}\partial\!\!\!/S)\psi$ has the same conformal transformation law as $\partial\!\!\!/\psi$.]

Remark 2.3. Clearly, the Lagrangian (2.19), and hence, the equations of motion are invariant under the discrete γ_5 transformation $\psi\to i\gamma_5\psi$, $\bar{\psi}\to i\bar{\psi}\gamma_5$; ($J^N\to J^N$, $A_N\to A_N$, $S\to S$). Moreover, the equations of motion imply that the axial current $J_5^{\mu}=ie:\bar{\psi}\gamma_{\mu}\gamma_5\psi:$ is also conserved; this allows us to decouple the left and right handed Weyl spinors in the Dirac equation.

C) Gauge Properties. Relation to a Non-Local Gauge Fixing

The action integral $\int Ld^4x$, and hence the entire set of Eqs. (2.18), (2.15), (2.20), (2.21) are invariant under the (restricted) gauge transformations

$$\begin{aligned}&\psi(x)\to e^{ie(\Lambda(x)+bS(x))}\Psi(x)\,(\bar{\psi}\to\bar{\psi}e^{-ie(\Lambda+bS)})\,,\qquad A_{\mu}\to A_{\mu}+\partial_{\mu}\Lambda\,,\\&A_{-}\to A_{-}+bq\,,\qquad S\to S+C\quad\text{for}\quad\Box^2\Lambda(x)=0(=\partial_{\mu}b=\partial_{\mu}C)\,.\end{aligned} \tag{2.23}$$

The above gauge freedom allows us to introduce a more general conformal transformation law for the basic fields which still respects the equations of motion. We shall use, in particular, a pure imaginary anomalous dimension for the spinor field (cf. [F7] and [B3(b)]):

$$\begin{aligned}U(\varrho)\psi(x)U(\varrho)^{-1}&=\varrho^{\frac{3}{2}+ie\delta}\psi(\varrho x)\,,\\U(\varrho)\bar{\psi}(x)U(\varrho)^{-1}&=\varrho^{\frac{3}{2}-ie\delta}\bar{\psi}(\varrho x)\,.\end{aligned} \tag{2.24}$$

It leads to a special conformal transformation law (for h_c: $x^\mu \to {}'x^\mu = \omega(x^\mu + c^\mu x^2)$, where $\omega^{-1} = 1 + 2xc + c^2x^2$) that differs from the standard one, $\psi \to {}'\psi$, by a gauge transformation:

$$U(h_c)\psi U(h_c)^{-1} = \omega^{ie\delta}\,{}'\psi = \exp\{-ie\delta \ln(1+2cx+c^2x^2)\}\,{}'\psi\,; \tag{2.25a}$$

it should be accompanied by an additive gauge transformation of the "compensating" (transformed) 4-potential ${}'A_\mu$:

$$\begin{aligned} U(h_c)A_\mu U(h_c)^{-1} &= {}'A_\mu + \delta\partial_\mu \ln\omega \\ &= {}'A_\mu - 2\delta \frac{c_\mu + c^2 x_\mu}{1+2cx+c^2x^2}\,. \end{aligned} \tag{2.25b}$$

Note that $\Lambda(x) = \delta \ln\omega$ satisfies the restriction $\Box^2\Lambda = 0$ of (2.23) for $\omega^{-1}(x, c) > 0$.

We shall now demonstrate in a formal way (within the framework of the Feynman path integral – see, e.g., [P4]) that the non-conventional (A_-- and S-dependent) terms in the Lagrangian (2.19) play the role of a gauge fixing.

We start with a modified form of the corresponding part of the action

$$\begin{aligned} &I(A_\mu, J^\nu; A_-, S; a) \\ &= \frac{\alpha}{2} a^2 + \int \left\{ \frac{\beta}{8}(A_- - a)\Box^2(A_- - a) - \frac{1}{2}(A_- - a)\Box\partial A \right. \\ &\quad \left. + \frac{\upsilon}{2} S\Box^2 S - (A_- - a)\partial_\mu S J^\mu \right\} d^4x\,, \end{aligned} \tag{2.26}$$

where "a" is a constant dynamical variable ("the value of A_- at infinity"). We define the generating functional

$$\begin{aligned} &F_{h,H}(A_\mu, J^\nu) = N \int DS \int D(A_- - a) \int_{-\infty}^{\infty} da \\ &\quad \exp\{i(I(A_\mu, \ldots, a) + \int (h(x)A_-(x) + H(x)S(x))d^4x)\} \\ &= N_1 \int DS \exp\left\{ i \int \left(\frac{\upsilon}{2} S\Box^2 S + HS \right.\right. \\ &\quad \left.\left. - \frac{2}{\beta}\left(\frac{1}{2}\Box\partial A - J_+ - h\right) \frac{1}{\Box^2} \left(\frac{1}{2}\Box\partial A - J_+ - h\right)\right) d^4x - i\frac{\tilde{h}^2}{2\alpha} \right\}, \\ &\tilde{h} = \int h(x) d^4x\,, \end{aligned} \tag{2.27}$$

where $h(x)$ and $H(x)$ are external sources, $J_+ = -J^\mu\partial_\mu S$ [see (2.18b)], N and N_1 are (infinite!) normalization constants; $\frac{1}{\Box^2}$ is the integral operator defined by

$$\left(\frac{1}{\Box^2} f\right)(x) = \frac{i}{(4\pi)^2} \int \ln \frac{l^2}{(x-y)^2 + i0} f(y) d^4y\,. \tag{2.28}$$

[It is independent of l on the subspace $\mathscr{V}_0$ of functions f with vanishing integral; it appears reasonable to assume that the external source $H(x)$ belongs to $\mathscr{V}_0$.]

The average of (2.27) over the (local) gauge group gives

$$\begin{aligned}&\int F_{h,H}(A_\mu+\partial_\mu\Lambda, J^\nu)D\Lambda\\&=N_2 e^{-i\frac{\tilde{h}^2}{2\alpha}}\int DS e^{i\int(\frac{\upsilon}{2}S\Box^2 S+HS)d^4x}\\&=C\exp\left\{-\frac{i}{2}\left(\frac{\tilde{h}^2}{\alpha}+\frac{1}{\upsilon}\int H\left(\frac{1}{\Box^2}H\right)d^4x\right)\right\},\end{aligned}\tag{2.29}$$

where C is a constant. We can choose the normalization constant N (respectively, N_1 and N_2) in such a way that $C=1$. We see, in particular, that for vanishing sources

$$\int F_{0,0}(A_\mu+\partial_\mu\Lambda, J^\nu)D\Lambda=C(=1),\tag{2.30}$$

which is (according to [P4]) precisely the condition that the contribution (2.26) to the action is a gauge fixing term. Moreover, it follows from (2.29) that the expectation value of a time-ordered product of gauge invariant fields (say, $F_{\mu\nu}$ and J_λ) and a number of fields A_- and S factorizes into a product of the standard gauge invariant expectation values (of T-products of $F_{\mu\nu}$'s and J_λ's) with generalized free type τ-functions of A_- and S that can be evaluated from (2.29). We have

$$\begin{aligned}&\langle TJ^{\lambda_1}(x_1)\dots J^{\lambda_m}(x_m)F_{\mu_1\nu_1}(x_{m+1})\dots F_{\mu_n\nu_n}(x_{m+n})e^{i\int(h(x)A_-(x)+H(x)S(x))d^4x}\\&=\langle TJ^{\lambda_1}(x_1)\dots F_{\mu_n\nu_n}(x_{m+n})\rangle_0\\&\cdot\exp\left\{\frac{-i\tilde{h}^2}{2\alpha}+\frac{1}{2\upsilon(4\pi)^2}\int\int H(x)\ln\frac{l^2}{(x-y)^2+i0}H(y)d^4x d^4y\right\};\end{aligned}\tag{2.31}$$

$$\langle TA_-(x)A_-(y)\rangle_0=\frac{i}{\alpha},\tag{2.32a}$$

$$\langle TS(x)S(y)\rangle_0=\frac{1}{(4\pi)^2\upsilon}\ln\frac{(x-y)^2+i0}{l^2}.\tag{2.32b}$$

We see that the τ-functions of A_- are only real for pure imaginary α's [i.e. for a complex action (2.26)]. As we shall see [in Eq. (3.16) below] the 2-point function (2.32a) that is related by conformal invariance to the normalized transverse photon propagator corresponds to

$$\alpha=-8\pi^2 i.\tag{2.32c}$$

We shall take the above heuristic argument as a justification for postulating (2.31), (2.32).

3. Quantization

A) General Requirements. Ward Identities. Electron Propagator

The construction of a quantum field theory corresponding to a given set of classical field equations requires new postulates that include a choice of certain operator properties of the basic quantized fields (commutation relations, properties of vacuum expectation values), and a proper definition of the (renormalized) composite fields (products of local field operators) which appear in the non-linear terms of the field equations. Looking for a conformal QED, we replace the canonical commutation relations of the standard (perturbative) approach by a number of requirements the most important of which is the conformal invariance of renormalized Green functions. This basic requirement implies that we are looking for a renormalization group fixed point (at least in the case when elementary representations of the subgroup of dilations are involved, leading to power laws for vertex functions) – see, e.g. [S2, M3, A1, 2].

The conformal quantum postulates involve some freedom related to the choice of representations for renormalized Green functions. We shall start in this section with those basic properties, which, in our view, should take place in any conformal QED [related to the classical Lagrangian (2.19)] leaving the more debatable requirements to Sect. 3B. The postulates within each section are also arranged in decreasing generality.

We first assume that the classical field equations remain valid at the quantum level provided that the encountered composite operators are renormalized by using operator product expansions. We shall not attempt to formalize this requirement in general; its precise meaning in the context of the present model of massless QED will be made clear in Sect. 4. We shall write the equations for Euclidean (Schwinger) functions which inherit the properties of time-ordered (τ-)functions. They include differentiation of the θ-function and use the equal time commutation relations for fields (like J^μ and A_-) whose dimensions remain canonical in the interacting case. Therefore, we find it convenient to set our basic postulates for (Minkowski space) time-ordered products of local field operators.

Rather than devising clumsy general formulations we shall write down an implication of the Maxwell equation (2.7), (2.15) for the time-ordered product of two gauge invariant fields.

Example 3.1. If $F_{\mu\nu}$ and $\partial_\mu A_-$ transform under the non-decomposable 10-dimensional representation of SO(4,2), described in Sect. 2, then we demand the validity of the following quantized form of the Maxwell equation:

$$\begin{aligned}\frac{\partial}{\partial x_\varrho} T(F_{\lambda\varrho}(x)F_{\mu\nu}(y)) - \frac{1}{2}\Box_x T(\partial_\lambda A_-(x)F_{\mu\nu}(y)) \\ = (\eta_{\lambda\mu}\partial_\nu - \eta_{\lambda\nu}\partial_\mu)\delta(x-y) + T(J_\lambda(x)F_{\mu\nu}(y)). \qquad (3.1\text{a})\end{aligned}$$

The δ-function term is a covariant reflection of the equal-time commutation relations between $F_{\lambda 0}$ and $F_{\mu\nu}$.

Similarly, Eq. (2.15) applied to the Euclidean photon propagator should read

$$\mathscr{M}(\partial)^M{}_{N'}\langle A^{N'}(x)A_N(0)\rangle_E = \delta^M{}_N\delta(x) + \langle J^M(x)A_N(0)\rangle_E. \qquad (3.1\text{b})$$

We shall assume throughout that we are working with the covariant (T^*- or Wick-) time-ordered product. The extension of a relation like (3.1) to products of several fields uses the standard algebraic properties of T-products. These include anticommutativity for any pair of local Fermi fields and commutativity of a local Bose field with any other local field as well as an associativity property which we shall write down in a special case to appear in our analysis of the Ward identity (WI) below:

$$T(\Psi(x_1)T(\bar{\psi}(x_2)J^\mu(x_3))) = T(T(J^\mu(x_3)\psi(x_1))\bar{\psi}(x_2))$$
$$= T(\psi(x_1)\bar{\psi}(x_2)J^\mu(x_3)). \quad (3.2)$$

A central postulate of local QED is the *operator* WI

$$\frac{\partial}{\partial y^\mu}(T\psi(x)J^\mu(y)) = \delta(x^0-y^0)\,[J^0(y),\psi(x)] = -e\delta(x-y)\psi(x), \quad (3.3a)$$

$$\frac{\partial}{\partial y^\mu}(T\bar{\psi}(x)J^\mu(y)) = \delta(x^0-y^0)\,[J^0(y),\bar{\psi}(x)] = e\delta(x-y)\bar{\psi}(x). \quad (3.3b)$$

The equal-time current-field commutation, incorporated in (3.3), is the only trace of the canonical anticommutation relations $[\psi(t,\mathbf{x}),\psi^*(t,\mathbf{y})]_+ = \delta(\mathbf{x}-\mathbf{y})$ (for $J^0(x) = ie{:}\bar{\psi}(x)\psi(x){:} = e{:}\psi^*(x)\psi(x){:}$). (These anticommutation relations will not be assumed for interacting fields, which will be allowed to acquire an anomalous dimension, while the property (3.3), characteristic for a current-charge density, will be retained.)

As a consequence of (3.3) we have the following standard WI relating the Euclidean 3- and 2-point (Schwinger) functions

$$\frac{\partial}{\partial x_3^\mu}\langle\psi(x_1)\bar{\psi}(x_2)J^\mu(x_3)\rangle_E = ieS_\psi(x_{12})\,[\delta(x_{23})-\delta(x_{13})], \quad (3.4)$$

where

$$S_\psi(x_{12}) \equiv \langle\psi(x_1)\bar{\psi}(x_2)\rangle_E, \qquad x_{12}\equiv x_1-x_2. \quad (3.5)$$

[The difference in the i-factor between (3.3) and (3.4) comes from the relation $(d_4p)_M \leftrightarrow i(d_4p)_E$ between Minkowski and Euclidean momentum space volume elements for $p^0 \leftrightarrow ip_4$.]

Proposition 3.2. *The field equations* (2.15) *and the* WI (3.3) *together with the assumption that $A_-(x)$ behaves like a constant at infinity and that $\psi(x)$ has an appropriate asymptotic behaviour, consistent with its conformal properties, imply the relations*

$$\langle\psi(x_1)\ldots\psi(x_n)\bar{\psi}(y_1)\ldots\bar{\psi}(y_n)A_-(z)\rangle_E$$
$$= \frac{e}{8\pi^2}\langle\psi(x_1)\ldots\psi(x_n)\bar{\psi}(y_1)\ldots\bar{\psi}(y_n)\rangle_E\left(i\ln\frac{(y_1-z)^2\ldots(y_n-z)^2}{(x_1-z)^2\ldots(x_n-z)^2}+C_n\right). \quad (3.6)$$

Similarly,

$$\langle\psi(x_1)\bar{\psi}(x_2)A_-(x_3)A_-(x_4)\rangle_E$$
$$= S_\psi(x_{12})\left\{\left(\frac{e}{8\pi^2}\right)^2\left(i\ln\frac{x_{23}^2}{x_{13}^2}+C\right)\left(i\ln\frac{x_{24}^2}{x_{14}^2}+C\right)+\langle A_-(x_3)A_-(x_4)\rangle_E\right\}. \quad (3.7)$$

Proof. The divergence of $F^{\mu\nu}$ in the conformal Maxwell equation

$$\partial_\nu F^{\mu\nu} - \tfrac{1}{2}\Box\partial^\mu A_- = J^\mu \quad (3.8)$$

does not contribute to the local charge density (see [F2, S5]). It then follows from (3.3) and (3.8) that

$$\begin{aligned} \tfrac{1}{2}\Box_y^2 T(\psi(x)A_-(y)) &= ie\delta(x-y)\psi(x)\,, \\ \tfrac{1}{2}\Box_y^2 T(\bar{\psi}(x)A_-(y)) &= -ie\delta(x-y)\bar{\psi}(x)\,. \end{aligned} \tag{3.9}$$

Equations (3.6), (3.7) are an (Euclidean) integrated version of (3.9) that takes into account the asymptotic conditions and the identity $\Box^2 \ln \frac{l^2}{x^2} = (4\pi)^2\delta(x)$ (where $\Box$ is the Euclidean Laplacian).

Remark 3.3. Equation (3.9) corresponds to the following Minkowski space commutation relations:

$$\begin{aligned} [A_-(x), \psi(y)] &= \frac{e\varepsilon(x^0-y^0)}{2\pi}\theta(-(x-y)^2)\psi(y)\,, \\ [A_-(x), \bar{\psi}(y)] &= -\frac{e\varepsilon(x^0-y^0)}{2\pi}\theta(-(x-y)^2)\bar{\psi}(y)\,, \end{aligned} \tag{3.10}$$

where the coefficient in the right-hand side is related to the Pauli-Jordan commutator function $\mathscr{D}(x)$ by

$$\Box\frac{-\varepsilon(x^0)}{4\pi}\theta(-x^2) = \mathscr{D}(x) = \frac{\varepsilon(x^0)}{2\pi}\delta(x^2) = 2\pi i \int \varepsilon(p^0)\delta(p^2)e^{ipx}d_4p\,. \tag{3.11}$$

We shall first use in an essential way the conformal invariance in determining the Dirac propagator.

C1. *The 2n-point Schwinger functions $\langle\psi(x_1)\dots\psi(x_n)\bar{\psi}(y_1)\dots\bar{\psi}(y_n)\rangle_E$ are γ_5-invariant and invariant with respect to the tensor product of (2n) elementary Dirac-type representation of the Euclidean conformal group Spin(5,1) corresponding to a (real) dimension d.*

The general form of the conformal electron propagator (3.5) is given by (1.3).

Remark 3.4. The 2-point Wightman function

$$w_{\psi\psi^*}(x) = \frac{2\pi C_\psi}{\Gamma(\frac{5}{2}-d)\Gamma(d-\frac{3}{2})}\int(-p^2)_+^{d-5/2}\theta(p^0)(p^0-\gamma^0\boldsymbol{\gamma}\cdot\mathbf{p})e^{ipx}d_4p \tag{3.12}$$

related to (1.3) is positive definite for $d \geqq \frac{3}{2}$, $C_\psi > 0$. The vanishing of $w_{\psi\psi^*}$ for $d-\frac{5}{2}=0,1,\dots$ is related to the assumption that S_ψ is dilation invariant at these points. The anomalous dimension $d-\frac{3}{2}$ of the charged field is gauge dependent. A standard renormalization group analysis using second order calculations gives

$$d-\frac{3}{2} \approx \left(\beta - \frac{1}{8\pi^2 \upsilon}\right)\frac{e^2}{(4\pi)^2}\,. \tag{3.13}$$

(For a similar computation in the framework of conventional massless QED see [A2, E1, F6].) We shall not exploit here the WI's associated with the conserved axial current.

B) Conformal QED *with a Free Photon Propagator*

Proposition 3.5. *The general conformal invariant 2-point function* $\langle A^M(x)A_N(0)\rangle_E$ *for the* Spin(5,1)*-representation, characterized by the infinitesimal law* (2.4), *is a multiple of the free photon Schwinger function*

$$\Delta_A(x)=\left\langle \begin{pmatrix} A^\mu(x) \\ A_-(x) \end{pmatrix} (A_\nu(0)A_-(0)) \right\rangle_E = \int e^{ipx}\tilde{\Delta}_A(p)d_4p\,, \tag{3.14a}$$

where

$$\tilde{\Delta}_A(p)=\begin{pmatrix} \frac{1}{p^2}\left(\delta^\mu_\nu-(1-\beta)\frac{p^\mu p_\nu}{p^2}\right) & 2i\frac{p^\mu}{p^4} \\ -2i\frac{p_\nu}{p^4} & -2\pi^2\delta(p) \end{pmatrix} \tag{3.14b}$$

satisfies (for $\mathscr{M}$ *defined in* (2.14))

$$\mathscr{M}(ip)\tilde{\Delta}_A(p)=\mathbb{1}\,,\quad \text{so that}\quad \mathscr{M}(\partial)\Delta_A(x)=\delta(x)\,. \tag{3.15}$$

A proof of this statement based on the manifestly covariant formalism is given in Appendix B to [F7].

Remark 3.6. Equation (3.15) does not determine the coefficient to $\delta(p)$ in the right-hand side of (3.14b). One way to verify that our choice $(-2\pi^2)$ fits the requirement of conformal invariance is to exploit the infinitesimal law (2.4) and the condition $C_\mu|0\rangle=0$. The result is

$$\langle A_-(x)A_-(0)\rangle_0=-\frac{1}{8\pi^2}=\langle A_-(x)A_-(0)\rangle_E \tag{3.16}$$

in accord with (3.14).

Corollary 3.7. *Equation* (3.14) *together with the Euclidean Maxwell equations* (3.1b) *lead to trivial current-field and current-current 2-point functions:*

$$\langle J^M(x)A_N(0)\rangle_E=\mathscr{M}(\partial)^M{}_{M'}\Delta_A(x)^{M'}{}_N-\delta^M{}_N\delta(x)=0 \tag{3.17}$$

(where $J^-\equiv J_+$*) and hence*

$$\langle J^M(x)J_N(0)\rangle_E=0\,. \tag{3.18}$$

Remark 3.8. The current-current Wightman function evaluated in second order perturbation theory (using the free electron propagator (1.3) for $d=\frac{3}{2}$) is non-trivial and consistent with conformal invariance:

$$\langle J^\mu(x)J_\nu(0)\rangle_0=\frac{e^2}{\pi^4}\frac{r^\mu{}_\nu(x)}{(x^2+i0x^0)^3}=\frac{e^2}{12\pi^4}(\delta^\mu_\nu\Box-\partial^\mu\partial_\nu)\frac{1}{(x^2+i0x^0)^2} \tag{3.19a}$$

$$=\frac{e^2}{6\pi}\int\theta_+(p)(p^\mu p_\nu-p^2\delta^\mu_\nu)e^{ipx}d_4p\,,\quad (\theta_+(p)=\theta(p^0)\theta(-p^2)) \tag{3.19b}$$

$$r^\mu{}_\nu=\delta^\mu{}_\nu-2\frac{x^\mu x_\nu}{x^2}\,. \tag{3.20}$$

However, its Euclidean counterpart is only covariant under a non-decomposable (logarithmic) representation of the dilation subgroup. Indeed, defining the finite part of $\frac{1}{x^4}$ by

$$\left(\frac{1}{x^4}\right)_\ell = \lim_{\varepsilon \to +0} \frac{d}{d\varepsilon}\left(\frac{\varepsilon}{x^4}\left(\frac{x^2}{e^2}\right)^\varepsilon\right),$$

we find

$$\begin{aligned}\langle J^\mu(x) J_\nu(0)\rangle_E &= \frac{e^2}{12\pi^2}(\delta^\mu_\nu \Box - \partial^\mu \partial_\nu)\left(\frac{1}{x^4}\right)_\ell \\ &= \frac{e^2}{12\pi^2} \lim_{\varepsilon\to 0} \frac{d}{d\varepsilon}\left\{\frac{\Gamma(1+\varepsilon)}{\Gamma(2+\varepsilon)} \int \frac{p^\mu p_\nu - p^2 \delta^\mu_\nu}{(4\ell^2 p^2)^\varepsilon} e^{ipx} d_4 p\right\} \\ &= \frac{e^2}{12\pi^2} \int \left(\ln \frac{\mu^2}{p^2}\right)(p^\mu p_\nu - p^2 \delta^\mu_\nu) e^{ipx} d_4 p \end{aligned} \tag{3.21}$$

(where the "subtraction point" μ is proportional to $\frac{1}{\ell}$). More troublesome, the invariant 2-point function $\langle J^\mu(x) J_+(y)\rangle_0$ is inconsistent with current conservation in that case. It is nevertheless clear that the vanishing of the 2-point functions (3.17), (3.18) is representation dependent.

Equation (3.18) was adopted as a unique possibility in the conformal QED of Johnson-Baker-Adler (JBA) (see [J1, A1,2, B1]). Although we leave open the possibility that (a modified) conformal invariance is still consistent with the more realistic ansatz (3.19), we shall only explore in this paper the simplest possibility (3.18).

C2. *The photon propagator is invariant under the standard (non-decomposable) conformal law* (2.3), (2.4).

As a consequence of Proposition 3.5 and of Corollary 3.6, this postulate leads to a JBA-type model of conformal QED.

C3. *The Schwinger functions*

$$\langle \psi(x_1)\ldots\psi(x_m)\bar\psi(y_1)\ldots\bar\psi(y_m) J^{\mu_1}(z_1)\ldots J^{\mu_n}(z_n)\rangle_E$$

are conformal and γ_5-invariant with respect to the tensor product of the standard elementary representation for ψ (with dimension d) and J^μ (with dimension 3).

Corollary 3.9. *It follows from* (3.3), (3.10), *and* C3. *that*

$$\begin{aligned}&\langle \psi(x_1)\bar\psi(x_2) J^\mu(x_3)\rangle_E \\ &= ieC_\psi \frac{\Gamma(d+\frac{1}{2})}{\Gamma(\frac{5}{2}-d)}\left(\frac{4}{x_{12}^2}\right)^{d-\frac{3}{2}}\left\{\frac{B+1}{2} S_{3/2}(x_{13})\gamma^\mu S_{3/2}(x_{32}) \right.\\ &\left. + (B-1)\frac{\not{x}_{12}}{x_{12}^2} \Delta(x_{13}) \overleftrightarrow{\partial}{}^\mu_3 \Delta(x_{23})\right\}, \quad \left(\Delta(x) = \frac{1}{4\pi^2 x^2}\right),\end{aligned} \tag{3.22}$$

where B is an arbitrary (real) constant (see Eq. (IV.48) *of* [T1]*).*

Remark 3.10. It should be noted that the 3-point function

$$S_-(x_1,x_2;x_3)\equiv\langle\psi(x_1)\bar\psi(x_2)A_-(x_3)\rangle_E=S_\psi(x_{12})\left(\frac{ie}{8\pi^2}\ln\frac{x_{23}^2}{x_{13}^2}+C\right) \quad (3.23a)$$

(obtained from (3.6) for $n=1$) is *not* invariant under the standard transformation law (2.22), (2.4) for ψ and A_-. However, there exists a more general law which relates S_- and

$$S_\nu(x_1,x_2;x_3)\equiv\langle\psi(x_1)\bar\psi(x_2)A_\nu(x_3)\rangle_E \quad (3.23b)$$

with the longitudinal 3-point function

$$l_\nu(x_1,x_2;x_3)\equiv\langle\psi(x_1)\bar\psi(x_2)\partial_\nu S(x_3)\rangle_E, \quad (3.23c)$$

and

$$l_-(x_1,x_2)\equiv\langle\psi(x_1)\bar\psi(x_2)q\rangle_E=S_\psi(x_{12}). \quad (3.23d)$$

(In the last equation we have used the ansatz $\psi\bar\psi q=\psi\bar\psi$ – cf. (2.18a).) A compact form of this general law is given in terms of the conformal inversion $I_r x=\frac{x}{x^2}$ (cf. [T1, 2]). The Euclidean picture fields would transform as follows under I_r for elementary representations of the (extended) Spin(5,1):

$$\psi(x)\to\psi^{(r)}(x)=i\gamma_5\frac{\not x}{(x^2)^{d+1/2}}\psi(I_r x), \quad (3.24a)$$

$$\bar\psi(x)\to\bar\psi^{(r)}(x)=\bar\psi(I_r x)i\gamma_5\frac{x}{(x^2)^{d+1/2}}, \quad (3.24b)$$

$$A_-(x)\to A_-^{(r)}(x)=A_-(I_r x),\qquad A_\mu\to A_\mu^{(r)}(x)=\frac{r_{\mu\nu}(x)}{x^2}A^\nu(I_r x)\quad\text{etc.}, \quad (3.24c)$$

where $r^\mu_\nu(x)$ is given by (3.20). Only $l_-(x_1,x_2)$ is invariant under this elementary law:

$$l_-^{(r)}(x_1,x_2)\equiv\langle\psi^{(r)}(x_1)\bar\psi^{(r)}(x_2)q\rangle_E=l_-(x_1,x_2)\,(=S_v(x_{12})). \quad (3.25a)$$

The non-decomposable representation spanned by $\partial_\nu S$ and q implies the following *non-elementary* covariance property for l_ν:

$$l_\nu^{(r)}(x_1,x_2;x_3)+\frac{2x_{3\nu}}{x_3^2}l_-(x_1,x_2)=l_\nu(x_1,x_2;x_3). \quad (3.25b)$$

It is straightforward to verify that the pair of functions S_- (3.23a) and l_- (3.23c) satisfy the I_r-covariance law,

$$S_-^{(r)}(x_1,x_2;x_3)-\frac{ie}{8\pi^2}l_-(x_1,x_2)\ln\frac{x_1^2}{x_2^2}=S_-(x_1,x_2;x_3). \quad (3.26a)$$

The law for S_ν (proposed in [F7]) combines together the last two formulas:

$$S_\nu^{(r)}(x_1,x_2;x_3)-\frac{ie}{8\pi^2}l_\nu(x_1,x_2;x_3)\ln\frac{x_1^2}{x_2^2}+\frac{2x_{3\nu}}{x_3^2}S_-(x_1,x_2;x_3)$$
$$=S_\nu(x_1,x_2;x_3). \quad (3.26b)$$

C4. *The Schwinger functions of* $\psi, \bar{\psi}$; A_ν, A_-; $\partial_\nu S, q$ *are related among themselves via the conformal inversion according to* Eqs. (3.25), (3.26) *(and their extension to higher point Green functions).*

There is a more complicated transformation law for the fields which makes all 3-point functions of the 5-potential invariant. For the dilation and the I_r-inversion of Euclidean picture spinor fields it reads [S3, F7]

$$U(\varrho)\psi(x)U(\varrho)^{-1} = \varrho^{3/2} : \varrho^{ie\hat{\delta}}\psi(\varrho x):, \qquad U(\varrho)\bar{\psi}_{(x)}U(\varrho)^{-1} = \varrho^{3/2} : \varrho^{-ie\hat{\delta}}\bar{\psi}(\varrho x):, \tag{3.27a}$$

$$U(I_r)\psi(x)U(I_r)^{-1} = i\gamma_5 \frac{\not{x}}{x^4} : \left(\frac{1}{x^2}\right)^{ie\hat{\delta}} \psi(I_r x): \tag{3.27b}$$

while the 5-potential obeys the standard non-decomposable law,

$$U(I_r)A_-(x)U(I_r)^{-1} = A_-(I_r x)\,(= A_-^{(r)}(x)), \qquad U(I_r)A_\nu(x)U(I_r)^{-1} = A_\nu^{(r)}(x) + \frac{2x_\nu}{x^2} A_-(x). \tag{3.28}$$

Here the normal products are defined as usual:

$$:\varrho^{ie\hat{\delta}}\psi(x): = \varrho^{ie\delta^{(+)}}\psi(x)\varrho^{ie\delta^{(-)}}, \tag{3.29}$$

where

$$\delta^{(-)}|0\rangle = 0 = \langle 0|\delta^{(+)}, \qquad \delta^{(-)} = \delta^{(+)*}. \tag{3.30}$$

Although *local fields* either strictly anticommute (for a pair of Fermi fields) or strictly commute (otherwise) under the sign of an Euclidean expectation value, positive and negative frequency parts, in general, do not commute with the fields. Thus, in order to apply the operator law (3.27) to the evaluation of Schwinger functions, we should define the commutation relations of $\delta^{(\pm)}$ between each other and with the local fields.

We shall set

$$[\delta^{(-)}, \delta^{(+)}] = 0, \qquad [\delta^{(\mp)}, \psi(x)] = \pm \frac{ieb}{(4\pi)^2}\psi(x), \qquad [\delta^{(\pm)}, \bar{\psi}(x)] = \pm \frac{ieb}{(4\pi)^2}\psi(x), \tag{3.31a}$$

so that

$$\varrho^{-ie\delta^{(+)}}\psi(x)\varrho^{ie\delta^{(+)}} = \varrho^{-\frac{e^2 b}{(4\pi)^2}}\psi(x) \quad \text{etc.}, \tag{3.31b}$$

$$[\delta^{(\mp)}, A_-(x)] = \pm \frac{1}{8\pi^2} q, \qquad [\delta^{(\mp)}, A_\mu(x)] = \pm \frac{1}{8\pi^2}\partial_\mu S(x). \tag{3.32}$$

The electron propagator (1.3) is now reproduced from (3.26–3.28) provided that

$$d = \frac{3}{2} + \frac{e^2}{(4\pi)^2} b. \tag{3.33}$$

Comparing with (3.13), we see that b plays the role of an overall gauge parameter. Furthermore, if we require

$$\psi(x)\bar{\psi}(y)q = q\psi(x)\bar{\psi}(y) = \psi(x)\bar{\psi}(y) \tag{3.34}$$

(in accord with (2.18)), we rederive the expression (3.23a) for S_- (up to a normalization constant) from the invariance requirement under the above operator law. In order to reproduce also the covariance property (3.25b), we demand that

$$[\delta^{(\pm)}, \partial_\mu S(x)] = 0 = [\delta^{(\pm)}, q]\,. \tag{3.35}$$

Proposition 3.11. *Under the above assumptions the general conformal invariant 3-point Schwinger function of $\partial_\mu S$, consistent with* (2.16) *and* (3.34), *is*

$$\langle\psi(x_1)\bar{\psi}(x_2)\partial_\mu S(x_3)\rangle_E = S_\psi(x_{12})\left\{-\left(\frac{x_{13\mu}}{x_{13}^2}+\frac{x_{23\mu}}{x_{23}^2}\right)+\alpha\lambda_\mu\right\}. \tag{3.36}$$

Equations (3.32) *and* (3.36) *imply*

$$\begin{aligned} S_\mu(x_1, x_2; x_3) = {} & \frac{ie}{8\pi^2} S_\psi(x_{12})\left\{\left(\alpha\lambda_\mu - \frac{x_{13\mu}}{x_{13}^2} - \frac{x_{23\mu}}{x_{23}^2}\right)\ln\frac{x_{23}^2}{x_{13}^2} + A\lambda_\mu\right\} \\ & - iC_\psi e\frac{B+1}{8\pi^2}\left(d-\frac{1}{2}\right)\Delta_{d-1/2}(x_{12})\frac{\not{x}_{13}}{x_{13}^2}\gamma_\mu\frac{\not{x}_{23}}{x_{23}^2}, \end{aligned} \tag{3.37}$$

where the coefficient to the last term is chosen to agree with (3.22)*; here Δ_δ is given by* (1.2) *and*

$$\lambda_\mu = \frac{x_{13\mu}}{x_{13}^2} - \frac{x_{23\mu}}{x_{23}^2}\left(=\frac{1}{2}\partial_{3\mu}\ln\frac{x_{23}^2}{x_{13}^2}\right). \tag{3.38}$$

The proof of this proposition is straightforward.

C) Subspace of Physical 0-*Charge States*

Assumption C2 that leads to a free photon propagator allows an explicit construction of the physical subspace (of the indefinite metric space for the local gauge under consideration) at least in the zero-charge sector.

We define the submanifold $\mathscr{D}'$ of physical states of the field domain $\mathscr{D}$ as a maximal conformally invariant linear subspace of vectors Φ, Ψ of $\mathscr{D}$ such that

$$\langle\Phi|A_-(x)|\Psi\rangle = 0 = \langle\Phi|\partial_\nu S|\Psi\rangle = \langle\Phi|q|\Psi\rangle\,. \tag{3.39}$$

Remark 3.12. Equation (3.39) excludes from $\mathscr{D}'$ vectors of the form $A_-|0\rangle$ or $\partial_\nu S|0\rangle$ that would have violated the non-negativity of the inner product. The positivity violation for $A_-|0\rangle$ is manifest from (3.16); for $\partial_\nu S$ it is derived from the expression for the 2-point Wightman function (cf. (2.32))

$$\begin{aligned} \langle\partial_\mu S(x)\partial_\nu S(0)\rangle_0 &= \frac{r_{\mu\nu}(x)}{8\pi^2 v(x^2+i0x^0)} \\ &= \frac{2\pi}{v}\int\theta(p^0)p_\mu p_\nu\delta'(-p^2)e^{ipx}d_4p\,. \end{aligned} \tag{3.40}$$

Let $\mathscr{D}'_0$ be the part of the 0-charge sector of $\mathscr{D}'$, generated by the action to the vacuum of polynomials of local fields smeared by test functions with compact support.

Proposition 3.13. *$\mathscr{D}'_0$ is generated by polynomials of $F_{\mu\nu}$, $\partial_\nu A_-$ and $\Box^2 S$, the states $\partial_\nu A_-|0>$, $\partial_\nu F^{\mu\nu}|0>$ and $\Box^2 S|0>$ having zero norm.*

The proposition follows from the factorization postulate (2.31) and from the properties of (free) 2-point functions of the encountered (Bose) fields. (We note that Eq. (3.39) does not contradict our assumption (2.18a) because of the vanishing of the current matrix elements in $\mathscr{D}'$, which follows from (3.18) – see [S4].)

4. Renormalized Schwinger-Dyson Equations (S-D Eqs)

A) Equations Involving 2- and 3-Point Functions

We follow here the standard scheme of defining the renormalized S-D Eqs – see [S6, B5] (as well as [Z2]).

Let F be an arbitrary polynomial in the fields A_N, $\partial_\mu S$, ψ, $\bar\psi$ and their derivatives: Eqs. (2.15), (2.18b), (2.20), (2.21) imply the following set of renormalized S-D Eqs for Schwinger functions (since we only deal with Euclidean Green functions in this section, we skip the subscript E in the expectation values):

$$\partial\!\!\!/\langle\psi(x)F\rangle-\left\langle\frac{\partial F}{\delta\bar\psi}\right\rangle=ie\langle N(h\!\!\!/(x)\psi(x))F\rangle\,,\qquad h_\mu(x)=A_\mu(x)-A_-(x)\partial_\mu S(x)\,,\tag{4.1a}$$

$$(\partial^\mu\partial_\nu-\Box\delta^\mu_\nu)\langle A^\nu(x)F\rangle-\frac{1}{2}\partial^\mu\Box\langle A_-(x)F\rangle-\left\langle\frac{\delta F}{\delta A_\mu(x)}\right\rangle=\langle J^\mu(x)F\rangle\,,\tag{4.1b}$$

$$\frac{1}{2}\Box\partial_\mu\langle A^\mu(x)F\rangle-\frac{\beta}{4}\Box^2\langle A_-(x)F\rangle-\left\langle\frac{\delta F}{\delta A_-(x)}\right\rangle=-\langle N(\partial^\mu S(x)J_\mu(x))F\rangle\,,\tag{4.1c}$$

$$\upsilon\Box^2\langle S(x)F\rangle-\left\langle\frac{\delta F}{\delta S(x)}\right\rangle=\langle\partial_\mu N(A_-(x)J^\mu(x))F\rangle\tag{4.1d}$$

where

$$J_\mu(x)=ieN(\bar\psi(x)\gamma_\mu\psi(x))\,.\tag{4.1e}$$

The normal products $N(\dots)$ will be determined in a non-perturbative way as finite parts of Wilson expansions inserted in conformal invariant Schwinger functions. The remaining freedom will be restricted by the requirement of gauge invariance.

Proceeding to the study of the Dirac equation (4.1a) we shall use (and extend) some of the techniques of [F4, P1] (devised for the study of the 2-dimensional Thirring model). We define the smeared composite field $\psi_1(x,\varepsilon)$ by

$$\langle\psi_1(x,\varepsilon)F\rangle=ieZ_2(\varepsilon)\int d\Omega_z\langle h\!\!\!/(x+\varepsilon z)\psi(x)F\rangle-(Z_2(\varepsilon)-1)\partial\!\!\!/\langle\psi(x)F\rangle\,,\tag{4.2}$$

where $d\Omega_z$ is the normalized measure on S^3:

$$\int d\Omega_z f(z) = \frac{1}{\pi^2} \int \delta(z^2-1) f(z) d^4 z \,. \tag{4.2b}$$

The field $h_\mu = qA_\mu - A_- \partial_\mu S$ transforms "homogeneously" [i.e. under an elementary representation of Spin(5, 1)]. The factor q can be dropped [as in (4.1a)] when multiplied by a pair of conjugate spinor fields [see (3.34)].

The renormalization constant $Z_1(\varepsilon)$ does not appear, since we have set $Z_1 = Z_2$ as required by gauge invariance.

Setting $F = \bar{\psi}(y)$ we can rewrite Eq. (4.1a) in the form

$$\not{\partial}_x \langle \psi(x)\bar{\psi}(y)\rangle - \delta(x-y) = \lim_{\varepsilon\to 0} \langle \psi_1(x,\varepsilon)\bar{\psi}(y)\rangle \,. \tag{4.3}$$

(The δ-function, which fits the free field limit for $e \to 0$, is cancelled by a similar term in the right-hand side (RHS) in the case when Ψ acquires anomalous dimension.) The 3-point function $\langle h_\mu(x+\varepsilon z)\psi(x)\bar{\psi}(y)\rangle$ [which appears in (4.3) if we insert ψ_1 from (4.2)] is obtained from (3.37) by deleting the logarithmic term and replacing $\frac{1+B}{2}$ and A by two arbitrary constants g_1 and g_2. It is clear [taking into account the composite character of the second term in h_μ (4.1a)] that one should actually deal with the 4-point function

$$S_\mu^{(4)} = \langle A_-(x_1)\partial_\mu S(x_2)\psi(x_3)\bar{\psi}(x_4)\rangle$$

in the limit $x_{12} \to 0$.

The small ε behaviour of the spherically smeared 3-point function[2]

$$G(x,p;\varepsilon) = \int d^4 y \int d\Omega_z \langle \not{h}(x+\varepsilon z)\psi(x)\bar{\psi}(y)\rangle e^{-ipy} \tag{4.4}$$

involved in (4.3) is analysed in the Appendix. The final expression (A.7) contains two terms of the type appearing in the left-hand side of (4.3), where we insert the conformal invariant 2-point function (1.3). In this way we obtain the following relations between the anomalous dimension $\lambda = d - \frac{3}{2}$, the ψ-field renormalization constant Z_2 and the free parameters g_1 and g_2 in the conformal invariant expression for G:

$$\lambda \left(\equiv d - \frac{3}{2} \right) = \frac{e^2}{(4\pi)^2} (g_1 + g_2) \,, \tag{4.5}$$

$$Z_2(\varepsilon,\lambda) = z(\lambda) \left(\frac{\varepsilon^2}{4} \right)^\lambda \quad (Z_2(\varepsilon,0) = z(0) = 1 \quad \text{for} \quad C_\psi(0) = 1, e^2 = 0) \,, \tag{4.6a}$$

where the product of $z(\lambda)$ with the normalization constant C_ψ of the electron propagator (1.3) is

$$C_\psi z(\lambda) = \frac{1}{2 + (2g_1 - g_2)\dfrac{e^2}{(4\pi)^2}} \frac{\Gamma(3-\lambda)}{\Gamma(1+\lambda)} \,. \tag{4.6b}$$

---

2 Note that a similar analysis of the Dirac equation in the framework of [F3, P2] is hindered by the non-local character of the photon 3-point function

Remark 4.1. The Euclidean counterpart of the more complicated action (2.26) leads to a somewhat modified system of equations. In particular, a term $\langle a\not{\partial} S(x_1)\psi(x_1)\bar{\psi}(x_2)\rangle$ is added to the RHS of the Dirac equation (4.3). The new system includes one more equation, obtained by varying with respect to a:

$$(8\pi^2)\langle aF\rangle + \left\langle\frac{\delta F}{\delta a}\right\rangle + \int d^4x\left(\frac{1}{2}\Box\partial_\mu\langle A^\mu(x)F\rangle - \langle J_+(x)F\rangle\right) = 0\,.$$

(We have used that $\int\Box^2\langle A_-(x)F\rangle d^4x = 0$ according to (4.1a) and the Ward identities.) Combined with (4.1b) this equation leads to the relations

$$\langle aa\rangle = \langle aA_-\rangle = \langle A_-A_-\rangle\left(=-\frac{1}{8\pi^2}\right),$$

$$\langle aF\rangle = 0 \quad \text{if} \quad \left\langle\frac{\delta F}{\delta A_-}\right\rangle = 0 = \left\langle\frac{\delta F}{\delta a}\right\rangle.$$

As a result, the Dirac equation (4.3) remains unchanged.

Turning next to the definition of the current, we set

$$\begin{aligned}&\langle J^\mu(x,\varepsilon)F\rangle\\ &= ieZ_2(\varepsilon,\lambda)\int d\Omega_z\langle\bar{\psi}(x+\varepsilon z)\gamma^\mu\psi(x)F\rangle\\ &\quad-\langle\{\mathscr{D}^\mu_\nu(\varepsilon,\partial)A^\nu(x)+\mathscr{D}^\mu_1(\varepsilon,\partial)A_-(x)+\mathscr{D}^\mu_2(\varepsilon,\partial)S(x)\}F\rangle\,, \qquad (4.7a)\end{aligned}$$

where

$$\begin{aligned}\mathscr{D}^\mu_\nu &= \frac{a_1}{\varepsilon^2}\,\delta^\mu_\nu + a_2\partial^\mu\partial_\nu + a_3\Box\delta^\mu_\nu\,,\\ \mathscr{D}^\mu_i &= \frac{b_i}{\varepsilon^2}\,\partial^\mu + C_i\Box\partial^\mu\,, \qquad i=1,2\,.\end{aligned} \qquad (4.7b)$$

According to our assumption C2 about the conformal transformation properties of the photon propagator and Proposition 3.5, the charge renormalization constant Z_3 is finite and can, therefore, be set equal to 1. We should like to give meaning to the current $J^\mu(x)$ in the Maxwell equation (4.1b) as a limit for $\varepsilon\to 0$ of (4.7a). Taking the divergence of (4.1b) for $F = F(y) = A_\nu$, A_- or $\partial_\nu S$ and using current conservation, we find the free field relation

$$\frac{1}{2}\Box^2_x\langle A_-(x)F(y)\rangle + \frac{\partial}{\partial x^\mu}\left\langle\frac{\delta F}{\delta A_\mu(x)}\right\rangle = 0\,,$$

which is satisfied by the conformal invariant 2-point functions (3.14). It follows that momentum space 2-point functions satisfy

$$\begin{aligned}&\int e^{-ipy}\left(\delta^\mu_\sigma - \frac{p^\mu p_\sigma}{p^2}\right)\left\{p^2\langle A^\sigma(x)F(y)\rangle - \frac{\delta F(y)}{\delta A_\sigma(x)}\right\}d^4y\\ &= \lim_{\varepsilon\to 0}\int e^{-ipy}\langle J^\mu(x,\varepsilon)F(y)\rangle d^4y\,. \qquad (4.8)\end{aligned}$$

The left-hand side of this equation vanishes identically for our choice of the photon propagator. Before studying the implications of this fact to the RHS we shall make a few comments.

Remark 4.2. Finite subtractions appear, as discussed in [B5], in order to maintain the properties of the current functions dictated by gauge invariance which is destroyed by our regularization procedure. The type of the operators $\mathscr{D}_i$ is restricted by the asymptotic behaviour of the first term in (4.7a). More subtractions are possible in principle, e.g. cubic terms of the type $\mathscr{D}_{\mu\varrho\gamma\sigma} A^\varrho(x-\varepsilon z)A^\gamma(x)A^\sigma(x+\varepsilon z)$ [B5]. We neglect them since in our case they factorize in the average and do not give new structures.

Remark 4.3. The definition (4.7–4.8) of the current applies only if F is a polynomial of basic fields. If F includes composite operators, additional (contact) terms are needed in general (see, e.g., [Z2]). (This remark applies of course to all composite operators in the model.) We shall give only an example of functions involving two current operators:

$$\begin{aligned}&\langle J_\mu(x_1)J_\nu(x_2)F\rangle\\ &\quad=\lim_{\varepsilon\to 0}\{\langle J_\mu(x_1,\varepsilon)J_\nu(x_2)F\rangle-\int d\Omega_z[D^1_{\mu\nu\gamma\varrho}(\varepsilon,x_{12})\\ &\quad\cdot\langle A^\gamma(x_2+\varepsilon z)A^\varrho(x_2)F\rangle+D^2_{\mu\nu\gamma}(\varepsilon,x_{12})\langle A^\gamma_{(x_2)}F\rangle\\ &\quad+D^3_{\mu\nu}(\varepsilon,x_{12})\langle F\rangle]\}\,.\end{aligned}\tag{4.9a}$$

In particular, for $F=\mathbb{1}$, we have

$$D^1_{\mu\nu\gamma\varrho}=\gamma_1\eta_{\mu\gamma}\eta_{\nu\varrho}\delta(x_2)\,;\qquad D^3_{\mu\nu}=[\gamma_2\partial_\mu\partial_\nu+\gamma_3(\partial_\mu\partial_\nu-\Box\eta_{\mu\nu})]\delta(x_{12})\,.\tag{4.9b}$$

The asymptotic behaviour of the 3-point functions involved in the RHS of (4.8) as well as the contribution of the subtraction terms are evaluated in the Appendix. We write here only the expression for the 3-point function of A_μ [see (3.37) with $\alpha=0$]:

$$\begin{aligned}&ieZ_2(\varepsilon,\lambda)\int d\Omega_z e^{-ipy}\langle\bar\psi(x+\varepsilon z)\gamma_\mu\psi(x)A_\nu(y)\rangle d^4y\\ &=-\frac{e^2C_\psi z(\lambda)}{(2\pi)^2}\frac{\Gamma(\lambda+2)}{\Gamma(1-\lambda)}e^{-ipx}\\ &\quad\cdot\left\{\frac{2}{\varepsilon^2p^2}\eta_{\mu\nu}+2\frac{(2B+A)}{\varepsilon^2}\frac{p_\mu p_\nu}{p^4}+\left(\frac{1-A}{4}-\frac{1+B}{4}\right)\frac{p_\mu p_\nu}{p^2}\right.\\ &\quad\left.+\alpha_0\eta_{\mu\nu}+\left(\frac{1+B}{8}+\frac{1}{12}\right)\eta_{\mu\nu}\ln\frac{p^2\varepsilon^2}{4}\right\}+0(\varepsilon^2)\,.\end{aligned}\tag{4.10}$$

Here α_0 is a numerical constant [see (A.11)]. All terms in (4.10) except the last one are compensated by the contribution of the subtractions in (4.8) leading to some relations for the arbitrary constants in (4.7). These constants are completely determined when the other two equations in (4.8) (for $F=A_-$ or $\partial_\mu S$) are also taken into account. There is no way, however, to cancel the finite transverse last term in (4.10) unless

$$e^2z(\lambda)\frac{\Gamma(2+\lambda)}{\Gamma(1-\lambda)}C_\psi\left(\frac{1+B}{2}+\frac{1}{3}\right)=\frac{e^2}{2}\frac{(1-\lambda^2)(2-\lambda)}{2+(2g_1-g_2)\dfrac{e^2}{(4\pi)^2}}\left(B+\frac{5}{3}\right)=0\,,\tag{4.11}$$

i.e. either $B=-\frac{5}{3}$ or $e(1-\lambda^2)(2-\lambda)=0$.

Remark 4.4. Concerning the first possibility ($B=-\frac{5}{3}$), we notice that it is not available in the framework of [F3, P2, K1]; as compared to this framework we have a second independent transverse structure in the 3-point function (3.37). The above result can also be obtained from the Maxwell equation for $\langle A_\mu J_\nu \rangle$ [see (A.12)]. The finite contribution of the current 3-point function (3.22) to the Maxwell equation has been evaluated earlier in [M4] where a similar condition for B has been put forward. [It is, however, incorrectly assumed in [M4] that the conformal invariant current Green's function $\langle TJ_\mu(x)J_\nu(0)\rangle_0$ is proportional to the associate homogeneous distribution (3.21).] A number of the above formulas (as well as those of the Appendix) have been obtained in an inconclusive earlier attempt [D5] to tackle conformal QED[3]).

Regarding the current as an elementary conformal field we easily analyse also the last two equations (4.1c, d) for the various 2-point functions $\langle \partial_\mu S(x)F(y)\rangle$ and $\langle A_-(x)F(y)\rangle$ and prove (exploiting current conservation) their consistency with the (free) expressions (2.32), (3.14). The implications of the composite structure of J^μ (4.7) is further discussed in Sect. 4B.

B) Equations Involving 4-Point Functions

The Maxwell equations (4.1b) with $F\to F\psi\bar{\psi}$ and $F=A_\nu, \partial_\mu S$ or A_- together with the corresponding Ward identities imply the equations:

$$\frac{1}{2}\square^2\langle A_-(x_4)F(x_3)\psi(x_1)\bar{\psi}(x_2)\rangle+\left\langle \frac{\delta F(x_3)}{\delta A_\mu(x_4)}\psi(x_1)\bar{\psi}(x_2)\right\rangle$$
$$=ie[\delta(x_{14})-\delta(x_{24})]\langle F(x_3)\psi(x_1)\bar{\psi}(x_2)\rangle\,. \tag{4.12}$$

In agreement with Proposition 3.2 we have the following

Proposition 4.1. *Let F be either A_μ, $\partial_\mu S$ or A_-. Then*

$$\mathcal{S}_F=\langle A_-(x_4)F(x_3)\psi(x_1)\bar{\psi}(x_2)\rangle$$
$$=\left(C+\frac{ie}{8\pi^2}\ln\frac{x_{42}^2}{x_{41}^2}\right)\langle F(x_3)\psi(x_1)\bar{\psi}(x_2)\rangle$$
$$+\langle\psi(x_1)\bar{\psi}(x_2)\rangle\langle A_-(x_4)F(x_3)\rangle\,. \tag{4.13}$$

The last term is present only for $F=A_\mu$ or A_-.

One can check that no finite part of $\mathcal{S}_F$ survives when $x_2\to x_1$, $x_4\to x_1$. Hence there is no contribution to the R.H.S. of Eq. (4.1d) for $F=A_\mu$, $\partial_\mu S$ or A_-. This shows that the composite character of the current does not affect the conclusion of the preceding section.

We let $x_4\to x_3$ in (4.13) for $F=\partial_\mu S$ thus getting

$$\langle A_-(x_3)\partial_\mu S(x_3)\psi(x_1)\bar{\psi}(x_2)\rangle=\left(C+\frac{ie}{8\pi^2}\ln\frac{x_{32}^2}{x_{31}^2}\right)\langle\partial_\mu S(x_3)\psi(x_1)\bar{\psi}(x_2)\rangle\,. \tag{4.14}$$

If we choose $\alpha=0$ in the expression (3.36) for $\langle\partial_\mu S\psi\bar{\psi}\rangle$, (4.14) has the general form of the 3-point function (3.37) with $1+B=0=A$. Hence, as discussed in Sect. 4A, it

3 The model in [D5] reproduces the standard 3-point current function and an expression for $\langle A_\mu\psi\bar{\psi}\rangle$ which differs from (3.37) by purely longitudinal terms. However, the normal product in the Dirac equation requires an unconventional subtraction, which is avoided here

just cancels down the contribution of the inhomogeneous term in (3.37) in the Dirac equation. Then we finally have in (4.4)

$$g_1 = \frac{1+B}{2}, \qquad g_2 = A. \tag{4.15}$$

Combining (4.15) with (A.13), we obtain

$$\lambda\left(=d-\frac{3}{2}\right) = \frac{e^2}{(4\pi)^2}(\beta + \text{const}). \tag{4.16}$$

The fact that (4.14) does not contain the standard transverse part of (3.37) leads us to the following important conclusion.

Proposition 4.2. *The Maxwell equation* (4.1b) *with* $F = A_-(x_1)\partial_\mu S(x_3)$ *(or* $F = (A_\mu - h_\mu)(x_3)$*) is not satisfied unless*

$$eC_\psi z(\lambda)\frac{\Gamma(\lambda+2)}{\Gamma(1-\lambda)} = \frac{e(2-\lambda)(1-\lambda^2)}{2+(2g_1-g_2)\dfrac{e^2}{(4\pi)^2}} = 0. \tag{4.17}$$

The proof concerning the 3-point function (4.14) repeats the derivation of (4.10), (4.11).

Thus, the only way to cancel the $\log p^2$-term is to have either a trivial theory (with $e\psi = 0$ – see the discussion in the Introduction) or a special gauge in which the anomalous dimension has the (negative) value $\lambda \to -1$.

Appendix A. Small Distance Expansions of 3-Point Functions

Our objective in this appendix is to evaluate the leading terms for $\varepsilon \to 0$ of smeared 3-point functions of the type of $G(x, p; \varepsilon)$ (4.4). Using the identity

$$\frac{\not{x}}{x^2} = \lim_{\delta \to 0}\left(\frac{1}{2\delta}\not{\partial} x^{2\delta}\right)$$

we can write the general conformal invariant expression of the integrand in (4.4) as the limit of

$$\begin{aligned}&\langle \not{h}(x_3)\psi(x_1)\bar{\psi}(x_2)\rangle_\delta \\ &= \frac{ieC_\psi}{(2\pi)^2}\left\{-\frac{1}{2\delta}\left(g_1(2d-1)\frac{\not{x}_{13}}{x_{13}^2}\not{\partial}_3 + \frac{1}{2}g_2\not{\partial}_3\not{\partial}_1\right)\Delta_{d-1/2}(x_{12})\right. \\ &\quad \left.\cdot\left(\frac{x_{23}^2}{2}\right)^\delta + \frac{1}{2}g_2\frac{\not{x}_{13}}{x_{13}^2}S_d(x_{12})\right\} \quad \text{for} \quad \delta \to 0.\end{aligned} \tag{A.1}$$

Setting $x_1 = x$, $x_3 = x + \varepsilon z$, and taking the Fourier transform in $x_2 = y$, we obtain

$$\begin{aligned}G(x,p;\varepsilon) = {} & \frac{ieC_\psi e^{-ipx}}{(2\pi)^2 2^{1-\lambda}}\frac{\Gamma(\lambda+1)}{\Gamma(1-\lambda)}\lim_{\delta\to 0}\frac{1}{\delta}\int d\Omega_z\left\{g_1(\lambda+1)\frac{\not{z}}{\varepsilon^2 z^2}\not{\partial}_z\right. \\ & \left. + g_2\frac{1}{2\varepsilon^2}(i\varepsilon\not{\partial}_z\not{p} + \Box_z)\right\} I(-1-\delta, \lambda; \varepsilon z, p),\end{aligned} \tag{A.2}$$

where $\lambda = d - \frac{3}{2}$ is the anomalous dimension and

$$I(\kappa, \lambda; z, p)$$

$$= \frac{e^{ipx}}{(2\pi)^2} \int \left(\frac{2}{(x+z-y)^2}\right)^{\kappa+1} \left(\frac{2}{(x-y)^2}\right)^{\lambda+1} e^{-ipy} d^4y$$

$$= \frac{e}{\Gamma(\kappa+1)\,\Gamma(\lambda+1)} \left(\frac{p^2}{z^2}\right)^{\frac{1}{2}(\kappa+\lambda)} \int_0^1 du$$

$$\cdot \left(\frac{u}{1-u}\right)^{\frac{\kappa-\lambda}{2}} e^{-iupz\varepsilon} K_{\kappa+\lambda}(\sqrt{u(1-u)p^2z^2}) \tag{A.3}$$

(cf. Eqs. (F.1, 2) of Appendix F to [D4]); here K_ν is the modified Bessel function:

$$2K_\nu(\zeta) = \Gamma(\nu)\Gamma(1-\nu)\,(I_{-\nu}(\zeta) - I_\nu(\zeta)) \tag{A.4a}$$

$$I_\nu(\zeta) = \left(\frac{\zeta}{2}\right)^\nu \sum_{n=0}^{\infty} \frac{1}{n!\,\Gamma(n+\nu+1)} \left(\frac{1}{4}\zeta^2\right)^n. \tag{A.4b}$$

Retaining up to quadratic terms in the (p, z)-expansion of the integrand, we can write

$$I(\kappa, \lambda; z, p)$$

$$= \frac{\Gamma(\kappa+\lambda)\Gamma(1-\kappa)\Gamma(1-\lambda)}{\Gamma(1+\kappa)\Gamma(1+\lambda)\Gamma(2-\kappa-\lambda)} \left(\frac{2}{z^2}\right)^{\kappa+\lambda} \left\{1 - \frac{1-\lambda}{2-\kappa-\lambda} ipz\right.$$

$$- \frac{(1-\lambda)(2-\lambda)}{(2-\kappa-\lambda)(3-\kappa-\lambda)} \frac{(pz)^2}{2} + \frac{(1-\kappa)(1-\lambda)}{(1-\kappa-\lambda)(2-\kappa-\lambda)(3-\kappa-\lambda)}$$

$$\left.\cdot \frac{p^2z^2}{4} + O((pz)^3)\right\} + \frac{\Gamma(-\kappa-\lambda)}{\Gamma(\kappa+\lambda+2)} \left(\frac{p^2}{2}\right)^{\kappa+\lambda} \left\{1 - \frac{\kappa+1}{\kappa+\lambda+2} ipz\right.$$

$$\left. - \frac{(1+\kappa)(2+\kappa)}{(\kappa+\lambda+2)(\kappa+\lambda+3)} \frac{(pz)^2}{2} + \frac{(1+\kappa)(1+\lambda)\frac{p^2z^2}{4}}{(\kappa+\lambda+1)(\kappa+\lambda+3)} + O((pz)^3)\right\}. \tag{A.5}$$

Inserting (A.5) (with $z \to \varepsilon z$, $\kappa = -1 - \delta \to -1$) into (A.2) and using the identity

$$\int d\Omega_z \not{z} pz = \tfrac{1}{4}\not{p}, \tag{A.6}$$

we find

$$G(x, p; \varepsilon) = \frac{ieC_\psi e^{-ipx}}{(4\pi)^2 \lambda}$$

$$\cdot \left\{\left(\frac{4}{\varepsilon^2}\right)^\lambda \frac{\Gamma(2+\lambda)}{\Gamma(3-\lambda)} \left(2g_1 + \frac{2-\lambda}{1+\lambda} g_2\right) - p^{2\lambda}(g_1 + g_2)\right\}. \tag{A.7}$$

In order to relate the anomalous dimension λ to the charge e and to the parameters g_1 and g_2 and to evaluate the renormalization constant $Z_2(\varepsilon, \lambda)$, we insert (A.7) into the Fourier transform of the Dirac equation (4.3) (with $\psi_1(x, \varepsilon)$ given by (4.2).) The result

$$1=C_{\psi}\lim_{\varepsilon\to 0} Z_2(\varepsilon,\lambda)\left\{p^{2\lambda}\left[1-\frac{g_1+g_2}{\lambda}\frac{e^2}{(4\pi)^2}\right]\right.$$
$$\left.+\left(\frac{4}{\varepsilon^2}\right)^{\lambda}\frac{\Gamma(1+\lambda)}{\Gamma(3-\lambda)}\left(2g_1-g_2+2\frac{g_1+g_2}{\lambda}\right)\frac{e^2}{(4\pi)^2}\right\} \quad \text{(A.8)}$$

yields (4.5) and (4.6).

Proceeding now to the implication of the Maxwell equation (4.8) for $F(y)=A_\nu(y)$ or $A_-(y)$, we first replace the log-terms in the 3-point functions (3.23), (3.37) by

$$\log\frac{x_{23}^2}{x_{13}^2}=\lim_{\delta\to 0}\frac{1}{\delta}\left\{\left(\frac{2}{x_{13}^2}\right)^{\delta}-\left(\frac{2}{x_{23}^2}\right)^{\delta}\right\},$$

and use the general integration formula

$$\int d\Omega_z z_\mu z_\nu (zq)(zp)^{2\kappa-1}$$
$$=\frac{1}{2^{\kappa+2}}\left(\frac{p^2}{2}\right)^{\kappa-2}\left[\frac{p^2}{2}(pq\eta_{\mu\nu}+p_\mu q_\nu+p_\nu q_\mu)+(\kappa-1)p_\mu p_\nu(pq)\right]\frac{(2\kappa)!}{\kappa!(\kappa+2)!}. \quad \text{(A.9)}$$

In studying the small ε behaviour of

$$ieZ_2(\varepsilon,\lambda)\int d\Omega_z\int d^4y e^{-ipy}\langle\bar{\psi}(x+\varepsilon z)\gamma^\mu\psi(x)A_\nu(y)\rangle$$

and representing the Fourier transform of $\frac{\not{x}_{13}}{x_{13}^2}\gamma^\mu\frac{\not{x}_{23}}{x_{23}^2}$ in (3.37) as

$$(\not{x}-i\not{\partial}_p)\gamma^\mu(\not{x}+\varepsilon\not{z}-i\not{\partial}_p)\int e^{-ipy}\frac{d^4y}{(x-y)^2(x+\varepsilon z-y)^2},$$

we need the expansion

$$K_0(\zeta)=-I_0(\zeta)\ln\frac{\zeta}{2}+\sum_{k=0}^{\infty}\frac{\psi(k+1)}{(k!)^2}\left(\frac{\zeta^2}{4}\right)^k,\qquad \psi(k)=\frac{d}{dk}\ln\Gamma(\kappa),$$

which gives

$$I(\kappa,-\kappa;z,p)$$
$$=-\sum_{s=0}^{\infty}\frac{(-ipz)^s}{s!}\sum_{n=0}^{\infty}\left(\frac{p^2z^2}{4}\right)^n\frac{(1+\kappa)_{n+s}(1-\kappa)_n}{(n!)^2(2n+s+1)!}$$
$$\cdot\left[\ln\frac{p^2z^2}{2}+\psi(n+\kappa+s+1)+\psi(n+1-\kappa)\right.$$
$$\left.-\psi(2n+s+2)-2\psi(n+1)\right], \quad \text{(A.10)}$$

where $(a)_n=\frac{\Gamma(a+n)}{\Gamma(a)}$. A straightforward (though rather lengthy) computation leads to (4.10) with

$$\alpha_C=\frac{1+B}{8}\left(-\frac{1}{2}-\psi(1)\right)-\frac{1}{6}\left(\psi(1)+\frac{5}{3}\right). \quad \text{(A.11)}$$

We can use the general formula (A.5) instead of (A.10) to obtain (4.10). Then we need an intermediate analytic regularization $(1/x_{13}^2 x_{23}^2) \to (1/x_{13}^2 x_{23}^2)^{1+\delta}$. The operator $\lim_{\delta \to 0} \frac{d}{d\delta} \delta$ which does not change the initial expressions then recovers a finite part which differs from (4.10) by the value of α_0. The subtraction term involving the constant a_3 in (4.7) accounts for this arbitrariness. Similarly, we find using (A.10)

$$ie \int d\Omega_z \int d^4 y e^{-ipy} \langle \bar{\psi}(x_1 + \varepsilon z) \gamma_\mu \psi(x_1) J_\gamma(y) \rangle Z_2(\varepsilon, \lambda)$$
$$= -\frac{e^2 C_\psi z(\lambda)}{(2\pi)^2} \frac{\Gamma(2+\lambda)}{\Gamma(1-\lambda)} e^{-ip \cdot x_1} \left\{ \left[\alpha_0' + \left(\frac{1+B}{8} + \frac{1}{12} \right) \ln \frac{p^2 \varepsilon^2}{4} \right] \right.$$
$$\left. \cdot (\eta_{\mu\nu} p^2 - p_\mu p_\nu) + \eta_{\mu\nu} \frac{2}{\varepsilon^2} - \frac{p_\mu p_\nu}{6} \right\} + O(\varepsilon^2), \tag{A.12}$$

where α_0' is a numerical constant. Comparing (A.12) with (4.9a, b), one determines the constants γ_i and reproduces the relation (4.11). (The finite transverse part of (A.12) coincides with the second order expression for the photon self-energy.)

Evaluating the contribution of the subtraction terms in (4.9) for $F = A_\mu$, A_-. $\partial_\mu S$, we get the following values for the subtraction constants:

$$a_1 = -\frac{2e^2 C_\psi z(\lambda) \Gamma(2+\lambda)}{(2\pi)^2 \Gamma(1-\lambda)}, \quad a_2 = \frac{a_1}{2}\left(\frac{1}{6} + \alpha_0\right), \quad a_3 = \frac{a_1}{2} \alpha_0,$$
$$b_1 = -\frac{e^2 C_\psi z(\lambda)}{(2\pi)^2} (A - \beta + 1 + 2B) \frac{\Gamma(2+\lambda)}{\Gamma(1-\lambda)},$$
$$c_1 = -\frac{e^2 C_\psi z(\lambda)}{12(2\pi)^2} \left(A - \beta + 1 - \frac{3}{2}(1+B) - 6\alpha_0(1-\beta)\right) \frac{\Gamma(2+\lambda)}{\Gamma(1-\lambda)}, \tag{A.13}$$
$$b_2 = -\frac{e^2 C_\psi z(\lambda) \Gamma(2+\lambda) 2v (2\pi)^2}{(2\pi)^2 \Gamma(1-\lambda) ie} = -6C_2.$$

In dealing with 4-point functions, we only work in x-space and use in addition to (A.6) the relation (A.9).

Acknowledgements. We would like to thank Dr. V. Dobrev and Dr. A. Ganchev for making available their unpublished calculations and Dr. E. Nissimov for valuable discussions.

Note added in proof. Equation (2.21) [or (4.1d)] is consistent with the commutation relations (3.32), (3.35). To amend the situation we should add a trilinear term, $\frac{q}{4} A_-^2 \Box^2 S$, replacing the Lagrangian (2.19), (2.14) by

$$L = \frac{1}{2} F^{\mu\nu} \left(\frac{1}{2} F_{\mu\nu} - \partial_\mu A_\nu + \partial_\nu A_\mu \right) - \frac{q}{2} A_- \Box \partial A + \frac{\beta}{8} (\Box A_-)^2 + \frac{v}{2} (\Box S)^2$$
$$- \bar{\psi} [\not{\partial} - ie(\not{A} - A_- \not{\partial} S)] \psi + \frac{q}{4} A_-^2 \Box^2 S. \tag{1}$$

The Maxwell and the Dirac equations (2.7) and (2.20) remain unchanged while the gauge fixing equations for ∂A and S assume the form

$$\frac{q}{2}\Box\partial A+\partial_\mu S J^\mu=\frac{q}{2}A_-\Box^2 S;\qquad \upsilon\Box^2 S+\partial_\mu A_- J^\mu+\frac{q}{4}\Box^2 A_-^2=0. \tag{2}$$

The new set of equations has the virtue of being invariant with respect to the 1-parameter family of field transformations

$$A_-\to A_-+\lambda,\qquad A_\mu\to A_\mu+\lambda\partial_\mu S. \tag{3}$$

It is straightforward to verify [using (3.7) and (3.23a)] that

$$\begin{aligned}\frac{1}{4}\Box_3^2\langle\psi(1)\bar\psi(2)A_-^2(3)\rangle_0&=\frac{e^2}{8\pi^4}\left[\frac{x_{12}^2}{(x_{13}^2+iox_{13}^0)(x_{23}^2+iox_{23}^0)}\right]^2\langle\psi(1)\bar\psi(2)\rangle_0\\&=-\langle\psi(1)\bar\psi(2)J^\mu(3)\partial_\mu A_-(3)\rangle_0\end{aligned} \tag{4}$$

so that [according to (2)] $\upsilon\Box_3^2\langle\psi(1)\bar\psi(2)S(3)\rangle_0=0$ in agreement with (3.36).

The argument of Sect. 2C is not directly applicable to this modified Lagrangian, however a calculation based on a power series expansion in the new trilinear term indicates that it does not affect gauge invariant matrix elements.

References

[A1] Adler, S.: Short distance behaviour of quantum electrodynamics and an eigenvalue condition for α, Phys. Rev. **D5**, 3021–3047 (1972), Erratum, ibid. **D7**, 1948 (1973); Euclidean quantum electrodynamics on the 5-dimensional unit hypersphere. Phys. Rev. **D6**, 3445–3461 (1972), Erratum, ibid. **D7**, 3821 (1973)

[A2] Adler, S., Bardeen, W.: Quantum electrodynamics without photon self-energy parts: an application of the Callan-Symanzik scaling equations. Phys. Rev. **D4**, 3045–3054 (1971) Erratum, ibid. **D6**, 734 (1972)

[A3] Aizenman, M.: Geometric analysis of φ^4 fields and Ising models. Commun. Math. Phys. **86**, 1–40 (1982)

[B1] Baker, M., Johnson, K: Applications of conformal symmetry in quantum electrodynamics. Physica **96A**, 120–130 (1979)

[B2] Bateman, H.: The transformation of the electrodynamical equations. Proc. London Math. Soc. **8**, 223–264 (1910)

[B3] Bayen, F., Flato, M.: (a) Remarks on conformal space. J. Math. Phys. **17**, 1112–1114 (1976); (b) Conformal covariant QED revisited (to be published)

[B4] Binegar, B., Fronsdal, C., Heidenreich, W.: Conformal QED. J. Math. Phys. **24**, 2828–2846 (1983)

[B5] Brandt, R.: Gauge invariance in quantum electrodynamics. Ann. Phys. (N.Y.) **52**, 122–176 (1969)

[C1] Cunningham, E.: The principle of relativity in electrodynamics and an extension thereof. Proc. London Math. Soc. **8**, 77–98 (1909)

[D1] Dell'Antonio, G.F.: On dilation invariance and the Wilson expansion. Nuovo Cim. **12A**, 756–762 (1972)

[D2] Dirac, P.A.M.: Wave equations in conformal space. Ann. Math. **37**, 429–442 (1936)

[D3] Dobrev, V.K., Petkova, V.B., Petrova, S.G., Todorov, I.T.: Dynamical derivation of vacuum operator product expansion in Euclidean conformal quantum field theory. Phys. Rev. **D13**, 887–912 (1976)

[D4] Dobrev, V.K., Mack, G., Petkova, V.B., Petrova, S.G., Todorov, I.T.: Harmonic analysis. In: Lecture Notes in Physics, Vol. 63. Berlin, Heidelberg, New York: Springer 1977

[D5] Dobrev, V.K., Ganchev, A., Petkova, V.B.: Schwinger-Dyson equations in conformal quantum electrodynamics (1981) (unpublished)

[E1] Englert, F.: Self-consistent approach to the fine-structure constant. Nuovo Cim. **16A**, 557–572 (1973)

[F1] Ferrara, S., Gatto, R., Grillo, A.: Logarithmic scaling and spontaneous breaking. Phys. Lett. **42B**, 264–266 (1972)

[F2] Ferrari, R., Picasso, L., Strocchi, F.: Some remarks on local operators in quantum electrodynamics. Commun. Math. Phys. **35**, 25–38 (1974)

[F3] Fradkin, E.S., Kozhevnikov, A.A., Palchik, M.Ya., Pomeransky, A.A.: Conformal invariant formulation of quantum electrodynamics. Commun. Math. Phys. **91**, 529–541 (1983)

[F4] Fradkin, E.S., Palchik, M.Ya.: Recent developments in conformal invariant QFT. Phys. Rep. **44C**, 249–349 (1978)

[F5] Fröhlich, J.: On the triviality of $\lambda\varphi_d^4$ theories and the approach to the critical point in $d \underset{(-)}{\geq} 4$ dimensions. Nucl. Phys. **B200**, 281–296 (1982)

[F6] Fry, M.: Connection between the electron propagator and the Baker-Johnson function in conformal invariant quantum electrodynamics. Nuovo Cim. **31A**, 129–150 (1976)

[F7] Furlan, P., Petkova, V.B., Sotkov, G.M., Todorov, I.T.: Conformal quantum electrodynamics with a 5-potential, ISAS preprint 52/83/EP, Trieste; Conformal quantum electrodynamics and non-decomposable representations (to be submitted to Rivista Nuovo Cim.) ISAS preprint 51/84/EP, Trieste

[J1] Johnson, K., Willey, R., Baker, M.: Vacuum polarization in quantum electrodynamics. Phys. Rev. **163**, 1699–1715 (1967)

[K1] Krasnikov, N.V.: Is finite renormalization of electric charge possible? Phys. Lett. **126B**, 483–484 (1983)

[M1] Mack, G., Salam, Abdus: Finite component field representations of the conformal group. Ann. Phys. (N.Y.) **53**, 174–202 (1969)

[M2] Mack, G., Symanzik, K.: Currents, stress tensor and generalized unitarity in conformal invariant quantum field theory. Commun. Math. Phys. **27**, 247–281 (1972)

[M3] Mack, G., Todorov, I.T.: Conformal invariant Green functions without ultraviolet divergences. Phys. Rev. **D8**, 1787 (1973)

[M4] Mannheim, P.: Structure of the vertex function in finite quantum electrodynamics. Preprint LBL-3610, Berkeley (1974)

[M5] Matsubara, Y., Suzuki, T., Yotsuyanagi, I.: On a possible inconsistency of the φ^4 and the Yukawa theories in four dimensions, Kanazawa Univ. Preprint DPKU-8303 (1983)

[M6] Matsubara, Y., Suzuki, T., Yotsuyanagi, I.: Can $(\mathrm{QED})_4$ be a trivial theory? Kanazawa Univ. Preprint DPKU-8304 (1983)

[M7] Mayer, D.H.: Vector and tensor fields in conformal space. J. Math. Phys. **16**, 884–893 (1975)

[P1] Palchik, M.Ya.: On the operator product $j_\mu(x)\psi(x)$ in the Thirring model. Short Communications of the Lebedev Physics Institute, No. 4, 19–24 (1980) (in Russian)

[P2] Palchik, M.Ya.: A new approach to the conformal invariance problem in quantum electrodynamics. J. Phys. **A16**, 1523–1527 (1983)

[P3] Petkova, V.B., Sotkov, G.M.: Exceptional representations of the conformal group and applications, I and II. Bulg. J. Phys. **10**, 144–172, 256–277 (1983); The six-point families of exceptional representations of the conformal group, ISAS preprint 15/83/E.P. Trieste; Lett. Math. Phys. **8**, 217–226 (1984)

[P4] Popov, V.N.: Path Integrals in Quantum Field Theory and Quantum Statistical Mechanics (Atomizdat, Moscow 1976; in Russian). See, in particular, Chap. 3

[S1] Salam, Abdus, Strathdee, J.: Nonlinear realizations. II. Conformal symmetry. Phys. Rev. **184**, 1760–1768 (1969)

[S2] Schroer, B.: A necessary and sufficient condition for the softness of the trace of the energy-momentum tensor. Lett. Nuovo Cim. **2**, 867–872 (1972)

[S3] Sotkov, G., Stoyanov, D.: On the conformal invariance in quantum electrodynamics. J. Phys. **A13**, 2807–2816 (1980); Conformal quantization of electrodynamics. J. Phys. **A16**, 2817–2826 (1983)
[S4] Strocchi, F.: Extension of Jost and Schroer's theorem to quantum electrodynamics. Phys. Rev. **D6**, 1193–1195 (1972)
[S5] Strocchi, F., Wightman, A.S.: Proof of the charge superselection rule in local relativistic quantum field theory. J. Math. Phys. **15**, 2198–2224 (1974); Erratum, ibid. **17**, 1930–1931 (1976)
[S6] Symanzik, K.: Green's functions method and renormalization of renormalizable quantum field theory. In: Lectures in High Energy Physics, Jaksič, B. (ed.). Zagreb 1961. New York: Gordon and Breach, 1965; Euclidean quantum field theory. In: Coral Gables Conference on Fundamental Interactions at High Energy, Gudehus, T. et al. (eds.). New York: Gordon and Breach, 1969, pp. 19–31
[T1] Todorov, I.T., Mintchev, M.C., Petkova, V.B.: Conformal Invariance in Quantum Field Theory. Scuola Normale Superiore, Pisa 1978
[T2] Todorov, I.T.: Local field representations of the conformal group and their applications. ZiF-Bielefeld preprint (1984)
[Z1] Zaikov, R.P.: On conformal invariance in gauge theories; quantum electrodynamics. JINR preprint E2-83-28 Dubna (1983)
[Z2] Zavialov, O.I.: Renormalized Feynman Diagrams. Moscow: Nauka, 1979; in Russian

Communicated by G. Mack

Received July 11, 1984; in revised form August 28, 1984

Longitudinal Jet Cross Sections in Order α_s^2

B. Lampe and G. Kramer

II. Institut für Theoretische Physik der Universität Hamburg, Luruper Chaussee 149, D-2000 Hamburg 50, Federal Republic of Germany

Dedicated to the memory of Kurt Symanzik

Abstract. We calculate the cross section for $e^+e^- \to 3$ jets for longitudinally polarized virtual photons up to order α_s^2 in the quark-gluon coupling.

1. Introduction

The cross sections for the production of 2, 3, and 4 jets in e^+e^- annihilation as measured at PETRA and PEP have given us useful information about the quark-gluon dynamics as described by QCD [1]. So, for example, the 3-jet cross section has been used in various ways for determining the quark-gluon coupling constant in the perturbative region [2].

Considering also the orientation of the jets with respect to the direction of the incoming beam the 3-jet cross section depends in general on three independent cross sections σ_U, σ_L, σ_T, and σ_I, where U, L, T, and I label the polarization of the ingoing virtual photon [3]. These polarization dependent cross sections which fully determine the jet angular correlations with respect to the incoming electron beam in $e^+e^- \to q\bar{q}g$ are known up to order α_s [3]. So far only the angular averaged 3-jet cross section $\sigma_{U+L} = \sigma_U + \sigma_L$ has been measured. The statistical accuracy of the e^+e^--data is not yet sufficient for determining also σ_L, σ_T, and σ_I. We expect such measurements in the near future with higher statistics data coming from PETRA and PEP. These cross sections σ_L, σ_T, and σ_I are useful to test the spin structure of the $e^+e^- \to q\bar{q}g$ matrix element, to measure the spin of the gluon or to obtain independent measurements of the quark-gluon coupling constant. For the latter it is essential to know these cross sections at least up to order α_s^2 since the coupling can uniquely be defined through renormalization only in higher order. The cross section σ_{U+L} has already been calculated up to this order [4]. These calculations have shown that it is possible to define the 3-jet cross section σ_{U+L} in such a way that it is infrared finite up to order α_s^2 [5].

In an earlier paper [6] we have calculated the virtual α_s^2 contributions to the longitudinal cross section σ_L for $e^+e^- \to q\bar{q}g$. They are infrared singular. In this paper we describe the calculation of the real α_s^2 contributions to the 3-jet cross section σ_L in order to get the infrared finite cross section which can be compared to experiment.

In Sect. 2 we collect the Feynman diagrams which contribute up to order α_s^2 and give the basic formula from which σ_L is calculated. The method for calculating the real contributions is described in Sect. 3. The results are infrared divergent as are the virtual contributions which are taken over from [6]. The sum of these two contributions is finite. It gives us the final formulas for the longitudinal part of $e^+e^- \to q\bar{q}g$ which are collected in Sect. 4. To point out some universal features with respect to the polarization of the virtual photon we include also the formulas for σ_{U+L} calculated earlier [4].

2. Higher Order Corrections to $e^+e^- \to q\bar{q}g$

The diagrams contributing to

$$e^+(p_+)+e^-(p_-)\to q(p_1)+\bar{q}(p_2)+g(p_3) \tag{2.1}$$

(the symbols in brackets denote the momenta of the particles in the initial and final state respectively) in order α_s^2 are shown in Fig. 2a (the virtual corrections). In Fig. 2b are the diagrams for $e^+e^- \to q\bar{q}gg$ and $e^+e^- \to q\bar{q}q\bar{q}$ which contribute to 3 jets in the case when two of the four partons (qg, $\bar{q}g$, gg etc.) are degenerate. Then the cross section for $e^+e^- \to 3$ jets in order α_s^2 is obtained by multiplying the diagrams in Fig. 2a with the lowest order diagrams in Fig. 1 and adding the product of the diagrams in Fig. 2b integrated over the 3-jet region of the 4-particle phase space.

The higher order diagrams in Fig. 2 have UV as well as IR divergences. We regulate them by going to $4-2\varepsilon$ spacetime dimensions [7]. After renormalization the sum of virtual and real contributions is finite.

The cross section for the process (2.1) is ($q=p_+ + p_-$)

$$d\sigma = \frac{e^4}{2q^6} L^{\mu\nu} H_{\mu\nu} PS^{(3)}, \tag{2.2}$$

where $PS^{(3)}$ is the 3-particle phase space in $4-2\varepsilon$ dimensions. $L_{\mu\nu}$ is the lepton tensor and $H_{\mu\nu}$ is the hadron tensor, where final spin, colour and flavour together with the quark charge factors are averaged out. Because of current conservation, $q^\mu H_{\mu\nu}=0$, $H_{\mu\nu}$ in general depends on five structure functions. They can be obtained by expanding $L_{\mu\nu}$ into products of polarization vectors of the incoming virtual photon: $\varepsilon_\mu^\pm = 1/\sqrt{2}(0, \mp 1, -i, 0)$ for transverse polarization and $\varepsilon^{(0)}=(0,0,0,1)$ for longitudinal polarization in the system $\mathbf{q}=0$ and then specifying the z-direction in this system. Longitudinal cross sections σ_L are obtained by substituting $\frac{q^2}{3}\varepsilon_\mu^{(0)}\varepsilon_\nu^{(0)}$

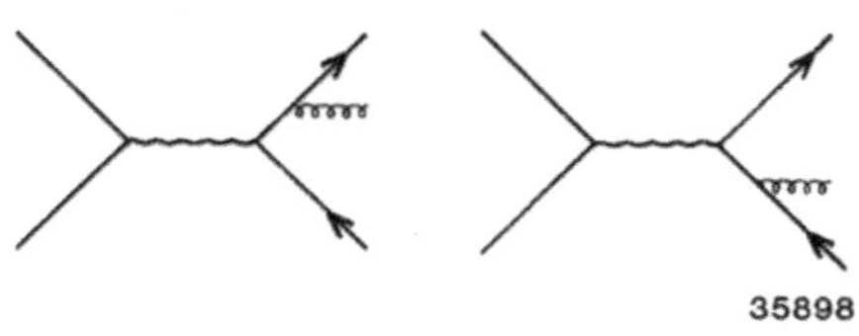

Fig. 1

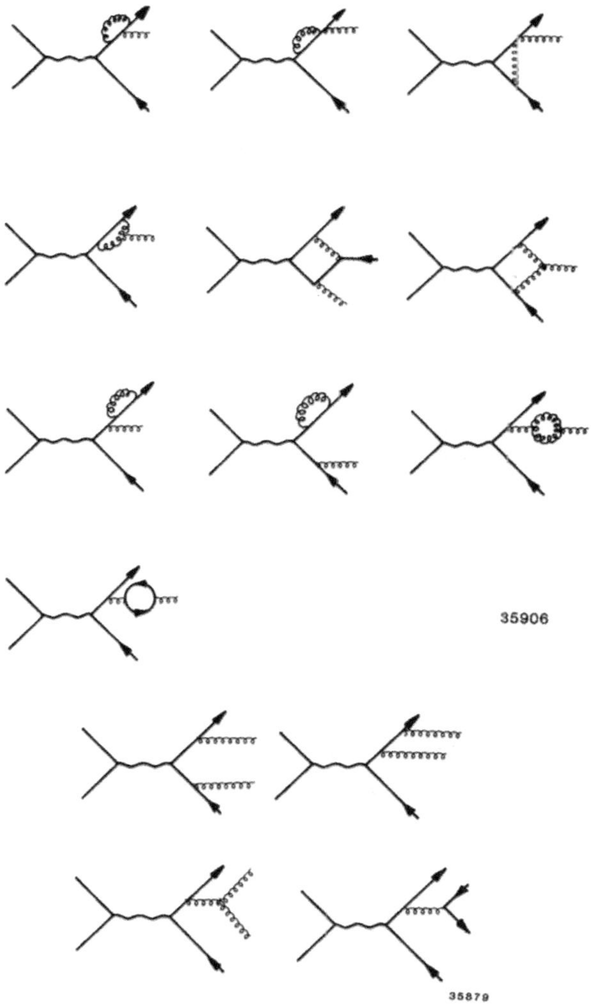

Fig. 2a and b

for $L_{\mu\nu}$ in (2.2). We have calculated $\sigma_L^{(1)}$ with z-axis parallel to $\mathbf{p}_1$ and $\sigma_L^{(3)}$ with z-axis parallel to $\mathbf{p}_3$. $\sigma_L^{(2)}$ with $z \| \mathbf{p}_2$ follows from $\sigma_L^{(1)}$ by interchanging $p_1 \leftrightarrow p_2$. Since $\varepsilon_\mu^{(0)} = p_{i\mu}/p_{i0} - q_\mu/q_0$ $(i=1,3)$ one has to calculate effectively $H_{\mu\nu} p_i^\mu p_i^\nu / p_{i0}^2$ with $i=1,3$. The total cross section $\sigma_{U+L} = \sigma_U + \sigma_L$ is obtained by integrating over angles

$$\int L_{\mu\nu} \frac{d\Omega}{4\pi} = \frac{q^2}{3}(-g_{\mu\nu} + q_\mu q_\nu / q^2), \tag{2.3}$$

where

$$L_{\mu\nu} = p_{+\mu} p_{-\nu} + p_{-\mu} p_{+\nu} - q^2/2\, g_{\mu\nu}. \tag{2.4}$$

3. Virtual and Real Corrections

All traces were calculated with the help of REDUCE [8] in $4-2\varepsilon$ dimensions. The sum of all graphs, either virtual (see Fig. 2a) or real (see Fig. 2b) consists of four distinct classes: (a) planar QED type graphs with (colour) group weight C_F^2, (b) non-planar QED type graphs with group weight $C_F(C_F-N_c/2)$, (c) QCD type graphs involving the three gluon vertex with group weight $C_F N_c$ and (d) graphs having a closed fermion loop with group weight $C_F\frac{N_f}{2}$. Here N_f is the number of flavours and $C_F=\frac{4}{3}$, $N_c=3$ and $T_r=\frac{N_f}{2}$ are the Casimir operators. We used the variables

$$y_{ij}=2p_ip_j/q^2\,, \tag{3.1}$$

$$y_{ijk}=y_{ij}+y_{jk}+y_{ik}\,, \quad i<j<k\,. \tag{3.2}$$

The calculation of the virtual contributions to $e^+e^-\to q\bar{q}g$ using the Gegenbauer expansion method has been described in detail in [6]. The result can be written in the following form

$$\frac{d^2\sigma_A}{dy_{13}dy_{23}}=\sigma^{(2)}\frac{\alpha_s(\mu^2)}{2\pi}C_F\left(\frac{4\pi\mu^2}{q^2}\right)^{\varepsilon}\frac{\theta(1-y_{13}-y_{23}-y_{12})}{\Gamma(2-\varepsilon)} \cdot (y_{13}y_{23}(1-y_{13}-y_{23}))^{-\varepsilon}T_A^{\mathrm{virtual}}. \tag{3.3}$$

Here

$$\sigma^{(2)}=\sigma_0\left(\frac{4\pi\mu^2}{q^2}\right)^{\varepsilon}\frac{\Gamma(2-\varepsilon)}{\Gamma(2-2\varepsilon)}, \qquad \sigma_0=\frac{4\pi\alpha^2}{3q^2}N_c\sum_{i=1}^{N_f}Q_i^2 \tag{3.4}$$

is the cross section for $q\bar{q}$ production in lowest order in $4-2\varepsilon$ dimensions. μ is an arbitrary parameter to define the coupling constant in such a way that it is dimensionless for arbitrary ε. The index A stands for $U+L$, L_1 or L_3. T_A^{virtual} is written as

$$T_A^{\mathrm{virtual}}=\frac{\Gamma(1-\varepsilon)}{\Gamma(1-2\varepsilon)}\left(\frac{4\pi\mu^2}{q^2}\right)^{\varepsilon}\frac{\alpha_s(\mu^2)}{2\pi}B_A(y_{13},y_{23})C_F \cdot\left\{C_F T_A^{C\,\mathrm{virtual}}+\frac{N_c}{2}T_A^{N\,\mathrm{virtual}}+\left(\frac{N_f}{3}-\frac{11}{6}N_c\right)T_A^{ct}\right\}. \tag{3.5}$$

We understand T_A to contain the renormalization counterterm

$$T_A^{ct}=1/\varepsilon \tag{3.6}$$

in the $\overline{MS}$ renormalization scheme [9]. B_A stands for the 3-jet distributions to order α_s:

$$B_{U+L}(y_{13},y_{23})=(1-\varepsilon)\left(\frac{y_{13}}{y_{23}}+\frac{y_{23}}{y_{13}}\right)+\frac{2y_{12}}{y_{13}y_{23}}-2\varepsilon\,,$$

$$B_{L1}(y_{13},y_{23})=2(1-\varepsilon)\frac{y_{12}}{(1-y_{23})^2}\,,$$

$$B_{L3}(y_{13},y_{23})=4\frac{y_{12}}{(1-y_{12})^2}\,. \tag{3.7}$$

The explicit formulas for $T_A^{C,\text{virtual}}$ and $T_A^{N,\text{virtual}}$ are rather lengthy and will not be given here. They can be looked up in our earlier work [6] and can be reconstructed from the final formulas for the cross section in (4.1). The singular pieces of $T_A^{C,\text{virtual}}$ and $T_A^{N,\text{virtual}}$ are rather simple and they are identical for $A=U+L$, L_1 and L_3:

$$T_{A,\text{div}}^{C,\text{virtual}} = -\frac{2}{\varepsilon^2} + \frac{1}{\varepsilon}(2\ln y_{12} - 3),$$

$$T_{N,\text{div}}^{N,\text{virtual}} = -\frac{2}{\varepsilon^2} + \frac{2}{\varepsilon}\ln(y_{13}y_{23}/y_{12}). \tag{3.8}$$

The fact that the divergent parts of $T_A^{C,\text{virtual}}$ and $T_A^{N,\text{virtual}}$ are equal for $A=U+L$, L_1, L_3, i.e. for the three different polarizations of the photon, is quite remarkable. It would be interesting to know whether this is true in general, i.e. also for the other cross sections σ_T and σ_I which have not been calculated yet. To point out this universal behaviour of the divergent parts is one of the reasons why we have included the total cross section σ_{U+L} as one of the cases in (3.3) and (3.5), although it was reported before.

We now come to the main part of this paper, the calculation of the real contributions to σ_{L_1} and σ_{L_3}. In order to produce physical cross sections one must add to the virtual contributions, given above, the infrared and collinear divergent parts of the four-parton cross sections $d\sigma_{q\bar{q}gg}$ and $d\sigma_{q\bar{q}q\bar{q}}$ integrated over a small phase space region which includes the singularities. To define this region one needs parameters to parametrize the boundary of the region inside which two partons are considered to be irresolvable, i.e. being one jet. For this we use an invariant mass constraint. The partons i and j are said to be irresolvable if the invariant mass squared $(p_i+p_j)^2 = 2p_ip_j$ is less than yq^2 (all partons being massless). This boundary has been used also for defining $d\sigma_{U+L}(y)$ in [4, second reference]. It has the advantage that it depends only on one parameter which makes it easier to study the dependence of $d\sigma_{3\text{-jet}}(y)$ on the boundary. We calculate $d\sigma_{3\text{-jet}}(y)$ only up to terms of order 1, i.e. we neglect term of order $y\ln y$ and higher. Therefore our results are applicable only for small $y \lesssim 0.05$ [1, 4]. To obtain these dominant terms we need only the most singular parts of the 4-parton matrix elements $\sim y_{ij}^{-1}$ which we take from our earlier work [4]. Of course, these matrix elements and the 4-parton phase space must be calculated in $4-2\varepsilon$ dimensions.

For the case that we integrate over the singularity y_{13}^{-1}, i.e. that parton 1 and 3 form one jet, the 4-parton phase space is [5]

$$\begin{aligned} PS^{(4)} = {} & \frac{q^4}{2^{11}\pi^3}\,\frac{S}{\Gamma(1-\varepsilon)\Gamma(2-2\varepsilon)}\left(\frac{4\pi}{q^2}\right)^{3\varepsilon} \\ & \cdot \int dy_{123}dy_{134}dy_{13}(y_{123}y_{134}-y_{13})^{-\varepsilon} \\ & \cdot (y_{13}+1-y_{123}-y_{134})^{-\varepsilon}y_{13}^{-\varepsilon}\theta(y_{13}) \\ & \cdot \theta(y_{123}y_{134}-y_{13})\theta(y_{13}+1-y_{123}-y_{134}) \\ & \cdot \int_0^1 dv\,v^{-\varepsilon}(1-v)^{-\varepsilon}\frac{1}{N_{\theta'}}\int_0^\pi d\theta'\sin^{-2\varepsilon}\theta'. \end{aligned} \tag{3.9}$$

S is a statistical factor, $N_{\theta'}$ is the normalization for the θ' integration. For defining the angles we take the system $z\|\mathbf{p}_2$ and $\mathbf{p}_1+\mathbf{p}_3=0$. Then θ and θ' are the polar angles of $\mathbf{p}_1$ in this system and

$$v=\tfrac{1}{2}(1-\cos\theta)=1-\frac{y_{23}}{y_{123}}. \tag{3.10}$$

It appears that this coordinate system is very convenient for doing the rather complicated integrations of the singular terms over the 3-jet region bounded by y. Since the longitudinal cross sections $d\sigma_{L_1}$ and $d\sigma_{L_3}$ are obtained from invariant quantities $H_{\mu\nu}p_1^\mu p_1^\nu$ and $H_{\mu\nu}p_3^\mu p_3^\nu$ respectively the choice of the coordinate system does not matter.

When calculating the contributions to $H_{\mu\nu}p_1^\mu p_1^\nu$ and $H_{\mu\nu}p_3^\mu p_3^\nu$ one must be careful concerning the correct definition of the labels "1" and "3". Integrating over the singularity $1/y_{13}$ means that $p_3\to 0$ or $p_3\| p_1$. Then the corresponding 3-jet momenta are $p_{\rm I}=p_1+p_3$, $p_{\rm II}=p_2$, and $p_{\rm III}=p_4$, so that actually one has to calculate $H_{\mu\nu}p_{\rm I}^\mu p_{\rm I}^\nu=H_{\mu\nu}(p_1+p_3)^\mu(p_1+p_3)=v$ etc. Only with this choice of 3-jet momenta for defining $d\sigma_{L_1}$ and $d\sigma_{L_3}$ one can achieve consistency, i.e. the cancellation of the $1/\varepsilon^2$ and $1/\varepsilon$ singularities. In the end all contributions are proportional to $B_A(y_{\rm I\,III}, y_{\rm II\,III})$ as they should.

Before presenting the results we make a few remarks concerning the region of integration in (3.7). For the graphs of type (c) with a gluon loop and the graphs of type (d) the singularity is $1/y_{34}$ instead of $1/y_{13}$. Then one uses the analogue of the phase space formula (3.9) in the system $\mathbf{p}_3+\mathbf{p}_4=0$ and integrates over $0\leqq y_{34}\leqq y$. Outside the 2-jet region $y\ll y_{134}$, y_{234}, and the integrations are rather simple. For the other graphs the v-integration is somewhat complicated, because one has singularities $1/y_{13}$ and $1/y_{23}$ which must be considered together. To avoid double counting in the infrared region, $p_3\to 0$, one integrates the $1/y_{13}$ terms over the region

$$2\int_0^1 dv-\int_{1-y/y_{\rm III}}^1 dv\,, \tag{3.11}$$

which then accounts correctly for the $1/y_{23}$ contribution.

After having done the θ'-integration we obtain the following expression for the integrand,

$$\begin{aligned}\text{Integrand}=B_A(y_{\rm I\,III}, y_{\rm II\,III})\bigg\{&C_F\frac{1}{y_{13}}\left(\frac{2v}{1-v}+(1-\varepsilon)(1-v)\right)\\ &+N_c\frac{2}{y_{34}}\left(\frac{v}{1-v}+v(1-v)\right)+\frac{N_f}{2}\frac{1}{y_{34}}\frac{v^2+(1-v)^2-\varepsilon}{1-\varepsilon}\bigg\}.\end{aligned} \tag{3.12}$$

A stands for $U+L$, L_1 and L_3. $y_{\rm I\,III}$ and $y_{\rm II\,III}$ are the 3-jet variables. Equation (3.12) must be integrated over (3.11) and $0\leqq y_{13}\leqq y$ in the case of the $1/y_{13}$ singular part and over the analogous regions in the case of the $1/y_{34}$ singular piece. It is remarkable that the curly bracket in (3.12) is universal for $A=U+L$, L_1 and L_3. For graphs of type (a) and (b), which do not depend on θ' this is obvious. In this case one encounters in the original expression within the traces always the following product

$$\ldots\bar{v}(p_1)\frac{\not\varepsilon(\not p_1+\not p_3)}{(p_1+p_3)^2}\ldots \tag{3.13}$$

where ε is the polarization vector of the gluon which has momentum p_3. We decompose $p_3=p_3^{\|}+p_3^{\perp}$ with respect to p_1, i.e. $p_3^{\|}(p_3^{\perp})$ is in the direction of (perpendicular to) p_1. $p_3^{\perp}$ is small since parton 1 and 3 are assumed to form one jet and can be neglected. Then the expression (3.13) goes over into

$$\ldots\bar{v}(p_1)\frac{\not{\varepsilon}(\not{p}_1+\not{p}_3^{\|})}{2p_1p_3}\ldots=\frac{2\varepsilon(p_1+p_3^{\|})}{2p_1p_3}\ldots\bar{v}(p_1)\ldots, \tag{3.14}$$

so that (3.13) is reduced to a contribution equal to the lowest order diagrams in Fig. 1 multiplied by a universal function which is independent of the virtual photon polarization. These arguments do not go through for graphs of type (c) and (d). Indeed there one finds that the θ'-dependence of the integrand depends on A. Nevertheless we find the universal factor in (3.12) after integrating over θ'!

The final result is written again in the form (3.3)–(3.5) with "virtual" to be replaced by "real" and T_A^{ct} replaced by $T_A^{N_fN_c\,\mathrm{real}}$. Instead of the indices I, II, III to label the 3-jet variables we use again 1, 2, 3. Then we obtain

$$\begin{aligned}
T_A^{C\,\mathrm{real}}&=\frac{2}{\varepsilon^2}-\frac{1}{\varepsilon}(2\ln y_{12}-3)+7-4\zeta_2-2\ln^2 y\\
&\quad-3\ln y+4\ln y\ln y_{12}-\ln^2 y_{12},\\
T_A^{N\,\mathrm{real}}&=\frac{2}{\varepsilon^2}-\frac{2}{\varepsilon}\ln(y_{13}y_{23}/y_{12})+\tfrac{4}{3}-4\zeta_2-\ln^2 y_{13}-\ln^2 y_{23}\\
&\quad+\ln^2 y_{12}-2\ln^2 y+4\ln y\ln y_{13}+4\ln y\ln y_{23}-4\ln y\ln y_{12},\\
T_A^{N_fN_c\,\mathrm{real}}&=-\frac{1}{\varepsilon}-\frac{5}{3}+\ln y.
\end{aligned} \tag{3.15}$$

This completes the calculation of the real contributions with the invariant mass constraint y.

4. Cross Sections

Adding (3.15) to (3.5) it is evident from (3.6) and (3.8) that the infrared and collinear singularities proportional to ε^{-2} and ε^{-1} cancel. The remainder in the sum gives the 3-jet cross sections for the three polarizations $A=U+L$, L_1, and L_3. They are written in the following form in which the lowest order contribution is also included:

$$\begin{aligned}
\frac{d^2\sigma_A}{dy_{13}dy_{23}}&=\sigma_0\frac{\alpha_s(q^2)}{2\pi}C_F\theta(1-y_{13}-y_{23})B_A(y_{13},y_{23})\\
&\quad\cdot\left[1+\frac{\alpha_s(q^2)}{2\pi}\left(C_FT_A^C+\frac{N_c}{2}T_A^N\right)+\left(\frac{N_f}{3}-\frac{11}{6}N_C\right)T_A^{N_fN_c}\right],
\end{aligned} \tag{4.1}$$

where

$$\begin{aligned}
T_A^C&=-1+\tfrac{2}{3}\pi^2-2\ln^2 y_{12}-2\ln^2 y+4\ln y\ln y_{12}-3\ln y+h_A,\\
T_A^N&=\tfrac{4}{3}+\tfrac{1}{3}\pi^2+2\ln^2 y_{12}-2\ln^2 y_{13}-2\ln^2 y_{23}-2r(y_{13},y_{23})\\
&\quad-2\ln^2 y+4\ln y\ln(y_{13}y_{23}/y_{12})+g_A,\\
T_A^{N_fN_c}&=-\tfrac{5}{3}+\ln y,
\end{aligned} \tag{4.2}$$

and the functions h_A and g_A appearing in (4.2) have the following form:

$$h_{U+L}=\hat{g}(y_{13},y_{23})/B_{U+L}(y_{13},y_{23}),$$

$$h_{L_1}=-\frac{2}{3}\pi^2+2\ln y_{23}+\frac{2y_{13}}{y_{13}+y_{23}}+2\ln y_{12}\left(\frac{y_{13}}{y_{13}+y_{23}}+\frac{y_{12}y_{23}}{(y_{13}+y_{23})^2}\right)$$

$$-2r(y_{12}, y_{13})+2\frac{y_{12}}{y_{13}}r(y_{12}, y_{23}),$$

$$h_{L_3}=-\frac{2}{3}\pi^2+2\ln y_{12}+\frac{y_{13}+y_{23}}{2y_{12}}-\frac{y_{13}^2}{2y_{12}(y_{12}+y_{13})}-\frac{y_{23}^2}{2y_{12}(y_{12}+y_{23})}$$

$$+\frac{1}{2}\ln y_{13}\left(\frac{y_{13}y_{23}}{(y_{12}+y_{23})^2}+\frac{2y_{23}}{y_{12}+y_{23}}\right)+\frac{1}{2}\ln y_{23}\left(\frac{y_{13}y_{23}}{(y_{12}+y_{13})^2}+\frac{2y_{13}}{y_{12}+y_{13}}\right)$$

$$-r(y_{12}, y_{13})-r(y_{12}, y_{23}),$$

$$g_{U+L}=(\tilde{g}(y_{13}, y_{23})-\hat{g}(y_{13}, y_{23}))/B_{U+L}(y_{13}, y_{23}), \tag{4.3}$$

$$g_{L_1}=-\frac{1}{3}\pi^2+1-\frac{2y_{13}}{y_{13}+y_{23}}-2\ln y_{12}\left(\frac{y_{13}}{y_{13}+y_{23}}+\frac{y_{12}y_{13}}{(y_{13}+y_{23})^2}\right)$$

$$+2r(y_{12}, y_{13})-2\frac{y_{12}}{y_{13}}r(y_{12}, y_{23}),$$

$$g_{L_3}=-\frac{1}{3}\pi^2-2\ln y_{12}+r(y_{12}, y_{13})+r(y_{12}, y_{23})-\frac{y_{13}}{y_{12}+y_{23}}\ln y_{13}$$

$$-\frac{y_{23}}{y_{12}+y_{13}}\ln y_{23},$$

with $r(x, y)$, $\hat{g}(y_{13}, y_{23})$ and $\tilde{g}(y_{13}, y_{23})$ defined as in [6]:

$$r(x, y)=\ln x\ln y-\ln x\ln(1-x)-\ln y\ln(1-y)-L_2(x)-L_2(y)+\tfrac{1}{6}\pi^2,$$

$$\hat{g}(y_{13}, y_{23})=4\ln y_{12}\left[\frac{2y_{12}}{y_{13}+y_{23}}+\frac{y_{12}^2}{(y_{13}+y_{23})^2}\right]$$

$$+\ln y_{13}\left[\frac{4y_{12}}{y_{12}+y_{23}}+\frac{2y_{13}}{y_{12}+y_{23}}-\frac{y_{13}y_{23}}{(y_{12}+y_{23})^2}\right]$$

$$+\ln y_{23}\left[\frac{4y_{12}}{y_{12}+y_{13}}+\frac{2y_{23}}{y_{12}+y_{13}}-\frac{y_{13}y_{23}}{(y_{12}+y_{23})^2}\right]$$

$$-2r(y_{12}, y_{13})\frac{y_{12}^2+(y_{12}+y_{23})^2}{y_{13}y_{23}}$$

$$-2r(y_{12}, y_{23})\frac{y_{12}^2+(y_{12}+y_{13})^2}{y_{13}y_{23}}+\frac{y_{12}}{y_{12}+y_{13}}$$

$$+\frac{y_{12}}{y_{12}+y_{23}}+\frac{4y_{12}}{y_{13}+y_{23}}-\frac{y_{12}}{y_{13}}-\frac{y_{12}}{y_{23}}-\frac{y_{13}}{y_{23}}-\frac{y_{23}}{y_{13}}, \tag{4.4}$$

$$\tilde{g}(y_{13}, y_{23})=\ln y_{13}\left[4\frac{y_{12}+y_{13}}{y_{12}+y_{23}}-\frac{y_{13}y_{23}}{(y_{12}+y_{23})^2}\right]$$

$$+\ln y_{23}\left[4\frac{y_{12}+y_{23}}{y_{12}+y_{13}}-\frac{y_{13}y_{23}}{(y_{12}+y_{13})^2}\right]$$

$$+\frac{y_{12}}{y_{12}+y_{13}}+\frac{y_{12}}{y_{12}+y_{23}}+\frac{y_{12}}{y_{13}}+\frac{y_{12}}{y_{23}}+\frac{y_{13}}{y_{23}}+\frac{y_{23}}{y_{13}}.$$

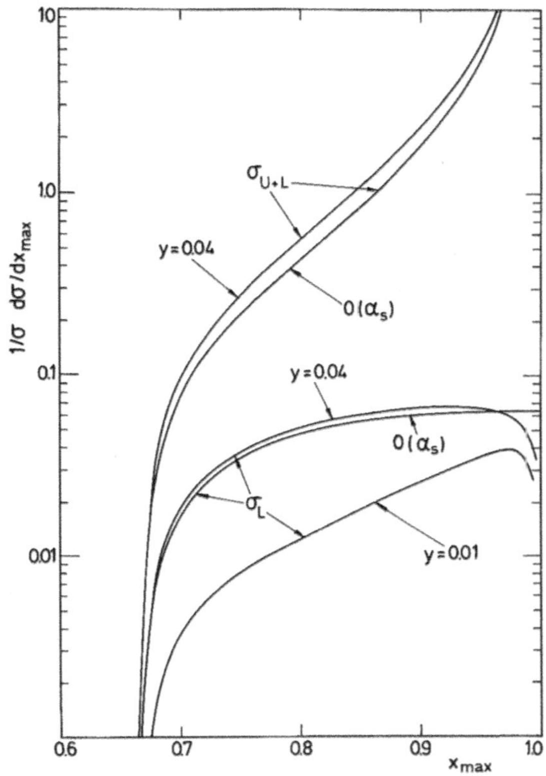

Fig. 3. Three-jet cross sections $1/\sigma d\sigma_{U+L}dx_{\max}$ and $1/\sigma d\sigma_L/dx_{\max}$ for $y=0.04$ and $y=0.01$ together with the Born cross section $(O(\alpha_s))$ as a function of $x_{\max}$ for $\alpha_s=0.16$

These formulas contain all results which are known up to now about 3-jet cross sections with invariant mass resolution. The result for $d^2\sigma_{U+L}$ agrees with our earlier calculations [4]. We have included it here since it makes it easier to program all three cross sections simultaneously when used for analysing experimental data.

In order to get an idea about the amount of corrections originating from the higher order terms we have plotted in Fig. 3 $1/\sigma d\sigma_L/dx_{\max}$ as a function of $x_{\max}$ for $y=0.04$ and $y=0.01$ together with the $O(\alpha_s)$ curve. $x_{\max}$ is the 3-parton thrust equal to the maximum of x_1, x_2, and x_3 where $x_i=1-y_{jk}$ (ijk cyclic). We have $N_f=5$ and $\alpha_s=0.16$, which is a realistic value obtained from analysis of experimental distributions of σ_{U+L} with $O(\alpha_s^2)$ corrections included [2]. We see that for $y=0.04$ the $O(\alpha_s^2)$ corrections to the $x_{\max}$ distribution are quite small. In average they amount to an increase of 6% if compared to the $O(\alpha_s)$ curve. The larger deviation above $x_{\max}=0.95$ is in the two-jet region and is caused by the cut-off $y=0.04$. For $y=0.01$ the corrections are very large. But $y=0.01$ lies outside the perturbative region as was already observed in connection with σ_{U+L} [4]. For comparison the $x_{\max}$ distribution $1/\sigma d\sigma_{U+L}/dx_{\max}$ is also shown for $y=0.04$ and in

$O(\alpha_s)$. Here the correction caused by the $O(\alpha_s^2)$ terms are somewhat larger, near 25% in average.

The results given here are valid only as long as terms $O(y)$ can be neglected. So one must choose y small enough for using these formulas. In case one wants results for larger y's where our approximation might be questionable one always can choose first a small y and correct for the difference numerically by adding the 4-jet contribution in the desired y bin with two partons lying in this bin averaged over.

Appendix

In our earlier paper [6] we discovered the following printing errors: In (4.10) the second line reads

$$\frac{1}{\varepsilon}(2\ln y_{12}-3)-8+\frac{2}{3}\pi^2-\ln^2 y_{12}.$$

In (4.19) I_2 and I_3 are:

$$I_2=N_c\left(\ln^2\frac{y}{y_{12}}-\ln^2\frac{y}{y_{13}}-\ln^2\frac{y}{y_{23}}-r(y_{13},y_{23})-\frac{11}{6}\ln y+\frac{67}{18}\right.$$
$$\left.+\frac{1}{6}\pi^2-\frac{y}{y_{12}}\ln\frac{y^2}{y_{12}}+\frac{y}{y_{13}}\ln\frac{y^2}{y_{13}}+\frac{y}{y_{23}}\ln\frac{y^2}{y_{23}}\right),$$
$$I_3=\frac{N_f}{2}\left(\tfrac{2}{3}\ln y-\tfrac{10}{9}\right).$$

References

1. Kramer, G.: Theory of jets in electron-positron annihilation. Springer tracts in modern physics, Vol. 102. Berlin, Heidelberg, New York: Springer 1984
2. Wolf, G.: The determination of α_s in e^+e^- annihilation. Talk presented an the XIVth Intern. Symposium on multiparticle dynamics. Lake Tahoe, CA (June 1983) DESY-report 83-096 (1983)
3. Kramer, G., Schierholz, G., Willrodt, J.: QCD predictions for jets in e^+e^- annihilation: angular correlations and asymmetries. Phys. Lett. **78** B, 249 (1978), Erratum Phys. Lett. **80** B, 433 (1979)
4. Fabricius, K., Kramer, G., Schierholz, G., Schmitt. I.: Higher order perturbative QCD calculation of jet cross sections in e^+e^- annihilation. Z. Physik C **11**, 315 (1982)
   Gutbrod, F., Kramer, G., Schierholz, G.: Higher order QCD corrections to the three-jet cross section: bare versus dressed jets. Z. Physik C **21**, 235 (1984)
5. For the infrared finiteness of the sum of 3- and 4-jet cross section in the limit of zero jet resolution see: Ellis, R.K., Ross, D.A., Terrano, A.E.: The perturbative calculation of jet structure in e^+e^- annihilation. Nucl. Phys. B **178**, 421 (1982)
6. Lampe, B., Kramer, G.: Application of Gegenbauer integration method to e^+e^- annihilation. Physica Scripta **28**, 585 (1983). Some misprints in this paper are corrected in the appendix
7. 't Hooft, G., Veltman, M.: Regularization and renormalization of gauge fields. Nucl. Phys. B **44**, 189 (1972)
8. Hearn, A.C.: REDUCE 2, a system and language for algebraic manipulation. Comm. ACM **14**, 511 (1971)
9. 't Hooft, G., Veltman, M.: Dimensional regularization and the renormalization group. Nucl. Phys. B **61**, 455 (1973)

Communicated by G. Mack

Received August 30, 1984

Convergent Perturbation Expansions for Euclidean Quantum Field Theory

Gerhard Mack and Andreas Pordt

II. Institut für Theoretische Physik der Universität Hamburg, D-2000 Hamburg 50, Federal Republic of Germany

Dedicated to the memory of Kurt Symanzik

Abstract. Mayer perturbation theory is designed to provide computable convergent expansions which permit calculation of Greens functions in Euclidean quantum field theory to arbitrary accuracy, including "nonperturbative" contributions from large field fluctuations. Here we describe the expansions at the example of 3-dimensional $\lambda\phi^4$-theory (in continuous space). They are not essentially more complicated than standard perturbation theory. The nth order term is expressed in terms of $O(n)$-dimensional integrals, and is of order λ^k if $4k-3 \leqq n \leqq 4k$.

1. Introduction

Mayer perturbation theory is designed to provide computable convergent expansions which permit calculation of Greens functions in Euclidean quantum field theory to arbitrary accuracy, including "nonperturbative contributions" from large field fluctuations. Their n^{th} order term is given by $O(n)$-dimensional integrals, as is the case in standard perturbation theory. In principle such expansions have a chance of converging for asymptotically free theories – including superrenormalizable ones – for problems where a small coupling constant is effective. Some models will require a more sophisticated treatment of the large field region, though. In this paper we describe the expansions in their simplest form at the example of weakly coupled massive $\lambda\phi^4$-theory in $\nu=3$ dimensions. Part I of the paper presents the main ideas and constructions. It describes the expansions in an elementary way, and discusses the relation to standard perturbation theory: The sum of all terms in the Mayer expansion up to order $4n$ equals the sum of all renormalized Feynman diagrams up to order n, plus a computable correction of higher order or nonperturbative origin. The raison d'être of Mayer expansions is their computability and convergence. Here we concentrate on computability. We plan to present estimates and discuss convergence properties in a subsequent article (Part II).

Mayer expansions [i.e., iterative solutions of Kirkwood Salsburg or Mayer Montroll equations, in place of Schwinger Dyson equations] were introduced into quantum field theory in Symanzik's celebrated paper on Euclidean quantum field theory [1]. Brydges and Federbush [2] used Mayer expansions to obtain an effective action for a Coulomb gas. The Mayer expansions on the staggered lattice that we are going to present here are equivalent to iterated Mayer expansions using successively lower lattice cutoffs (cf. end of Sect. 2). Iterated Mayer expansions with Pauli-Villars cutoff were developped by Göpfert and Mack in [3, 4], and applied to prove permanent confinement of static quarks in 3-dimensional U(1) lattice gauge theory. The work presented here is a continuation of the program started there, but the paper will be selfcontained. The expansions presented here are examples of what is called a *phase cell cluster expansion* in constructive field theory. Phase cell cluster expansions were introduced by Glimm and Jaffe [5] and further developped by Magnen and Sénéor [6] and Battle and Federbush [7, 8]. Our convergence proof (in preparation) combines elements of the analysis of Magnen and Sénéor, and of [4], with a UV-stability argument (the mass counterterm is dominated by the quartic interaction, cf. below). Elements of the exact renormalization group approach of Kupiainen and Gawedzki [9] are also incorporated[1].

The space on which the theory lives originally [continuum $\mathbb{R}^\nu$ or cubic lattice $(a\mathbb{Z})^\nu$ of lattice spacing a] will be called the *base space*. The generating function $\ln Z(\psi)$ of the connected free-propagator-amputated Euclidean Greens functions depends on a field ψ on base space. Its derivatives at $\psi=0$ give the Greens functions, see Appendix A. In Mayer perturbation theory they are computed from an expansion of the form

$$\ln Z(\psi) = \sum_{n \geqq 0} \int_{x_1} \dots \int_{x_n \in \Lambda} \tilde{\mathcal{M}}(x_1, \dots, x_n | \psi) . \tag{1.1a}$$

The resulting expansions for Greens functions become particularly simple when suitable normalization conditions are imposed, see Eqs. (1.16a), (1.19) below. The leading term $n=0$ is given by

$$\tilde{\mathcal{M}}(\emptyset | \psi) = -\frac{\lambda}{4!} \int_{\text{base}} \psi(z)^4 . \tag{1.1b}$$

The integrations over x_i in the later terms are actually summations over all points x of a *staggered lattice* Λ that is superimposed on the base space (see Fig. 1). The staggered lattice $\Lambda = (\Lambda_0, \Lambda_1, \Lambda_2, \dots)$ consists of a finite or infinite sequence [disjoint union] of lattices Λ_j of decreasing lattice spacing a_j. For instance, $a_j = L^{-j} a_0$, where L is a suitably chosen integer >1 and a_0 is the physical length scale[2]. A point $x \in \Lambda$ is an element of Λ_j for some $j \geqq 0$. In the presence of a lattice cutoff a, there is a finest lattice Λ_N with lattice spacing $a_N = a$. For a continuum theory $N = \infty$ – i.e. the sequence is infinite – and the staggered lattice admits as maps a semigroup of scale transformations by integer powers of L^{-1} in addition to

1 In recent announcements, Feldman, Magnen, Rivasseau, and Sénéor promise an improved version of the expansions of [6], with application to the infrared behavior of massless lattice $\lambda\phi^4$-theory in 4 dimensions [10], and Kupiainen and Gawedzki describe how they handle the same model [11]. This supports the belief that asymptotically free renormalizable models are not necessarily intrinsically more difficult than superrenormalizable ones

2 We believe that $L=2$ is an optimal choice yet

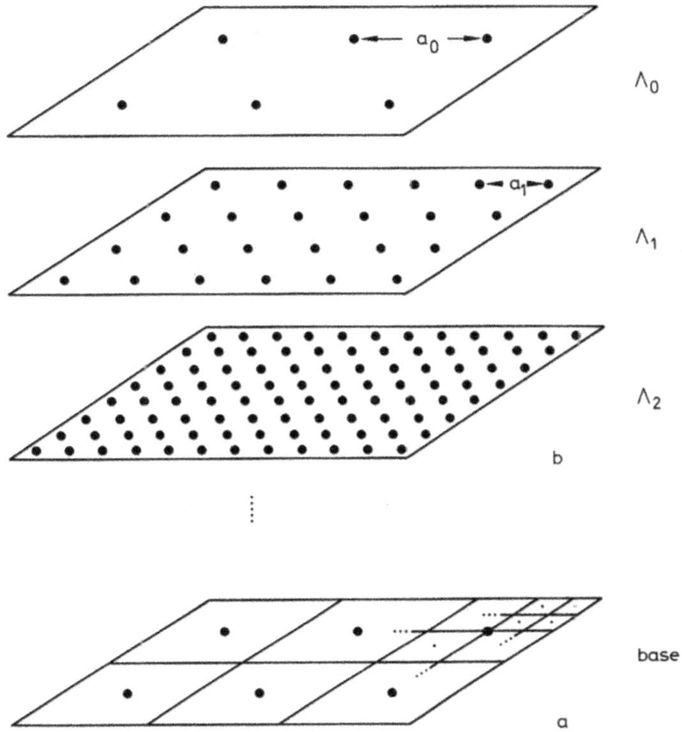

Fig. 1a and b. *The staggered lattice.* A point $x \in \Lambda_j$ of the staggered lattice may be regarded as a cube of side length a_j in base space. The cubes are positioned in such a way that smaller cubes fit into larger ones (i.e. do not intersect their boundaries). **a** base space with lattices of side length $a_j = L^{-j}a_0$ superimposed, $j=0, 1, 2, \ldots$. **b** Every cube in one of the lattices that is superimposed on base space is represented by a dot in this pictorial representation of the staggered lattice $\Lambda = \Lambda_0 + \Lambda_1 + \Lambda_2 + \ldots$. The dots in layer Λ_j represent cubes of side length a_j. (Drawing for 2 dimensions)

the group of translations by integer multiples of a_0. Integration over the staggered lattice includes a summation over length scales j. In ν dimensions

$$\int_{x\in\Lambda} (\ldots) = \sum_{j=0}^{N} a_j^{\nu} \sum_{x\in\Lambda_j} (\ldots). \tag{1.2}$$

The *augmented Mayer amplitudes* $\tilde{\mathcal{M}}(x_1, \ldots, x_n|\psi)$ are finite sums of products of Mayer amplitudes $\mathcal{M}(x_{i_1}, \ldots, x_{i_k}|\psi)$ with $k \leqq n$, multiplied with combinatorial coefficients. The general formula is given in Eqs. (3.6) of Sect. 3. The *Mayer amplitudes* $\mathcal{M}$ are defined for finite subsets $X = \{x_1, \ldots, x_n\} \subset \Lambda$ of the staggered lattice. That is, they are symmetrical in their arguments and defined only when all their arguments are distinct. They are defined by considering auxiliary theories in which the interaction has been switched off, in a suitable way, outside finite subsets X of the staggered lattice. For 3-dimensional $\lambda\phi^4$-theory this is done as follows.

The interaction of the model is $V(\phi)-\delta V(v|\phi)$,

$$V(\phi)=\frac{\lambda}{4!}\int_{\text{base}}\phi(z)^4\,,$$
$$\delta V(v|\phi)=\delta e(v)+\int_{\text{base}}\tfrac{1}{2}\delta m^2(v|z)\phi(z)^2\,. \tag{1.3}$$

It is expedient to consider the counterterms as functions of the free propagator v of the theory. In our model [and, more generally, in asymptotically free theories], the counterterms are known a priori. For instance

$$\delta m^2(v|z)=\tfrac{1}{2}\lambda v(z,z)-\frac{1}{3!}\lambda^2\int_{z'\in\text{base}}v(z,z')^3\,. \tag{1.4}$$

There is a similar expression for the field independent ("vacuum energy") counterterm δe. Alternatively, δe may be fixed by imposing normalization conditions. One may also speed up UV-convergence by making *oversubstractions*, i.e. include wave function and coupling constant renormalization counterterms, and higher order terms in (1.4), if one wishes to. When the propagator v is translation invariant, then δm^2 is independent of z. Both counterterms vanish when $v=0$.

One splits the field ϕ into pieces φ^j that are constant on cubes of side length a_j. The points $x\in\Lambda_j$ of the staggered lattice may be identified with cubes in the base space of side length a_j. As a result, the pieces φ^j specify a field on the staggered lattice, viz. $\varphi(x)=\varphi^j(z)$ for $z\in x$, $x\in\Lambda_j$. The decomposition reads

$$\phi(z)=\int_{x\in\Lambda}\mathscr{A}(z,x)\varphi(x)\equiv\mathscr{A}\varphi(z)\quad\text{for}\quad z\in\text{base}\,. \tag{1.5}$$

The kernels $\mathscr{A}$ are determined by a *choice of block spin* (and the free propagator v). To be specific, let us choose the block spin on the lattice Λ_j of lattice spacing a_j as

$$\Phi^j(x)=\underset{z\in x}{av}\,\phi(z)\equiv a_j^{-v}\int_{z\in x}\phi(z)\quad\text{for}\quad x\in\Lambda_j\,. \tag{1.5'}$$

Then the kernels $\mathscr{A}$ are the same as in the work of Kupiainen and Gawedzki (except for the presence of a mass) [9], see Appendix B. There is a corresponding split of the free propagator

$$v(z_1,z_2)=\iint_{x_1,x_2\in\Lambda}\mathscr{A}(z_1,x_1)\,v\,(x_1,x_2)\mathscr{A}(z_2,x_2)\quad\text{for}\quad z_1,z_2\in\text{base}\,. \tag{1.6}$$

The propagator $v(x_1,x_2)$ on the staggered lattice vanishes unless x_1 and x_2 are in the same layer. Both $\mathscr{A}(z,x)$ and $v(x_1,x_2)$ decay exponentially with distance [between z and $x\subset$ base, or between x_1 and x_2] with decay length[3] a_j if $x\in\Lambda_j$, $x_{1,2}\in\Lambda_{j+1}$, see [9]. The decay of the propagator on Λ_0 is determined by the mass. The propagator v is positive semi-definite.

The interaction may be considered as a function of the field φ on the staggered lattice. We set

$$\mathfrak{V}(\psi,\varphi)=V(\psi+\phi)-\frac{\lambda}{4!}\int_{\text{base}}\psi(z)^4\,,$$
$$\delta\mathfrak{V}(v|\psi,\varphi)=\delta V(v|\psi+\phi)\,, \tag{1.7}$$

3 This implies that the propagator of φ^j has infrared cutoff a_{j-1}^{-1}, while a UV-cutoff a_j^{-1} is supplied by the lattice Λ_j

when ϕ and φ are related by Eq. (1.5). [The external field ψ on base space is argument of the generating function for amputated Greens functions.]

While it is custumary to call V the interaction, the reader is advised to think of interaction as being mediated by propagators. The interaction is switched off by putting the propagator equal to zero outside X. Given a subset X of the staggered lattice, let χ_X be its characteristic function

$$\chi_X(x)=\begin{cases}1 & \text{for} \quad x\in X\\ 0 & \text{otherwise},\end{cases}$$

and define X-dependent propagators by

$$\begin{aligned} v_X(x_1,x_2)&=\chi_X(x_1)v(x_1,x_2)\chi_X(x_2),\\ \upsilon_X(z_1,z_2)&=\iint_{x_1,x_2\in\Lambda}\mathscr{A}(z_1,x_1)v_X(x_1,x_2)\mathscr{A}(z_2,x_2).\end{aligned} \tag{1.8}$$

Partition functions $Z_1(X|\psi)$ for finite subsets $X\subset\Lambda$ are defined using as free field measure the Gaussian measure $d\mu_{v_X}(\varphi)$ with propagator [covariance] v_X. Because the propagator vanishes outside X this measure is supported on fields $\varphi=\chi_X\varphi$ that vanish outside X. Therefore the integral is effectively n-dimensional when X contains n points [cf. Eq. (3.15) of Sect. 3]. The interaction $\mathfrak{V}$ is as before, but the counterterms are adjusted by substituting υ_X for υ in the argument of $\delta\mathfrak{V}$.

$$Z_1(X|\psi)=\int d\mu_{v_X}(\varphi)\exp[-\mathfrak{V}(\psi,\varphi)+\delta\mathfrak{V}(\upsilon_X|\psi,\varphi)]. \tag{1.9a}$$

Our definitions are such that

$$Z_1(\emptyset|\psi)=1. \tag{1.9b}$$

The generating function for the amputated Green functions is obtained as an "∞-volume limit" $X\nearrow\Lambda$ (see end of Appendix B)

$$\ln Z(\psi)=-\frac{\lambda}{4!}\int_{\text{base}}\psi(z)^4+\ln Z_1(\Lambda|\psi)\qquad(+\text{const}). \tag{1.10}$$

This switchoff of interaction and induced counterterms on part of the staggered lattice is the central idea of this paper. It is the key to computability and simplicity. The adjustment of the counterterms becomes natural when the interaction is viewed as mediated by propagators. It is also essential in order to maintain stability. In and close to the continuum the bare mass squared is negative. The mass counterterm must therefore be dominated by the quartic interaction. The basic inequality is

$$\sup_{\phi}\int_{\text{base}}\left[-\frac{\lambda}{4!}\phi(z)^4+\tfrac{1}{2}\delta m^2(z)\phi(z)^2\right]\leqq 6\lambda^{-1}\int_{\text{base}}[\delta m^2(z)]^2. \tag{1.11}$$

When the counterterm is switched off outside X, $\lambda^{-1}\int[\delta m^2]^2$ becomes of the order of $\lambda\cdot$ "volume of X" and its exponential can be controlled by powers of λ in suitable units.

Let us return to the computation of Mayer amplitudes $\mathscr{M}$. They are defined by requiring validity of the following relations for arbitrary finite nonempty subsets X

of the staggered lattice Λ. Set

$$A(X|\psi)=n!\,a(x_1)^\nu\ldots a(x_n)^\nu \mathscr{M}(x_1,\ldots,x_n|\psi)+\delta_{1,n}$$

for

$$X=\{x_1,\ldots,x_n\}\,,\quad \text{with}\quad a(x)\equiv a_j \quad \text{for}\quad x\in\Lambda_j\,. \tag{1.12a}$$

The dimension $\nu=3$ in our model. We demand

$$Z_1(X|\psi)=\sum_{X=\Sigma Y}\prod_Y A(Y|\psi)\quad \text{for all finite } X\,. \tag{1.12b}$$

Summation is over all partitions of X into disjoint nonempty subsets Y. Equations (1.9b), (1.12b) exhibit $Z_1(X)$ as partition function of a polymer system with activities $A(Y)$ (see Sect. 2). It is well known that Eq. (1.12b) has a converse which determines the activity $A(Y)$ uniquely, given the partition functions $Z_1(X)$ for $X\subseteqq Y$.

$$A(Y|\psi)=\sum_{n\geqq 1}(-1)^{n-1}(n-1)!\sum_{Y=\Sigma_1^n X_i}\prod_i Z_1(X_i|\psi)\,. \tag{1.12c}$$

For practical calculations (and for deriving estimates) one uses other more convenient formulae for the Mayer amplitudes in place of Eq. (1.12c). They will be described in the following sections. It is now clear, however, that the Mayer amplitudes $\mathscr{M}(x_1,\ldots,x_n|\psi)$ are computable in terms of $\leqq n$-dimensional integrals, because expression (1.9a) is an integral over n real variables $\varphi(x_i)$ when $X=\{x_i\}$ has n points.

Let us note, for comparison, that standard perturbation theory can also be put on a staggered lattice. This can be done Feynman graph by Feynman graph. Consider for instance a nth order graph with 4-leg vertices only. The Feynman integrand in coordinate space is a product of propagators $v(z_i,z_j)$, multiplied with $(-\lambda)^n$ and possibly with a combinatorial factor. Split the propagator as in Eq. (1.6) and imagine that the z_j-integration over base space are done. This produces nonlocal vertex functions on the staggered lattice,

$$\mathscr{V}_4(x_1,\ldots,x_4)=-\frac{\lambda}{4!}\int_{z\in\text{base}}\mathscr{A}(z,x_1)\ldots\mathscr{A}(z,x_4)\,,\qquad \text{etc.} \tag{1.13}$$

The Feynman integral becomes an integral over $4n$ arguments x_i in the staggered lattice. It can be rewritten as a sum of "integrals" over $k\leqq n$ arguments y_i that assume distinct values in Λ (cf. Sect. 4).

Now we will explain, briefly, how our renormalization procedure with X-dependent counterterms works. We begin with a perturbation theoretic illustration. For simplicity we will ignore tadpoles and the $O(\lambda)$ piece of the mass counterterm that cancels them. To order λ^2 the generating function $\ln Z(\psi)$ is a polynomial in ψ, and its quadratic term in ψ is given by

$$\ln Z(\psi)=\iint_{z_1,z_2\in\text{base}}\psi\ldots$$ 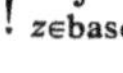 $$\ldots\psi-\int_{z_1\in\text{base}}\psi\ldots \underset{z_1}{\times}\ldots\psi+\ldots \tag{1.14a}$$

in graphical notation, with

$$\underset{z_1}{\times}=\int_{z_2\in\text{base}}\psi\ldots$$ 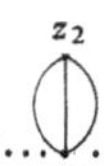 $$\ldots\psi\,. \tag{1.14b}$$

Inserting the expression for the counterterm gives

$$\ln z(\psi) = \iint_{z_1, z_2 \in \text{base}} \left[\psi \ldots \right.$$ 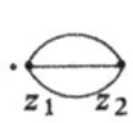 $$\ldots \psi - \psi \ldots$$ 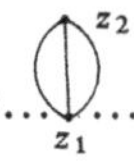 $$\left. \ldots \psi \right] \quad (1.15a)$$

$$= \frac{(-\lambda)^2}{3!} \iint_{z_1, z_2} \psi(z_1)[\psi(z_2) - \psi(z_1)] v(z_1, z_2)^3 . \quad (1.15b)$$

The counterterm differs from the main term by reversal of sign and a "shift of the field" ψ from z_2 to z_1. Such "shift of the field" is also the basic operation in the recent work of Gallavotti and Nicoló [12] on perturbation theoretic renormalization. In coordinate space language, the logarithmic UV-divergence of the self mass in continuous space appears in the guise of an ill defined product of distributions $v(z_1, z_2)^3$ which contains a divergent piece proportional $\delta(z_1 - z_2)$ [13]. Expression (1.15) is a well defined distribution, however, for smooth fields ψ, because of the zero of the factor $[\psi(z_2) - \psi(z_1)]$ at $z_2 = z_1$.

To put the renormalized Feynman integral (1.15a) on the staggered lattice, we perform the split (1.6) of the propagator v in the factor $v(z_1, z_2)^3$ in expression (1.15b). This amounts to splitting the main term and the counterterm simultaneously, and combining the two terms piecewise in the obvious fashion. The split introduces integrations over six variables x_1^a, $x_2^a (a = 1, 2, 3)$ on the staggered lattice, and we imagine that the two integrations of z_1, z_2 over the base space are done first. [They converge because restricted propagators $v_X(z_1, z_2)$ are less singular at z_1, z_2 than $v(z_1, z_2)$, by positivity.] After the split, UV-divergences would show up as divergences in the sums over scales j as $j \to \infty$. Such sums are implied in the integrations (1.2) over the staggered lattice. Consider now a term in the integrals (actually sums) over Λ which involves a propagator $v(x_1, x_2)$ with x_1, $x_2 \in \Lambda_j$. It appears in the combination $\mathscr{A}(z_1, x_1) v(x_1, x_2) \mathscr{A}(z_2, x_2) \equiv \hat{v}(z_1, z_2)$. Because of the decay properties of kernels $\mathscr{A}$ and v, the restricted propagator $\hat{v}$ becomes exponentially small with distance between z_1 and z_2 with decay length a_j. Suppose now that the field ψ is smooth on length scale a_k for some $k < j$. The zero of the factor $[\psi(z_2) - \psi(z_1)]$ in expression (1.15b) will then give rise to a suppression factor a_j / a_k. This suppression factor is enough to eliminate the logarithmic divergence in the sum over length scales $j \to \infty$ that would be present if the mass counterterm had not been included in (1.15).

Self mass graphs (1.15a) may also appear as subgraphs of larger graphs. In this case the factor $\psi(z_1)\psi(z_2)$ is replaced by some function $f(z_1, z_2)$ that is determined by the rest of the graph. Since potential UV-divergences would show up as divergences in the sum over length scales a_j as $j \to \infty$ (where a_j is the decay rate of the fastest decaying restricted propagator in the subgraph) one needs only consider the case that $f(z_1, z_2)$ is smooth on some length scale a_k, with $k < j$. (k is determined by the restricted propagators that appear in the rest of the graph.)[4] The effect of the counterterm is to substitute $f(z_1, z_2) - f(z_1, z_1)$ for $f(z_1, z_2)$. This gives rise to a convergence producing suppression factor a_j / a_k as before.

---

4 In the perturbative renormalization theory of Gallavotti and Nicoló [12] ordering of lines by frequency j replaces the Zimmermann forests [13]

Our discussion of renormalization has been perturbative so far. Our Mayer amplitudes may be thought of as given by low order Feynman graphs (split as described above) plus corrections. The corrections turn out not to produce new UV-divergences because they have explicit extra factors a_j. In conclusion, renormalization suppresses dangerous contributions to *individual* Mayer amplitudes $\mathcal{M}(x_1, \ldots, x_n|\psi)$ with some arguments x_r in very fine lattices Λ_j (j large). This works in such a way that the sums over length scales that have to be done as part of the integrations in (1.1) become convergent.

In the analysis of Magnen and Sénéor [6], part of the UV-convergence factors L^{-j} is used up to control combinatorial factors. They arise because a large cube $x' \in \Lambda_k$ may be coupled to many smaller cubes x, $x' \supset x \in \Lambda_j$, $j > k$.

We conclude our introductory discussion of renormalization with a remark about *renormalization conditions*. The X-dependent mass- and vacuum energy counterterms δm^2 and δe are chosen in such a way that the following renormalization conditions are approximately fulfilled, for all finite nonempty subsets $X = \{x_1, \ldots, x_n\}$ of Λ and all $z \in$ base.

$$\mathcal{M}(x_1, \ldots, x_n|\psi = 0) = 0\,, \tag{1.16a}$$

$$\int_{z' \in \mathrm{base}} \frac{\delta^2}{\delta\psi(z')\delta\psi(z)} \mathcal{M}(x_1, \ldots, x_n|\psi = 0) = 0\,. \tag{1.16b}$$

The renormalization conditions (1.16) are not "too many" because there are counterterms $\delta e(X)$ and $\delta m^2(X|z)$ to be fixed for arbitrary $X \subset \Lambda$ and $z \in$ base.

Exact validity of Eq. (1.16a) may be enforced by fixing the constant term $\delta e(X)$ in the action such that

$$Z_1(X|\psi = 0) = 1 \quad \text{for all } X\,. \tag{1.17}$$

If this is done, then the expansions for the Greens functions become particularly simple. The full, free-propagator-amputated Greens functions (including disconnected parts) are given by

$$\begin{aligned} G(\underline{z}_1, \ldots, \underline{z}_k) &= Z(\Lambda|\psi = 0)^{-1} \frac{\delta^k}{\delta\psi(z_1)\ldots\delta\psi(z_k)} Z(\Lambda|\psi = 0)\,, \\ Z(X|\psi) &\equiv Z_1(X|\psi) \exp\left[-\frac{\lambda}{4!}\int_{\mathrm{base}} \psi(z)^4\right]. \end{aligned} \tag{1.18}$$

Performing the differentations in expression (1.12b) one obtains with Eqs. (1.12a), (1.16a), and (1.17),

$$\begin{aligned} G(\underline{z}_1, \underline{z}_2) = \sum_{n \geq 1} \int_{x_1 \ldots x_n \in \Lambda} \!\!\ldots\! \int \Bigg[& \frac{\delta^2}{\delta\psi(z_1)\delta\psi(z_2)} \mathcal{M}(x_1, \ldots, x_n|\psi = 0) \\ & + \sum_{k=1}^{n-1} \binom{n}{k} \frac{\delta}{\delta\psi(z_1)} \mathcal{M}(x_1, \ldots, x_k|\psi = 0) \frac{\delta}{\delta\psi(z_2)} \mathcal{M}(x_{k+1}, \ldots, x_n|\psi = 0) \Bigg] \end{aligned} \tag{1.19}$$

and similar formulae for the higher n-point functions. In Eq. (1.19) the second term in [ ] is actually zero by symmetry.

The free-propagator-amputated Greens functions $G(\underline{x}_1, \ldots, \underline{x}_n)$ are not 1-particle irreducible, and neither are the Mayer amplitudes $\mathscr{M}$. 1-particle irreducible Mayer amplitudes $\mathscr{M}^{(1i)}$ can be introduced by the following defining relation. We write it for activities A which are related to Mayer amplitudes $\mathscr{M}$ by Eq. (1.12a).

$$A(X|\psi) = A^{(1i)}(X|\psi) + \sum_{X=X_1+X_2} \iint_{z_1,z_2 \in \text{base}} \iint_{\substack{x_1\in X_1 \\ x_2 \in X_2}} \frac{\delta A^{(1i)}(X_1|\psi)}{\delta\psi(z_1)} \cdot \mathscr{A}(z_1,x_1)v(x_1,x_2)\mathscr{A}(z_2,x_2)\frac{\delta A(X_2|\psi)}{\delta\psi(z_2)}. \tag{1.20}$$

Summation is over partitions of X in two nonempty subsets X_1 and X_2. Renormalization conditions (1.16) carry over to the one particle irreducible amplitudes $\mathscr{M}^{(1i)}$ (and also to $\tilde{\mathscr{M}}$).

One may want to impose also the second renormalization condition (1.16b) exactly. Description of such an alternative renormalization scheme is outside the scope of the present paper.

Our method may also be used to compute *effective actions in the sense of Wilson* [15], and corresponding Boltzmann factors, for lattice cutoffs a_j. They depend on a block spin Φ^j on Λ_j. Set

$$\Lambda_{[>j]} = \sum_{k>j} \Lambda_j \quad \text{and} \quad \Phi^j(x) = 0 \quad \text{for} \quad x \notin \Lambda_j. \tag{1.21}$$

Then $\ln Z(\Lambda_{[>j]}|\mathscr{A}\Phi^j)$ and $Z(\Lambda_{[>j]}|\mathscr{A}\Phi^j)$, which are determined by Mayer amplitudes $\mathscr{M}(X|\psi)$ with $X \subset \Lambda_{[>j]}$, are almost the desired quantities. ["Almost" because one needs to correct for the fact that the "partial bare mass" $m_0^2(z) = m^2 - \delta m^2(v_{\Lambda[>j]}|z)$ need not be exactly invariant under continuous translations.] This is of interest for theories which are not weakly coupled at the physical length scale a_0. We hope to come back to this problem in a future paper. Ultimately one may hope to design a *Monte Carlo procedure for continuum theories* in which *correction terms* in the effective Boltzmann factor are *picked randomly*.

The plan of the rest of the paper is as follows: In Sect. 2 we recall results of the theory of polymer systems, and formulae for and properties of truncated expectation values. In Sect. 3 we show how the Mayer amplitudes can be computed as truncated expectation values of so-called "molecular activities" B that are determined by the interaction V and the kernels $\mathscr{A}$ (i.e. propagator v and block spin). Formulae for and properties of these molecular activities are discussed. The result is summarized in Theorem 3.3. Section 4 discusses the relation with perturbation theory. It includes a pedagogical discussion of the relation between standard perturbation theory and Mayer expansions on a simple (as opposed to staggered) lattice. In Sect. 5 the trivial, but still instructive, model of $\lambda\phi^4$-theory on a lattice with only one point is considered.

2. Theory of Polymer Systems

Following Gruber and Kunz [16], a polymer system on a finite or countable set Λ is specified as follows. The elements of Λ are called sites. Certain finite nonempty subsets P of Λ are declared to be polymers. Among them are the monomers which consist of a single site $x \in \Lambda$. To every polymer P, a (real) activity $A(P)$ is assigned. The partition function $Z(X)$ for an arbitrary finite subset X of Λ is defined as

$$Z(X) = \sum_{X=\Sigma P_i} \prod_{P_i} A(P_i) . \tag{2.1}$$

Here and in the following, we write $\sum$ and $+$ for union of disjoint sets. The sum in Eq. (2.1) runs over all partitions of X into polymers. In particular

$$Z(\emptyset) = 1 . \tag{2.1'}$$

In our applications, the partition functions will be guaranteed to be positive, but the activities $A(P)$ for polymers other than monomers are not necessarily nonnegative. An example of a polymer system is described in Fig. 2. By definition (2.1), the partition functions satisfy the following recursion relation, for any $x \in X$

$$Z(X) = \sum_{\substack{Y \\ Y \subsetneq X, x \in Y}} A(Y) Z(X-Y) . \tag{2.2}$$

They are known as Kirkwood-Salsburg equations. Equation (2.1) may be regarded as the iterative solution of (2.2) with initial condition (2.1′). Only a finite number of iteration steps is needed to obtain $Z(X)$ for a finite set X.

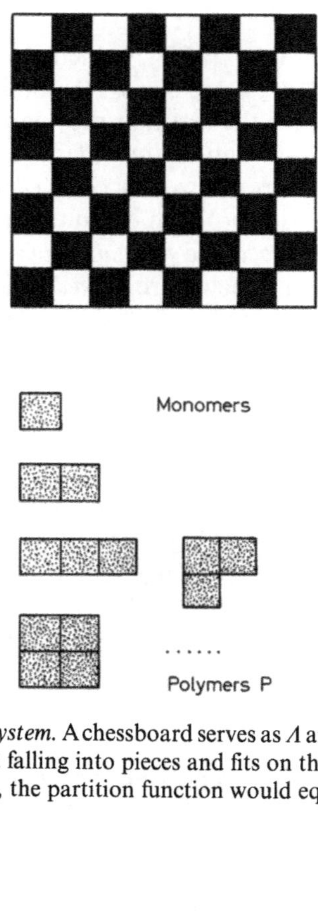

Fig. 2. Example of a *polymer system*. A chessboard serves as Λ and any union of squares that can be cut out of cardboard without falling into pieces and fits on the board is a polymer. Choosing as activities $A(P) = \exp[-\beta|P|]$, the partition function would equal $e^{-64\beta} \cdot$ (# partitions of Λ into polymers)

It will be important for us that Eq. (2.1) can be inverted. That is, the activity $A(P)$ is uniquely determined by the partition functions $Z(X)$ for $X \subseteqq P$. Indeed it is obvious from Eq. (2.1) that $A(P)=Z(P)-\sum\prod$ (activities of smaller polymers). Therefore the activities can be recursively determined. The result is well known [17]

$$A(P)=\sum_{n=1}^{|P|}(-1)^{n-1}(n-1)!\sum_{P=\Sigma_1^n X_i}\prod_i Z(X_i). \tag{2.3}$$

The inner sum runs over partitions of P into nonempty subsets X_i, and $|P|$ is the number of sites in P.

A function $Z(\cdot)$ of finite subsets $X \subset \Lambda$ may be regarded as a partition function of a polymer system on Λ if it obeys Eq. (2.1'). Arbitrary finite subsets of Λ must be admitted as polymers for general $Z(\cdot)$. Since validity of Eq. (2.1) follows from (2.2) and (2.1'), it suffices to check validity of Eqs. (2.2) for some choice of $x=x(X)$ in order to verify that some $A(\cdot)$ are indeed the activities of the polymer system with partition function $Z(\cdot)$. The monomer activities $A(\{x\})=Z(\{x\})$ are positive if the partition functions are positive.

A central result of the theory of polymer systems is a series expansion for the free energies $\ln Z(X)$, together with a sufficient condition for its convergence. We state it first for finite sets X. Given X, a *cluster* $\mathfrak{Q}$ is a nonempty collection of not necessarily distinct polymers with the property that the following graph $\gamma(\mathfrak{Q})$ is connected. Draw a vertex for every polymer P in $\mathfrak{Q}$, and a line joining P to P' if $P \cap P' \neq \emptyset$. It is customary to write

$$\mathfrak{Q}=(P_1^{n_1}, \ldots, P_k^{n_k})$$

if $\mathfrak{Q}$ contains k distinct polymers P_i with multiplicities $n_i \geqq 1$. One defines reduced activities by

$$\bar{A}(P)=A(P)\Big/\prod_{x\in P}A(\{x\}). \tag{2.4}$$

The expansion formula reads

$$\ln Z(X)=\sum_{x\in X}\ln A(\{x\})+\sum_{\mathfrak{Q}}\mathfrak{a}(\mathfrak{Q})\prod_{P\in\mathfrak{Q}}\bar{A}(P). \tag{2.5}$$

The sum over $\mathfrak{Q}$ runs over all clusters of nontrivial polymers on X. Nontrivial polymers are, for now, all polymers other than monomers. $\mathfrak{a}(\mathfrak{Q})$ are combinatorial coefficients which are given by

$$\mathfrak{a}(\mathfrak{Q})=\sum_{C\subseteqq\gamma(\mathfrak{Q})}(-1)^{l(C)}\Big/\prod_i n_i!. \tag{2.6}$$

Summation is over all connected subgraphs C of the graph $\gamma(\mathfrak{Q})$ that was mentioned above, and $l(C)$ is the number of lines in C. Alternative expressions and bounds on $\mathfrak{a}(\mathfrak{Q})$ can be found in the literature [18].

For real $\xi>1$ one defines

$$\mathfrak{B}_X(\xi)=\sup_{x\in X}\frac{1}{\xi}\left[1+\sum_{\substack{P\\ x\in P\subseteqq X,\,|P|\geqq 2}}|\bar{A}(P)|\xi^{|P|}\right]. \tag{2.7}$$

The expansion (2.5) for the free energy $\ln Z(X)$ converges if $\mathfrak{B}_X(\xi)<1$ for some $\xi>1$ [16].

By differentiating Eq. (2.5) one obtains an expansion for the so-called "reduced correlation functions" of the polymer system

$$\varrho_X(Y) = \frac{Z(X-Y)}{Z(X)} = \frac{\partial}{\partial A(Y)} \ln Z(X)\,, \quad (Y \subsetneqq X)\,. \tag{2.8}$$

If Λ is an infinitely extended lattice, then these expansions continue to converge in the infinite volume limit $X \nearrow \Lambda$, if $\mathfrak{B}_\Lambda(\xi) < 1$ for some $\xi > 1$. Moreover,

$$|\varrho_X(Y)| \leqq [1 - \mathfrak{B}_X(\xi)]^{-1} \xi^n \prod_{x \in Y} A(\{x\})^{-1} \quad \text{with} \quad n = |Y| \tag{2.9}$$

for all $X \subsetneqq \Lambda$. For a more precise statement of these results, see [16]. For the renormalized Mayer expansions on the staggered lattice one uses these convergence results only at $\psi = 0$. One can avoid the need to use them (in the discussion of Greens functions) altogether by imposing the renormalization condition (1.16a), viz. $\mathcal{M}(X|\psi = 0) = 0$. This can be seen from Eq. (1.19). Expansion (2.5) for $\ln Z$ serves then merely as a bookkeeping device for combinatorial factors in equations that could also be obtained by differentiating Eq. (1.12b) for Z. [In the expression for the connected amputated n-point Greens function that is obtained by differentiating expansion (2.5) with respect to the external field ψ at $\psi = 0$, all terms with $\mathfrak{Q}$ containing more than n polymers will vanish as a result of the renormalization condition (1.16a). There results a finite sum for finite X.] The point is that the convergence conditions mentioned above are much more restrictive than the condition for convergence of the sum (1.19) – in particular they remain nontrivial for finite X.

It is convenient to introduce the notion of a *polymer system with empty sites*. Denote its activities by $M(P)$. Its partition function is defined as

$$Z(X) = 1 + \sum_{\substack{Y \\ \emptyset \neq Y \subsetneqq X}} \sum_{Y = \Sigma P_i} \prod_i M(P_i)\,. \tag{2.10}$$

Evidently this is equal to the partition function of a polymer system proper with activities

$$A(\{x\}) = 1 + M(\{x\})\,, \quad A(P) = M(P) \quad \text{for} \quad |P| \geqq 2\,. \tag{2.11}$$

We may identify with a polymer system proper in still another way. Split each site of the original set Λ into two. Polymers of the original system, including its monomers $\{x\}$, shall occupy double sites. In addition we introduce a monomer with activity 1 which occupies only one site (either half of a double site). Applying the expansion formula (2.5) to this system we obtain

$$\ln Z(X) = \sum_{\mathfrak{Q}} \mathfrak{a}(\mathfrak{Q}) \prod_{P \in \mathfrak{Q}} M(P)\,. \tag{2.12}$$

Summation is now over all clusters containing arbitrary polymers P (of the original system). The expansion will converge if

$$\sup_{x \in X} \frac{1}{\xi} \left[1 + \sum_{\substack{P \\ x \in P}} |M(P)| \xi^{2|P|} \right] < 1 \tag{2.13}$$

for some $\xi > 1$. Monomers are included in the sum over P.

Let us now consider polymer systems whose activities depend on a field φ on Λ

$$\mathscr{Z}(X|\varphi) = \sum_{X=\Sigma P} \prod_P B(P|\varphi). \tag{2.14}$$

It will be required that the *activity* $B(P|\varphi)$ *depends only on the restriction of* φ *to* P. We wish to consider Gaussian integrals of such partition functions. To avoid confusion with quantities that will appear later on, the polymers of the polymer system (2.14) will be called *molecules*, with *molecular activity* $B(\cdot|\varphi)$. Let $d\mu_v(\varphi)$ be the Gaussian measure with covariance v (= normalized free field measure with propagator v) and consider

$$Z(X) = \int d\mu_v(\varphi)\mathscr{Z}(X|\varphi) \equiv \langle \mathscr{Z}(X|\varphi)\rangle_v. \tag{2.15}$$

The symbol $\langle\cdot\rangle_v$ stands for expectation value with respect to the Gaussian measure $d\mu_v$. Evidently $Z(\emptyset)=1$. We may therefore regard $Z(X)$ as partition functions of a polymer system

$$Z(X) = \sum_{X=\Sigma P} \prod_P A(P). \tag{2.16}$$

The activities A are uniquely determined by the partition functions Z as we know. If $\{M_1, \ldots, M_n\}$ is a finite set of disjoint molecules, we may consider the truncated Gaussian expectation value[5]

$$\langle B(M_1|\varphi); B(M_2,\varphi); \ldots; B(M_n|\varphi)\rangle_v \equiv \left\langle \prod_{i=1}^{n} [B(M_i|\varphi);]\right\rangle_v. \tag{2.17}$$

Truncated expectation values obey the defining relation

$$\left\langle \prod_{a\in I} B_a(\varphi)\right\rangle_v = \sum_{I=\Sigma J} \prod_J \left\langle \prod_{b\in J} [B_a(\varphi);]\right\rangle_v. \tag{2.18}$$

Summation is over partitions of the index set $I=\{1, \ldots, n\}$ into nonempty subsets J. Note that this formula has the character of a polymer representation too, with the ordinary expectation values on the left-hand side playing the role of partition functions $Z(I)$. The truncated expectation values are the corresponding activities.

It follows from Eqs. (2.15) and (2.18) that the activities A in the polymer representation (2.16) of the partition functions $Z(X)$ are given by truncated expectation values

$$A(P) = \sum_{P=\Sigma M} \left\langle \prod_M [B(M|\varphi);]\right\rangle_v. \tag{2.19}$$

These relations can be transcribed into the language of polymer systems with empty sites. We leave this to the reader.

Truncated Gaussian expectation values admit a tree formula which expresses them as integrals over auxiliary real variables $s_i = 0\ldots 1$ of ordinary Gaussian expectation values. To state it we introduce some notation [19, Sect. 2.4].

An n-tree is a map

$$\eta: [2, \ldots, n] \to [1, \ldots, n-1] \quad \text{with} \quad \eta(i) < i. \tag{2.20}$$

5 We use the customary "semicolon notation" for truncated expectation values

It specifies a tree graph with vertices $1 \ldots n$, root 1, and links $(i, \eta(i))$. Given an $(n-1)$-tupel $s=(s_1, \ldots, s_{n-1})$ of real variables in the interval $0 \ldots 1$, one sets

$$f(\eta|s)=\prod_{a=2}^{n} s_{a-2}s_{a-3}\ldots s_{\eta(a)}, \qquad ds=ds_1 \ldots ds_{n-1}. \tag{2.21}$$

Empty products which arise when $\eta(a)=a-1$ (in particular for $a=2$) are read as 1.

Given a set of n disjoint molecules $M_i \subset \Lambda$, let χ_i be the characteristic function of M_i (viz. $\chi_i(x)=1$ for $x \in M_i$, and $=0$ otherwise) and define the interpolating covariance $v[\pi, s]$ by

$$v[\pi, s]=\sum_i \chi_i v \chi_i + \sum_{1 \leqq i<j \leqq n} s_i \ldots s_{j-1}[\chi_{\pi(i)} v \chi_{\pi(j)} + \chi_{\pi(j)} v \chi_{\pi(i)}]. \tag{2.22}$$

It depends on a permutation π of $1 \ldots n$, besides the $n-1$ tupel of real variables s. We assume that products of activities $B(M_a|\varphi)$ are integrable with respect to the Gaussian measure $d\mu_v(\varphi)$, and differentiable.

Proposition 2.1. *Let $M_1, \ldots, M_n$ be disjoint subsets of Λ and suppose that the activities $B(M_a|\varphi)$ depend only on the restriction of φ to M_a. Then their truncated Gaussian expectation values admit the following representation in terms of ordinary Gaussian expectation values*

$$\left\langle \prod_{a=1}^{n} [B(M_a|\varphi);] \right\rangle_v = \sum_\eta \sum_\pi{}' \int ds\, f(\eta|s) \cdot \left\langle \prod_{b=2}^{n} \left[\iint_{\substack{x \in M_{\pi(b)} \\ y \in M_{\pi(\eta(b))}}} \frac{\delta}{\delta\varphi(x)} v(x,y) \frac{\delta}{\delta\varphi(y)} \right] \prod_{c=1}^{n} B(M_c|\varphi) \right\rangle_{v[\pi,s]}.$$

Summation over η is over all n-trees. Summation over π is over all $(n-1)!$ permutations of $1 \ldots n$ with $\pi(1)=1$. Integration over the $n-1$ variables s_i runs from 0 to 1.

This proposition is a standard result of constructive field theory [2, 19, 20]. It can be proven either by the method of partial integration of Gaussian measures of Glimm and Jaffe [19], or by making a Fourier transform in φ and using the techniques of Mayer expansions for classical gases [20, 3].

Remark. This proposition covers the special case that no two factors $B(M_a|\varphi)$ depend on the field $\varphi(x)$ at the same point x. A tree formula exists also in the general case where this condition is not satisfied. It can be used, for instance, to derive a *tree formula for the sum of all (unrenormalized) connected n^{th} order Feynman integrals*, on a lattice or in the continuum. As is well known, the sum of all connected n^{th} order vacuum Feynman diagrams in a theory with interaction Lagrangian $\lambda\mathscr{L}_I$ equals

$$\frac{(-\lambda)^n}{n!} \int_{z_1} \ldots \int_{z_n} \langle \mathscr{L}_I(\varphi(z_1)); \ldots; \mathscr{L}_I(\varphi(z_n)) \rangle_v.$$

The general tree formula is obtained with the help of the following trick which reduces the general case to the special case covered by Proposition 2.1.

The covariance operator v in a Gaussian measure is permitted to have zero eigenvalues. The simplest example is the 1-dimensional Dirac δ-measure $d\mu_0(\xi)$

$=\delta(\xi)d\xi$, which is a Gaussian measure with covariance 0. Consider the Gaussian measure for a field $\phi(x,i)$ on $\Lambda\times\ldots\times\Lambda$ (n copies $i=1\ldots n$) with covariance

$$\tilde{v}(x,i,y,j)=v(x,y). \tag{2.23}$$

It equals

$$d\mu_{\tilde{v}}(\phi)=d\mu_v(\phi(\cdot,1))\prod_{j=2}^{n}\prod_{x\in\Lambda}\delta(\phi(x,j)-\phi(x,1))d\phi(x,j). \tag{2.24}$$

This can be verified by checking that $\langle\phi(x,i)\phi(y,j)\rangle_{\tilde{v}}=\tilde{v}(x,i,y,j)$. Set

$$\tilde{B}(M_i|\phi)=B(M_i|\phi(\cdot,i)). \tag{2.25}$$

Evidently, for any $I\subseteqq[1,\ldots,n]$,

$$\left\langle\prod_{i\in I}B(M_i|\varphi)\right\rangle_v=\left\langle\prod_{i\in I}\tilde{B}(M_i|\phi)\right\rangle_{\tilde{v}}. \tag{2.26}$$

No two activities $\tilde{B}(M_i|\phi)$ depend on the field ϕ at the same point (x,i), and the proposition can therefore be applied to find a formula for the truncated expectation value $\langle\prod[B(M_i|\varphi);]\rangle_v$ in terms of a Gaussian measure for a field on $\Lambda\times\ldots\times\Lambda$. □

Finally we wish to consider the case that the Gaussian measure $d\mu_v(\varphi)$ is represented as convolution of Gaussian measures with covariances v^i by splitting the propagator v[6]. If

$$v=\sum_{i=0}^{N}v^i\equiv v^{[\leqq N]}\quad\text{with}\quad v^i\geqq 0, \tag{2.27}$$

then

$$\int d\mu_v(\varphi)f(\varphi)=\int d\mu_{v^0}(\varphi^0)\ldots d\mu_{v^N}(\varphi^N)f\left(\sum_{i=0}^{N}\varphi^i\right). \tag{2.28}$$

Gaussian measures are only well defined for propagators v that are kernels of positive semidefinite operators. Therefore the split of the propagator must be such that v^i have this property, viz. $v^i\geqq 0$ (as operators).

Let us first give the result for a split of a propagator v into two. The general case will then be handled by iterating the result.

Proposition 2.2. *Let $v=u+w$ with $v\geqq 0$, $u\geqq 0$ and $w\geqq 0$ as operators. Then the truncated expectation values obey the following relations*

$$\left\langle\prod_{a\in I}[B_a(\varphi);]\right\rangle_v=\sum_{I=\Sigma J}\left\langle\prod_J\left[\left\langle\prod_{b\in J}[B_b(\Phi+\zeta);]\right\rangle_w;\right]\right\rangle_u. \tag{2.29}$$

The expectation value $\langle\cdot\rangle_w$ is computed with Gaussian measure $d\mu_w(\zeta)$, and the expectation value $\langle\cdot\rangle_u$ with $d\mu_u(\Phi)$. Summation is over partitions of the finite set I into nonempty subsets $J\subseteqq I$.

The proof of this proposition will be given below, at the end of this section.

6 The material in the rest of this section is not needed for pedestrian calculations

Let us now iterate Eq. (2.29) to treat the split (2.27) of the propagator and corresponding split of the field in (2.28). Write

$$\begin{aligned} v \equiv v^{[\leqq N]} &= v^N + v^{[\leqq N-1]} = \ldots = v^{[>N-\ell]} + v^{[\leqq N-\ell]}, \\ v^{[\leqq N-1]} &= v^{N-1} + v^{[\leqq N-2]} \\ &\vdots \end{aligned} \tag{2.30a}$$

and

$$\begin{aligned} \varphi \equiv \varphi^{[\leqq N]} &= \varphi^N + \varphi^{[\leqq N-1]} = \ldots = \varphi^{[>N-\ell]} + \varphi^{[\leqq N-\ell]}, \\ \varphi^{[\leqq N-1]} &= \varphi^{N-1} + \varphi^{[\leqq N-2]} \\ &\vdots \end{aligned} \tag{2.30b}$$

The labelling of the propagators shall be such that v^N propagates the highest frequencies, and v^0 the lowest. (It was the other way around in [4].) Following [3, 4] we define ℓ-vertices with constituents $i \in I$ and associated vertex functions inductively, for a given index set $I \subsetneqq N$. A *0-vertex* is a single element $a \in I$ which is its own constituent. The associated vertex function is

$$\sigma_a^0(\varphi^{[\leqq N]}) = B(M_a|\varphi^{[\leqq N]}). \tag{2.31}$$

Higher vertices are defined inductively. An *ℓ-vertex* α' is a finite collection $\{\alpha\}$ of $(\ell-1)$-vertices, no two of which share a constituent. $a \in I$ is a constituent of α' if it is a constituent of one of the $(\ell-1)$-vertices $\alpha \in \alpha'$. We write $a \Subset \alpha'$ in this case, and $C(\alpha')$ for the set of all constituents of α'. The vertex functions are defined by the *recursion relation*

$$\sigma_{\alpha'}^{\ell}(\varphi^{[\leqq N-\ell]}) = \left\langle \prod_{\alpha \in \alpha'} [\sigma_\alpha^{\ell-1}(\varphi^{[\leqq N-\ell]} + \varphi^{N-\ell+1});] \right\rangle_{v^{N-\ell+1}}. \tag{2.32}$$

The truncated expectation value is computed with Gaussian measure $d\mu_{v^{N-\ell+1}}(\varphi^{N-\ell+1})$. It follows from Proposition 2.2 by induction that

$$\left\langle \prod_{b \in J} [B(M_b|\psi + \varphi^{[>N-\ell]});] \right\rangle_{v^{[>N-\ell]}} = \sum_{\substack{\alpha \\ C(\alpha)=J}} \sigma_\alpha^\ell(\psi) \tag{2.33}$$

for all $\ell = 0, 1, \ldots, N$. Summation is over all ℓ-vertices α whose set of constituents is $C(\alpha)$. The special case $\ell = N+1$ gives

$$\left\langle \prod_{a \in I} [B(M_a|\varphi);] \right\rangle_v = \sum_{\substack{\alpha \\ C(\alpha)=I}} \sigma_\alpha^{N+1}(0). \tag{2.34}$$

Summation is over all $(N+1)$-vertices with set of constituents I. (The present definition of vertex functions differs by a combinatorial factor from [3, 4].) Following Gallavotti and Nicoló [12], the terms in the sums (2.33) or (2.34) [i.e. the $(N+1)$-vertex α] can be represented by tree diagrams. To avoid confusion with other kinds of trees that occurred before, we call them Gallavotti Nicoló trees (GN-trees).

The trees are drawn on ruled paper, with lines labelled by $k = -1, \ldots, N$. A ℓ-vertex α is represented by a point on line $N-\ell$. If the ℓ-vertex α consists of $(\ell-1)$-vertices $\alpha_1, \ldots, \alpha_k$, then the point α on line $N-\ell$ is linked to k points $\alpha_1, \ldots, \alpha_k$

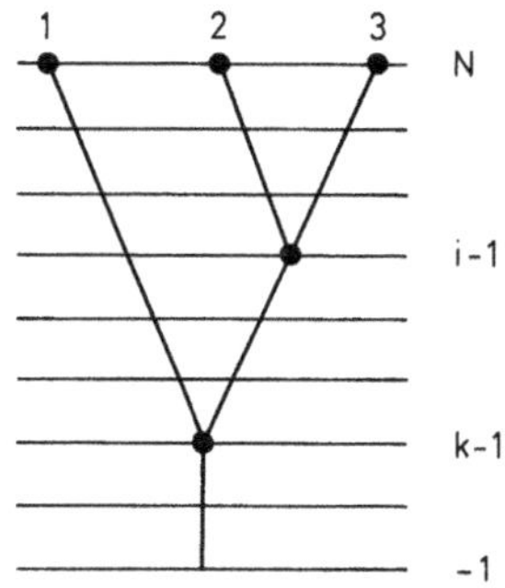

Fig. 3. Example of a *Gallavotti-Nicoló tree*. The tree shown in the figure represents the expectation value

$$\langle\langle B(M_1,\varphi);\langle B(M_2,\varphi);B(M_3,\varphi)\rangle_{v^i}\rangle_{v^k}$$

assuming M_1, M_2, M_3 consist of points of Λ_i and Λ_k only

on line $N-\ell+1$. Then the procedure is repeated. If the $(\ell-1)$-vertex α_1 consists of $(\ell-2)$-vertices $\alpha_{11},\ldots,\alpha_{1m}$, it is linked to m points α_{1i} on line $N-\ell+2$, etc. If α had constituent set I then the procedure ends with the drawing of points on line N representing 0-vertices, one for each constituent = element of I.

In this way, a one to one correspondence is established between ℓ-vertices and GN trees whose root is on line $N-\ell$, and whose tips of branches on line N are labelled by the elements of the constituent set $C(\alpha)$. In our applications it happens often that the activities $B(M_a|\varphi)$ are all independent of some φ^j so that the integration $\int d\mu_{vj}(\varphi^j)$ is trivial. In this case the GN-tree cannot branch on line $j-1$, and we erase the vertices on this line. An example is shown in Fig. 3.

These formulae embody the essence of the iterated Mayer expansions of [3, 4] in a succinct way. They are also the starting point of the perturbative renormalization theory of Gallavotti and Nicoló [12]. The truncated expectation values in the right-hand side of recursion relation (2.32) can be expressed in a tree formula by Proposition 2.1 (and remark following it).

It remains to prove Proposition 2.2. Since truncated expectation values are uniquely determined by ordinary expectation values by the defining relation (2.18), it suffices to show that

$$\left\langle\prod_{a\in H}B_a(\varphi)\right\rangle_{w+u}=\sum_{H=\Sigma K}\prod_K\left\{\sum_{K=\Sigma J}\left\langle\prod_J\left[\left\langle\prod_{b\in J}[B_b(\Phi+\zeta);]\right\rangle_w\right];\right\rangle_u\right\}. \tag{2.35}$$

By Eq. (2.28) the left-hand side equals

$$\left\langle\left\langle\prod_{a\in H}B_a(\Phi+\zeta)\right\rangle_w\right\rangle_u.$$

By the defining relation (2.18) of truncated expectation values this equals

$$\sum_{H=\Sigma J}\left\langle\prod_J\left\langle\prod_{b\in J}[B_b(\Phi+\zeta);]\right\rangle_w\right\rangle_u.$$

Using the defining relation (2.18) of truncated expectation values again to reexpress the outer ordinary expectation value, this becomes

$$\sum_{H=\Sigma J}\sum_{\{J\}=\Sigma\mathscr{K}}\prod_{\mathscr{K}}\left\langle\prod_{J\in\mathscr{K}}\left[\left\langle\prod_{b\in J}[B_b(\Phi+\zeta);]\right\rangle_w;\right]\right\rangle_u. \tag{2.36}$$

The inner sum is over partition of the set $\{J\}$ (of subsets of H) into disjoint subsets $\mathscr{K}$. Thus, $\mathscr{K}$ is a collection of sets J. Set $K = \sum_{J \in \mathscr{K}} J$. Then

$$\sum_{H=\Sigma J} \sum_{\{J\}=\Sigma \mathscr{K}} \prod_{\mathscr{K}} \left\langle \prod_{J \in \mathscr{K}} \cdot \right\rangle = \sum_{H=\Sigma K} \prod_{K} \left\{ \sum_{K=\Sigma J} \left\langle \prod_{J} \cdot \right\rangle \right\}$$

by the distributive law. Upon inserting this, expression (2.36) becomes equal to the right-hand side of Eq. (2.35). q.e.d.

3. Expansions on the Staggered Lattice

Now we consider $\lambda\phi^4$-theory in $\nu=3$ dimensions, in the continuum or on a cubic lattice of arbitrarily small lattice spacing $a \equiv a_N$. This space, on which the theory lives originally, will be called the base space. Integration over the base space is written as

$$\int_{z \in \text{base}} \equiv \begin{cases} \int_{\mathbb{R}^3} d^3 z & \text{in the continuum} \\ a^3 \sum_{z \in (a\mathbb{Z})^3} & \text{on the lattice}. \end{cases} \tag{3.1}$$

The generating function for amputated Greens functions, the Mayer amplitudes etc., will depend on an external field ψ on base space.

We put the theory on a staggered lattice Λ as described in the introduction. When the interaction and induced counterterms are switched off outside a finite subset X of Λ, the partition function becomes

$$Z(X|\psi) \equiv Z_1(X|\psi) \exp\left[-\frac{\lambda}{4!} \int_{\text{base}} \psi(z)^4 \right] = \int d\mu_{v_X}(\varphi) \mathscr{Z}(X|\psi + \mathscr{A}\varphi). \tag{3.2}$$

Integration is over fields φ that vanish outside X, so that the integral has as many dimensions as X has points, and

$$\mathscr{Z}(X|\psi) = \exp\left\{ \int_{z \in \text{base}} \left[-\frac{\lambda}{4!} \psi(z)^4 + \tfrac{1}{2} \delta m^2(v_X|z) \psi(z)^2 \right] + \delta e(X) \right\}. \tag{3.3}$$

The mass counterterm is given by Eq. (1.4), and the vacuum energy counterterm $\delta e(X) \equiv \delta e(v_X)$ can be determined so that

$$Z(X|\psi = 0) = 1. \tag{3.4}$$

The factor $\exp[-\lambda \int \psi^4/4!]$ is extracted in (3.2) in order to assure that $Z_1(\emptyset|\psi) = 1$, as is necessary for partition functions of polymer systems by Eq. (2.1').

The Mayer amplitudes $\mathscr{M}$ are related to the partition functions.

$$Z_1(X|\psi) = \sum_{X=\Sigma Y} \prod_{Y} A(Y|\psi), \tag{3.5a}$$

$$A(X|\psi) = \delta_{1,n} + M(X|\psi), \tag{3.5b}$$

$$M(X|\psi) = n!\, a(x_1)^\nu \ldots a(x_n)^\nu \mathscr{M}(x_1, \ldots, x_n|\psi), \quad (\nu = 3) \tag{3.5c}$$

for $X = \{x_1, \ldots, x_n\}$ with $a(x) = a_j$ for $x \in \Lambda_j$. A are the activities of the polymer system whose partition functions are Z_1, and M are the activities of the corresponding polymer system with empty sites. Arbitrary finite nonempty subsets of the staggered lattice are admitted as polymers. According to the discussion in Sect. 2, activities are uniquely determined by partition functions for polymer systems. Therefore the Mayer amplitudes are well defined for distinct arguments $x_1, ..., x_n$. They are symmetrical in these arguments.

It is useful to introduce also augmented Mayer emplitudes $\tilde{\mathcal{M}}(x_1, ..., x_n|\psi)$. They are defined for arbitrary, not necessarily distinct, values of their arguments, and symmetrical in these arguments. If all arguments x_i are distinct

$$\tilde{\mathcal{M}}(x_1, ..., x_n|\psi) = \mathcal{M}(x_1, ..., x_n|\psi) \quad (x_i \text{ distinct}). \tag{3.6a}$$

The general definition is as follows. Consider clusters $\mathbb{Q}$ of polymers $P \subset \Lambda$ (including monomers) as defined in Sect. 2. Let $\operatorname{supp}\mathbb{Q}$ be the disjoint union of $P \in \mathbb{Q}$ – i.e. a point $x \in \Lambda$ is contained in $\operatorname{supp}\mathbb{Q}$ with a certain multiplicity $n(x)$. It equals the number of polymers $P \in \mathbb{Q}$ which contain x. Now define, for X = finite family of not necessarily distinct elements x_i of Λ

$$\tilde{M}(X|\psi) = \sum_{\substack{\mathbb{Q} \\ \operatorname{supp}\mathbb{Q} = X}} \mathfrak{a}(\mathbb{Q}) \prod_{P \in \mathbb{Q}} M(P|\psi), \tag{3.6b}$$

$$n!\, a(x_1)^\nu \ldots a(x_n)^\nu \tilde{\mathcal{M}}(x_1, \ldots, x_n|\psi) = \tilde{M}(X|\psi) \prod_{\substack{\text{distinct} \\ x \in X}} n(x)! \tag{3.6c}$$

for $X = \{x_1, \ldots, x_n\}$. The combinatorial factors $\mathfrak{a}(\mathbb{Q})$ were defined by Eq. (2.6). Equation (3.6a) is a special case of (3.6b, c) because $\mathbb{Q}$ consists of a single polymer X if $\operatorname{supp}\mathbb{Q}$ contain no point with multiplicity $\geqq 2$, and $\mathfrak{a}(\mathbb{Q}) = 1$ in this case. In general, the total number of arguments $x \in P$ in the amplitudes M on the right-hand side of (3.6b) is n, and the sum over clusters $\mathbb{Q}$ is a finite sum.

It follows from Eqs. (2.12) and (3.6b) that

$$\ln Z_1(X|\psi) = \sum_X \tilde{M}(X|\psi) = \sum_{n \geqq 1} \int_{x_1} \ldots \int_{x_n \in X} \tilde{\mathcal{M}}(x_1, \ldots, x_n|\psi) \tag{3.7}$$

in the sense of formal power series in the amplitudes $\mathcal{M}(\ldots)$. The sum is actually convergent if the Mayer amplitudes satisfy suitable bounds. For the computation of Greens functions this is not needed when the normalization condition (3.4) is imposed, and ψ may be considered as infinitesimal in this case. See the discussion in Sect. 2.

Now we turn to the computation of activities A (or Mayer amplitudes $\mathcal{M}$). Write φ_X for a field that vanishes outside X and set

$$\mathcal{Z}(X|\psi + \mathcal{A}\varphi_X) = \mathcal{Z}_1(X|\psi, \varphi_X) \exp\left[-\frac{\lambda}{4!} \int_{\text{base}} \psi(z)^4\right] \tag{3.8}$$

so that $\mathcal{Z}_1(\emptyset|\psi, \cdot) = 1$. We regard $\mathcal{Z}_1(X|\psi, \varphi_X)$ as partition function of a polymer system.

$$\mathcal{Z}_1(X|\psi, \varphi_X) = \sum_{X = \Sigma P} \prod_P B(P|\psi, \varphi_P) \quad \text{with} \quad \varphi_P(x) = \begin{cases} \varphi_X(x) & \text{for } x \in P \subseteqq X \\ 0 & \text{otherwise}. \end{cases} \tag{3.9}$$

The polymers of this polymer system will be called molecules, and their activities B will be called molecular activities. Arbitrary finite nonempty subsets of Λ are admitted as molecules. A pedestrian way to compute the molecular activities from the given interaction (3.3) uses the inversion formula (2.3) viz.

$$B(P|\psi,\varphi_P)=\sum_{n\geqq 1}(-1)^{n-1}(n-1)!\sum_{P=\Sigma_1^n X_i}\mathscr{Z}_1(X_i|\psi,\varphi_{X_i})\,. \tag{3.10}$$

A more convenient formula will be derived later on.

By inserting Eqs. (3.8), (3.9) into (3.2), $Z_1(X|\psi)$ becomes expressed as a Gaussian expectation of a partition function $\mathscr{Z}_1(X,\cdot\cdot)$ of a polymer system. According to the discussion in Sect. 2 [Eq. (2.19)] this implies that the corresponding activities are obtained as truncated expectation values

$$A(X|\psi)=\sum_{X=\Sigma P}\left\langle\prod_P[B(P|\psi,\varphi_P);]\right\rangle_{vX}. \tag{3.11}$$

The Mayer amplitudes are given in terms of A's by Eqs. (3.5b, c). The truncated expectation values can be expressed in terms of ordinary expectation values by the tree formula, Proposition 2.1. Finally we note again that the resulting Gaussian integrals are n-dimensional integrals if X has n points. Let us elaborate on this point.

Let X be a subset containing n elements of Λ, and suppose that the function F of the field φ depends only on $\varphi(x)$ for $x\in X$. Set

$$\varphi_x=a_j^d\varphi(x)\quad\text{for}\quad x\in X\cap\Lambda_j,\quad\text{with}\quad d=\tfrac{1}{2}(v-2)=\tfrac{1}{2}, \tag{3.12}$$

and consider the $n\times n$ matrix

$$\mathbb{V}=(\mathbb{V}_{xy})_{x,y\in X}$$

$$\mathbb{V}_{xy}=a_j^{2d}v(x,y)\quad\text{for}\quad x\in X\cap\Lambda_j \tag{3.13}$$

$\mathbb{V}_{xy}$ is zero unless both x and y are in the same layer $X\cap\Delta_j$. By hypothesis

$$F(\varphi)=f(\{\varphi_x\}_{x\in X})\,. \tag{3.14}$$

In this notation

$$\int d\mu_v(\varphi)F(\varphi)=(\det 2\pi\mathbb{V})^{-\frac{1}{2}}\int\left[\prod_{x\in X}d\varphi_x\right]e^{-\frac{1}{2}\Sigma_{x,y\in X}\varphi_x(\mathbb{V}^{-1})_{xy}\varphi_y}f(\{\varphi_z\})\,. \tag{3.15}$$

The right-hand side of this formula is an n-dimensional integral as promised. Equation (3.15) can be verified by using the standard formula for the Fourier transform of a Gaussian measure

$$\int d\mu_v(\varphi)e^{i(q,\varphi)}=e^{-\frac{1}{2}(q,vq)}\,. \tag{3.16}$$

Instead of using the explicit formula (3.10) for the molecular activities B in terms of the interaction, they can also be determined from a recursion relation which we will now describe. Let us set

$$\mathscr{Z}^Y(X|\psi+\mathscr{A}\varphi_X)=\mathscr{Z}(X+Y|\psi+\mathscr{A}\varphi_X)\quad\text{for}\quad X\cap Y=\emptyset\,. \tag{3.17}$$

The quantity $\mathscr{Z}_2^Y(X)=\mathscr{Z}^Y(X)/\mathscr{Z}^Y(\emptyset)$ obeys $\mathscr{Z}_2^Y(\emptyset)=1$ and may therefore be regarded as partition function of a polymer system. Denote the corresponding activities by $B^Y(P|\psi,\varphi_P)$. They obey the Kirkwood-Salsburg equations [for any

$x = x(P) \in P$]

$$\mathscr{Z}^Y(X|\psi + \mathscr{A}\varphi_X) = \sum_{\substack{P \\ x \in P \subsetneqq X}} B^Y(P|\psi, \varphi_P)\mathscr{Z}^Y(X - P|\psi + \mathscr{A}\varphi_{X-P}). \tag{3.18}$$

The molecular activities in which we are actually interested (for now) are

$$B(P|\psi, \varphi_P) = B^{\emptyset}(P|\psi, \varphi_P).$$

Let us write $P + x$ in place of $P \cup \{x\}$ if $x \notin P$, and φ_x in place of $\varphi_{\{x\}}$. φ_P is short for $\chi_P \varphi$, χ_P = characteristic function of $P \subset \Lambda$.

Proposition 3.1. *The generalized molecular activities obey the following recursion relation for* $w \notin P + Y$, $x \in P$

$$B^Y(P + w|\psi, \varphi_{P+w}) = \sum_{\substack{R \\ x \in R \subsetneqq P}} B^Y(P - R + w|\psi, \varphi_{P-R+w})$$
$$\cdot [B^{Y+w}(P|\psi + \mathscr{A}\varphi_w, \varphi_P) - B^Y(P|\psi, \varphi_P)],$$

and the initial condition

$$B^Y(\{x\}|\psi, \varphi_x) = \mathscr{Z}(Y + x|\psi + \mathscr{A}\varphi_x)/\mathscr{Z}(Y|\psi).$$

The notation $\mathscr{A}\varphi$ *was introduced in Eq.* (1.5)*; the other quantities were introduced earlier in this section.*

The proof of Proposition 3.1 is given in Appendix C. This recursion relation can be solved by methods similar to those of Sect. 2. We shall defer their further analysis to Part II and record here only one result.

Corollary 3.2. *If* P *consists of* n *points,* $4k - 3 \leqq n \leqq 4k$, *then*

$$B(P|\psi, \varphi_P) = \delta_{1,n} + O(\lambda^k).$$

It follows from Eq. (3.11) that the activities $A(X|\psi)$ share this property. Let us summarize the results obtained so far as

Theorem 3.3. *Let the molecular activities* $B(P|\psi, \varphi_P) \equiv B^{\emptyset}(P|\psi, \varphi_P)$ *be determined by the interaction Boltzmannian*

$$\mathscr{Z}(X|\psi) = \exp\left\{\delta e(X) + \int_{z \in \text{base}} \left[-\frac{\lambda}{4!}\psi(z)^4 + \tfrac{1}{2}\delta m^2(v_X|z)\psi(z)^2 \right]\right\}$$

via Eqs. (3.8), (3.10) *or Proposition 3.1. Then the Mayer amplitudes are expressed as truncated expectation values as follows. Let* $X = \{x_1, \ldots, x_n\}$. *Then*

$$n!\, a(x_1)^v \ldots a(x_n)^v \mathscr{M}(x_1, \ldots, x_n|\psi) = A(X|\psi) - \delta_{1,n},$$

$$A(X|\psi) = \sum_{X = \Sigma P} \left\langle \prod_P [B(P|\psi, \varphi_P);] \right\rangle_{v_X}. \tag{3.11}$$

The truncated expectation value is expressed in terms of n*-dimensional Gaussian integrals by Proposition 2.1 and Eq.* (3.15). *If* $4k - 3 \leqq n \leqq 4k$ *then*

$$\mathscr{M}(x_1, \ldots, x_n|\psi) = O(\lambda^k).$$

The generating function for connected free-propagator-amputated Greens functions is expressed in terms of Mayer amplitudes by Eqs. (1.1) and (3.6). Alternatively, the full (disconnected) free-propagator-amputated Greens functions may be obtained from Eqs. (1.19).

Let us note that the tree formula (Proposition 2.1) involves also integrations over $n-1$ real variables $s_i = 0\ldots1$, and summation over $n-1$ integers $\eta(a) = 1\ldots a-1$ (apart from symmetrization π). They replace summation over all Feynman graphs in standard perturbation theory. The summation over order n and arguments $x_1, \ldots, x_n$ in Eqs. (1.1) or (1.19) for the Greens functions replace the sum over orders and integrals over coordinate space in standard perturbation theory.

If one wishes to, one can compute the truncated expectation value (3.11) integrating out the fields φ^j on Λ_j in sequence, beginning with the highest frequencies j. How this is done is explained at the end of Sect. 2. The result is given by a sum of terms that are labelled by Gallavotti-Nicoló trees. The tree formula (Proposition 2.1) can be used to do the Gaussian integration in the recursion relation for vertex functions, Eq. (2.32).

Of course, the truncated expectation values (3.11) can also be evaluated without recourse to the tree formula (Proposition 2.1) by using their standard expression in terms of ordinary expectation values, and Eqs. (3.8), (3.10) can be used in place of the recursion relation (Proposition 3.1).

In Eq. (3.11) the renormalization cancellation occurs between different terms in the sum over partition $X = \sum P$ of X. In Part II we will show how to change this by giving the same s-dependence to the propagators that appear as arguments of counterterms and to the covariance of the Gaussian measure in the tree formula.

4. Relation with Perturbation Theory

The Mayer amplitudes may be expanded in formal (asymptotic) power series in the coupling constant λ. Theorem 3.3 asserts that $\mathcal{M}(x_1, \ldots, x_n|\psi)$ is of order λ^k if $4k-3 \leqq n \leqq 4k$. It follows from their definition (3.6) that the same is then also true for the augmented Mayer amplitudes $\tilde{\mathcal{M}}(x_1, \ldots, x_n|\psi)$. Therefore, if we truncate the series (1.1) for the generating function of the connected free-propagator-amputated Greens functions after the term $n = 4k$, the result will be correct to order λ^k. In other words, the result differs from the sum of all Feynman diagrams up to order k by higher order and nonperturbative contributions.

In order to deepen the reader's understanding of this fact we will now indulge in a pedagogical exercise. We will examine ν-dimensional $\lambda\phi^4$-theory without any counterterms on a lattice Λ of lattice spacing a, and its Mayer expansions on this simple lattice. This may be regarded as a simple special case of the general one – the staggered lattice has only one layer, and the counterterms are zero.

We will exhibit an explicit relation between the Mayer amplitudes, expanded in a formal power series in λ, and the Feynman amplitudes. Stated in words it says that the Mayer amplitude $\mathcal{M}(x_1, \ldots, x_n|\psi)$ equals the sum of all "point connected" Feynman diagrams whose arbitrarily many vertices occupy the n distinct points $x_1, \ldots, x_n$ of the lattice Λ. A Feynman diagram with given positions of its vertices is called *point connected* if it is connected or becomes connected when vertices that are positioned at the same site of the lattice are identified.

The partition function at zero external field for the simplified model is

$$Z(\Lambda|0)=\int d\mu_v(\phi)\exp\left[-\frac{\lambda}{4!}\int_{z\in\Lambda}\phi(z)^4\right]. \tag{4.1}$$

In perturbation theory, the partition function is obtained as a sum of coordinate space integrals associated with arbitrary not necessarily connected vacuum Feynman diagrams.

$Z(\Lambda|0)=\sum$ (disconnected vacuum Feynman diagrams)

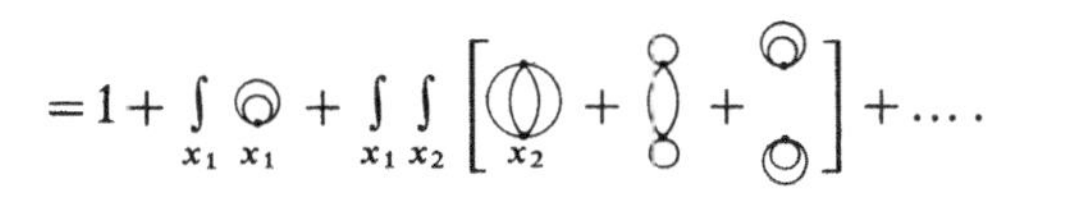

(4.2)

We adopt the convention that Feynman integrands associated with arbitrary diagrams are to be evaluated according to the standard rules, including combinatorial factors. The combinatorial factors coming from symmetries under permutations of disconnected pieces imply, for instance

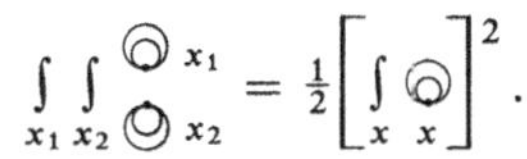

(4.3)

As a result of these combinatorial factors, the series of $Z(\Lambda|0)$ exponentiates to a sum of integrals associated with connected vacuum diagrams,

$\ln Z(\Lambda|0)=\sum$ (connected vacuum diagrams)

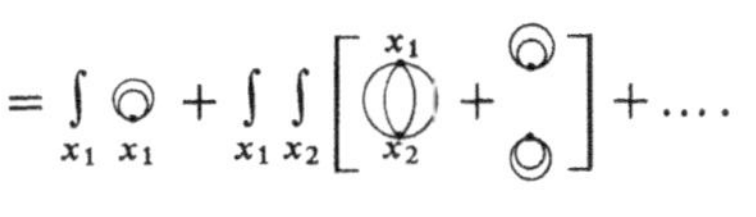

(4.4)

Now we perform a reordering and partial resummation. In (4.2) we collect all the terms in the combined sum, over graphs and over n-tuples of points $(x_1, \ldots, x_k)$ with arbitrary $k \geqq n$, whose vertices $\{x_i\}_{i=1\ldots k}$ occupy the same set $Y=\{y_1, \ldots, y_n\} \subseteqq \Lambda$ of n distinct lattice points.

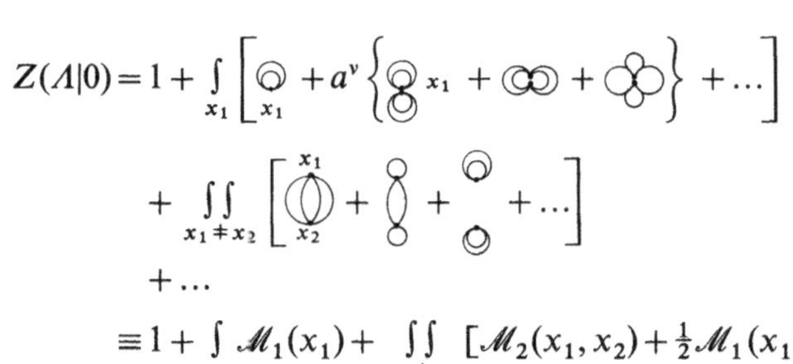

$$\equiv 1+\int_{x_1}\mathcal{M}_1(x_1)+\iint_{x_1 \neq x_2}[\mathcal{M}_2(x_1,x_2)+\tfrac{1}{2}\mathcal{M}_1(x_1)\mathcal{M}_1(x_2)]+\ldots \tag{4.5}$$

etc., $\mathcal{M}_n(x_1, \ldots, x_n)$ is the sum of all point connected vacuum Feynman diagrams whose vertices are positioned at n distinct sites $y_1, \ldots, y_n$ of the lattice. There are infinitely many such diagrams because arbitrarily many vertices may be positioned on the same site x. In effect we are summing up diagrams with arbitrarily many propagators joining any two (identical or distinct) sites $y_1, \dots, y_n$ of the lattice.

By construction, $\mathcal{M}_n(x_1, \dots, x_n)$ will be of order λ^n as $\lambda \to 0$. An expansion for $\ln Z(\Lambda|0)$ is obtained from the expansion for $Z(\Lambda|0)$ by multiplying $\mathcal{M}_n$ with ξ^n, expanding in powers of ξ and setting $\xi = 1$ in the end. This expansion will include correction terms that come from the constraints $x_1 \neq x_2$ etc. in the sum for $Z(\Lambda|0)$ ("excluded volume effect").

$$\begin{aligned} \ln Z(\Lambda|0) = & \int_{x_1} \mathcal{M}_1(x_1) + \iint_{x_1 \neq x_2} \mathcal{M}_2(x_1, x_2) + \dots \\ & - \tfrac{1}{2} a^\nu \int_{x_1} \mathcal{M}_1(x_1)^2 - \dots \\ & \pm \dots \\ = & \int_{x_1} \tilde{\mathcal{M}}_1(x_1) + \iint_{x_1 x_2} \tilde{\mathcal{M}}_2(x_1, x_2) + \dots \end{aligned} \tag{4.6}$$

with

$$\begin{aligned} \tilde{\mathcal{M}}_1(x_1) &= \mathcal{M}_1(x_1) \\ \tilde{\mathcal{M}}_2(x_1, x_2) &= \begin{cases} \mathcal{M}_2(x_1, x_2) & \text{if} \quad x_1 \neq x_2 \\ -\tfrac{1}{2} \mathcal{M}_1(x_1)^2 & \text{if} \quad x_1 = x_2 \,. \end{cases} \end{aligned} \tag{4.7}$$

The whole discussion can be repeated for $Z(\Lambda|\psi)$ with general external field ψ (see Appendix A). One starts from

$$Z(\Lambda|\psi) = 1 + \int_{x_1}$$
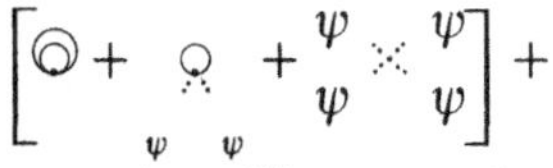
$$+ \dots . \tag{4.8}$$

The Mayer amplitudes $\mathcal{M}_n(x_1, \dots, x_n)$ will now depend on ψ through $\psi(x_1), \dots, \psi(x_n)$. The general form of the expansion for $Z(\Lambda|\psi)$ is

$$\begin{aligned} Z(\Lambda|\psi) = 1 + & \sum_{N \geq 1} \sum_k \sum_{\substack{\{\ell_j\}_1^k \\ \Sigma \ell_j = N}} c(\{l_j\}) \underbrace{\int \dots \int}_{\substack{x_1 \quad x_n \\ \text{distinct}}} \mathcal{M}_{\ell_1}(x_1, \dots, x_{\ell_1}) \\ & \cdot \mathcal{M}_{\ell_2}(x_{\ell_1 + 1}, \dots, x_{\ell_1 + \ell_2}) \dots \mathcal{M}_{\ell_k}(\dots x_N) \,. \end{aligned} \tag{4.9}$$

The combinatorial factor is

$$c(\{\ell_j\}) = \prod_{\ell = 1}^{\infty} \frac{1}{m_\ell !} \tag{4.10}$$

if m_ℓ of the integers $\ell_j \geq 1$ equal ℓ.

Let us state the relation between Mayer amplitudes and connected Feynman diagrams in general terms.

Consider $n \times \ell$ matrices $k^{(\ell)} = (k_{ij}^{(\ell)})$ whose entries are nonnegative integers. By a permutation of rows or columns, such a matrix can be brought to block form. We call it *irreducible* if it consists of only one block and no row or column is identically 0. We write $\mathcal{F}(y_1, \dots, y_k)$ for the sum of all connected k^{th} order vacuum diagrams with vertices positioned at $y_1, \dots, y_k$. We abbreviate

$$\mathcal{F}(\underbrace{x_1, \dots, x_1}_{m_1}, x_2 \dots, \underbrace{x_n, \dots, x_n}_{\substack{m_n \\ \text{arguments}}}) \equiv \mathcal{F}(x_1^{m_1}, \dots, x_n^{m_n}) \,. \tag{4.11}$$

We shall not indicate the dependence on the external field ψ explicitly.

Proposition 4.1.

$$\mathscr{M}_n(x_1,\dots,x_n)=\frac{1}{n!}\sum_{\ell\geqq 1}\sum_{k^{(\ell)}}\frac{\prod_{j=1}^{\ell}\left(\sum_{i=1}^{n}k_{ij}\right)!}{\prod_{j=1}^{\ell}\prod_{i=1}^{n}k_{ij}!}\frac{1}{\ell!}\prod_{j=1}^{\ell}\mathscr{F}(x_1^{k_{1j}},\dots,x_n^{k_{nj}}) \tag{4.12}$$

as formal power series in λ. The inner summation is over all irreducible $n\times\ell$ matrices $k^{(\ell)}$ whose entries are nonnegative integers.

Proof. Write

$$Z(\Lambda|\psi)=\int d\mu_v(\phi)\prod_{x\in\Lambda}e^{-\lambda\mathscr{V}(\varphi_x)},$$

where $\mathscr{V}$ depends on ψ, and $\varphi_x=a^d\varphi(x)$, $d=\frac{1}{2}(v-2)$. Both the Mayer amplitude and the Feynman amplitude may be expressed as truncated expectation values

$$\mathscr{M}(x_1,\dots,x_n)=\frac{a^{-vn}}{n!}\left[\left\langle\prod_{i=1}^{n}[e^{-\lambda\mathscr{V}(\varphi_{x_i})};]\right\rangle_v-\delta_{1,n}\right] \tag{4.13}$$

and

$$\mathscr{F}(y_1,\dots,y_k)=\frac{1}{k!}\left\langle\prod_{j=1}^{k}[-\lambda\mathscr{V}(\varphi_{y_j});]\right\rangle_v.$$

In the first expression one expands

$$e^{-\lambda\mathscr{V}(\varphi_x)}=\sum_{\ell=0}^{\infty}\frac{1}{\ell!}[-\lambda\mathscr{V}(\varphi_x)]^{\ell}.$$

Then one expresses the resulting "partially truncated" expectation value of a product of factors $-\lambda\mathscr{V}(\varphi_.)$ in terms of completely truncated expectation values with the help of the following

Lemma 4.2.

$$\left\langle\prod_{i=1}^{n}[F_i(\varphi_{y_i})^{n_i};]\right\rangle_v$$

$$=\sum_{\ell\geqq 1}\sum_{\substack{k^{(\ell)}\\ \sum_{j=1}^{\ell}k_{ij}^{(\ell)}=n_i}}\frac{\prod_{i=1}^{n}n_i!}{\prod_{i=1}^{n}\prod_{j=1}^{\ell}k_{ij}^{(\ell)}!}\frac{1}{\ell!}\prod_{j=1}^{\ell}\left\langle\prod_{i=1}^{n}[F_i(\varphi_{y_i});]^{k_{ij}^{(\ell)}}\right\rangle_v. \tag{4.14}$$

Summation is over irreducible $n\times\ell$ matrices $k^{(\ell)}$ as above which satisfy indicated constraint.

Sketch of the Proof of Lemma 4.2. One considers first the ordinary expectation value

$$\langle F_1(\varphi_{y_1})^{n_1}\dots F_n(\varphi_{y_k})^{n_k}\rangle.$$

Using the defining relation (2.18) of truncated expectation values, this can be expressed as a truncated expectation value of k factors $F_i(\varphi_{y_i})^{n_i}$ and also as a truncated expectation value of $N=\sum n_i$ factors $F_i(\varphi_{y_i})$. It suffices to show that the two expressions become equal when Eq. (4.14) is inserted into the first. This is so because of uniqueness of the activities of a polymer system, cf. remark after Eq. (2.18). After some combinatorics it is found that they are indeed equal. Details are spelled out in [21]. □

We mentioned in the introduction that standard Feynman perturbation theory may be put on a staggered lattice, graph by graph. The new graphs will involve nonlocal vertices which can connect different layers, and propagators that link only points in the same layer. Using this, the results of this section could be extended to the Mayer amplitudes on a staggered lattice. We think, however, that they are good enough as they stand to give the general idea.

To make sure that there is no misunderstanding, let us emphasize one fact at the end of this perturbation theoretical section. Mayer amplitudes are *not* defined as divergent sums of Feynman amplitudes. They are *defined* as perfectly meaningful truncated expectation values like (4.13). These well defined expressions admit an asymptotic power series expansion in λ whose coefficients are Feynman amplitudes.

5. The Special Case of a Lattice with only One Point

The partition function for $\lambda\phi^4$-theory on a lattice $\Lambda=\{0\}$ that consists of a single point is given by a convergent 1-dimensional integral

$$Z(\{0\}|\psi)=(2\pi v)^{-1/2}\int_{\mathbb{R}} d\varphi \exp\left[-\tfrac{1}{2}v^{-1}\varphi^2-\frac{\lambda}{4!}(\varphi+\psi)^4+\tfrac{1}{2}\delta m^2(\varphi+\psi)^2+\delta e\right],$$
$$\lambda>0,\ v>0,\ \text{real}. \qquad (5.1)$$

Its perturbation expansion in powers of λ will be divergent no matter how δm^2 and δe depend on λ, because the integral makes no sense for $\lambda<0$, and is therefore not analytic at $\lambda=0$ [23].

On the other hand, a Mayer expansion for this model amounts to "do nothing". There is only a single Mayer amplitude, which is given by a convergent 1-dimensional integral, and

$$Z(\{0\}|\psi)=\mathcal{M}(0|\psi)\exp\left[-\frac{\lambda}{4!}\psi^4\right]. \qquad (5.2)$$

Greens functions are obtained by differentiation at $\psi=0$. They do not involve divergent sums.

Appendix A.
Generating Function for Free-Propagator-Amputated Greens Functions

In a theory with free propagator v, the connected, free-propagator-amputated Greens functions $G_c(\underline{z}_1, \ldots, \underline{z}_n)$ are related to the corresponding unamputated Greens function $G_c(z_1, \ldots, z_n)$ by

$$G_c(z_1, \ldots, z_n)=\int_{z_1'} \ldots \int_{z_n'} v(z_1, z_1')\ldots v(z_n, z_n')G_c(\underline{z}_1', \ldots, \underline{z}_n') \quad (n\geqq 3) \qquad (A.1)$$

and

$$G_c(z_1, z_2)=v(z_1, z_2)+\iint_{z_1' z_2'} v(z_1, z_1')v(z_2, z_2')G_c(\underline{z}_1', \underline{z}_2')=G(z_1, z_2).$$

Amputation is indicated by underlining the argument. The Greens functions can be obtained by differentiation from generating functions

$$G_c(z_1, \ldots, z_n) = \frac{\delta^n}{\delta J(z_1) \ldots \delta J(z_n)} G(J)|_{J=0}, \tag{A.2}$$

$$G_c(\underline{z}_1, \ldots, \underline{z}_n) = \frac{\delta^n}{\delta \psi(z_1) \ldots \delta \psi(z_n)} F(\psi)|_{\psi=0}. \tag{A.3}$$

The above relations correspond to the following relation between generating functions

$$F(\psi = vJ) = G(J) - \tfrac{1}{2} J v J \tag{A.4}$$

in obvious notation. In a theory with interaction Lagrangian V, the generating function $G(J)$ is the logarithm of the partition function in the presence of a source

$$e^{G(J)} = \int d\mu_v(\phi)\, e^{-V(\phi) + J\phi}. \tag{A.5}$$

The free field measure is

$$d\mu_v(\phi) = (\det 2\pi v)^{-1/2}\, e^{-\frac{1}{2}\phi v^{-1}\phi} \mathscr{D}\phi. \tag{A.6}$$

It follows that

$$\begin{aligned} d\mu_v(\phi - \psi) &= e^{-\frac{1}{2}\psi v^{-1}\psi + \phi v^{-1}\psi} d\mu_v(\phi) \\ &= e^{-\frac{1}{2}JvJ + \phi J} d\mu_v(\phi) \quad \text{if} \quad \psi = vJ. \end{aligned} \tag{A.7}$$

Let us define

$$Z(\psi) = \int d\mu_v(\phi)\, e^{-V(\phi + \psi)}. \tag{A.8}$$

We show that

$$F(\psi) = \ln Z(\psi) \tag{A.9}$$

obeys the relation (A.4).

Making a shift of the field in (A.8) and using Eq. (A.7) we obtain

$$\begin{aligned} Z(\psi) &= \int d\mu_v(\phi)\, e^{-\frac{1}{2}JvJ + \phi J} e^{-V(\phi)} \\ &= e^{G(J) - \frac{1}{2}JvJ} \quad \text{if} \quad \psi = vJ. \end{aligned} \tag{A.10}$$

Making a shift in the field, we must verify that boundary conditions at infinity are preserved. They are preserved if ψ falls to zero at infinity. This will be the case for locally supported source J if v is a massive propagator. Equation (A.10) agrees with the desired relation (A.4). q.e.d.

The connected free-propagator-amputated Greens functions are not 1-particle irreducible, and must not be confused with the so-called vertex-functions ≡ full-propagator-amputated 1-particle irreducible Greens functions whose generating function is related to $G(J)$ by a Legendre transform.

Appendix B. Block Spins, Propagators, and Kernels $\mathscr{A}$

We will follow the procedure of Kupiainen and Gawedzki [9] to carry out the split (1.5), (1.6) of the field ϕ and propagator $v>0$ on base space. No summation will be implied by repeated indices.

Let us assume that there is a finest lattice Λ_N= base space. Results for continuum base space may be obtained by letting $N\to\infty$. As we have mentioned in the main text, the points $x\in\Lambda_j$ may be considered as blocks of side length a_j in base space.

Given a field $\phi=\phi_N$ on base space, we define the blockspin $\Phi^j(x)$ on Λ_j as block average of ϕ

$$\Phi^j(x)=\underset{z\in x}{av}\,\phi(z)\equiv C^j\phi(x)\,. \tag{B.1}$$

Accordingly, the free propagators [covariances] v and u_j of the [Gaussian] fields ϕ and Φ^j are related by

$$u_j=C^j v C^{j*}>0\,. \tag{B.2}$$

$*$ is the adjoint with respect to the scalar products (,) that are furnished by integration over base space and Λ_j, respectively. The field ϕ may be split into a part that is determined by the block spin Φ^j, and a fluctuation field ζ^{j+1} on base space that has zero block average.

$$\phi=\mathscr{A}^j\Phi^j+\zeta^{j+1}\quad\text{with}\quad C^j\zeta^{j+1}=0\,. \tag{B.3}$$

In the special case $j=N$ this holds with $\zeta^{N+1}=0$, $\mathscr{A}^N=1$. The kernel $\mathscr{A}^j$ is chosen as

$$\mathscr{A}^j=vC^{j*}u_j^{-1}\,, \tag{B.4}$$

so that the two pieces on the right-hand side are orthogonal with respect to the scalar product $\langle\ ,\ \rangle$ that is specified by the inverse propagator

$$\langle f,g\rangle=(f,v^{-1}g)\,.$$

Evidently

$$C^j\mathscr{A}^j=1\,. \tag{B.5}$$

We will define the pieces φ^k of the field ϕ in such a way that

$$\zeta^j=\sum_{k\geqq j}\mathscr{A}^k\varphi^k\,. \tag{B.6}$$

Expressed in words, the fluctuation field ζ^j in the sum of the parts of the field φ with "frequency" $k\geqq j$ (appropriately transported to base space with kernels $\mathscr{A}^k$). For $j=N+1$ this relation holds with $\varphi^{N+1}=0$. Insert (B.6) into (B.3) to obtain

$$\phi=\mathscr{A}^j\Phi^j+\sum_{k>j}\mathscr{A}^k\varphi^k\,. \tag{B.7}$$

Suppose this relation holds for some j. We determine φ^j so that it holds for $j-1$.

To this end we define the operation $C_{j-1,j}$ which averages functions on Λ_j over cells of Λ_{j-1}. Evidently, definition (B.1) implies

$$\Phi^{j-1}=C_{j-1,j}\Phi^j\,. \tag{B.8}$$

We split the field Φ^j into a piece that is determined by its block average Φ^{j-1} and a part φ^j on Λ_j whose block average vanishes,

$$\Phi^j = \mathscr{A}_{j,j-1}\Phi^{j-1} + \varphi^j, \tag{B.9}$$

with

$$C_{j-1,j}\varphi^j = 0. \tag{B.10}$$

The kernel $\mathscr{A}_{j,j-1}$ is chosen as

$$\mathscr{A}_{j,j-1} = u_j C^*_{j-1,j} u^{-1}_{j-1}, \tag{B.11}$$

so that

$$\mathscr{A}^j \mathscr{A}_{j,j-1} = \mathscr{A}^{j-1}. \tag{B.12}$$

Inserting Eqs. (B.9) and (B.12) into (B.7), we see that (B.7) remains true for $j-1$, as promised.

Equation (B.9), together with definitions (B.1) of the block spins and (B.11), (B.2) of the kernel $\mathscr{A}_{..}$, defines the pieces φ^j of the field for all $j \geqq 1$ in such a way that (B.7) holds for all j. For $j=0$ this equation is of the form (1.5) if we set, finally

$$\varphi^0 = \Phi^0. \tag{B.13}$$

$\mathscr{A}(z,x)$ is the integral kernel of $\mathscr{A}^j$ if $x \in \Lambda_j$, viz.

$$\mathscr{A}^j f(z) = \int_{x \in \Lambda_j} \mathscr{A}(z,x) f(x). \tag{B.14}$$

It remains to determine the free propagator [covariance] v of φ^j. Regard φ^j as a function of ϕ. We may then use the definition (B.9), (B.13) of φ^j to compute

$$v(x,y) = \langle \varphi^j(x)\varphi^k(y)\rangle \quad \text{for} \quad x \in \Lambda_j,\ y \in \Lambda_k. \tag{B.15}$$

Using the relation (B.8) and the known propagator u_j of Φ^j, it turns out to be of the form

$$v(x,y) = \delta_{jk} v^j(x,y)$$

with

$$\begin{aligned} v^j &= u_j - u_j C^*_{j-1,j} u^{-1}_{j-1} C_{j-1,j} u_j \quad \text{for} \quad j \geqq 1, \\ v^0 &= u_0. \end{aligned} \tag{B.16}$$

v is positive semidefinite by definition (B.15). It follows from Eq. (B.15) and (1.15), viz.

$$\phi = \sum_{j=0}^{N} \mathscr{A}^j \varphi^j \tag{B.17}$$

that the relation (1.6) between propagators holds. That is

$$v = \sum_j \mathscr{A}^j v^j \mathscr{A}^{j*} \equiv \mathscr{A} v \mathscr{A}^*. \tag{B.18}$$

Decay properties of kernels $\mathscr{A}$ and propagators v were studied in [9, 22]. Finally we should transcribe Gaussian measures. We claim that

$$\int d\mu_v(\phi) F(\phi) = \int d\mu_v(\varphi) F(\mathscr{A}\varphi). \tag{B.19}$$

It suffices to verify this for $F(\phi)=e^{i(q,\phi)}$. Using the known Fourier transformation (3.16) of Gaussian measures, one finds

$$\int d\mu_v(\phi)\, e^{i(q,\phi)} = e^{-\frac{1}{2}(q,vq)} = e^{-\frac{1}{2}(q,\mathscr{A}v\mathscr{A}^*q)},$$

while

$$\int d\mu_v(\varphi)\, e^{i(q,\mathscr{A}\varphi)} = e^{-\frac{1}{2}(\mathscr{A}^*_q, v\mathscr{A}^*_q)} = e^{-\frac{1}{2}(q,\mathscr{A}v\mathscr{A}^*_q)}. \quad \text{q.e.d.}$$

Using Eq. (B.19) it follows from the formula (A.8) for the generating function $\ln Z(\psi)$ of connected Greens functions, and definitions (1.7), (1.9a) that the relation (1.10) for $\ln Z(\psi)$ holds for finite volume and cutoff.

Appendix C. Proof of the Recursion Relation (Proposition 3.1)

We start from the Kirkwood-Salsburg Eq. (3.18) which defines the generalized molecular activities B^Y. Iterating it once we obtain, for $w \notin X+Y$

$$\mathscr{Z}^Y(X+w|\psi+\mathscr{A}\varphi_{X+w}) = \sum_{\substack{P \\ x\in P\subsetneqq X}} C^Y(P+w|\psi,\varphi_{P+w})\,\mathscr{Z}^Y(X-P|\psi+\mathscr{A}\varphi_{X-P}) \tag{C.1}$$

with

$$C^Y(P+w|\psi,\varphi_{P+w}) = B^Y(P+w|\psi,\varphi_{P+w}) + \sum_{\substack{Q \\ x\in Q\subsetneqq P}} B^Y(Q|\psi,\varphi_Q)\,B^Y(P-Q+w|\psi,\varphi_{P-Q+w}). \tag{C.2}$$

$C^Y(\ldots)$ can be determined recursively from Eq. (C.1). Therefore, two expressions for $C^Y(\ldots)$ which both satisfy Eq. (C.1) must necessarily be equal. By definition (3.17) we have

$$\mathscr{Z}^Y(X+w|\psi+\mathscr{A}\varphi_{X+w}) = \mathscr{Z}^{Y+w}(X|\psi'+\mathscr{A}\varphi_X), \qquad \psi'=\psi+\mathscr{A}\varphi_w.$$

Inserting the Kirkwood-Salsburg Eq. (3.18) for the right-hand side gives

$$\mathscr{Z}^Y(X+w|\psi+\mathscr{A}\varphi_{X+w}) = \sum_{\substack{P \\ x\in P\subsetneqq X}} B^{Y+w}(P|\psi',\varphi_P)\,\mathscr{Z}^{Y+w}(X-P|\psi'+\mathscr{A}\varphi_{X-P}).$$

But

$$\begin{aligned}\mathscr{Z}^{Y+w}(X-P|\psi'+\mathscr{A}\varphi_{X-P}) &= \mathscr{Z}^Y(X-P+w|\psi+\mathscr{A}\varphi_{X-P+w}) \\ &= \sum_{\substack{Q \\ w\in Q\subsetneqq X-P+w}} B^Y(Q|\psi,\varphi_Q)\,\mathscr{Z}^Y(X-P+w-Q|\psi+\mathscr{A}\varphi_{X-P+w-Q})\end{aligned}$$

by the Kirkwood-Salsburg equation. Inserting this and setting

$$P+Q-w=R \quad \text{so that} \quad x\in R\subsetneqq X,\ Q=R-P+w,\ P\subsetneqq R,$$

we obtain

$$\mathscr{Z}^Y(X+w|\psi+\mathscr{A}\varphi_{X+w}) = \sum_{\substack{R \\ x\in R\subsetneqq X}} \left\{ \sum_{\substack{P \\ x\in P\subsetneqq R}} B^{Y+w}(P|\psi',\varphi_P)\,B^Y(R-P+w|\psi,\varphi_{R-P+w}) \right\} \cdot \mathscr{Z}^Y(X-R|\psi+\mathscr{A}\varphi_{X-R}).$$

This is another equation of the form (C.2). Equating the coefficient in { } to $C^Y(R+w|\ldots)$ produces the recursion relation (Proposition 3.1).

Acknowledgements. This work was begun while one of the authors (G.M.) was at the Institute for Advanced Studies in Princeton, NJ. He would like to thank S. Adler and R. Dashen for their hospitality, and the Deutsche Forschungsgemeinschaft for a travel grant. He would also like to thank P.K. Mitter for the invitation to a topical meeting on renormalization group problems in Paris in spring 1984, and its participants for many stimulating discussions. Finally we would like to thank G. Gallavotti and F. Nicoló for providing a copy of their paper [12] before publication.

References

1. Symanzik, K.: Euclidean quantum field theory. I. Equations for a scalar model. J. Math. Phys. **7**, 510 (1966)
2. Brydges, D.: A rigorous approach to Debye screening in dilute classical Coulomb systems. Commun. Math. Phys. **58**, 313 (1978)
   Brydges, D., Federbush, P.: Debye screening. Commun. Math. Phys. **73**, 197 (1980)
3. Göpfert, M., Mack, G.: Iterated Mayer expansions for classical gases at low temperatures. Commun. Math. Phys. **81**, 97 (1981)
4. Göpfert, M., Mack, G.: Proof of confinement of static quarks in 3-dimensional $U(1)$ lattice gauge theory for all values of the coupling constant. Commun. Math. Phys. **82**, 545 (1982)
5. Glimm, J., Jaffe, A.: Positivity of the φ_3^4-Hamiltonian. Fortschr. Phys. **21**, 327 (1973)
6. Magnen, J., Seneor, R.: The infrared behavior of $(\nabla\phi)_3^4$ theory. Ann. Phys. **152**, 130 (1984). See also: Phase space cell expansion and Borel summability of the Euclidean φ_3^4-theory. Commun. Math. Phys. **56**, 237 (1977)
7. Battle, G., Federbush, P.: A phase cell cluster expansion for Euclidean field theories. Ann. Phys. **142**, 95 (1982); A phase cell cluster expansion for a hierarchial ϕ_3^4-model. Commun. Math. Phys. **88**, 263 (1983)
8. Battle, G.: A new combinatoric estimate for cluster expansions. Commun. Math. Phys. **94**, 133 (1984)
   Battle, G., Federbush, P.: A note on cluster expansions, tree graph identities, extra $1/N!$ factors!!! Lett. Math. Phys. **8**, 55–57 (1984)
9. Kupiainen, A., Gawedzki, K.: A rigorous block spin approach to massless lattice theories. Commun. Math. Phys. **77**, 31 (1980); Block spin renormalization group for dipole gas and $(\nabla\phi)^4$. Ann. Phys. **147**, 198 (1983)
10. Feldman, J., Magnen, J., Rivasseau, V., Seneor, R.: Methodes pour la theorie constructive des champs renormalisables asymptotiquement libres, preprint école polytechnique, Palaiseau, June 1984
11. Kupiainen, A., Gawedzki, K.: Massless lattice $\lambda\phi^4$-theory: A nonperturbative control of a renormalizable model. Preprint IHES Bures-sur-Yvette, June 1984
12. Gallavotti, G., Nicoló, F.: Renormalization theory in four dimensional scalar fields. Rutgers preprint, spring 1984
13. Bogolubov, N.N., Shirkov, D.V.: Introduction to the theory of quantized fields. New York: Interscience 1959
14. Zimmermann, W.: Convergence of Bogolubov's method of renormalization in momentum space. Commun. Math. Phys. **15**, 208 (1969)
15. Wilson, K.: Renormalization group and strong interactions. Phys. Rev. D **3**, 1818 (1971)
16. Gruber, C., Kunz, H.: General properties of polymer systems. Commun. Math. Phys. **22**, 133 (1971). See also: Gallavotti, G., Miracle Solé, S.: Correlation functions of a lattice system. Commun. Math. Phys. **7**, 274 (1968)
17. Ruelle, D.: Statistical mechanics. New York: Benjamin 1969
18. Seiler, E.: Gauge theories as a problem in constructive quantum field theory and statistical mechanics. Lecture Notes in Physics, Vol. 159. Berlin, Heidelberg, New York: Springer 1982
    Fredenhagen, K., Marcu, M.: Charged states in $\mathbb{Z}_2$ gauge theory. Commun. Math. Phys. **92**, 81 (1983) (Appendix)
    Göpfert, M.: Das Peierls Argument für $Z(2)$-Gitter-Eichtheorien. Diplomarbeit Hamburg 1979 (unpublished)

Malyshev, V.A.: Uniform cluster expansions for lattice models. Commun. Math. Phys. **64**, 131 (1979);
Münster, G.: High temperature expansion for the free energy of vortices and the string tension in lattice gauge theory. Nucl. Phys. B **180** [FS2], 23 (1981)
19. Glimm, J., Jaffe, A.: Quantum physics. Berlin, Heidelberg, New York: Springer 1981
20. Brydges, D., Federbush, P.: A new form of the Mayer expansion in classical statistical mechanics. J. Math. Phys. **19**, 2604 (1978)
21. Pordt, A.: Mayer-Entwicklungen für Euklidische Gitterfeldtheorie: Konvergenzeigenschaften und Beziehung zur Störungstheorie. Diplomarbeit Hamburg 1984
22. Balaban, T.: Regularity and decay of lattice Green's functions. Commun. Math. Phys. **89**, 571 (1983)
23. Dyson, F.: Divergence of perturbation theory in quantum electrodynamics. Phys. Rev. **85**, 631 (1952)

Communicated by A. Jaffe

Received August 31, 1984; in revised form September 21, 1984

Renormalization of the Higgs Model: Minimizers, Propagators and the Stability of Mean Field Theory *

Tadeusz Bałaban **, John Imbrie, and Arthur Jaffe

Lyman Laboratory of Physics, Harvard University, Cambridge, MA 02138, USA

Dedicated to the memory of Kurt Symanzik

Abstract. We study the effective actions $S^{(k)}$ obtained by k iterations of a renormalization transformation of the U(1) Higgs model in $d=2$ or 3 space-time dimensions. We identify a quadratic approximation $S_Q^{(k)}$ to $S^{(k)}$ which we call mean field theory, and which will serve as the starting point for a convergent expansion of the Green's functions, uniformly in the lattice spacing. Here we show how the approximations $S_Q^{(k)}$ arise and how to handle gauge fixing, necessary for the analysis of the continuum limit. We also establish stability bounds on $S_Q^{(k)}$, uniformly in k. This is an essential step toward proving the existence of a gap in the mass spectrum and exponential decay of gauge invariant correlations.

Table of Contents

---

* Supported in part by the National Science Foundation under Grant PHY 82-03669

** Current address: Department of Mathematics, University of Michigan, Ann Arbor, MI 48109, USA

1. The Model

We study the Higgs-type model of a U(1) gauge field u coupled to a complex scalar field ϕ with a quartic self interaction. The total action functional is given on the unit lattice as

$$S(u,\phi)=\sum_p e(\varepsilon)^{-2}(1-\mathrm{Re}\,u(p))+\tfrac{1}{2}\sum_b |(D_u\phi)_b|^2+\sum_x P(\phi_x)+E\,. \tag{1.1}$$

The relation between the ε-lattice and unit-lattice actions is described in the Cargèse lectures [1], as well as a general introduction to mass generation in this model. This approximation with Reu is called the Wilson approximation to the continuum action.

The lattice points $x\in Z^d$ are endpoints of directed, unit length lattice bonds $b=(b_-,b_+)$ which bound unit squares (plaquettes) denoted by p. The gauge field u: bonds $\to$ U(1), and

$$u(p)\equiv\prod_{b\in\partial p}u_b\,.$$

Furthermore, we assume $b^{-1}=(b_+,b_-)$ and $u_b^{-1}=u_{b^{-1}}$. The covariant derivative $D_u\phi$ is defined by $(D_u\phi)_b=u_b\phi_{b_+}-\phi_{b_-}$, and the coupling constant $e(\varepsilon)$ is $e(\varepsilon)=e\varepsilon^{(4-d)/2}$. We consider $d=2$ or 3 space-time dimensions, in which case the self interaction is

$$\mathscr{P}(\xi)=\lambda(\varepsilon)\,|\xi|^4-\tfrac{1}{4}(m^2+\delta m^2)\varepsilon^2|\xi|^2\,, \tag{1.2}$$

where $\lambda(\varepsilon)=\lambda\varepsilon^{4-d}$. The mass shift is consistent with perturbation theory with

$$\delta m^2(\varepsilon)=\begin{cases}O(\ln\varepsilon^{-1}), & d=2\,,\\ O(\varepsilon^{-1}), & d=3\,.\end{cases} \tag{1.3}$$

The constant $E=E(e,\lambda,\varepsilon,\Lambda)$ is chosen to include vacuum energy normalization and renormalization counterterms from perturbation theory.

The continuum limit of the model with the action (1.1) has not yet been proved to exist, but our methods should lead to that result. The continuum limit of another action, the so-called Gaussian action, has been extensively studied. The Gaussian action is obtained by replacing Re$u(p)$ in (1.1) with its quadratic approximation in Lie algebra variables. In two dimensions the existence theorem for the Gaussian action was established in [2]. In three dimensions and with a massive gauge field, the fundamental stability bounds were proved in [3] and the existence theorems based on these bounds are deduced in [4]. The infinite volume limit in these papers was analyzed using correlation inequalities, a method which is well suited for establishing existence theorems but which has not yet led to results on the nature of the spectrum in the resulting model. A lattice cutoff version of our spectral results is given in [5], where we also discuss the connection between Gaussian and Wilson actions on a lattice.

Our program is to study the Wilson action using a cluster expansion method. In this way we expect to both prove the existence of the theory and to establish properties of the particle spectrum. We expect that the Wilson action yields the same limiting continuum field theory as the Gaussian action. Hence the resulting invariance properties of the ground state and the properties of the particle spectrum which are consequences of the cluster expansion will also extend to the Gaussian model constructed previously. The Wilson action furthermore gives a natural setting for a possible generalization of these results to a nonabelian gauge model.

The basic strategy to gain control over the spectrum is to establish exponential decay of gauge invariant correlations. We develop a method of steepest descent to study integrals of the form

$$\int \psi e^{-S}\mathcal{D}u\mathcal{D}\phi / \int e^{-S}\mathcal{D}u\mathcal{D}\phi\,, \tag{1.4}$$

where ψ is a gauge invariant functional of u and ϕ. Thus we need to define an expansion $S(u,\phi)=S(u_{cl},\phi_{cl})+\dots$ in terms of a classical field (u_{cl},ϕ_{cl}) and a fluctuation field $(uu_{cl}^{-1},\phi-\phi_{cl})$. We obtain this expansion in a sequence of steps, by studying block spin renormalization transformations of $S(u,\phi)$.

For most of this paper, we are concerned with identifying $S(u_{cl},\phi_{cl})$ and determining its properties. In particular, we find a quadratic action S_Q which is close to $S(u_{cl},\phi_{cl})$. The transformation $S\to S_Q$ is an approximation to our renormalization transformation. We study the quadratic terms which arise by iteration of the renormalization transformation, yielding $S_{Q_1}, S_{Q_2}, \dots, S_{Q_k}$, and establish uniform stability estimates (strictly positive lower bounds) on the S_Q's. These quadratic forms dominate the action in the "small field" region. In other papers we estimate the corrections due to large fields and show that they do not affect the basic picture. The end result is exponential decay of gauge invariant correlations, and a mass gap for both the Higgs particle and the photon.

Let us return to the action (1.1). The formal picture depends upon relating the action S to quadratic actions S_Q. Note that we can define the plaquette field

$$f_p=(ie(\varepsilon))^{-1}\ln u(p)\,,$$

so for small f_p,

$$\sum_p e(\varepsilon)^{-2}(1-\mathrm{Re}\,u(p))=\sum_p \tfrac{1}{2}f_p^2+O(e(\varepsilon)^2)\,.$$

The error term vanishes pointwise as $\varepsilon\to 0$ for $d<4$, and shows that the action (1.1) is the formally correctly scaled unit lattice action to study the continuum limit

$\varepsilon \to 0$. In this paper we establish stability properties of various quadratic approximations to S. We use these later to establish properties of integrals (1.4).

2. Block Fields and Scaling

We study block spin renormalization of the Wilson action. This involves defining averages of u and ϕ in lattice blocks, and performing the integral (1.4) conditioned by the block averages taking given values. In physics terminology we integrate high momentum degrees of freedom while fixing low momentum degrees of freedom in each block. If the blocks have side length L, a small positive integer, then we obtain functions on an LZ^d lattice. Repeating this process leads us to study lattices $Z^d, LZ^d, L^2Z^d, \ldots, L^kZ^d, \ldots$. Since after each step it is often convenient to scale the LZ^d lattice back to the unit lattice, we consider lattices with spacing a, with

$$a=\varepsilon, \ldots, L^{-k}, \ldots, L^{-1}, 1, L. \tag{2.1}$$

The natural L^2 inner product for spacing a comes from the norm (for scalar functions on the a-lattice),

$$\|\phi\|_a^2 = \sum_{i\in T_a} |\phi_i|^2 a^d. \tag{2.2}$$

Here T_a denotes a lattice with spacing a and periodic (toroidal) boundary conditions. Generally we take $a=1$, L, or L^{-k}, the latter arising from k-fold scalings of an L-lattice to a unit lattice, following a block spin transformation. We also use lattices

$$T_a^{(k)} = T_{aL^{-k}} \cap aZ^d. \tag{2.3}$$

Averaging operators map functions on the T_a lattice to functions on the T_{La} lattice. We specify the map from the unit-lattice to the L-lattice, but this definition extends naturally from an a-lattice to an La-lattice. We assume this extension in dealing with powers of averaging operators, etc.

Consider the example of a scalar field. Imbed the unit lattice in an L-lattice of block $B(y)$. Here $y=Ln$ denotes a corner of a block, and $n\in Z^d$. Then $B(y)$ consists of points $x=(x_1, \ldots, x_d)$ such that

$$Ln_j \leqq x_j < L(n_j+1), \quad j=1,2,\ldots,d. \tag{2.4}$$

Let Γ denote a path (contour) composed of bonds, and define

$$u(\Gamma)=\prod_{b\in\Gamma} u_b. \tag{2.5}$$

Contours Γ_1, Γ_2 can be composed if the endpoint of Γ_2 coincides with the starting point of Γ_1, so $u(\Gamma_1 \circ \Gamma_2)=u(\Gamma_1)u(\Gamma_2)$.

We require a set of standard contours Γ_{yx} from corners y of blocks $B(y)$ to points $x\in B(y)$. Define Γ_{yx} as the path obtained by following the coordinate axis one, then axis two, ..., etc., in going from x to y. For example, if $x\in B(y)$, then

$$x-y=\sum_{i=1}^{d} n_i e_i,$$

where e_i denote unit vectors in the ith coordinate directions [and $n_i \geqq 0$ as $x \in B(y)$]. The path Γ_{yx} is the path of length n_1 in the direction $-e_1$ followed by n_2 bonds in direction $-e_2$, etc. Let Γ^{-1} denote the reverse of the path Γ. For a point y' which is a corner of an adjacent block, let $\Gamma_{yy'}$ denote the L-lattice bond from y to y'.

The average $Q\phi$ of ϕ is defined by

$$(Q\phi)_y = L^{-d} \sum_{x \in B(y)} u(\Gamma_{yx})\phi_x . \tag{2.6}$$

An important property of the average is that Q commutes with gauge transformations. In other words, let h denote a map from the unit lattice to $U(1)$. Then h defines the gauge transformation

$$\phi_y \to h(y)\phi_y \equiv \phi_y^h, \qquad u_b \to h(b_-)h(b_+)^{-1}u_b = u_b^h . \tag{2.7}$$

Clearly

$$(Q\phi)^h = Q\phi^h . \tag{2.8}$$

A further property of Q is that

$$QQ^* = I , \tag{2.9}$$

where Q^* is the adjoint in the scalar product (2.2). Thus Q^* maps functions on the a-lattice into functions on the $L^{-1}a$ lattice. Furthermore Q^*Q is an orthogonal projection. In other words Q is a partial isometry.

One choice of gauge is especially good for algebraic operations, namely axial gauge. In axial gauge we choose h in order to set $u_b = 1$ for every b which occurs in some Γ_{yx}. In other words, we specify a maximal tree $T(y)$ in each block $B(y)$ composed of bonds in any Γ_{yx}, $x \in B(y)$.

In the axial gauge, $Q\phi$ reduces to the ordinary average of ϕ, and Q^*Q is the projection onto ϕ which are constant on blocks $B(y)$.

Let us next consider the average of a gauge field u. If $b' = yy'$ is a bond on the L-lattice, then $\Gamma_{yy'} = b'$ and we average over contours $\Gamma_{xx'}$ which are translates of $\Gamma_{yy'}$, i.e. $x - x' = y - y'$. More specifically, define Q on gauge fields on the unit lattice by

$$(Qu)_{yy'} = u(\Gamma_{yy'}) \exp\left[L^{-d} \sum_{x \in B(y)} \ln u(\Gamma_{yx} \circ \Gamma_{xx'} \circ \Gamma_{y'x'}^{-1} \circ \Gamma_{y'y})\right] . \tag{2.10}$$

In (2.10) choose the logarithm so that

$$-\pi \leqq \arg \ln u < \pi . \tag{2.11}$$

With this choice, the discontinuity of $\ln u$ occurs at $u = -1$, namely the region of largest action. This is also the region of maximum suppression of probability, and hence the discontinuity will be seen not to affect integrals (1.4) in an important manner.

The average (2.10) is highly nonlinear. We also desire a linear average, defined on Lie algebra variables. Suppose that we can define

$$A_b = (ie(\varepsilon))^{-1} \ln u_b . \tag{2.12}$$

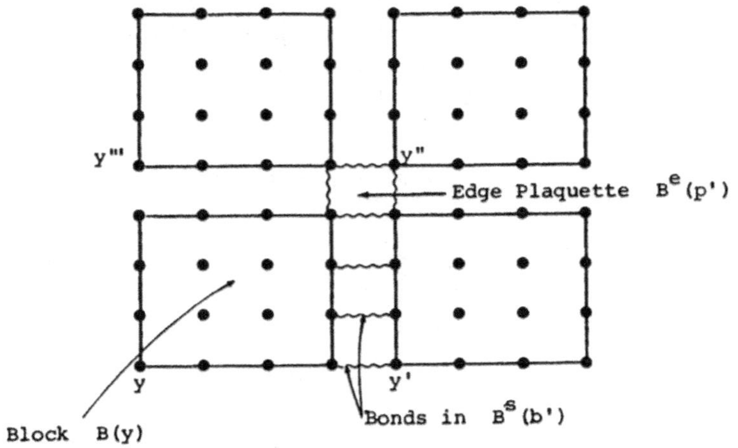

Fig. 1. Block variables for gauge fields, drawn in $d=2$. Four blocks are illustrated. Let $B(y)$ denote a block with $L=4$. The L-lattice bonds illustrated are $b'=yy'$, $y'y''$, $y''y'''$, and $y'''y'$. These four L-bonds bound a plaquette p'. The corresponding edge plaquette $B^e(p')$ is indicated. The bond b' has associated surface bonds $B^s(b')$ which connect $B(y)$ and $B(y')$. A few bonds connecting distinct blocks are indicated by wavey lines

For $b'=(b'_-, b'_+)$ a (directed) L-lattice bond from b'_- to b'_+, we define the average

$$(QA)_{b'} = L^{-(d+1)} \sum_{x\in B(b'_-)} \sum_{b\in\Gamma_{xx'}} A_b, \tag{2.13}$$

where $\Gamma_{xx'}$ is the special contour from x to x' and xx' is the parallel transport of the bond b' to start at x. Thus the bonds b which enter the sum (2.13) range over the interior of the two L-blocks $B(b'_-)$ and $B(b'_+)$, as well as the (surface) bonds connecting these blocks.

The natural inner product on functions on bonds is

$$\langle A, B\rangle_a = \sum_b a^d A_b B_b. \tag{2.14}$$

We do not explicitly compute the adjoint Q^*. However, we now study related averages on surface bonds which connect adjacent blocks and on edge plaquettes which intersect four blocks.

Define the set of surface bonds $B^s(b')$ for an L-lattice bond b' as follows:

$$B^s(b') = \{b : b=(b_-, b_+), b_- \in B(b'_-), b_+ \in B(b'_+)\} \tag{2.15}$$

(see Fig. 1). Associated with the surface is the average

$$(Q^s A)_{b'} = L^{-(d-1)} \sum_{b\in B^s(b')} A_b. \tag{2.16}$$

Then

$$(Q^{s*}A)_b = \begin{cases} LA_{b'}, & \text{if } b\in B^s(b'), \\ 0, & \text{otherwise}, \end{cases} \tag{2.17}$$

using the inner product (2.14). Also

$$Q^s Q^{s*} = LI, \quad Q^{s*}Q^s = LP^s, \tag{2.18}$$

where P^s is the projection onto configurations which are constant on surface bonds and zero on interior bonds. Furthermore, by (2.13),

$$QQ^{s*}=I. \tag{2.19}$$

Finally, we define an edge average Q^e associated with lattice plaquettes. The natural scalar product for functions on lattice plaquettes is

$$\langle f,g\rangle_a=\sum_p a^d f_p g_p, \tag{2.20}$$

where p is a plaquette on the a-lattice. Let p' denote an L-lattice plaquette. The set of edge plaquettes $B^e(p')$ are defined as those unit plaquettes p which (i) are parallel to p', (ii) whose bonds touch four distinct L-blocks, and (iii) such that these L-blocks contain the corners of p' (see Fig. 1). The edge average Q^e is defined by

$$(Q^e f)(p')=L^{-(d-2)}\sum_{p\in B(p')} f(p). \tag{2.21}$$

Then

$$(Q^{e*}f)(p)=\begin{cases}L^2 f(p'), & \text{if } p\in B^e(p'),\\ 0, & \text{otherwise}.\end{cases} \tag{2.22}$$

Also

$$Q^eQ^{e*}=L^2 I, \quad Q^{e*}Q^e=L^2P^e, \tag{2.23}$$

where P^e is the orthogonal projection onto configurations which are constant on edge plaquettes and zero elsewhere.

We also need the k-fold averaging operators $Q_k\equiv(Q)^k$, etc. Especially important are $Q_k^e\equiv(Q^e)^k$ and $Q_k^s=(Q^s)^k$ which satisfy

$$Q_k^eQ_k^{e*}=L^{2k}I=\eta^{-2}I, \quad Q_k^sQ_k^{s*}=L^kI=\eta^{-1}I, \tag{2.24}$$

where $\eta=L^{-k}$.

Finally we consider a scale transformation $\mathscr{S}_{L^{-1}}$ which maps L-lattice fields to unit lattice fields. The canonical scaling for scalar fields

$$(\mathscr{S}_{L^{-1}}\phi)_x=L^{-(d-2)/2}\phi_{Lx}. \tag{2.25}$$

For gauge fields, the L-lattice action

$$\sum_p L^d\langle f_p,\sigma f_p\rangle$$

can be written in terms of a unit lattice action with

$$(\mathscr{S}_{L^{-1}}f)_p=L^{d/2}f_{Lp}, \tag{2.26}$$

so for a quadratic form σ,

$$\sum_{p\in T_L} L^d\langle f_p,\sigma f_p\rangle=\sum_{p\in T_1}\langle(\mathscr{S}_{L^{-1}}f)_p,\sigma(\mathscr{S}_{L^{-1}}f)_p\rangle. \tag{2.27}$$

Then the ε-lattice action

$$S_\varepsilon=\sum_{p\in T_\varepsilon} e^{-2}\varepsilon^{d-4}(1-\mathrm{Re}\,u(p))+\tfrac{1}{2}\sum_{b\in T_\varepsilon}\varepsilon^d|D_u\phi|^2+\sum_{x\in T_\varepsilon}\varepsilon^d\mathscr{P}_\varepsilon(\phi_x)+E, \tag{2.28}$$

where $\mathscr{P}_\varepsilon(\xi)=\lambda|\xi|^4-\frac{1}{4}(m^2+\delta m^2)|\xi|^2$ is related to the unit lattice action (1.1) by

$$S_\varepsilon=S(\mathscr{S}_\varepsilon\phi,\mathscr{S}_\varepsilon u). \tag{2.29}$$

3. The First Renormalization Step

The renormalization transformation R defines a mapping

$$R: S \to S^{(1)}, \tag{3.1}$$

where S, $S^{(1)}$ are both unit lattice actions. This transformation involves two steps:

(i) Integrate high momentum degrees of freedom while fixing block spin averages.

(ii) Rescale the resulting L-lattice action to the unit lattice.

Thus we can write

$$R = \mathscr{S}_L \mathscr{T}, \tag{3.2}$$

where $\mathscr{T}$ is defined by

$$(\mathscr{T} e^{-S})(v, \psi) = \int e^{-S(u,\phi)} \delta_{Ax}(u) \delta(v/Qu) \delta_H(\psi - Q\phi) \mathscr{D}u \mathscr{D}\phi. \tag{3.3}$$

In (3.3) we specify the axial gauge with $\delta_{Ax}(u)$ which sets $u_b = 1$ on the set of bonds chosen as follows: Within each L-block $B(y)$ with corner y in the L-lattice, let $T(y)$ denote the tree composed of unit bonds in Γ_{yx}. Here x ranges over $B(y)$, and $T(y)$ is maximal – in the sense that adjoining any additional unit bond b in $B(y)$ to $T(y)$ makes $T(y)$ multiply connected. Then

$$\delta_{Ax}(u) = \prod_{y \in T_L} \prod_{b \in T(y)} \delta(u_b). \tag{3.4}$$

Let $\mathscr{G}_0$ denote the subgroup of gauge transformations which are specified by h's in (2.7) which are constant on each L-block $B(y)$. These gauge transformations preserve the axial gauge and affect only bonds on the unit lattice which connect different blocks. They generate the gauge group of the integral $\mathscr{T}(\exp - S)$.

The factor $\delta(v/Qu)$ is a delta function which specifies that the average gauge field in each block $B(y)$ is v. Thus for b' an L-lattice bond,

$$\delta(v/Qu) = \prod_{b' \in T_L} \delta(v_{b'}/(Qu)_{b'}). \tag{3.5}$$

The factor δ_H is an approximate delta function which specifies that the block averages of the scalar (Higgs) field equal ψ. In particular, if y denotes an L-lattice site,

$$\delta_H(\psi - Q\phi) = \prod_{y \in T_L} [(a/2\pi) \exp(-(1/2) a |\psi - Q(u)\phi|^2)]. \tag{3.6}$$

The integral (3.2) therefore has the normalization property

$$\int \mathscr{T} e^{-S} \mathscr{D}v \mathscr{D}\psi = \int e^{-S} \mathscr{D}u \mathscr{D}\phi, \tag{3.7}$$

where $\mathscr{D}v\mathscr{D}\psi$ is a measure corresponding to (3.3) but on the L-lattice.

Having defined the renormalization transformation, we now analyze its behavior for small fields. This will yield a Gaussian approximation to the integral (3.1) and a corresponding quadratic form for $S^{(1)}$. The quadratic form has the structure

$$S_Q^{(1)} = \tfrac{1}{2} \langle f^{(1)}, \sigma_1 f^{(1)} \rangle + \tfrac{1}{2} \langle \psi, \Delta_1(u_1) \psi \rangle, \tag{3.8}$$

which we now derive and analyze. In particular we will establish strict positivity of σ_1 and $\Delta_1(u_1)$ in (3.8).

We begin the study of (3.1) by rewriting the Wilson action contribution to $S(u, \phi)$. First define a unit lattice field u_b' with block averages equal to 1. In this way we can write the resulting u_b' in terms of a vector potential which fluctuates about zero. Let

$$(u')_b = \begin{cases} u_b, & \text{if } b \notin B^s(b'), \\ u_b v_{b'}^{-1}, & \text{if } b \in B^s(b') \end{cases} \tag{3.9}$$

be given in terms of the averaging operator Q of (2.10), where

$$v_{b'} = (Qu)_{b'}. \tag{3.10}$$

It then follows that

$$(Qu')_{b'} = 1 \tag{3.11}$$

for all L-lattice bonds b'. Now define A_b' by

$$(u')_b = \exp(ie(\varepsilon) A_b'). \tag{3.12}$$

Then the Wilson action can be written

$$\sum_{p \in T_1} e(\varepsilon)^{-2}(1 - \mathrm{Re}\, u(p)) = \tfrac{1}{2} \sum_{p \in T_1} |\partial A' + L^{-d/2} Q^{e*} F|^2 + O(e(\varepsilon)^2). \tag{3.13}$$

Here we define the L-lattice field $F(p')$ in terms of $v(p')$, where

$$v(p') \equiv \prod_{b' \in \partial p'} v_{b'} \equiv \exp[ie(L\varepsilon) F(p')]. \tag{3.14}$$

The factor $L^{-d/2}$ in (3.13) then arises from combining the factor L^{-2} from (2.22) with $L^{-(d-4)/2} = e(\varepsilon)/e(L\varepsilon)$.

The quadratic form in (3.13) can be reexpressed by translating to its minimum. By the Appendix, with $\alpha = \partial$ and $B = L^{-d/2} Q^{e*} F$, we have

$$A_{cl} = -L^{-d/2} C \partial^* Q^{e*} F. \tag{3.15}$$

Here $C = (\partial^* \partial)^{-1}$, restricted to the subspace of A satisfying both $QA = 0$ and the axial gauge condition. Then by Corollary A2, and

$$A' = A + A_{cl}, \tag{3.16}$$

$$\tfrac{1}{2} \sum_{p \in T_1} |\partial A' + L^{-d/2} Q^{e*} F|^2 = \tfrac{1}{2} \sum_{p \in T_1} |\partial A|^2 + \tfrac{1}{2} \langle F, \Sigma F \rangle, \tag{3.17}$$

where

$$\Sigma = L^{-d} Q^e (I - \partial C \partial^*) Q^{e*}. \tag{3.18}$$

The fluctuation field A lives on the unit lattice, and will be integrated. For small fluctuations we use this Gaussian approximation to yield a perturbative expression for the integral. Large fluctuations will lead to small probabilities.

The quadratic form $\langle F, \Sigma F \rangle$ provides the formula for the background field and it will come outside the integral, since the average fields v and hence F are fixed. This form will be scaled to the unit lattice. This scaling absorbs the factor L^{-d}, so

$$\mathcal{S}_L \langle F, \Sigma F \rangle = \langle f^{(1)}, \sigma_1 f^{(1)} \rangle, \tag{3.19}$$

where

$$\sigma_1 = Q^e(I - \partial C \partial^*)Q^{e*}, \quad f^{(1)}(p) = \frac{1}{ie(L\varepsilon)} \ln v(p). \tag{3.20}$$

Here Q^{e*} maps the unit lattice to the L^{-1} lattice, and $\partial C \partial^*$ is an L^{-1} lattice operator.

Define the unit lattice field

$$u^{(1)} = u \exp(-ie(\varepsilon)A), \tag{3.21}$$

with A the fluctuation field defined by (3.16). Then after scaling back to the L^{-1} lattice,

$$u_1 \equiv \mathcal{S}_L^* u^{(1)}$$

is the gauge field on the L^{-1} lattice on which the unit lattice renormalized action depends.

Let us now direct attention to the boson part of the integral (3.1). We expand

$$S(u, \phi) + \tfrac{1}{2} a \|\psi - Q(u)\phi\|^2 \tag{3.24}$$

to identify the quadratic terms in ϕ. Using (3.23), we also expand u about its background value u_1, yielding additional contributions from A which are accompanied by (small) factors $O(e(\varepsilon))$. Thus the leading terms in the Higgs action are

$$\tfrac{1}{2} a \|Q(u^{(1)})\phi - \psi\|^2 + \tfrac{1}{2}\langle \phi, \Delta_{u^{(1)}}\phi \rangle, \tag{3.25}$$

with all corrections included in the potential $V_1(A, \phi, f_1)$ which have coefficients $O(\varepsilon)$.

Define

$$G = (-\Delta_{u^{(1)}} + aQ(u^{(1)})^* Q(u^{(1)}))^{-1}. \tag{3.26}$$

The quadratic form (3.25) is minimized by

$$\phi_{cl} = -aGQ(u')^*\psi. \tag{3.27}$$

See Appendix A. Writing $\phi = \phi_{cl} + \phi_{fl}$, expand (3.24) into background and fluctuation parts, yielding

$$\tfrac{1}{2}\langle \phi_{fl}, G^{-1}\phi_{fl}\rangle + \tfrac{1}{2} a \langle \psi, [I - aQGQ^*]\psi \rangle. \tag{3.28}$$

In (3.28) the field ϕ_{fl} is a unit lattice field but ψ is an L-lattice field. The scaling to the unit lattice will rewrite the second term in (3.28) in terms of a unit lattice field $\psi^{(1)} = \mathcal{S}_L^*\psi$. Let $G_1 = \mathcal{S}_L^* G \mathcal{S}_L = G_1(u_1)$ denote the scaled Green's function, which is expressed in terms of the L^{-1} lattice gauge field u_1.

4. Form of the kth Effective Action

In this section we state the quadratic or mean field approximation to the action which is produced after k renormalization transformations. We define the basic objects and give some detailed formulas for the quadratic action. Basically the action is the sum of two parts: a gauge field action and a scalar field part which is quadratic in ϕ and depends on an average value of the gauge field. There are further interaction terms between the gauge field and the scalar field, but these interactions yield small corrections to our mean field action and are not considered here.

We begin with a discussion of the gauge field terms. The scalar field is more straightforward and is postponed to the final subsection. The formulas which we give here are justified in Sect. 6, when we show that renormalization of the kth effective action yields the quadratic form for the $(k+1)^{st}$ effective action. The action has the form

$$S_k(v,\psi)=\tfrac{1}{2}\langle f^{(k)},\sigma_k f^{(k)}\rangle+\tfrac{1}{2}\langle\psi,\Delta_k(u_k)\psi\rangle+\text{interaction terms}, \tag{4.1}$$

where the interaction terms will be considered in detail in a later paper.

4.1. The Axial Gauge Propagator and Minimizer

We require an axial gauge propagator to express the effective action. Define $G_{k,Ax}$ by the functional integral

$$\exp(\tfrac{1}{2}\langle J,G_{k,Ax}J\rangle)=Z_{k,Ax}^{-1}\int \mathscr{D}A\,\delta(Q_kA)\,\delta_{k,Ax}(A)\exp(-\tfrac{1}{2}\|\partial A\|^2+\langle A,J\rangle). \tag{4.1.1}$$

Here the normalization factor $Z_{k,Ax}$ can be computed by setting $J=0$. The averaging operator Q_k is the k-fold composition of the 1-step averaging operators Q for bond variables, $Q_k=(Q)^k$. It follows that Q_k is given by the formula (2.13), where L is replaced by L^k. Finally, the axial gauge is fixed by the delta function

$$\delta_{k,Ax}(A)\equiv\prod_{j=0}^{k-1}\delta_{Ax}(Q_jA). \tag{4.1.2}$$

The jth factor in the product acts on the $L^j\eta=L^{j-k}$ lattice. It sets bond field averages Q_jA to zero on contours Γ_{yx} defined in the $L^j\eta$ lattice in the same way that the axial gauge was fixed in Sect. 3 on contours in the unit lattice.

The reader may wonder whether ∂ has zero modes on the subspace of gauge fields satisfying $Q_kA=0$ and satisfying the axial gauge condition. Such zero modes do not occur, and as a consequence the integral (4.1.1) is convergent also for noncompact gauge fields. A proof of this fact for $k=1$ follows by considering on each plaquette in a lattice block the condition $\partial A=0$ and the axial gauge condition. This shows that A can be nonzero only on bonds connecting different blocks, and it must be constant on the bonds connecting two given blocks. The condition $QA=0$ then ensures that A is everywhere zero. Similar, elementary reasoning yields an inductive proof for $k>1$, but we leave out the details.

The propagator $G_{k,Ax}$ itself is the second moment of the measure $\exp(-\frac{1}{2}\|\partial A\|^2)$, restricted to the space of gauge fields which satisfy the kth axial gauge condition and condition on the kth average fields vanishing. We do not know a simple operator theoretic definition of $G_{k,Ax}$, and we generally establish the properties of the propagator by appealing to the definition (4.1.1).

Another aspect of the action $S_G(A)=\frac{1}{2}\|\partial A\|^2$ concerns what configurations minimize $S_G(A)$ subject to the constraints of a gauge condition and a condition that the average field Q_kA equals a given value B. We introduce a transformation

$H_{k,Ax}$ which maps B into such a minimizing configuration for axial gauge, and we call $H_{k,Ax}$ the *axial gauge minimizer*. Explicitly,

$$H_{k,Ax}B = Z_{k,Ax}(B)^{-1} \int \mathscr{D}A\delta(Q_kA - B)\delta_{k,Ax}(A)A\exp(-\tfrac{1}{2}\|\partial A\|^2), \quad (4.1.3)$$

where

$$Z_{k,Ax}(B) = \int \mathscr{D}A\delta(Q_kA - B)\delta_{k,Ax}(A)\exp(-\tfrac{1}{2}\|\partial A\|^2). \quad (4.1.4)$$

Note that by definition

$$Q_kH_{k,Ax}B = B. \quad (4.1.5)$$

4.2. The Quadratic Action and Plaquette Fields

We begin with the definition of the basic quadratic form σ_k. Again we use a functional integral definition

$$\exp(-\tfrac{1}{2}\langle f, \sigma_k f\rangle) = Z_{k,Ax}^{-1} \int \mathscr{D}A\delta(Q_kA)\delta_{k,Ax}(A)\exp(-\tfrac{1}{2}\|\partial A - Q_k^{e*}f\|^2). \quad (4.2.1)$$

Here f is any real (Lie algebra) valued field defined on unit lattice plaquettes.

Clearly σ_k is related to the propagator $G_{k,Ax}$ of Sect. 4.1. Comparing (4.2.1) with (4.1.1) we find

$$\sigma_k = Q_k^e(I - \partial G_{k,Ax}\partial^*)Q_k^{e*} = \eta^{-2}I - Q_k^e\partial G_{k,Ax}\partial^* Q_k^{e*}. \quad (4.2.2)$$

Here we have used $Q_k^eQ_k^{e*} = \eta^{-2}$, see (2.24). Note that if $k=1$, then $G_{k,Ax} = C$ and (4.2.2) agrees with (3.21).

The right side of (4.2.2) involves the combination $\partial G_{k,Ax}\partial^*$ which is invariant under a change of gauge. Hence the gauge chosen to define σ_k is irrelevant, and the quadratic form σ_k is gauge invariant. We use this property in Sect. 7 when we derive an explicit momentum-space representation for σ_k. Using that representation we also establish a uniform, positive lower bound

$$0 < c \leqq \sigma_k. \quad (4.2.3)$$

The quadratic part of the effective action is obtained by evaluating σ_k on a special field configuration $f^{(k)}$ related to the gauge field after k renormalization steps. Let v be defined for unit lattice bonds, with values in the group U(1). Define the plaquette field $f^{(k)}$ on the unit lattice by

$$f^{(k)} = (ie_k)^{-1}\ln v(\partial p), \quad p \in T_1^{(k)}, \quad \text{where} \quad e_k = e(L^k\varepsilon)^{(4-d)/2}. \quad (4.2.4)$$

The quadratic part of the action for the gauge field is then

$$\tfrac{1}{2}\langle f^{(k)}, \sigma_k f^{(k)}\rangle. \quad (4.2.5)$$

We also use a field f_k which lives on the $\eta = L^{-k}$ lattice and is defined by

$$f_k = (I - \partial G_{k,Ax}\partial^*)Q_k^{e*}f^{(k)}. \quad (4.2.6)$$

Thus $f_k(p)$ is a function with $p \in T_\eta$. The action (4.2.5) also can be written

$$\tfrac{1}{2}\langle f^{(k)}, \sigma_k f^{(k)}\rangle = \tfrac{1}{2}\|f_k\|^2. \quad (4.2.7)$$

This identity is a consequence of the fact that $\partial G_{k,Ax}\partial^*$ is a projection operator. Alternatively, we compute from (4.1.1) and (4.2.1) that

$$\exp(-\tfrac{1}{2}\langle f,\sigma_k f\rangle)=\exp(\tfrac{1}{2}\langle\partial^* Q_k^{e*}f, G_{k,Ax}\partial^* Q_k^{e*}f\rangle)\exp(-\tfrac{1}{2}\|Q_k^{e*}f\|^2),$$

from which (4.2.7) follows.

4.3. Quadratic Action for Curls

The quadratic form σ_k simplifies on curls, namely for fields f of the form $f=\partial B$. In particular

$$\langle\partial B,\sigma_k\partial B\rangle=\langle B,\Delta_k B\rangle, \tag{4.3.1}$$

which defines an action Δ_k. To give a functional integral representation for Δ_k, we note that

$$\begin{aligned}\exp(-\tfrac{1}{2}\langle\partial B,\sigma_k\partial B\rangle)&=Z_{k,Ax}^{-1}\int\mathscr{D}A\delta(Q_kA)\delta_{k,Ax}(A)\exp(-\tfrac{1}{2}\|\partial A-Q_k^{e*}\partial B\|^2)\\&=Z_{k,Ax}^{-1}\int\mathscr{D}A\delta(Q_kA-B)\delta_{k,Ax}(A)\exp(-\tfrac{1}{2}\|\partial A\|^2)\\&=\exp(-\tfrac{1}{2}\langle B,\Delta_k B\rangle).\end{aligned} \tag{4.3.2}$$

Here we have performed a translation of A on surface bonds $A\to A+Q_k^{s*}B$. We use the fact that the axial gauge condition is independent of the surface bonds, we use $\partial Q_k^{s*}=Q_k^{e*}\partial$, and we also use (2.19).

The action Δ_k yields the unit lattice propagator $C^{(k)}$, defined as follows:

$$\exp(\tfrac{1}{2}\langle J,C^{(k)}J\rangle)=(Z^{(k)})^{-1}\int\mathscr{D}B\delta(QB)\delta_{Ax}(B)\exp(-\tfrac{1}{2}\langle B,\Delta_k B\rangle+\langle B,J\rangle). \tag{4.3.3}$$

The actions Δ_k and propagators $C^{(k)}$ were studied by Balaban [6II] who established that $C^{(k)}$ is well defined and is bounded in norm

$$\|C^{(k)}\|\leqq c \tag{4.3.4}$$

uniformly in k. Furthermore $C^{(k)}$ has a kernel which decays exponentially, uniformly in k.

$$|C^{(k)}(x,y)|\leqq a\exp(-b|x-y|). \tag{4.3.5}$$

4.4. Landau Gauge Propagator and Minimizer

The axial gauge has many useful properties, especially in providing convenient definitions of Green's functions, and for establishing many algebraic identities in Sect. 5. However, the axial gauge does not provide either good regularity or good decay properties for the Green's functions. These are much better in Landau gauge.

Landau gauge is defined for our lattice theory in a complicated way, because of the restrictions on gauge transformations. We introduce a gauge fixing function $\mathscr{G}(\partial^*A)$. The exact form of $\mathscr{G}$ is important for the proof of estimates. An appropriate choice is

$$\mathscr{G}(\partial^*A)=\exp(-\tfrac{1}{2}\|\partial^*A\|^2)/(\int d\lambda\delta(Q'\lambda)\exp(-\tfrac{1}{2}\|\partial^*A-\Delta\lambda\|^2)), \tag{4.4.1}$$

where Q' denotes the ordinary average over k-blocks. Such a Faddeev-Popov type choice leads to the formula for the Landau gauge minimizer

$$H_k B = Z_k(B)^{-1} \int \mathcal{D}A \delta(Q_k A - B) \mathcal{G}(\partial^* A) A \exp(-\tfrac{1}{2}\|\partial A\|^2), \qquad (4.4.2)$$

where

$$Z_k(B) = \int \mathcal{D}A\, \delta(Q_k A - B) \mathcal{G}(\partial^* A) \exp(-\tfrac{1}{2}\|\partial A\|^2). \qquad (4.4.3)$$

We could define a well-behaved, Landau gauge propagator G_k by a formula similar to (4.4.1), but with $\mathcal{G}(\partial^* A)$ replacing $\delta_{k,Ax}(A)$. However, we do not use this propagator in our consideration of the Higgs effect. Rather, we define a propagator $\mathcal{D}_k$ in terms of minimizers H_j and the covariance operators $C^{(j)}$ of Sect. 4.3. Let

$$\mathcal{D}_k \equiv \sum_{j=0}^{k-1} H_j C^{(j)} H_j^* . \qquad (4.4.4)$$

Here the transformations H_j are now defined by the same formulas as (4.4.2); however, they act on the $\eta = L^{-k}$ lattice instead of the L^{-j} lattice and they involve j^{th} order averages Q_j. Similarly the propagators $C^{(j)}$ are defined by the formula (4.3.3) but on the $L^j\eta = L^{j-k}$ lattice, rather than on the unit lattice. They still involve one-step averages Q.

We remark that $\mathcal{D}_k$, $G_{k,Ax}$, and G_k are related by change of gauge formulas, some of which we establish in Sect. 5.

4.5. Bond Fields

The first renormalization step in Sect. 3 yielded an effective action $S^{(1)}(v, \psi)$ depending on a unit lattice field v, through the L^{-1} lattice field u_1. We need to generalize this picture to the k-step action $S^{(k)}$. After k renormalization steps we again have as independent variables the gauge field v defined on the unit lattice. We introduce a generalization of the configuration u_1 denoted u_k. To begin, we have to define a transformation Q_k^{s*} from unit-lattice, group-valued configurations v to η-lattice, group-valued configurations.

We can represent an arbitrary v as

$$v_b = \exp(ie_k B_b), \qquad (4.5.1)$$

where $B_b = (ie_k)^{-1} \ln v_b$, and the branch of the logarithm was chosen in (2.11). Then we extend the definition Q_s^* from the Lie algebra to group variables v by the definition

$$(Q_k^{s*} v)_b = \exp((ie_k \eta Q_k^{s*} B)_b), \qquad b \in T_\eta . \qquad (4.5.2)$$

This raises the question of what happens if we choose another branch of the logarithm. The independence of the resulting transformation on this choice follows from the equivalent definition:

$$(Q_k^{s*} v)_b = \begin{cases} 1 & \text{if } b \text{ is strictly contained in a } k\text{-block} \\ & \text{(both endpoints belong to the block)} \\ v_c & \text{if the } \eta\text{-lattice bond } b \text{ belongs to the corridor} \\ & \text{of bonds connecting the two blocks } B^k(c_-) \text{ and} \\ & B^k(c_+)\text{. Here } c \text{ is a unit lattice bond.} \end{cases} \qquad (4.5.3)$$

Now we define

$$(u_k)_b = (Q_k^{s*}v)_b \exp[-ie_k\eta(\mathcal{D}_k\partial^* Q_k^{e*} f^{(k)})_b]. \tag{4.5.4}$$

This field configuration u_k is the minimal configuration for the approximate action, up to a gauge transformation. It serves as the background field (external gauge field) in the expressions for the scalar field action and propagators.

4.6. The Scalar Field Action

The fundamental propagator $G_k(u_k)$ for the scalar field in this theory can also be defined by a functional integral,

$$\exp(\tfrac{1}{2}\langle h, G_k(u_k)h\rangle) = Z_k(u_k)^{-1}\int \mathcal{D}\phi \exp[-\tfrac{1}{2}\|D_{u_k}\phi\|^2 - \tfrac{1}{2}a_k\|Q_k(u_k)\phi\|^2 + \langle\phi, h\rangle]. \tag{4.6.1}$$

Since no restrictions on ϕ occur in the Gaussian integral (4.6.1), we can also write

$$G_k(u_k) = [-\Delta_{u_k} + a_k Q_k^*(u_k)Q_k(u_k)]^{-1}, \tag{4.6.2}$$

where

$$-\Delta_{u_k} = D_{u_k}^* D_{u_k}. \tag{4.6.3}$$

The coefficients a_k are produced by iterating one-step renormalization transformations which use a constant a in the Gaussian. This yields $a_k = a(1-L^{-2})(1-L^{-2k})^{-1}$ in the k-step transformation [3].

The quadratic form which arises for the scalar field is $\langle\psi, \Delta_k(u_k)\psi\rangle$, where ψ is the unit lattice scalar field and where

$$\Delta_k(u_k) = a_k I - a_k^2 Q_k(u_k)G_k(u_k)Q_k^*(u_k). \tag{4.6.4}$$

The scalar field action depends on the unit lattice field ψ through the η-lattice minimizer ψ_k, where

$$\psi_k = a_k G_k(u_k)Q_k^*(u_k)\psi.$$

The other propagators such as $C^{(k)}(u_k)$ are defined as in [3] with the only change that the background gauge field is u_k.

5. Relations Among Minimizers and Propagators

5.1. Change of Gauge for Minimizers

In this section we discuss relations among the operators H and G. We begin by relating $H_{k,Ax}$ and H_k. Recall that H_kB is the configuration which minimizes the action $\|\partial A\|^2$, subject to a gauge condition as well as the restriction $Q_kA = B$ on field averages. We claim that $H_{k,Ax}B$ and H_kB differ by a gauge transformation,

$$H_{k,Ax}B - H_kB = \partial\lambda. \tag{5.1.1}$$

We show that λ is an explicit, linear function of H_kB. In order to specify λ, we choose a sequence of points $x_0, \ldots, x_k$, where $x = x_0$, $y = x_k$, and where

$$x \in B^j(x_j), \quad x_j \in T_{L^j\eta}^{(j)}, \tag{5.1.2}$$

that is

$$x_j \in B(x_{j+1}). \tag{5.1.3}$$

The contour $\Gamma_{x_j,x}$ runs from x to x_1 in $B(x_1)$, from x_1 to x_2 in $B(x_2)$, etc.

Proposition 5.1.1. *The relation* (5.1.1) *holds with*

$$\lambda(x) = -\sum_{j=0}^{k-1} L^{j-k}\left[(Q_jH_kB)(\Gamma_{x_{j+1},x_j}) - \sum_{x'\in B(x_{j+1})} (L^{-d}Q_jH_kB)(\Gamma_{x_{j+1},x'})\right]. \tag{5.1.4}$$

Here we use $(Q_jH_kB)(\Gamma) = \sum_{b\in\Gamma} (Q_jH_kB)(b)$, *with no factor of the lattice spacing for* Γ.

Proof. The definition of $H_{k,Ax}B$ is

$$H_{k,Ax}B = Z_{k,Ax}(B)^{-1}\int dA\delta(Q_kA-B)\delta_{k,Ax}(A)\exp(-\tfrac{1}{2}\|\partial A\|^2)A, \tag{5.1.5}$$

where

$$\delta_{k,Ax}(A) = \prod_{j=0}^{k-1} \delta_{Ax}(Q_jA). \tag{5.1.6}$$

Furthermore,

$$H_kB = Z_k(B)^{-1}\int dA\delta(Q_kA-B)\mathcal{G}(\partial^*A)\exp(-\tfrac{1}{2}\|\partial A\|^2)A, \tag{5.1.7}$$

where $\mathcal{G}(\partial^*A)$ is a gauge fixing term, and where both integrals are normalized so that without the final factor A they equal 1. We use a variant of the Faddeev-Popov procedure and insert the factor 1 in the integrand (5.1.5), expressed as

$$1 = Z_k^{-1}\int d\lambda\delta(Q_k'\lambda)\mathcal{G}(\partial^*A-\Delta\lambda).$$

Therefore,

$$\begin{aligned} Z_{k,Ax}(B) &= \int dA\delta(Q_kA-B)\delta_{k,Ax}(A)\exp(-\tfrac{1}{2}\|\partial A\|^2) \\ &= Z_k^{-1}\int dAd\lambda\,\delta(Q_k'\lambda)\mathcal{G}(\partial^*A-\Delta\lambda)\delta(Q_kA-B)\delta_{k,Ax}(A)\exp(-\tfrac{1}{2}\|\partial A\|^2). \end{aligned}$$

Making the gauge transformation $A\to A+\partial\lambda$, and using $Q_kA\to Q_kA+Q_k\partial\lambda = Q_kA+\partial Q_k'\lambda = Q_kA$, on account of the δ function yields

$$\begin{aligned} Z_{k,Ax}(B) &= Z_k^{-1}\int dA\mathcal{G}(\partial^*A)\delta(Q_kA-B)\exp(-\tfrac{1}{2}\|\partial A\|^2)\int d\lambda\,\delta(Q_k'\lambda)\delta_{k,Ax}(A+\partial\lambda) \\ &= Z_k^{-1}\int dA\mathcal{G}(\partial^*A)\delta(Q_kA-B)\exp(-\tfrac{1}{2}\|\partial A\|^2) = Z_k^{-1}Z_k(B). \end{aligned} \tag{5.1.8}$$

If we perform the same steps leading to (5.1.8) in (5.1.5), we obtain by (5.1.8)

$$\begin{aligned} H_{k,Ax}B = Z_k(B)^{-1}&\int dA\delta(Q_kA-B)\mathcal{G}(\partial^*A)\exp(-\tfrac{1}{2}\|\partial A\|^2) \\ &\cdot\int d\lambda\delta(Q_k'\lambda)\delta_{k,Ax}(A+\partial\lambda)(A+\partial\lambda). \end{aligned} \tag{5.1.9}$$

In order to evaluate (5.1.9), we solve the system of linear equations for λ which result from the delta functions $\delta(Q_k'\lambda)$ and $\delta_{k,Ax}(A+\partial\lambda)$. Using this value of $\lambda = \lambda(A)$, we have

$$H_{k,Ax}B = Z_k(B)^{-1}\int dA\,\delta(Q_kA-B)\mathcal{G}(\partial^*A)\exp(-\tfrac{1}{2}\|\partial A\|^2)(A+\partial\lambda(A)).$$

We now study the equations which determine $\lambda(A)$. The axial gauge fixing conditions in (5.1.9) entails

$$(Q_jA)(\Gamma_{z,z'}) = -(L^j\eta)^{-1}[(Q_j'\lambda)(z') - (Q_j'\lambda)(z)],$$

where z and z' are endpoints of a bond in $B(x_{j+1})$ which lies on an axial gauge fixing contour. Since every point in $B(x_{j+1})$ lies on some such contour, we can sum these identities to show that for all $x' \in B(x_{j+1})$,

$$(Q_jA)(\Gamma_{x_{j+1},x'}) = -(L^j\eta)^{-1}[(Q_j'\lambda)(x') - (Q_j'\lambda)(x_{j+1})]. \qquad (5.1.10)$$

Summing over $x' \in B(x_{j+1})$ yields

$$(Q_j'\lambda)(x_{j+1}) = (Q_{j+1}'\lambda)(x_{j+1}) + L^j\eta Q'(Q_jA)(\Gamma_{x_{j+1}},\cdot). \qquad (5.1.11)$$

We use the fact that $(Q_j'\lambda)(x_{j+1})$ is constant in this sum, and the final Q' acts on "·". Substituting (5.1.11) in (5.1.10) with $x' = x_j$ then gives

$$(Q_j'\lambda)(x_j) = (Q_{j+1}'\lambda)(x_{j+1}) - L^j\eta[(Q_jA)(\Gamma_{x_{j-1},x_j}) - Q'(Q_jA)(\Gamma_{x_{j+1},\cdot})]. \quad (5.1.12)$$

The identities (5.1.12) hold for $j = 0, 1, \dots, k-1$. Setting $j = 0$ we have

$$\lambda(x) = (Q_0'\lambda)(x_0) = (Q'\lambda)(x_1) - \eta[A(\Gamma_{x_1,x}) - (Q'A)(\Gamma_{x_1},\cdot)].$$

Furthermore, summing (5.1.12) from $j = 0$ to $j = k-1$, the factors $(Q_j'\lambda)(x_j)$ cancel except for $j = 0$ and $j = k$. Thus

$$\lambda(x) = (Q_k'\lambda)(x_k) - \sum_{j=0}^{k-1} L^j\eta[(Q_jA)(\Gamma_{x_{j+1},x_j}) - Q'(Q_jA)(\Gamma_{x_{j+1},\cdot})].$$

Using the δ function for $Q_k'\lambda$, and $\eta = L^{-k}$ gives

$$\lambda(x) = \lambda(A,x) = -\sum_{j=0}^{k-1} L^{j-k}[(Q_jA)(\Gamma_{x_{j+1},x_j}) - Q'(Q_jA)(\Gamma_{x_{j+1},\cdot})]. \quad (5.1.13)$$

We now notice that $\lambda = \lambda(A)$ is a linear function of A. Thus we can write $\lambda(A,x) = \lambda(H_kB,x) - \lambda(A - H_kB,x)$. The term $\lambda(H_kB,x)$ is independent of A and comes outside the A integral. Thus

$$\begin{aligned} H_{k,Ax}B = H_kB + \partial\lambda(H_kB) + Z_k(B)^{-1}\int dA\,\delta(Q_kA - B)\mathcal{G}(\partial^*A) \\ \cdot \exp(-\tfrac{1}{2}\|\partial A\|^2)\partial\lambda(A - H_kB). \qquad (5.1.14) \end{aligned}$$

Since H_kB minimizes the quadratic form $\|\partial A\|^2$ subject to the restrictions imposed by $\delta(Q_kA - B)$ and $\mathcal{G}$, the term in (5.1.14) involving $\partial\lambda(A - H_kB)$ is linear in $A - H_kB$ and hence its integral vanishes. Thus $H_{k,Ax}B = H_kB + \partial\lambda(H_kB)$, and the proof of Proposition 5.1.1 is complete.

5.2. Decomposition of Axial Gauge Green's Functions

Let $C^{(k)}$ denote the unit lattice propagator defined in (4.3.3). Here we give a simple relation between $C^{(k)}$ and the axial gauge propagator $G_{k,Ax}$ of (4.1.1). This relation is similar to the definition (4.4.4) of the Landau gauge propagator $\mathcal{D}_k$ in terms of the $C^{(j)}$. The algebraic identity for $G_{k,Ax}$ will be very useful, especially in computing the form of the action after renormalization, even though the axial gauge propagator has poor regularity and decay properties.

Proposition 6.2. *The $G_{k,Ax}$ satisfy the recursion relation*

$$G_{k,Ax}=H_{k-1,Ax}C^{(k-1)}H^*_{k-1,Ax}+G_{k-1,Ax}, \tag{5.2.1}$$

with the solution:

$$G_{k,Ax}=\sum_{j=0}^{k-1} H_{j,Ax}C^{(j)}H^*_{j,Ax}. \tag{5.2.2}$$

Proof. Clearly (5.2.1) and (5.2.2) are equivalent. We prove (5.2.1) starting from the definition of $G_{k,Ax}$, namely (4.1.1). Write $Q_k=QQ_{k-1}$, and expand the integrand of (4.1.1) using

$$\delta(Q_kA)\delta_{k,Ax}(A)=\int \mathscr{D}B\,\delta(QB)\delta(Q_{k-1}A-B)\delta_{Ax}(B)\delta_{k-1,Ax}(A).$$

Then translate the integrand with respect to A to the minimum of the quadratic form $\frac{1}{2}\|\partial A\|^2-\langle A,J\rangle$, under the restriction $Q_{k-1}A=B$ and the appropriate axial gauge. The minimum is achieved at $H_{k-1,Ax}B$, so we write

$$A=A'+H_{k-1,Ax}B. \tag{5.2.3}$$

Inserting this and using (4.1.5) gives

$$\begin{aligned}\exp(\tfrac{1}{2}\langle J,G_{k,Ax}J\rangle)=Z^{-1}&\int \mathscr{D}B\delta(QB)\delta_{Ax}(B)\\ &\cdot\exp(-\tfrac{1}{2}\|\partial H_{k-1,Ax}B\|^2+\langle B,H^*_{k-1,Ax}J\rangle)\\ &\cdot\int \mathscr{D}A'\delta(Q_{k-1}A')\delta_{k-1,Ax}(A')\\ &\cdot\exp(-\tfrac{1}{2}\|\partial A'\|^2+\langle A',J\rangle).\end{aligned} \tag{5.2.4}$$

The A' integral yields $Z_{k-1,Ax}\exp(\frac{1}{2}\langle J,G_{k-1,Ax}J\rangle)$, while the B integral gives

$$ZZ^{-1}_{k-1,Ax}\exp(\tfrac{1}{2}\langle H^*_{k-1,Ax}J,C^{(k-1)}H^*_{k-1,Ax}J\rangle). \tag{5.2.5}$$

Thus (5.2.2) holds and the proposition follows.

Now that we have related H_k and $H_{k,Ax}$, we can find the relation between $\mathscr{D}_k$ and $G_{k,Ax}$. In fact we need only the compositions $\mathscr{D}_k\partial^*$ and $G_{k,Ax}\partial^*$, since only these products occur.

Proposition 5.2.2. *There is a gauge transformation D such that*

$$G_{k,Ax}\partial^*-\mathscr{D}_k\partial^*=\partial D. \tag{5.2.6}$$

Explicitly

$$D=\sum_{j=0}^{k-1}\lambda_j(H_jC^{(j)}H_j^*\partial^*), \tag{5.2.7}$$

where λ_j is the function of Proposition 5.1.1 with k set equal to j. Here we write the operator identity, rather than the identity $G_{k,Ax}\partial^*B=\mathscr{D}_k\partial^*B+\partial DB$ for configurations.

Proof. We use three facts: the formula (5.2.1), the fact that

$$H^*_{j,Ax}\partial^*=H^*_j\partial^*, \tag{5.2.8}$$

is gauge invariant, and formulas (5.1.1), (5.1.4) for the factor $H_{j,Ax}-H_j$ in the expression

$$G_{k,Ax}\partial^* - \mathscr{D}_k\partial^* = \sum_{j=0}^{k-1} (H_{j,Ax}-H_j)C^{(j)}H_j^*\partial^* . \tag{5.2.9}$$

Substitution then yields (5.2.7).

Remark 1. A consequence of the proposition is

$$\partial G_{k,Ax}\partial^* = \partial\mathscr{D}_k\partial^* . \tag{5.2.10}$$

Remark 2. The gauge field quadratic form $\langle f^{(k)}, \sigma_k f^{(k)}\rangle$ can now be written entirely as a function of the η-lattice gauge field u_k. In fact

$$\langle f^{(k)}, \sigma_k f^{(k)}\rangle = \sum_{p\in T_\eta} \eta^d |f_k(p)|^2 , \tag{5.2.11}$$

where

$$f_k(p) = (ie_k\eta^2)^{-1} \ln u_k(\partial p) . \tag{5.2.12}$$

This follows from the definition (4.5.4) of u_k, giving

$$u_k(\partial p) = \exp[ie_k\eta^2 Q_k^{e*}f^{(k)} - ie_k\eta^2\partial\mathscr{D}_k\partial^* Q_k^{e*}f^{(k)}] .$$

From this and (5.2.10) we get

$$(ie_k\eta^2)^{-1} \ln u_k(\partial p) = Q_k^{e*}f^{(k)} - \partial G_{k,Ax}\partial^* Q_k^{e*}f^{(k)} ,$$

coinciding with (4.2.6).

5.3. *Minimizers and Green's Functions*

To help define the fluctuation field in the $(k+1)^{\text{st}}$ step, we need one more identity, namely

$$H_{k,Ax}B = Q_k^{s*}B - G_{k,Ax}\partial^* Q_k^{e*}B . \tag{5.3.1}$$

This identity follows by inspecting the definition (4.1.3) after the translation $A = A' + Q_k^{s*}B$. Use the facts $Q_kQ_k^{s*} = I$ and $\partial Q_k^{s*}B = Q_k^{e*}\partial B$ to obtain

$$H_{k,Ax}B = Z_{k,Ax}(B)^{-1} \int \mathscr{D}A'\delta(Q_kA')\delta_{k,Ax}(A')(A' + Q_k^{s*}B)\exp(-\tfrac{1}{2}\|\partial A' + Q_k^{e*}\partial B\|^2) .$$

The second term yields $Q_k^{s*}B$. In the first term we translate to the minimum of the quadratic form which is $G_{k,Ax}\partial^* Q_k^{e*}\partial B$. We see this by integrating.

6. Renormalization Transformations Produce the Effective Action as Claimed

In this section we study the renormalization of the quadratic part of the kth action S_k in (4.1). This action depends on the plaquette field $f^{(k)}$, the (background) bond field u_k and the scalar field ψ. Under renormalization we prove that $S_k \to S_{k+1}$. We exhibit this transformation by writing S_k as a sum of two terms: the first is S_{k+1}, but scaled from the unit lattice to the L-lattice. The second term is the contribution of the fluctuation field on the unit lattice. The fluctuation field will be integrated in performing the renormalization transformation, while the remaining terms scale back to give S_{k+1} on the unit lattice. In (4.1) we ignore the small corrections due to interactions and concentrate on the quadratic terms. Hence the result of this section is to justify our (otherwise ad hoc) choice of σ_k and $\Delta_k(u_k)$ as the mean field quadratic forms for the kth effective action.

In accord with the notation of Sect. 3 we denote the variables in S_k as u and ϕ which take their values on the unit lattice. Thus we study

$$\int \mathscr{D}u\mathscr{D}\phi\, \delta(v/Qu)\delta_{Ax}(u)\exp[-\tfrac{1}{2}a\|\psi-Q(u_k)\phi\|^2-S_k(u,\phi)]. \tag{6.1}$$

In this integral we perform a preparatory translation (the first translation in Sect. 3) to isolate the average gauge field on the kth length scale. Let v denote the average gauge field on the L-lattice, and on the unit lattice let

$$u=u'Q^{s*}v\,, \qquad u'=\exp(ie_kB')\,. \tag{6.2}$$

The plaquette field

$$f^{(k)}(p)=(ie_k)^{-1}\ln u(\partial p)\,, \qquad p\in T_1\,, \tag{6.3}$$

then decomposes as

$$f^{(k)}(p)=(ie_k)^{-1}\ln u'(\partial p)+(ie_k)^{-1}\ln(Q^{s*}v)(\partial p)=(\partial B')(p)+L^{-d/2}(Q^{e*}f^{(k+1)})(p)\,, \tag{6.4}$$

which is valid as long as $u'(p)$ and $(Q_k^{s*}v)(p)$ are close to 1. The new field $f^{(k+1)}$ is an L-lattice field, but Q_k^{e*} maps it back (to edge plaquettes) in the unit lattice.

The integral (6.1) will be studied in detail in later papers by dividing it into small and large field regions, defined by some restrictions on the range of ϕ and u. In this paper we are concerned with the small field regions which give the dominant contributions to the integral. In this region the identity (6.4) holds.

We will also see that the perturbative terms in the action are unaffected by the restriction to small fields, since these restrictions become trivial in the limit of zero coupling constants. In particular, this means that in the calculation of leading terms in the action S_{k+1} we can assume that the integrations over fields in (6.1) are restricted only by the delta functions which appear in the definition of the renormalization transformation.

6.1. $\sigma_k=\sigma_{k+1}+$ *Fluctuation Form*

The goal here is to show that

$$\langle f^{(k)},\sigma_k f^{(k)}\rangle=\mathscr{S}_L^{-1}\langle f^{(k+1)},\sigma_{k+1}f^{(k+1)}\rangle+\langle B,\Delta_k B\rangle\,. \tag{6.1.1}$$

In other words, the quadratic form σ_k in S_k which occurs in the integrand (6.1) can be decomposed into the sum of two independent forms: the $(k+1)$-step quadratic form scaled to the L-lattice and the quadratic form for the fluctuation field B. Hence the fluctuation field can be integrated and, after rescaling to the unit lattice, we obtain the desired $(k+1)$-step gauge field quadratic form. Using (6.4),

$$\begin{aligned}\langle f^{(k)},\sigma_k f^{(k)}\rangle=&\langle\partial B',\sigma_k\partial B'\rangle+2L^{-d/2}\langle\partial B',\sigma_k Q^{e*}f^{(k+1)}\rangle\\&+L^{-d}\langle f^{(k+1)},Q^e\sigma_k Q^{e*}f^{(k+1)}\rangle\,.\end{aligned} \tag{6.1.2}$$

Use definition (4.3.1) of Δ_k and the representation (4.2.2) for σ_k. In addition, note that $Q^e Q^e_k = Q^e_{k+1}$. Then

$$\langle f^{(k)}, \sigma_k f^{(k)}\rangle = \langle B', \Delta_k B'\rangle + 2L^{-d/2}\langle (I - \partial G_{k,Ax}\partial^*) Q^{e*}_k \partial B', Q^{e*}_{k+1} f^{(k+1)}\rangle$$
$$+ L^{-d}\langle Q^{e*}_{k+1} f^{(k+1)}, (I - \partial G_{k,Ax}\partial^*) Q^{e*}_{k+1} f^{(k+1)}\rangle . \qquad (6.1.3)$$

By (5.3.1), $H_{k,Ax} = Q^{s*}_k - G_{k,Ax}\partial^* Q^{e*}_k \partial$. Apply ∂ to this identity, and use the gauge invariance statement

$$\partial H_{k,Ax} = \partial H_k \qquad (6.1.4)$$

and the identity $\partial Q^{s*}_k = Q^{e*}_k \partial$. It follows that

$$\partial H_k = (I - \partial G_{k,Ax}\partial^*) Q^{e*}_k \partial . \qquad (6.1.5)$$

Substituting (6.1.5) in (6.1.3) gives

$$\langle f^{(k)}, \sigma_k f^{(k)}\rangle = \langle B', \Delta_k B'\rangle + 2L^{-d/2}\langle B', H^*_k \partial^* Q^{e*}_{k+1} f^{(k+1)}\rangle$$
$$+ L^{-d}\langle Q^{e*}_{k+1} f^{(k+1)}, (I - \partial G_{k,Ax}\partial^*) Q^{e*}_{k+1} f^{(k+1)}\rangle . (6.1.6)$$

Consider the quadratic form (6.1.6) as a function of B' and subject to the axial gauge and averaging restrictions of the integral (6.1). The minimum of the form is given by Proposition A3 of the Appendix, namely

$$B' = -L^{-d/2} C^{(k)} H^*_k \partial^* Q^{e*}_{k+1} f^{(k+1)} . \qquad (6.1.7)$$

The proposition is applicable with $\mathscr{H}_0$ the subspace of axial gauge configurations defined by the kth axial gauge delta function $\delta_{k,Ax}(B')$ and the delta function $\delta(QB')$. The covariance C of the proposition then agrees with $C^{(k)}$ defined in (4.3.3), and this is also the covariance in (6.1.7).

Furthermore, the translation

$$B' = B - L^{-d/2} C^{(k)} H^*_k \partial^* Q^{e*}_{k+1} f^{(k+1)} \qquad (6.1.8)$$

and the decomposition (A 16) yield

$$\langle f^{(k)}, \sigma_k f^{(k)}\rangle$$
$$= \langle B, \Delta_k B\rangle + L^{-d}\langle Q^{e*}_{k+1} f^{(k+1)}, (I - \partial G_{k,Ax}\partial^* - \partial H_k C^{(k)} H^*_k \partial^*) Q^{e*}_{k+1} f^{(k+1)}\rangle . \qquad (6.1.9)$$

Using (6.1.4) and the representation (5.1.15), we can rewrite (6.1.9) as

$$\langle f^{(k)}, \sigma_k f^{(k)}\rangle = \langle B, \Delta_k B\rangle + \mathscr{S}_L^{-1}\langle f^{(k+1)}, \sigma_{k+1} f^{(k+1)}\rangle ,$$

where the scaling $\mathscr{S}_L$ absorbs the factor L^{-d} and where

$$\sigma_{k+1} = Q^e_{k+1}(I - \partial G_{k+1,Ax}\partial^*) Q^{e*}_{k+1} .$$

This verifies the inductive hypothesis.

6.2. $u_k = u_{k+1} \cdot$ Fluctuation (Modulo Gauge Transformation)

We now investigate the effect of the two translations defined in Sect. 6.1 on the field u_k. This allows us to separate a fluctuation field and a gauge transformation which then leave u_{k+1}. Rewrite the definition of u_k (with u in place of v), so

$$u_k = Q_k^{s*} u \exp(-ie_k \eta \mathscr{D}_k \partial^* Q_k^{e*} f^{(k)}) . \tag{6.2.1}$$

By (6.2) we have

$$Q_k^{s*} u = (Q_{k+1}^{s*} v) e^{ie_k \eta Q_k^{s*} B'} , \tag{6.2.2}$$

so substituting this and (6.3) into (6.2.1) we obtain

$$u_k = Q_{k+1}^{s*} v \exp[ie_k \eta Q_k^{s*} B' - ie_k \eta (\mathscr{D}_k \partial^* Q_k^{e*} \partial B') - ie_k \eta L^{-d/2} (\mathscr{D}_k \partial^* Q_{k+1}^{e*} f^{(k+1)})] . \tag{6.2.3}$$

We have the identity $G_{k,Ax} \partial^* = \mathscr{D}_k \partial^* + \partial D$ to replace $\mathscr{D}_k \partial^*$ by $G_{k,Ax} \partial^*$. Then we apply again the identity (5.3.1) to replace the first two terms in the exponential by $ie_k \eta H_{k,Ax} B'$. This takes care of the first translation.

The second translation (6.1.5) then yields

$$\begin{aligned} u_k = Q_{k+1}^{s*} v \exp[&ie_k \eta H_{k,Ax} B - ie_k \eta L^{-d/2} H_{k,Ax} C^{(k)} H_k^* \partial^* Q_{k+1}^{e*} f^{(k+1)} \\ &- ie_k \eta L^{-d/2} (\mathscr{D}_k \partial^* Q_{k+1}^{e*} f^{(k+1)})] \cdot (\text{gauge trans}) . \end{aligned} \tag{6.2.4}$$

We make another gauge transformation to replace the axial gauge minimizers by Landau gauge minimizers, and we combine the second and third term into an operator $\mathscr{D}_{k+1}$. After rescaling, the $L^{-d/2}$ factor is absorbed and we have

$$u_k = e^{ie_k \eta H_k B} u_{k+1} \text{ (the gauge transformation)}, \tag{6.2.5}$$

where the gauge transformation is generated by

$$\exp[ie_k D(Q_k^{e*} \partial B') + ie_k \lambda_k (H_k B) - ie_k L^{-d/2} \lambda_k (H_k C^{(k)} H_k \partial^* Q_{k+1}^{e*} f^{(k+1)})] . \tag{6.2.6}$$

6.3. Scalar Field Renormalization

In Sect. 6.2 we established the general decomposition (6.2.5) for u_k,

$$u_k = u_{k+1} e^{ie_k \eta H_k B} e^{i\eta \partial \omega} , \tag{6.3.1}$$

where ω denotes the gauge transformation (6.2.6). This gauge transformation does not influence the gauge field quadratic form σ_k, since this form is fully gauge invariant. More generally, expressions involving only the gauge field are gauge invariant, so we must discuss the scalar field part of the action.

In that case, the form $\Delta_k(u_k)$ does depend on the gauge transformation ω. Our procedure for constructing S_k has the general invariance

$$S_k(u_k e^{-i\eta \partial \lambda}, e^{i\lambda} \phi) = S_k'(u_k, \phi) \tag{6.3.2}$$

for a gauge transformation λ. While we do not prove this invariance here, it is clear in the case of the quadratic forms for which we write explicit formulas.

The scalar field integral has the form

$$\int \mathscr{D}\phi \exp\{-[S_k(u_k, \phi) + \tfrac{1}{2} a \|\psi - Q(u_k)\phi\|^2]\} . \tag{6.3.3}$$

If we apply the decomposition (6.3.1) for u_k, and the gauge invariance (6.3.2), then (6.3.3) becomes

$$\int \mathscr{D}\phi \exp(-[S_k(u_{k+1} e^{ie_k \eta H_k B}, \phi) + \tfrac{1}{2} a \|e^{-i\omega}\psi - Q(u_{k+1} e^{ie_k \eta H_k B})\phi\|^2]) . \tag{6.3.4}$$

Here we use the gauge invariance of the measure $\mathscr{D}\phi$. The resulting integral still depends on ω. However, the integral of (6.3.4) over ψ is independent of ω, since $\mathscr{D}\psi$ is also gauge invariant.

We can either write the new action in terms of $\psi' = e^{-i\omega}\psi$, or use this gauge invariance and the fact that we are only interested in integrals of (6.3.4) over ψ, times other gauge invariant functions of ψ (as observables). We take the latter point of view and ignore ω.

Now we are in exactly the same situation as [3]. We expand the whole integrand with respect to $H_k B$ and perform the fluctuation field integral over boson fields as in that paper. It reproduces the $(k+1)$ quadratic form for the action.

7. Positivity and Localization of the Effective Action

Positivity bounds on σ_k and $\Delta_k(u_k)$ are a necessary step in proving that the renormalization transformation preserves stability of the effective action. The main difficulty centers on the quadratic form σ_k, which arises from the gauge field action. In particular, σ_k comes from the Wilson form of the action. Unlike the "Gaussian" action for the gauge field used in earlier studies of the U(1) Higgs model, the Wilson action has rather subtle positivity properties. We study this question first, yielding the main result of this section, $0 < c \leqq \sigma_k$. We then finish by stating minor generalizations of known estimates which apply to the other operators H_k, $\mathscr{D}_k$, and $\Delta_k(u_k)$ introduced in this paper.

7.1. Positivity of σ_k

Theorem 7.1.1. *There exists a constant $c > 0$, independent of k, such that*

$$c \leqq \sigma_k \,. \tag{7.1.1}$$

The strategy of the proof is to give an explicit formula for σ_k, which we then analyze in detail. Since we study periodic boundary conditions, σ_k is translation invariant. Thus it is natural to study σ_k as a multiplication operator $\sigma_k(p)$ in the Fourier transform representation. (Note that we now use p to denote a momentum variable, rather than a plaquette variable.) The momenta p have d components p_i, and $|p_i| \leqq \pi$.

We introduce a tensor notation with functions $f_{\mu\nu}$ on plaquettes given as antisymmetric functions on coordinate axes μ, ν which indicates plaquette orientation. They also depend on a lattice position. Then

$$\langle f^{(k)}, \sigma_k f^{(k)} \rangle = \int_{-\pi}^{\pi} \dots \int_{-\pi}^{\pi} \langle \tilde{f}^{(k)}(p), \sigma_k(p) \tilde{f}^{(k)}(p) \rangle \, dp \,, \tag{7.1.2}$$

where the momentum p inner product is defined by

$$\langle \tilde{f}(p), \sigma_k(p) \tilde{f}(p) \rangle = \sum_{\mu, \nu, \lambda, \kappa} \bar{\tilde{f}}_{\mu\nu}(p) \sigma_{k;\mu\nu\lambda\kappa}(p) \tilde{f}_{\lambda\kappa}(p) \,. \tag{7.1.3}$$

In order to study $\sigma_k(p)$ we need standard momentum space operations such as derivatives ∂ or Laplacians Δ. Since we have two special lattice scales, the unit lattice leading to momenta $|p_i| \leqq \pi$ and the $\eta = L^{-k}$ lattice leading to momenta

$|p_i| \leqq \pi/\eta = \pi L^k$, we require operators on both scales. We use a superscript (1) to denote the unit lattice scale.

With these conventions, the derivative ∂ can be represented by a diagonal $d \times d$ matrix $\partial(p)$ with eigenvalues

$$\partial_\mu(p) = (\exp(i\eta p_\mu) - 1)/\eta . \tag{7.1.4}$$

The unit lattice derivative $\partial^{(1)}(p)$ is a similar matrix with eigenvalues

$$\partial_\mu^{(1)}(p) = \exp(ip_\mu) - 1 . \tag{7.1.5}$$

The corresponding Laplacians are

$$\Delta(p) = \mathrm{Tr}\,\partial(p)^* \partial(p), \qquad \Delta^{(1)}(p) = \mathrm{Tr}\,\partial^{(1)}(p)^* \partial^{(1)}(p). \tag{7.1.6}$$

We require the operator $V(p)$ defined by

$$V(p) = \partial^{(1)}(p')\partial(p)^{-1} . \tag{7.1.7}$$

Here and below we use the notation $p_i' \in [-\pi, \pi]$ and

$$p' = p \bmod 2\pi , \tag{7.1.8}$$

so p' denotes a unit lattice momentum. The eigenvalues of $V(p)$ are $v_\mu(p) = \partial_\mu^{(1)}(p')/\partial_\mu(p)$. In terms of $V(p)$ we also define

$$u(p) = \det V(p) \tag{7.1.9}$$

and

$$\varphi_\mu(p') = \sum_{\substack{l \in 2\pi Z^d \\ |l_i| \leqq \pi/\eta}} |u(p'+l) v_\mu(p'+l)|^2 \Delta(p'+l)^{-1} . \tag{7.1.10}$$

In terms of these functions we can express the averaging operators Q_k^e, etc. For example

$$(Q_k^e f)^{\sim}_{\mu\nu}(p') = \sum_l u(p'+l) \tilde{f}_{\mu\nu}(p'+l)\,(v_\mu(p'+l) v_\nu(p'+l))^{-1} . \tag{7.1.11}$$

The basic object we wish to study is σ_k, defined in (4.2.2),

$$\sigma_k = Q_k^e(I - \partial G_{k,Ax} \partial^*) Q_k^{e*} . \tag{7.1.12}$$

The axial gauge Green's function $G_{k,Ax}$ has a very complicated momentum space structure. Thus we use gauge invariance of $\partial G_{k,Ax}\partial^*$ to replace it by the Landau gauge operator $\partial G_k \partial^*$. Here we emphasize that G_k is a slightly different operator from $\mathscr{D}_k$ introduced in Sect. 4. The transformation G_k is defined by replacing the gauge-fixing delta function $\delta_{k,Ax}(A)$ in (4.1.1) by the Landau-gauge fixing function $\mathscr{G}(\partial^* A)$ of (4.4.1). The operator G_k was given in the momentum representation in [6I, Eqs. (1.83) and (1.84)]. Starting from this expression, one can derive the following formulas for $\sigma_k(p)$ by straightforward, algebraic manipulation:

We express σ_k as a sum of two terms

$$\sigma_k = \tau_1 + \tau_2 . \tag{7.1.13}$$

Here τ_1 vanishes on curls. Thus if $f = \partial B$, then $\tau_1 f = 0$. Explicitly

$$\tau_{1,\mu\nu\lambda\kappa}(p') = \tfrac{1}{2} \sum_l \left(\frac{|u|^2}{\bar{v}_\mu \bar{v}_\nu v_\lambda v_\kappa} \left[\delta_{\mu\lambda} - \frac{\partial_\mu \bar{\partial}_\lambda}{\Delta} \right] \left[\delta_{\nu\kappa} - \frac{\partial_\nu \bar{\partial}_\kappa}{\Delta} \right] \right) (p'+l), \tag{7.1.14}$$

and

$$\tau_{2,\mu\nu\lambda\kappa}(p') = \left(a_\mu \bar{a}_\lambda (\varphi_\nu \varphi_\kappa)^{-1/2} \left[\delta_{\nu\kappa} - \frac{\partial_\nu^{(1)} \overline{\partial_\kappa^{(1)}}}{\sqrt{\varphi_\nu \varphi_\kappa}} \left(\sum_\varrho |\partial_\varrho^{(1)}|^2 / \varphi_\varrho \right)^{-1} \right] \right) (p') . \tag{7.1.15}$$

In the formula for $\tau_1(p')$, each term on the right side is evaluated at momentum $p'+l$. In the formula for τ_2, the averaging only occurs in

$$a_\mu(p') = \partial_\mu^{(1)}(p') \sum_l \left(\left| \frac{u}{v_\mu} \right|^2 \frac{1}{\Delta} \right) (p'+l) . \tag{7.1.16}$$

In both τ_1 and τ_2 the expressions inside brackets [ ] are projection operators. Thus σ_k has the general form of a sum of projection operators, or tensor products of projection operators, sandwiched between averaging operators.

The fact that τ_1 vanishes on curls can be established as follows: The general form of τ_1 in configuration space is evident from (7.1.11), (7.1.14) namely

$$\tau_1 = Q_k^e (I - P_\partial) Q_k^{e*} , \tag{7.1.17}$$

where P_∂ denotes the orthogonal projection onto curls. (If P_1 projects one-forms onto gradients, then $P_1 \otimes P_1$ projects two forms onto curls.) Furthermore $Q_k^{e*}\partial = \partial Q_k^{s*}$, so

$$\tau_1 \partial = Q_k^e (I - P_\partial) \partial Q_k^{s*} = 0 . \tag{7.1.18}$$

Let $\mathscr{K}(p)$ denote the space of momentum p lattice one-forms B. Write the two-forms $\mathscr{H}(p)$ as a sum of curls and an orthogonal complement,

$$\mathscr{H}(p) = \partial \mathscr{K}(p) + (\partial \mathscr{K}(p))^\perp . \tag{7.1.19a}$$

Here we now introduce the scalar product,

$$(f, g) = \sum_{\mu,\nu} \overline{f_{\mu\nu}(p)} |v_\mu(p) v_\nu(p)|^{-2} g_{\mu\nu}(p) = \langle f, |V|^{-2} \otimes |V|^{-2} g \rangle , \tag{7.1.19b}$$

and the decomposition (7.1.19) is with respect to this inner product. This decomposition is natural in the study of $\sigma_k(p)$, since $\tau_1(p)\partial\mathscr{K}(p) = 0$.

Remark that there are strictly positive constants c_1, c_2 such that

$$c_1 \langle f, f \rangle \leqq (f, f) \leqq c_2 \langle f, f \rangle ,$$

so the corresponding norms are equivalent. In fact

$$|v_\mu(p)| = \left[\frac{\sin(p_\mu/2)}{(p_\mu/2)} \right] \left[\frac{(p_\mu \eta/2)}{\sin(p_\mu \eta/2)} \right] ,$$

and since for $|x| \leqq \pi/2$,

$$\frac{2}{\pi} \leqq \frac{\sin x}{x} \leqq 1 ,$$

it follows that for $|p_i| \leqq \pi$,

$$\tfrac{2}{\pi} \leqq |v_\mu(p)| \leqq \tfrac{\pi}{2} . \tag{7.1.20}$$

Proposition 7.1.2. *Suppose there exists* $\varepsilon > 0$, *and* $\tau_0(p) \leqq \tau_1(p)$ *such that for* $|p_j| \leqq \pi$,

$$\varepsilon \leqq \tau_0(p) \restriction (\partial \mathscr{K}(p))^\perp \quad \text{and} \quad \varepsilon \leqq \tau_2(p) \restriction \partial \mathscr{K}(p) . \tag{7.1.21}$$

Then Theorem 7.1.1 holds.

Proof. It is sufficient to show that there is a constant $c > 0$ such that

$$c \leqq \sigma_k(p) \tag{7.1.22}$$

for all $|p_j| \leqq \pi$. The inequality

$$|a_\mu(p)/\varphi_\nu(p)^{1/2}| \leqq c_2 \tag{7.1.23}$$

is an extension of (7.1.20). As a consequence,

$$\|\tau_2\| \leqq M , \tag{7.1.24}$$

where M is a constant independent of p and k.

Let us now decompose $\sigma_k(p)$ according to (7.1.19). For $f = \partial B + f^\perp \in \mathscr{H}(p)$, write

$$\begin{aligned} \langle f, \sigma_k f \rangle &= \langle f, \tau_1 f \rangle + \langle f, \tau_2 f \rangle \\ &= \langle f^\perp, \tau_1 f^\perp \rangle + \langle f, \tau_2 f \rangle \geqq \langle f^\perp, \tau_0 f^\perp \rangle + \langle f, \tau_2 f \rangle \\ &\geqq \varepsilon \| f^\perp \|^2 + \langle f, \tau_2 f \rangle . \end{aligned} \tag{7.1.25}$$

Here we use $\tau_1 \partial = 0$ and the hypothesis on τ_1. The idea is to bound $\langle f, \tau_2 f \rangle$ using the positivity of τ_2. In fact, $0 \leqq \tau_2$ is evident from (7.1.15). Then for any $\delta \in [0, 1]$,

$$\langle f, \sigma_k f \rangle \geqq \varepsilon \| f^\perp \|^2 + \delta \langle f, \tau_2 f \rangle . \tag{7.1.26}$$

Note that by (7.1.24) and the positivity of τ_2,

$$\begin{aligned} \langle f, \tau_2 f \rangle &= \langle \partial B, \tau_2 \partial B \rangle + \langle \partial B, \tau_2 f^\perp \rangle + \langle f^\perp, \tau_2 \partial B \rangle + \langle f^\perp, \tau_2 f^\perp \rangle \\ &\geqq \langle \partial B, \tau_2 \partial B \rangle - 2M \| \partial B \| \, \| f^\perp \| . \end{aligned}$$

For any $\lambda > 0$, we then have

$$\langle f, \tau_2 f \rangle \geqq \langle \partial B, \tau_2 \partial B \rangle - (M\lambda)^2 \| \partial B \|^2 - \lambda^{-2} \| f^\perp \|^2 .$$

Use the lower bound (7.1.20) on $\tau_2 \restriction \partial \mathscr{K}$, and choose λ so $(M\lambda)^2 = \varepsilon/2$. Thus

$$\langle f, \tau_2 f \rangle \geqq \tfrac{1}{2} \varepsilon \| \partial B \|^2 - 2M^2 \varepsilon^{-1} \| f^\perp \|^2 .$$

Inserting this in (7.1.26) yields

$$\langle f, \sigma_k f \rangle \geqq (\varepsilon - 2M^2 \delta \varepsilon^{-1}) \| f^\perp \|^2 + \tfrac{1}{2} \delta \varepsilon \| \partial B \|^2 \geqq \tfrac{1}{2} \varepsilon \| f^\perp \|^2 + \tfrac{1}{2} \delta \varepsilon \| \partial B \|^2 , \tag{7.1.27}$$

where the last inequality follows for any $\delta < (\varepsilon/2M)^2 \leqq 1$. Let $\delta > 0$ and $c \equiv \frac{1}{2} \delta \varepsilon > 0$. Then

$$\langle f, \sigma_k f \rangle \geqq c(\| f^\perp \|^2 + \| \partial B \|^2) = c \| f \|^2 ,$$

as desired and completes the proof of the proposition.

Let us now prove the lower bound (7.1.21) on $\tau_0(p)$. Introduce the operator $\tau_0(p)$ obtained by restricting the sum in (7.1.14) to $l = 0$. Clearly each term in the sum over l is a nonnegative operator. Hence

$$\tau_0(p) \leqq \tau_1(p). \tag{7.1.28}$$

Note that because $f^\perp$ is defined with respect to the inner product (7.1.19b), τ_0 simplifies on $f^\perp$. In particular the ∂ terms vanish and

$$\langle f^\perp, \tau_0 f^\perp \rangle = \tfrac{1}{2}|u(p)|^2 \sum_{\mu,\nu} |f^\perp_{\mu\nu}(p)/(v_\mu(p)v_\nu(p))|^2. \tag{7.1.29}$$

From (7.1.20) we then infer

$$\varepsilon \leqq \tau_0 \restriction (\partial\mathscr{K})^\perp,$$

for some $\varepsilon > 0$, which with (7.1.28) yields the desired lower bound

$$\varepsilon \leqq \tau_1 \restriction (\partial\mathscr{K})^\perp. \tag{7.1.30}$$

Next we consider the bound on $\tau_2 \restriction \partial\mathscr{K}$. But since $\tau_1\partial = 0$,

$$\langle \partial B, \tau_2 \partial B \rangle = \langle \partial B, \sigma_k \partial B \rangle = \langle B, \Delta_k B \rangle, \tag{7.1.31}$$

using the definition (4.3.1) of Δ_k. An explicit formula for Δ_k shows that

$$\langle \partial B, \sigma_k \partial B \rangle = \frac{1}{2}\left(\sum_{\varrho} |\partial^{(1)}_\varrho|^2/\varphi_\varrho\right)^{-1} \sum_{\mu,\nu} |(\partial B)_{\mu\nu}|^2/(\varphi_\mu \varphi_\nu).$$

which is bounded below by $\varepsilon\|\partial B\|^2$ (see also [6I, 8]). This completes the proof of Theorem 7.1.1.

7.2. *Regularity and Decay of Propagators and Minimizers*

We are interested in Landau gauge propagators and minimizers, because they have good regularity and decay. The key objects are H_k and $C^{(k)}$, while estimates on $\mathscr{D}_k$ follow by (4.4.4). The minimizer H_k can be expressed as an integral kernel. For $x \in T_\eta$,

$$(H_k B)_\mu(x) = \sum_{y \in T_1^{(k)}, \nu} H_{k,\mu\nu}(x; y) B_\nu(y). \tag{7.2.1}$$

The kernel $H_{k,\mu\nu}(x, y)$ and its gradient decay exponentially. In particular there exists $\delta > 0$ and for $0 \leqq \alpha < 1$ a constant $M = M(\alpha) < \infty$ such that for $|x - x'| \leqq 1$,

$$|H_{k,\mu\nu}(x, y)| + |\nabla H_{k,\mu\nu}(x, y)| + |x - x'|^{-\alpha} |\nabla H_{k,\mu\nu}(x, y) - \nabla H_{k,\mu\nu}(x', y)| \leqq M e^{-\delta|x-y|}. \tag{7.2.2}$$

This inequality is a consequence of Proposition 1.2 and the representation (1.103) of [6I].

The unit lattice propagator $C^{(k)}$ also has exponential decay,

$$|C^{(k)}_{\mu\nu}(x, y)| \leqq M e^{-\delta|x-y|}, \tag{7.2.3}$$

for $x, y \in T_1^{(k)}$. This inequality follows from the bound (2.157) in [6II] and from the general theorem on unit lattice operators in [7]. In fact this inequality also holds for propagators with Dirichlet boundary conditions outside a domain Λ, uniformly in Λ.

The gauge transformation λ in (5.1.1) is bounded and depends on B through an exponentially decaying kernel D_k:

$$\lambda(x) = (D_k B)(x), |D_k(x,b)| \leqq M e^{-\delta \operatorname{dist}(x,b)}. \tag{7.2.4}$$

This estimate follows from (5.1.4) and (7.2.2).

The operators $\mathscr{D}_k$ have the same properties as the operators G_k in [6I], Proposition 1.2, with exponential decay but singularities on the diagonal. These properties follow from (4.4.4) and the above estimates on H_k, $C^{(k)}$.

7.3. *Positivity of $\Delta_k(u_k)$*

The scalar field quadratic form $\Delta_k(u_k)$ depends on the background field u_k. Hence we can only establish stability properties for $\Delta_k(u_k)$ with some restriction on u_k. In particular, let us assume that for the unit lattice field v,

$$|v(\partial p) - 1| \leqq e_k \not{p}(e_k), \tag{7.3.1}$$

where $\not{p}(e_k) = (1 + \ln e_k^{-1})^{\not{p}}$. Then the stability estimate can be stated in two forms. For constants $\gamma > 0$, $\alpha > 0$, $M < \infty$,

$$\langle \phi, \Delta_k(u_k)\phi \rangle \geqq \gamma \sum_{b \in T_1^{(k)}} |u_k(b)\phi(b_+) - \phi(b_-)|^2 - M e_k^{2-\alpha} \sum_{x \in T_1^{(k)}} |\phi(x)|^2. \tag{7.3.2}$$

The second form of the inequality substitutes v_b for $u_k(b)$ in the covariant derivative of ϕ. These inequalities can be proved by an extension of the proofs of [7]. The propagators arising from $\Delta_k(u_k)$, under the restriction (7.3.1) on the gauge field, also satisfy the regularity and decay estimates of [7]. In order to remain within the framework of this reference, we remark that by change of gauge u_k can be transformed in a local region Λ into a configuration of the form $\exp[ie_k \eta A]$, where A is smooth and small. In axial gauge for the configuration v in a domain $\Lambda' \supset \Lambda$ we can substitute $\mathscr{D}_k \partial^* = G_{k,Ax} \partial^* + \partial D$ in the formula for u_k, we use (5.3.1) to replace this by a minimizer in axial gauge. Then we use (5.1.1) to return the minimizers to Landau gauge. This is a local procedure since $f^{(k)}$ can locally be represented as a curl.

Appendix. Quadratic Forms

In the analysis of this paper we often wish to minimize a quadratic form on a Hilbert space $\mathscr{H}$, possibly subject to a constraint. We collect here a couple of elementary properties which we use, and which could serve as the start of a more general study of related problems.

Since we work with lattice fields, our space $\mathscr{H}$ is finite dimensional and all forms are bounded. Of course we may obtain limiting unbounded forms as $\dim \mathscr{H} \to \infty$ (e.g., as in infinite-volume or zero-lattice-spacing limit). Such forms are analyzed as limits of finite-dimensional approximations.

As a first example, let $0 \leqq \Delta = \alpha^* \alpha$ be a self-adjoint transformation on $\mathscr{H}$ and let $B \in \mathscr{H}$. Define the quadratic form

$$h(A) = \tfrac{1}{2} \|\alpha A + B\|^2. \tag{A1}$$

Let P denote the projection onto Range(α).

Proposition A1. Range $\alpha^* = \text{Domain } \Delta^{-1}$, *so* $P = \alpha\Delta^{-1}\alpha^*$. *Furthermore*

$$h(A) \geqq h(A_{cl}) = \tfrac{1}{2}\langle B, (I-P)B\rangle, \tag{A2}$$

where

$$A_{cl} = -\Delta^{-1}\alpha^* B \tag{A3}$$

is the unique minimum of (A1).

Proof. Note that Range $\alpha^* = (\text{Kernel } \alpha)^\perp$ and Kernel $\alpha = $ Kernel Δ. By the spectral theorem $\mathscr{H} = \text{Kernel}(\Delta) \oplus \text{Range}(\Delta)$, so Range $\alpha^* = $ Range $\Delta = $ Domain Δ^{-1}. Therefore $P = \alpha\Delta^{-1}\alpha^*$ exists, $P^2 = P = P^*$, and $P\alpha = \alpha$. Thus P is the projection onto Range α. Writing

$$h(A) = \tfrac{1}{2}\|\Delta^{1/2}A + \Delta^{-1/2}\alpha^* B\|^2 + \tfrac{1}{2}\langle B, (I-P)B\rangle$$

yields (A2). Since $h(A)$ is convex, the minimum is unique, to complete the proof.

If we translate the quadratic form h, using the change of variables

$$A = A' + A_{cl},$$

then as a consequence of $(I-P)\alpha = 0$, and $h(A_{cl}) = \tfrac{1}{2}\langle B, (I-P)B\rangle$, $h(A)$ decomposes into a sum of independent quadratic forms. We state

Corollary A2. *With* $h(A)$ *given by* (A1) *and* $A = A' + A_{cl}$ *defined by* (A3),

$$h(A) = \tfrac{1}{2}\langle A', \Delta A'\rangle + h(A_{cl}). \tag{A4}$$

We now specialize to the case where $\mathscr{H} = L_2(\mathbb{R}^N, dx)$, with dx Lebesgue measure. We consider the Green's function C of Δ. Assuming $0 < \Delta$, then we have two convenient representations for C. First

$$C = \Delta^{-1}, \tag{A5}$$

by which we express C as the operator inverse of Δ. Secondly, we can express C as a moment of a Gaussian measure, namely

$$\exp(\tfrac{1}{2}\langle B, CB\rangle) = Z^{-1} \int_{\mathscr{H}} \exp(-\tfrac{1}{2}\langle x, \Delta x\rangle + \langle x, B\rangle)dx. \tag{A6}$$

We are often interested, however, in the case of operators Δ with zero modes. Thus Δ has an inverse only when restricted to a subspace $\mathscr{H}_0 \subset \mathscr{H}$ which does not contain zero modes. Likewise an integral such as (A6) converges only when restricted to such a subspace. In gauge theories it is often natural to choose various subspaces $\mathscr{H}_0$, and furthermore Δ is generally not diagonal under the decomposition $\mathscr{H} = \mathscr{H}_0 \oplus \mathscr{H}_0^\perp$. Thus we are led to study quadratic forms Δ constrained to act on $\mathscr{H}_0$.

Define the transformation C on $\mathscr{H}$ by

$$\exp(\tfrac{1}{2}\langle B, CB\rangle) = Z^{-1} \int_{\mathscr{H}_0} \exp(-\tfrac{1}{2}\langle x, \Delta x\rangle + \langle x, B\rangle)dx. \tag{A7}$$

Clearly C is self-adjoint in the inner product $\langle\ ,\ \rangle$. Furthermore, define

$$\Delta_0 = \Delta \upharpoonright \mathscr{H}_0. \tag{A8}$$

In general $\mathscr{H}_0 \neq \mathscr{H}$ and $C \upharpoonright \mathscr{H}_0 \neq \Delta_0^{-1}$. Thus we ask

(1) How are Δ_0 and C related?

(2) What is the minimizer of the quadratic form $\frac{1}{2}\langle x, \Delta_0 x\rangle - \langle x, B\rangle$ for $x \in \mathscr{H}_0$?

Let us make the following assumptions: Let V be given and

$$I \leqq \Delta_0 \quad \text{on} \quad \mathscr{H}_0, \tag{A9}$$

$$I \leqq \Delta + V \quad \text{on} \quad \mathscr{H}, \tag{A10}$$

where $\mathscr{H}_0 = \text{Null}\, V$. With these definitions, define

$$G = (\Delta + V)^{-1} \tag{A11}$$

and let (,) be the inner product

$$(x, y) \equiv \langle x, (\Delta + V) y\rangle. \tag{A12}$$

Then let P denote the projection of $\mathscr{H}$ onto $\mathscr{H}_0$, where P is orthogonal in the inner product (A12).

Proposition A3. *Under the assumptions above,*

$$C = PG, \tag{A13}$$

so $\text{Range}\, C \subset \mathscr{H}_0$. *Furthermore*

and

$$C\Delta_0 C = C, \tag{A14}$$

$$HB = CB = Z^{-1} \int_{\mathscr{H}_0} \exp(-\tfrac{1}{2}\langle x, \Delta x\rangle + \langle x, B\rangle) x\, dx. \tag{A15}$$

is the minimum configuration of

$$\tfrac{1}{2}\langle x, \Delta x\rangle - \langle x, B\rangle, \qquad x \in \mathscr{H}_0.$$

Thus with $x = y + CB$,

$$\tfrac{1}{2}\langle x, \Delta x\rangle - \langle x, B\rangle = \tfrac{1}{2}\langle y, \Delta y\rangle - \tfrac{1}{2}\langle B, CB\rangle, \qquad x \in \mathscr{H}_0. \tag{A16}$$

Proof. For $x \in \mathscr{H}_0$, note that

$$\tfrac{1}{2}\langle x, \Delta_0 x\rangle - \langle x, B\rangle = \tfrac{1}{2}\langle x, (\Delta + V) x\rangle + \langle x, (\Delta + V) GB\rangle.$$

Thus (A7) can be written

$$\exp(\tfrac{1}{2}\langle B, CB\rangle) = Z^{-1} \int_{\mathscr{H}_0} \exp(-\tfrac{1}{2}\langle x - PGB, (\Delta + V)(x - PGB)\rangle + \tfrac{1}{2}\langle PGB, (\Delta + V) PGB\rangle) dx, \tag{A17}$$

where we have used

$$\langle PGB, (\Delta + V)x\rangle = (PGB, x) = (GB, Px) = (GB, x) = \langle GB, (\Delta + V)x\rangle. \tag{A18}$$

Thus (A17) equals

$$\exp(\tfrac{1}{2}\langle PGB, (\Delta + V) PGB\rangle) = \exp(\tfrac{1}{2}\langle GB, (\Delta + V) PGB\rangle) = \exp(\tfrac{1}{2}\langle B, PGB\rangle).$$

Thus $\langle B, CB\rangle = \langle B, PGB\rangle$, and (A13) holds as claimed. To verify (A14), note that

$$C\Delta_0 C = PG\Delta_0 PG = PG(\Delta + V)PG = P^2 G = PG = C.$$

Next note that (A 17) displays CB as the minimum of $\frac{1}{2}\langle x, \Delta x\rangle - \langle x, B\rangle$, since linear terms in the fluctuation $x - PGB$ do not occur. Also from (A 18) we have for $x \in \mathscr{H}_0$,

$$\langle CB, \Delta x\rangle = \langle B, x\rangle, \tag{A 19}$$

so the integral in (A 15) equals CB. Finally, for $x \in \mathscr{H}_0$, so by (A 19), $\langle CB, \Delta y\rangle \equiv \langle B, y\rangle$. It follows that

$$\begin{aligned}\tfrac{1}{2}\langle x, \Delta x\rangle - \langle x, B\rangle &= \tfrac{1}{2}\langle y, \Delta y\rangle + \langle \Delta y, CB\rangle + \tfrac{1}{2}\langle CB, \Delta CB\rangle - \langle y, B\rangle - \langle CB, B\rangle \\ &= \tfrac{1}{2}\langle y, \Delta y\rangle - \tfrac{1}{2}\langle B, CB\rangle,\end{aligned}$$

which is (A 16).

References

1. Bałaban, T., Imbrie, J., Jaffe, A.: Exact renormalization group for gauge theories, 1983 Cargese Lectures, Progress in Gauge Field Theories, pp. 79–104. G. 't Hooft et al. (ed.). New York: Plenum Press 1984
2. Brydges, D., Fröhlich, J., Seiler, E.: On the construction of quantized gauge fields. I. General results. Ann. Phys. **121**, 227–284 (1979). II. Convergence of the lattice approximation. Commun. Math. Phys. **71**, 159–205 (1980). III. The two-dimensional abelian Higgs model without cutoffs. Commun. Math. Phys. **79**, 353–400 (1981)
3. Bałaban, T.: $(\text{Higgs})_{2,3}$ quantum fields in a finite volume. I. A lower bound. Commun. Math. Phys. **85**, 603–626 (1983). II. An upper bound. Commun. Math. Phys. **86**, 555–594 (1982)
4. King, C.: The U(1) Higgs model. I. The continuum Limit, HUTMP 84/B167. II. The infinite volume limit. HUTMP 84/B168
5. Bałaban, T., Brydges, D., Imbrie, J., Jaffe, A.: The mass gap for Higgs models on a unit lattice. Ann. Phys. **158**, 281–319 (1984)
6. Bałaban, T.: Propagators and renormalization transformations for lattice gauge theories. I. Commun. Math. Phys. **95**, 17–40 (1984). II. Commun. Math. Phys. **96**, 223–250 (1984)
7. Bałaban, T.: Regularity and decay of lattice Green's functions. Commun. Math. Phys. **89**, 571–597 (1983)
8. Federbush, P.: A phase cell approach to Yang-Mills theory. I. Small field modes. University of Michigan preprint

Communicated by A. Jaffe

Received October 26, 1984

GPSR Compliance

The European Union's (EU) General Product Safety Regulation (GPSR) is a set of rules that requires consumer products to be safe and our obligations to ensure this.

If you have any concerns about our products, you can contact us on ProductSafety@springernature.com

In case Publisher is established outside the EU, the EU authorized representative is:

Springer Nature Customer Service Center GmbH
Europaplatz 3
69115 Heidelberg, Germany

Batch number: 10704763

Printed by Printforce, the Netherlands